AF342056

FIBER LASER

Edited by **Mukul Chandra Paul**

Fiber Laser

http://dx.doi.org/10.5772/60690
Edited by Mu

Contributors

Vladimir Kalashnikov, Sergey Sergeyev, Maryam Eilchi, Parviz Parvin, Kavintheran Thambiratnam, Harith Ahmad, Mukul Chandra Paul, Chien-Hung Yeh, Vasily Spirin, Andrei Fotiadi, Patrice Mégret, Sergey Kobtsev, Sergey Smirnov, Sergey Kukarin, Guolu Yin, Xin Wang, Shuqin Lou, Yiping Wang, Mohd Afiq Bin Ismail, Sulaiman Wadi Harun, Ci-Ling Pan, Alexey Zaystev, Yi-Jing You, Chih-Hsuan Lin, Zhijun Yan, Chengbo Mou, Yishan Wang, Jianfeng Li, Zuxing Zhang, Xianglian Liu, Lin Zhang, Kaiming Zhou, Catherine Wandera, B. N. Upadhyaya, Ricardo Ivan Alvarez Tamayo, Dan Savastru, Roxana Savastru, Sorin Miclos, Ion Lancranjan, Patricia Popoola, Olanrewaju Adesina, Olawale Fatoba, Sisa Pityana, Olawale Popoola, Gabriel Farotade

CBS, Edition **2017**

Published by InTech

Janeza Trdine 9, 51000 Rijeka, Croatia

Notice

Statements and opinions expressed in the chapters are these of the individual contributors and not necessarily those of the editors or publisher. No responsibility is accepted for the accuracy of information contained in the published chapters. The publisher assumes no responsibility for any damage or injury to persons or property arising out of the use of any materials, instructions, methods or ideas contained in the book.

Publishing Process Manager Iva Simcic

Technical Editor InTech DTP Team

Cover Designer InTech Design Team

Additional hard copies can be obtained from orders@intechopen.com

Fiber Laser, Edited by Mukul Chandra Paul
p. cm.
ISBN 978-953-51-2257-9

Contents

Preface

Fiber laser is an advanced field of modern science entering in all branches of science. This field continues to vastly expend with state-of-the-art developments across the entire spectrum of scientific, military, medical, industrial, and commercial applications ranging from spectroscopy to material cutting, welding, and marking. Light plays an important role in modern human's life. We cannot think our human life environment without light. Nowadays, fiber laser is an important light-based technology in the area of photonics. I am fully satisfied with the outcome made through the collective efforts of the wonderful team of all authors, who have gathered and shared their knowledge to make a complete shape of this book. The authors are researchers, academics, PhD students, and professionals in the field of fiber lasers and industries.

This book attempts to give various aspects of different fiber lasers as well as some of the recent developments in this field.

The plan of the book is as follows:

The first chapter of the book describes the self-pulsing dynamics and its elimination in different ytterbium (Yb)-doped continuous wave (CW) fiber laser configurations because of the development of high-power Yb-doped CW and pulsed fiber lasers for various material processing applications. The generation of truly high-power CW output from Yb-doped CW fiber lasers by pumping with high-power CW laser diodes is not so easy without amplitude modulations due to self-pulsing phenomenon as these random self-pulses may have very high peak power to cause catastrophic damage of fiber laser components and thereby inhibit generation of high-power CW output.

The second chapter of the book discusses the Q-switched fiber lasers' operating principle for active technique in terms of operation characteristics and demonstrates the experimental analysis of the pulsed fiber lasers by the active Q-switched technique using erbium (Er)/Yb-doped fiber as a gain medium. Then, it discusses about the experimental analysis of principal characteristics of single-wavelength operation of the fiber laser and cavity loss adjustment method for dual-wavelength laser operation. **Chapter 3** emphasizes on carbon-based saturable absorbers, namely graphene and carbon nanotubes (CNT), and their unique electronic band structures and optical characteristics. The methods of incorporating these carbon-based saturable absorbers into fiber laser cavity for generating passive Q-switched and mode-locked fiber lasers are demonstrated. **Chapter 4** presents a detailed analysis of the properties of double-scale pulses (also called noise-like pulses (NLPs) and femtosecond clusters) generated in fiber lasers and their promising applications. **Chapter 5** describes the experimental techniques and methods for passive stabilization of single-frequency fiber Brillouin lasers by employing the self-injection locking phenomenon. Then, it demonstrates that this locking phenomenon delivers a significant narrowing of the pump laser linewidth and generates the Stokes wave with linewidth of about 0.5 kHz. **Chapter 6** discusses the dual-wavelength fiber lasers (DWFLs), which provide a simple and cost-effective approach for the optical generation of microwave (MHz) and terahertz (THz) radiation using narrow-band filters and comb filters for microwave radiation generation, as well as the use of DWFLs with diethylaminosulfur tetrafluoride or $LiNbO_3$ crystals for generating THz radiation. **Chapter 7** is on tunable single-, dual-, and multiwavelength fiber laser based on twin core fiber (TCF). Then, it demonstrates the importance of tunable fiber lasers because their emission wavelength can be systematically tuned within a certain spectral range, which allows using a single-laser source instead of several sources. In this chapter, TCF-based filters, which work as the wavelength selective element, are introduced into the ring cavity to implement tunable single-, dual-, and multiwavelength fiber lasers. The authors have emphasized the tuning mechanism and the tuning characteristics of the tunable fiber lasers. **Chapter 8** demonstrates the reflective semiconductor optical amplifier (RSOA)-based fiber laser architectures for multiwavelength and wavelength-tunable operations by introducing an L-band multiwavelength laser, a C-band RSOA and a linear cavity formed by a fiber coupler, a polarization controller, and a reflected fiber mirror. In this chapter, the authors have demonstrated a multiwavelength laser source using a C-band RSOA with dual-ring fiber cavity and investigated a wavelength-tunable fiber ring laser architecture using the RSOA and SOA. **Chapter 9** discusses the interdisciplinary concept of dissipative soliton in fiber lasers. In this chapter, the authors have surveyed briefly about different mode-locking techniques as well as experimental realizations of dissipative soliton fiber lasers with an emphasis on their energy scalability. **Chapter 10** covers the generation and amplification of medium- and high-energy NLPs with Yb-doped optical fibers. In this chapter, the authors have demonstrated supercontinuum (SC) generation techniques, where NLPs serve as the pump source. **Chapter 11** reviews the recent achievements of 45° tilted fiber grating (TFG) in all fiber laser systems, including the theory, fabrication, and characterization of 45° TFGs and 45° TFGs-based ultra fast fiber laser systems working in different operating regimes at the wavelength of 1, 1.5, and 2 μm. **Chapter 12** discusses the role of simulation methods as tools for analysis of low and medium average power fiber laser operated in passively Q-switched and/or mode-locking regimes into the design of various applications, such as materials microprocessing of sensor applications. In this chapter, the authors have briefly presented the two main mathematical methods that used to analyze solid laser oscillators in passive optical Q-switching regime: the coupled rate equations approach and the iterative approach. The presented numerical simulation methods are validated by comparing with the experimental results. **Chapter 13** discusses the amplifying parameters in rare earth–doped optical fiber laser amplifiers under both CW and pulse conditions through a comprehensive analysis using the set of coupled propagation rate equations based on the atomic energy structure of dopant, as well as the absorption and emission cross sections. Then, it described the dependence of gain and saturation properties on the pump power, dopant concentration, and fiber length mainly due to the dominant effect of the overlapping factor of dual-clad fibers. **Chapter 14** describes the heat generation and the removal of fiber lasers particularly for high-power or high-energy operation. This chapter investigated the purpose of figuring out heat dissipation necessities for different parts of the active gain media and providing effective cooling procedures, thermal loading, as well as longitudinal and transverse temperature profiles of dual-clad fibers for both inside/outside of the doped fiber core based on considerable numerical analysis. The authors have shown that chilling mechanisms are very efficient ways to dissipate heat in high-power regimes. **Chapter 15** discusses the laser surface alloying (LSA) method for material processing, which uses the high-power density available from defocused laser beam to melt reinforcement powders and a part of the underlying substrate. This chapter described a three-dimensional (3D) mathematical model to obtain insights on the behavior of laser-melted pools subjected to various processing parameters. **Chapter 16** discusses laser surface modification techniques to improve the mechanical and terminological properties of materials. The selection of appropriate surface modification technique is required to prevent catastrophic failure of materials in the industry due to wear degradation phenomena where laser surface modification techniques have been established by researchers to improve the mechanical and terminological properties of titanium alloy material. This chapter demonstrates laser surface cladding and its processing parameters coupled with the oxidation, wear, and corrosion performances of laser-modified titanium material. **Chapter 17** reviews laser engineering net shaping method in the area of development of functionally graded materials for aero engine application. LENS™ offers a great deal in rapid prototyping, repair, and fabrication of 3D dense structures with superior properties in comparison with traditionally fabricated structures. **The final chapter** focuses on application of fiber lasers in material processing and describes the maximum processing speeds, maximum processing depths, and resulting cut edge quality governed by a number of parameters related to the laser system, work piece specification, and the cutting process. The effects of the processing parameters in the cutting of thick section stainless steel, mild steel, and medium-section aluminum along with optimization of the processing parameters for enhancement of the cut edge quality have been reported.

In summary, this 'Fiber Laser' book discusses some of the important topics that have made tremendous impact in the growth of science and technology.

Dr. Mukul Chandra Paul
Principal Scientist
Fiber Optics and Photonics Division
Central Glass and Ceramic research Institute
India

Chapter 1

Self-pulsing Dynamics in Yb-doped Fiber Lasers

B.N. Upadhyaya

Additional information is available at the end of the chapter

http://dx.doi.org/10.5772/62087

Abstract

There is worldwide considerable interest in the study and development of high-power Yb-doped continuous wave (CW) and pulsed fiber lasers for various material processing applications. Although it appears to be trivial to generate high-power CW output from Yb-doped CW fiber lasers by pumping with high-power CW laser diodes; however, it is not so easy to generate truly CW output without amplitude modulations due to self-pulsing phenomenon. The observation of random self-pulses overriding CW output has been reported by several authors. These random self-pulses may have very high peak power to cause catastrophic damage of fiber laser components and thereby inhibit generation of high-power CW output. This chapter describes self-pulsing dynamics and its elimination in different Yb-doped CW fiber laser configurations.

Keywords: Yb-doped fiber, self-pulsing, saturable absorption

1. Introduction

High-power Yb-doped double-clad fiber lasers have recently attracted considerable attention due to its advantages such as single-mode operation, all-fiber integration, high efficiency, compactness, no misalignment sensitivity, robustness, and efficient heat dissipation due to large surface area to volume ratio.[1,2] Yb-doped fibers provide wide absorption band in the range of ~800–1064 nm, and it can provide lasing in the range of ~974–980 and ~1010–1160 nm. Thus, Yb-doped fiber lasers are unique laser sources for a wide range of applications. Hanna et al. reported tuning range from 1010 to 1162 nm with a total span of 152 nm.[3] In view of this, there has been a growing interest in high-power Yb-doped fiber lasers as a potential replacement for bulk solid-state lasers in many applications. There are reports on the development of up to 2 kW of single-transverse-mode fiber laser and its commercial availability.[1] However, its efficiency and configuration is not reported in the literature. Record brightness from Yb-doped fiber laser with output power of up to 10 kW at 1070 nm of output wavelength has also

been reported by IPG Photonics.[1] This laser is the most intense CW laser of any kind with maintenance-free operation, a compact footprint, and ultra-long operational life. Such an intense laser can perform a variety of material processing and defense applications. For the generation of very high output power from fiber lasers, either only oscillator or master oscillator power amplifier (MOPA) configurations are normally used. However, the output power from Yb-doped fiber lasers is mainly limited by nonlinearities, thermal effects, fiber fuse effect, and optical damage of fiber-optic components used in oscillator or MOPA configuration.[4]

At a first glance, it is normally felt that it is an easy task to generate CW output from Yb-doped fiber lasers, but it is not so easy to achieve purely CW output having highly stable output without modulations as achieved in conventional lamp-pumped or diode-pumped solid-state lasers. It is normally expected to achieve a CW output from fiber lasers under CW pumping conditions. However, in several CW pumped rare-earth-doped fiber lasers with different resonator configurations and pumping geometries, modulations in the output and self-pulsations over-riding CW level has been reported.[5–13] Two types of self-pulsations have been reported in fiber lasers, they are as follows: (a) sustained self-pulsing (SSP) and (b) self-mode locking (SML).[7] In case of SSP, emission of high-intensity pulses at irregular intervals is observed, whereas in the case of SML, pulse spiking in the output or laser signal modulation is observed at the round trip time of the resonator cavity. Several possible origins have been suggested for self-pulsing in the literature. These include re-absorption of laser signal in the weakly pumped or un-pumped section of the rare-earth-doped fiber,[11–14] ion-pairing acting as a saturable absorber,[8–10] relaxation oscillations due to inversion in the gain medium and population of photons in the cavity,[10] resonator steady-state conditions and Q-factor of the cavity,[5,15] interaction between laser signal and population inversion,[16] pump noise as the source of self-pulsing.[17,18] Further, cascaded stimulated Brillouin scattering (SBS), distributed Rayleigh scattering, and the set of other nonlinear effects (SRS (stimulated Raman scattering); SPM (self-phase modulation); XPM (cross-phase modulation); and FWM (four wave mixing)) [19–22] have also been suggested as the sources of self-modulation and self-pulsing in different rare-earth-doped fiber lasers. Control on self-pulsing to achieve narrow pulses at regular interval and with enhanced Q-switching has also been reported.[23] Substantial effort has been put up by several authors to reduce self-pulsing. Suppression of self-pulsing has been carried out by using a low transmission output coupler giving a high Q-cavity,[5] using unidirectional fiber ring cavity,[5,7,24] increasing round trip time in the cavity by splicing a long section of matched passive fiber and thereby changing the relaxation oscillation dynamics,[25] uniform bidirectional pumping,[26] suppressing relaxation oscillation frequency component,[27] preventing rapid depletion of gain by resonant pumping near the lasing wavelength and thereby minimizing relaxation oscillations,[28,29] electronic feedback to the pump laser for shifting the gain and its phase.[30] In addition, the use of the narrow passband of a $\lambda/4$-shifted fiber Bragg grating (FBG) structure in a ring cavity to limit the number of longitudinal cavity modes [31] and the use of fast saturable gain of a semiconductor optical amplifier within the fiber laser resonator [32] have also been reported to suppress self-pulsing.

2. Effect of laser resonator on self-pulsing dynamics

The experimental setup to show the effect of laser resonator on self-pulsing dynamics consisted of a Yb-doped double-clad fiber having a core diameter of 10 μm and an inner-clad diameter of 400 μm with core and inner-clad numerical aperture (NA) of 0.075 and 0.46, respectively. This Yb-doped fiber had clad-pump absorption of 0.8 dB/m at 975 nm, and it had an octagonal inner-clad geometry to avoid excitation of skew modes. A fiber-coupled laser diode of 20 W output power and center wavelength of 975 nm was used to pump 18 m length of the above-mentioned Yb-doped fiber. The Yb-doped fiber was perpendicularly cleaved at both the ends to sustain higher damage thresholds. Study of three different Fabry–Perot resonator configurations as shown in Figures 1(a), 1(b), and 1(c) was carried out.[15] Figure 1(a) shows the forward-pumping configuration with high finesse in which a dichroic mirror with ~100% reflectivity at signal wavelength and high transmission at pump wavelength is kept in between the two lenses used for coupling pump light in to the Yb-doped fiber and the cleaved end with ~4% Fresnel reflection at the farther fiber end of the doped fiber acts as the output coupler. Figure 1(b) shows the backward-pumping configuration with high finesse in which two dichroic mirrors have been used, one for ~ 100% feedback of the signal and another at an angle of 45° for taking laser output from pump input end of fiber. In this case, the perpendicularly cleaved pump input end of the doped fiber with ~4% Fresnel reflection for laser signal acts as the output coupler. Figure 1(c) shows the low-finesse fiber laser resonator configuration; in this case, the cleaved ends with ~4% Fresnel reflection from both the fiber ends act as Fabry–Perot cavity mirrors. The dichroic mirrors used in these configurations are highly transmitting in the range of 960–980 nm and highly reflecting (≈98%) in the wavelength range of 1064–1140 nm. Output power was measured after filtering out the unabsorbed pump power using a thermal power meter. Maximum output power of 10.75 W was achieved with a slope efficiency of ~73% in the backward-pumping configuration at an input pump power of 17.2 W.

However, in the case of low-finesse cavity configuration of Figure 1(c), it was observed that the output power from both the ends ceases to increase beyond 1.8 W and starts fluctuating due to appearance of strong random self-pulsing over-riding CW output. With further increase in the pump power, an increase in the fluctuation about the average output power was observed and peak power of these random pulses were also found to increase. At an input pump power of 8 W, Figure 2(a) shows the observed random self-pulsing, and Figure 2(b) shows the expanded view of one of the random self-pulses with observed pulse duration of less than 25 ns. As the pulses are random in time domain with variation in frequency duration, and their peak powers are also not constant, the measured average power keeps fluctuating.

There is report on the occurrence of random self-pulsing in a V-groove pumping configuration in the case of "bad cavity" (R_1=100%, R_2=4% or R_1=4% and R_2=4%) and no such random self-pulsing in the case of "good cavity"(R_1=100%, R_2=80%) by Hideur et al.[5] They proposed that random self-pulsing behavior in the case of bad cavity may arise due to re-absorption of the laser signal in the un-pumped part of the fiber, and the occurrence of SRS and SBS, whereas in the case of a good cavity, the device behaves as a laser with saturable absorber and Brillouin and Raman effects do not occur.[11,12] However, it is clear that intracavity intensity in the case

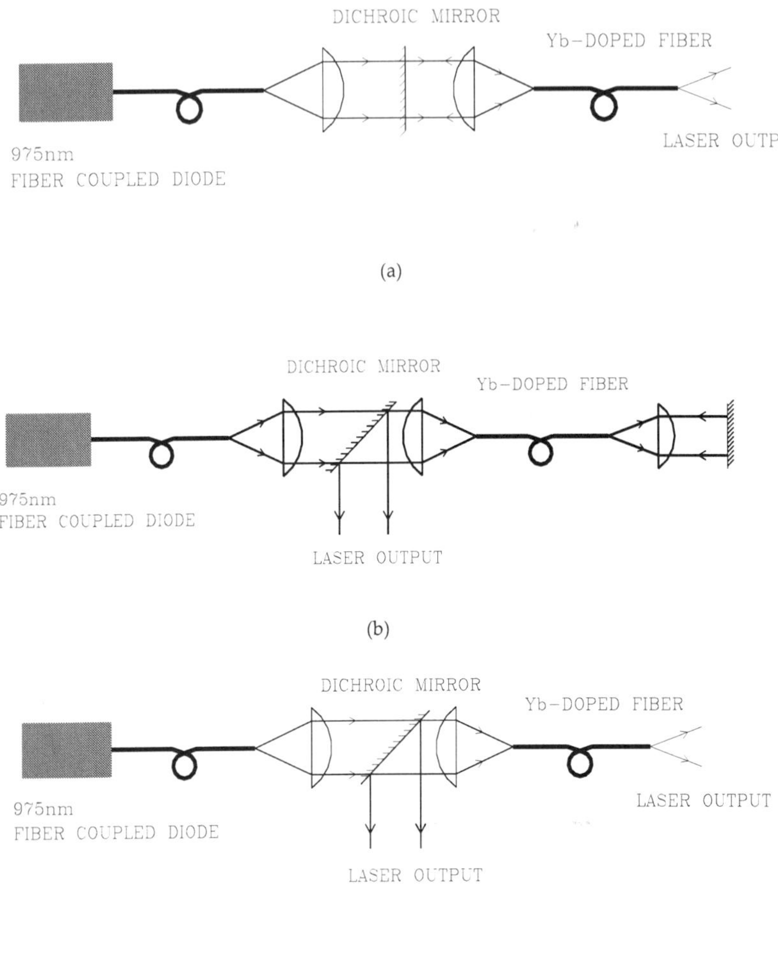

Figure 1. (a) High-finesse forward-pumping configuration with output power taken from the farther end of Yb-doped fiber, (b) high-finesse backward-pumping configuration with output power taken from the pumping end of the Yb-doped fiber using the 45° tilted dichroic mirror, and (c) low-finesse resonator configuration with laser output from both the ends.[15]

of a good-cavity will be much larger as compared to that in the case of a bad-cavity; therefore, the nonlinear SRS and SBS processes should occur at a much earlier value in the case of a good-cavity than in a bad-cavity, which contradicts their proposal. It has been shown by Fotiadi et al.[21] that mechanism of self-pulsing involves distributed backscattering in the form of Rayleigh scattering (RS) and SBS. At the beginning of the cycle, the pump provides a buildup of the population inversion in the gain medium and the gain increases gradually. However, CW lasing is completely inhibited due to the suppression of feedback by the imposed angle at the fiber ends and feedback into the cavity is provided mainly by the distributed RS and SBS that leads to generation of gigantic irregular random self-pulses. In cases of good-cavity or high-finesse cavity and bad cavity or low-finesse cavity, where feedback is not suppressed by angle cleaving of fiber ends and dominant feedback is through fiber-end facets or feedback mirrors, self-pulsing behavior is completely different.

We noted the threshold powers for occurrence of random self-pulsing in three different configurations for the investigation of random self-pulsing behavior in low- and high-finesse cavities. In cases of high-finesse resonator configurations R_1=98%, R_2=4% (forward pumping) and R_1=98%, R_2=80% (good-cavity), the random self-pulsing thresholds were found to be 6.2 and 12.7 W, respectively, whereas in the case of low-finesse resonator configuration (R_1=4%, R_2=4%), the self-pulsing threshold was found to be 2.8 W. Further, in the case of forward-pumping configuration, the output power was found to vary linearly with input pump power and the self-pulsing effect was very weak. Whereas in the case of low-finesse resonator configuration, output started fluctuating beyond 1.8 W and self-pulsing was found to be very strong. The higher threshold for occurrence of random self-pulsing in case of good-cavity and lower thresholds for bad-cavity configurations show that mirror reflectivities also play an important role in the onset of random self-pulsing behavior. Occurrence of a peak in the gain vs length curve could cause any pump-induced noise or distributed backscattered noise in the form of RS and SBS to build up and emit in the form of strong random pulses. Thus, in order to avoid random self-pulsing in CW fiber lasers, gain peaking along the fiber length should be avoided by uniform pumping along the fiber length with dual end pumping or multipoint pumping or by choosing appropriate cavity mirror reflectivities.

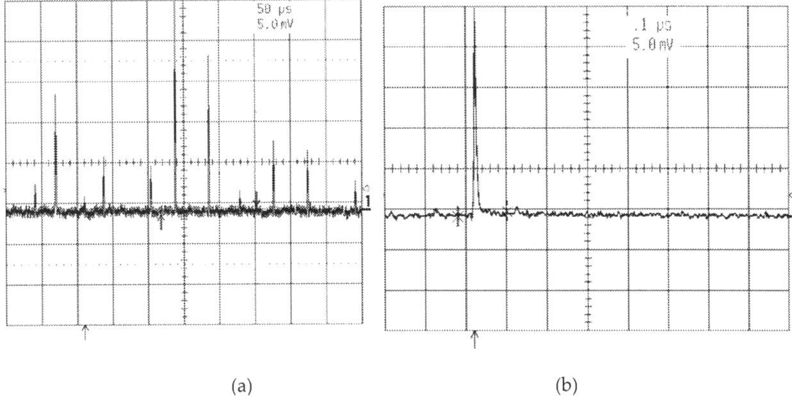

(a)   (b)

Figure 2. (a) Output of the fiber laser, showing random self-pulses in the case of low-finesse cavity of Figure 1(c), for an input pump power of 8 W, and (b) an expanded oscilloscope trace of one of the random self-pulses from Yb-doped fiber laser.[15]

Figure 3 shows average output spectrum in the case of low-finesse cavity configuration of Figure 1(c) before and after the onset of random self-pulsing behavior. This spectrum shows the presence of nonlinear SRS effect in the presence of random self-pulsing with first-order Stokes peak at a wavelength separation of 50 nm from the laser line. However, we could not confirm the occurrence of SBS due to smaller SBS shift and limited resolution of the optical spectrum analyzer.

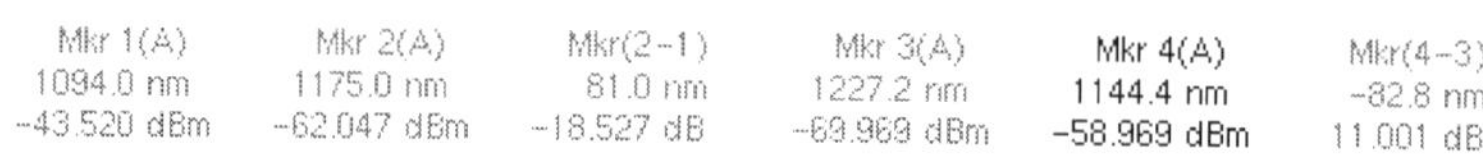

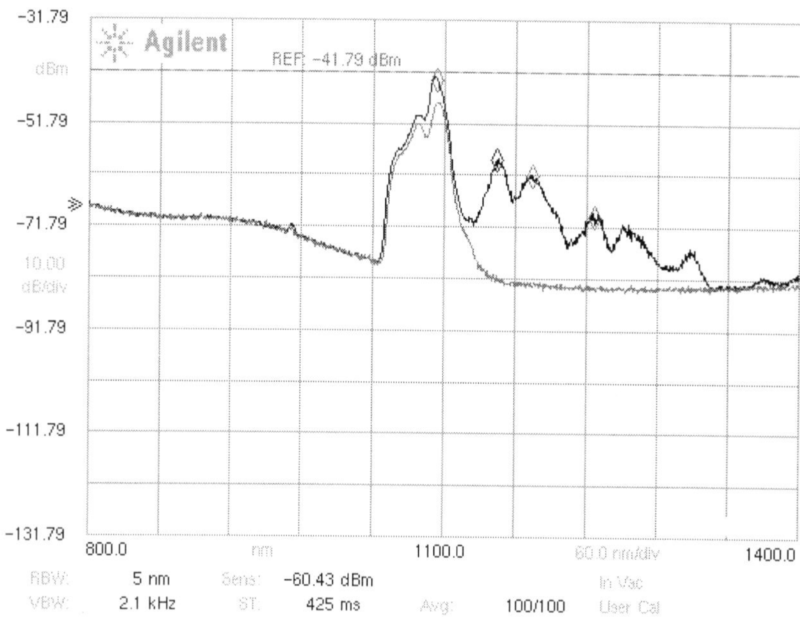

Figure 3. Output spectrum in case of low-finesse cavity of Figure 1(c); the lower trace shows spectrum before onset of random self-pulsing at 2.5 W of input pump power, and the upper trace shows output spectrum after onset of strong random self-pulsing at 8 W of input pump power.[15]

3. Effect of fiber length and laser linewidth on self-pulsing dynamics

Length of the doped fiber used for the fiber laser setup and the laser linewidth also plays an important role in self-pulsing dynamics. Root cause for the onset of self-pulsing is relaxation oscillation at low pump powers and saturable absorption along the weakly pumped portion of the fiber at the farther end, whereas nonlinear SRS and SBS effects are responsible for an increase in the peak power of these relaxation oscillation pulses. In fiber lasers with single-end or double-end pumping with wide- or narrow laser linewidths, most of the pump power is absorbed near the pump input end of the doped fiber; hence, there is always a weakly pumped region at the farther end of the doped fiber in the case of single-end pumping configuration and at the middle portion of the doped fiber in the case of double-end pumping configuration. The presence of weakly pumped region is responsible for signal re-absorption along the fiber length, and thereby causes self-pulsation due to "saturable absorption effect" as in the case of conventional passively Q-switched lasers. However, due to the "distributed nature" of saturable absorption of signal along the fiber length, self-pulsing is random in nature with variations in the frequency and time duration of self-pulses. With increase in input pump power, there is an increase in the frequency of these self-pulses, which is clear from the observed sequence of self-pulses under different resonator and pump conditions, and is a typical characteristic of passively Q-switched lasers. Gain uniformity increases with further increase in pump power and there is a corresponding reduction in saturable absorption of the signal along the fiber due to the absence of weakly pumped region, which eliminates self-pulsation at higher pump inputs. Depending on the signal intensity and excited state absorption of signal in the doped fiber, passively mode-locked pulses with a characteristic separation equal to the round-trip time may also appear in the self-pulse envelope, as in conventional bulk "passively Q-switched lasers having simultaneous mode locking". If the doped fiber length is long and laser linewidth is also narrow, it may happen that SBS threshold is reached, which will further enhance self-pulses. If the fiber length is long but the linewidth is not narrow, it may result in generation of SRS and thereby enhance self-pulses.[33–36]

4. Self-pulsing dynamics of high-power Yb-doped fiber laser with bulk mirror resonator

The experimental setup of Yb-doped fiber laser with bulk mirror resonator consisted of a large mode area Yb-doped double-clad fiber with a core diameter of 20 µm and an inner-clad diameter of 400 µm. The outer clad is made of fluoroacrylate polymer and had a diameter of 550 µm. Core and inner cladding NA's are 0.060 and 0.46, respectively. Inner-clad pump absorption at 975 nm is 1.7 dB/m. Thus, a length of 15 m of Yb-doped fiber was selected to have an efficient pump absorption of ~25 dB with minimum leakage of the input pump beam. An octagonal shape inner cladding of this fiber avoids excitation of skew modes. Yb-doped fiber ends were cleaved perpendicularly and coiling of the active fiber was done on a metallic mandrel to remove heat load from it. Fourteen number of laser diodes at 975 nm with an output power of 30 W and pigtail fiber core diameter of 200 µm and 0.22 NA were used to pump from both the ends of the active fiber using 7:1 multimode fiber-optic pump combiners. All the diodes were maintained at an operating temperature of 25 °C for the entire range of its output power using water-cooled heat sinks for mounting. Fiber output of seven such diodes were fusion-spliced individually with seven matched pump input ports of multimode fiber-optic pump combiner using GPX-3400 fusion splicing workstation from M/s Vytran. All the splice joints were optimized and a maximum transmission of ~86% was achieved. A cumulative loss of 14% is due to mismatch between fibers of diode pigtail and pump combiner fiber input ports along with insertion loss from each input port of fiber-optic pump combiner. Pump combiner output port had a core diameter of 400 μm and an NA of 0.46. Two such diode pump modules were made as shown in Figure 4(a), and these were used to pump from both ends of the active Yb-doped double-clad fiber using two 7:1 pump combiners. Pump beam at 975 nm from output port of each pump combiner was initially collimated using a 20 mm focal length lens, and then, it was imaged at the active fiber ends using another 20 mm focal length lens. Temperature-controlled metallic V-grooves were used to hold the ends of the Yb-doped fiber to prevent any possible thermal damage to the gain fiber coating by the heat generated in the gain fiber due to nonradiative emission processes or by means of any over filled pump or signal power. In order to have signal feedback in laser oscillator, a dichroic mirror with high reflectivity (HR) of ~100% in broadband from 1040 to 1100 nm for normal incidence and high transmission (HT) at 975 nm has been placed at one end of the Yb-doped double-clad fiber between the two lenses. Resonator mirrors are formed by this mirror along with the other cleaved end of the Yb-doped fiber providing 4% Fresnel reflection. Another dichroic mirror with HR in a broadband from 1040 to 1100 nm at 25° angle of incidence and HT at 975 nm has been placed between the two lenses to have laser output from the resonator. Figure 4(a) shows a schematic of the experimental setup, and Figure 4(b) shows an image of Yb-doped double-clad fiber-end face having octagonal inner cladding using a microscope.

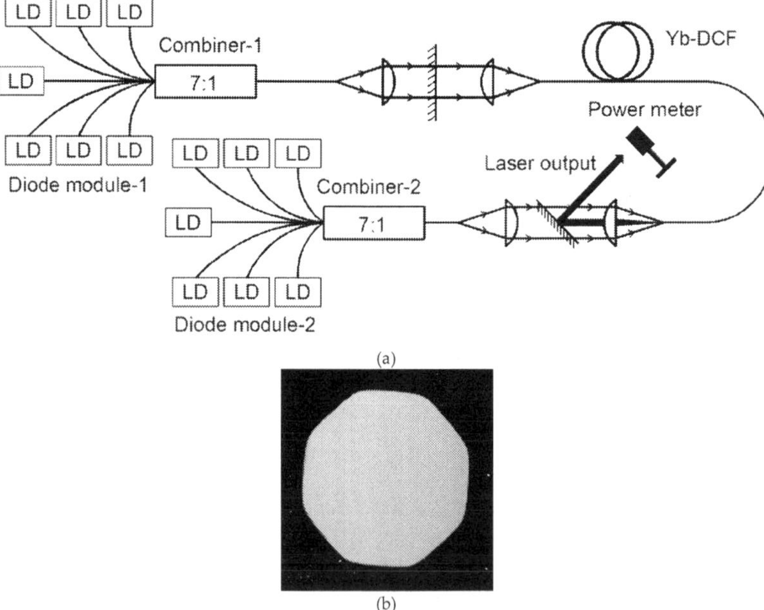

Figure 4. (a) A schematic view of 165 W of Yb-doped CW fiber laser, and (b) an image of Yb-doped double-clad fiber-end face with octagonal inner cladding using microscope.

From this experimental setup, an output power of 165 W was achieved at the combined maximum input pump power of 316 W from both the ends with a slope efficiency of 56.5% and an optical-to-optical conversion efficiency of 52%. Experimentally measured value of threshold pump power was 20 W. Theoretical estimate of threshold pump power was 9.5 W, which is much lower than the measured value. This is due to the fact that at lower pump powers, input pump wavelength was near to 968 nm, which is far away from the peak pump absorption wavelength of 976 nm. Using a compatible passive fiber in the same setup, pump coupling efficiency from the output of pump combiner to the active fiber was measured to be 65%, which is very less and can be further improved to reduce pump power losses and achieve higher conversion efficiencies. Taking into account, the pump coupling losses, optical-to-optical conversion efficiency improves to a value of 80.5%, which is close to the maximum reported figure of 83% for high-power Yb-doped fiber lasers by Jeong et al.[2] Variation of output pump power as a function of input pump power is shown in Figure 5. It shows that there is linear variation in output power with respect to an increase in input pump power, and there is no saturation in the output laser power even at the maximum value of input pump power. This shows that laser output power is limited only by the input pump power. Figure 6 shows recorded output spectrum at the maximum laser output power of 165 W. The laser output spectrum is peaked at 1079.7 nm with spread from 1064.1 to 1100.1 nm and full width half maximum linewidth of ~7 nm. There is another peak near 975 nm in the output spectrum, which shows pump wavelength peak.[37]

Laser output from the Yb-doped fiber is emitted from the core of 20 μm diameter in a full-cone angle of 120 mrad. V-number of the active fiber is 3.5, which suggests that it may provide guidance of a total of six number of fiber modes. Thus, fiber laser output may be slightly multimoded. But, the Yb-doped fiber was coiled on a mandrel of 150 mm diameter to increase losses for higher order modes and thereby ensure single-mode operation. A laser beam profiler from M/s Spiricon was used to measure laser output beam profile and measured value of M^2 was found to be ~1.04, which shows a nearly diffraction-limited laser output beam. During experiments, it was found that relaxation oscillations override the CW signal output near threshold, and consequently, generate high peak power random self-pulses with pulse duration of the order of a few nanoseconds, which damages fiber ends or any other fiber laser component. This damage of fiber components is irreversible. Thus, it is of utmost importance to remove random self-pulsing, before proceeding for the generation of high-power output from Yb-doped fiber lasers. Fiber ends are also prone to damage by dust particles due to emission of high power from very small core area and the presence of high-power density at the fiber-end faces. Hence, it is also important to protect fiber ends by means of fiber end caps. It was found that if the mirrors are not aligned properly, it introduces higher loss in the resonator and results in the generation of self-pulses, which may even damage the mirror coatings due to high peak power density of self-pulses. Figure 7 shows generation of high peak power random self-pulses when the resonator mirrors are slightly misaligned. However, if the mirrors are aligned properly to generate maximum output power, there are no self-pulses in the output. Thus, intracavity loss has to be minimized to prevent generation of random self-pulses.

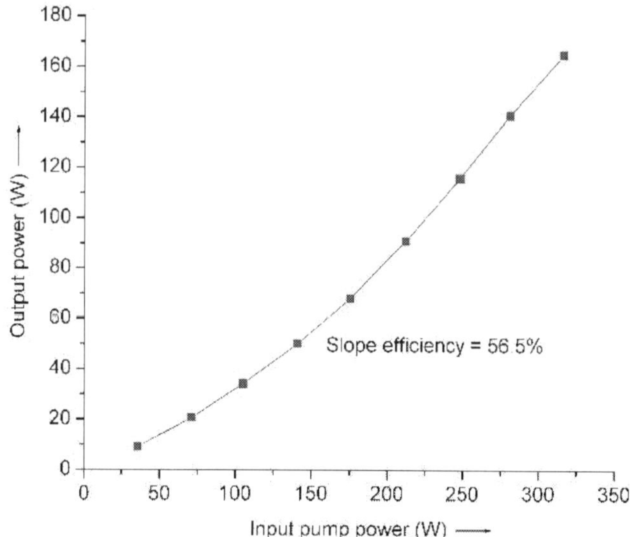

Figure 5. Variation of laser output power as a function of input pump power.

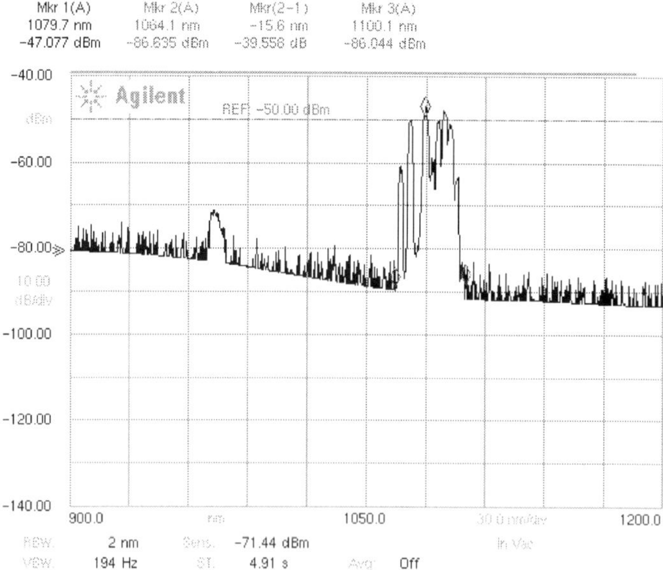

Figure 6. Output spectrum of Yb-doped CW fiber laser at the maximum output power of 165 W.

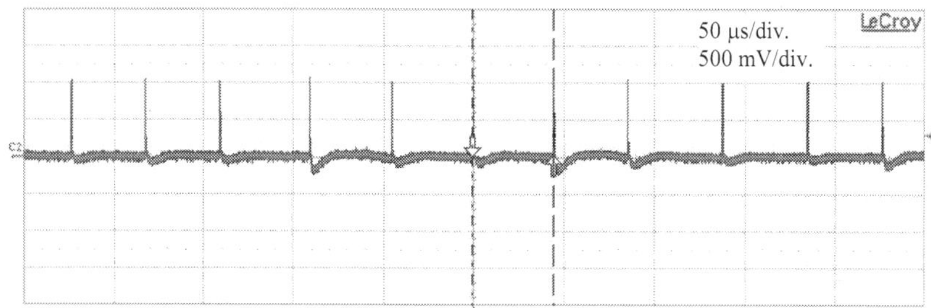

Figure 7. Self-pulses in the laser output with resonator mirrors slightly misaligned.

5. Self-pulsing dynamics of all-fiber oscillator configuration using FBG mirrors

Figure 8 shows a schematic view of 115 W of all-fiber Yb-doped CW fiber laser. In this all-fiber laser configuration, a Yb-doped double-clad fiber has been used as the gain medium having a core diameter of 20 µm and an inner-clad diameter of 400 µm. NAs of the core and inner clad are 0.075 and 0.46, respectively. Inner clad has an octagonal shape to avoid excitation of skew modes.

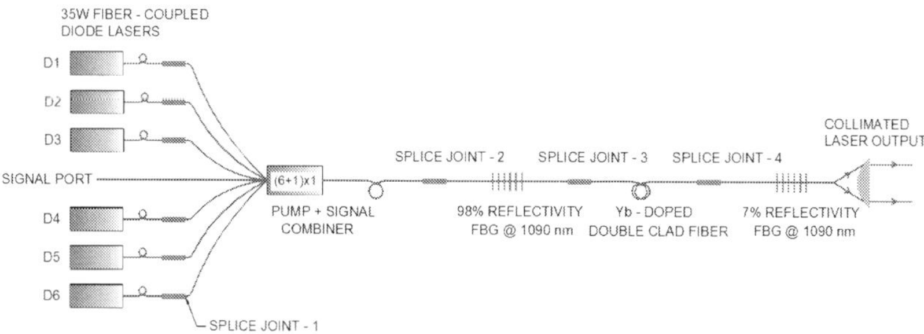

Figure 8. A schematic view of 115 W of all-fiber Yb-doped CW fiber laser.

Inner-clad pump absorption of the Yb-doped fiber at 975 nm is 1.7 dB/m. For efficient absorption of the pump beam, 10 m length of the active fiber has been used, which provides total pump absorption of ~17 dB. For pumping of Yb-doped fiber, a diode pump module of six fiber coupled diodes has been made. Each fiber coupled diode provides an output power of 35 W at 975 nm at 20 °C case temperature. Fiber coupled diode contains an optical fiber with a core diameter of 200 µm and an NA of 0.22. This diode-pump module has been spliced to (6+1)×1 fiber-optic signal and pump combiner. The core diameter of the fiber-optic pump combiner at

Figure 9. Optimized splice joint of 400 μm inner-clad diameter double-clad fiber with compatible fiber Bragg grating in double-clad fiber.

the output end is 400 μm with an NA of 0.46. Further, the output end of the fiber-optic pump combiner has been spliced to a FBG of ~98% reflectivity. This FBG is written in a compatible double-clad fiber, and it has a peak reflectivity at 1090 nm with a bandwidth of 0.2 nm. One end of Yb-doped fiber has been spliced to the other end of this HR FBG. Another FBG of ~7% reflectivity at 1090 nm has been spliced at the other end of Yb-doped fiber to make an all-fiber fiber laser oscillator. This 7% reflectivity FBG has also been written in a compatible double-clad fiber with a peak at 1090 nm and a bandwidth of 1.5 nm. The diode-pump module of six fiber-coupled diodes, pump combiner package, and FBGs are being cooled by means of water-cooled heat sinks and a closed loop water cooling unit at 20 °C of water temperature. A power supply unit has been used to control and operate all the six fiber-coupled laser diodes in series. The power supply unit provides voltage variations in the range of 0–15 V and current variations in the range of 0–100 Amp. Optimization of splice joint-1 of diode pump module with (6+1)×1 fiber-optic signal and pump combiner having 200 μm core diameter silica–silica fibers resulted in a transmission efficiency of 97%. After splice joint-1, the transmitted pump power from the pump combiner was 204 W at the maximum combined diode input pump power of 210 W. After splice joint-2 of the fiber-optic pump combiner with HR FBG mirror, transmission efficiency of 94% was achieved and the transmitted pump power from FBG was 192 W. Figure 9 shows optimized splice joint of 400 μm inner-clad diameter double-clad fiber with compatible FBG in double-clad fiber.[38]

Figure 10 shows variation of laser output power as a function of input pump power. It can be seen that there is no rollover even at the maximum output power due to thermal effects in Yb-doped double-clad fiber or any other fiber component. Threshold for lasing was about 17.0 W. An optical-to-optical conversion efficiency of 55% and a slope efficiency of 57% have been

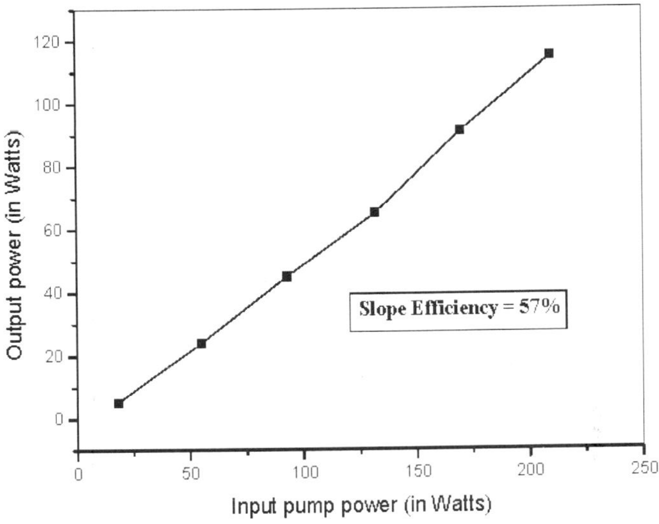

Figure 10. Variation of output laser power vs input pump power.

achieved in this all-fiber Yb-doped fiber laser oscillator configuration. Figure 11 shows wavelength spectrum of output laser signal at the maximum output power of 115 W. It shows signal peak at 1089.7 nm having a 3 dB linewidth of 0.64 nm. Major problems faced in this development were self-pulsing, optimization of splice joints, and heat removal. Self-pulsing was removed by minimization of intracavity losses and heat load from Yb-doped fiber was removed by tightly winding it on a copper spool, so that heat from fiber is conducted through copper spool. Intracavity loss is basically introduced by splice joints of FBGs with Yb-doped fiber. The splice joint loss was minimized by varying splice parameters such as fusion power, rate of fusion, argon flow rate, etc., using a commercial fusion splicing work station. This work station shows an estimated loss using image of splice joints. Although, it does not show an accurate data for splice loss, but a lower value of estimated loss indicates a better splice joint. Using this data, several splices were carried out by varying splice parameters and these joints were tested in resonator by measuring lasing threshold. A lower value of threshold was achieved with minimized estimated splice joint loss. During the testing of splice joints, it was found that laser output contained self-pulses and it became difficult to increase output power due to damage of FBG and fiber-end facets. It was found that with minimum value of lasing thresh-old and estimated loss, the self-pulses were removed and a truly CW output was observed without self-pulsing and no damage of fiber components took place. Figure 12(a) shows oscilloscope trace of self-pulsing amplitude with higher splice loss, and Figure 12 (b) shows oscilloscope trace with minimized splice loss. These traces were recorded using a 1 GHz photoreceiver and 1 GHz oscilloscope. It shows that self-pulse amplitude is minimized with reduction in intracavity splice losses. Laser output was emitted through the 20 µm core of doped fiber with a full-angle divergence of 150 mrad. V-number of the fiber is 4.425, and it can support about eight transverse modes. However, it was made to operate in single-transverse mode by winding it on a spool of 150 mm diameter, so that higher order modes are leaked out by bend induced loss. A laser cutting nozzle was made, which contains a 20 mm focal length lens for collimation and another 40 mm focal length lens for focusing to have a focused spot diameter of 40 µm. Using this cutting nozzle and oxygen as assist gas, micro-cutting of up to 2.5 mm thick stainless steel sheets and tubes was carried out with a kerf width of less than 200 µm.

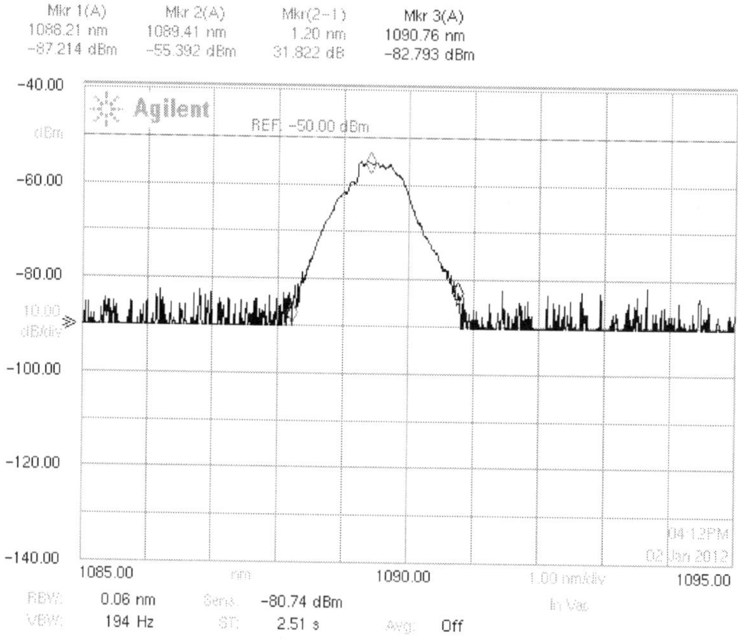

Figure 11. Wavelength spectrum of output laser signal.

6. Generation of 215 W of CW output power using all-fiber MOPA configuration and its self-pulsing dynamics

Figure 13 shows schematic of experimental setup of high-power 215 W of all-fiber Yb-doped CW fiber laser using MOPA configuration. In this all-fiber laser configuration, all-fiber oscillator as described in earlier in this chapter has been utilized. The output of oscillator was amplified by using another (6+1)×1 fiber-optic pump and signal combiner. From the oscillator stage, 115 W of output was achieved, which was further amplified to an output power of 215 W at the amplifier stage. As the laser is emitted from a very small (20 µm) core diameter of Yb-doped fiber, it is prone to damage by dust particles. Thus, at the exit end of amplifier, an end

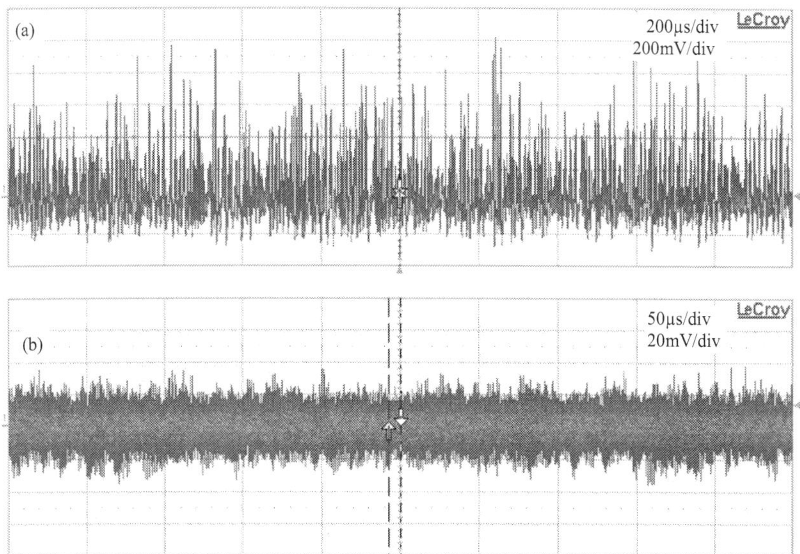

Figure 12. (a) Oscilloscope trace of self-pulses with higher splice loss, and (b) oscilloscope trace of laser output with minimized splice loss.

cap of 400 μm diameter was spliced to sustain higher damage thresholds. An optical-to-optical conversion efficiency of 55% and a slope efficiency of 57% have been achieved in this all-fiber Yb-doped fiber laser MOPA configuration. Output signal was peaked at 1089.7 nm having a 3 dB linewidth of 0.64 nm.[39]

As described in previous sections, major problems faced in this development were self-pulsing, optimization of splice joints, and heat removal. Self-pulsing, which is generation of high peak power ns-pulses even with CW pumping, results in catastrophic damage of fiber components and diode laser. It was removed by minimization of intracavity losses by minimization of losses through splice joints in the cavity. The splice joint loss was minimized by varying splice parameters such as fusion power, hot push, rate of fusion, argon flow rate, etc., using a fusion splicing work station. Heat load from Yb-doped fiber was efficiently removed by tightly winding it on a copper spool, so that heat from fiber is conducted through copper spool. Intracavity loss is basically introduced by splice joints of FBGs with Yb-doped fiber. Splice joints were also re-coated using low-index and high-index polymers for its protection. Figure 14 (a) and (b) shows the 2D and 3D beam profiles at the output of amplifier stage recorded using a Spiricon beam profiler, which shows that output is nearly diffraction limited with smooth variation in intensity. In this case also, if the slice joint losses are not minimized, it may result in higher intracavity losses, and consequently, result in high peak power self-pulses in the laser output. If not checked, these self-pulses may contain sufficient peak power to damage in-line fiber components and diode laser also.

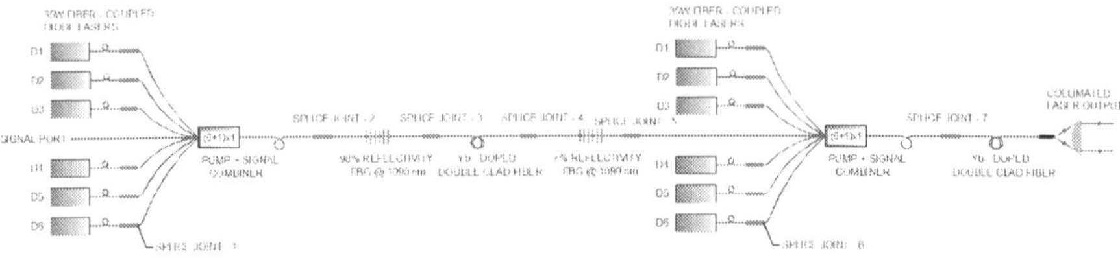

Figure 13. Schematic of experimental set up for MOPA configuration.

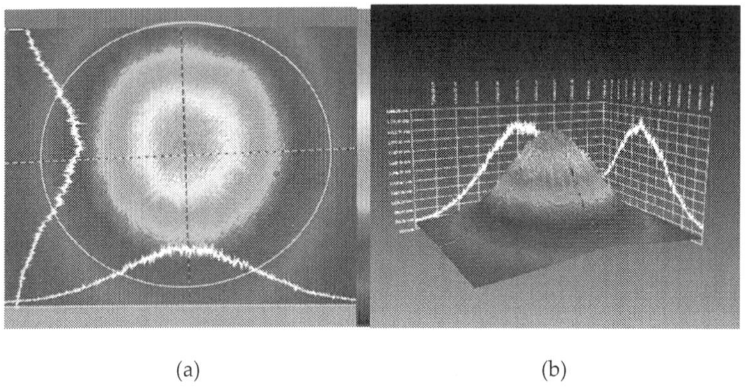

(a)   (b)

Figure 14. (a) 2D and (b) 3D beam profile recorded at the output of 215 W of Yb-doped CW fiber laser.

Thus, most of the fiber laser configurations suffer from self-pulsing. It is necessary to suppress self-pulsing in the output by a careful choice of doped fiber length, resonator configuration, and output spectrum with minimization of intracavity losses before proceeding towards high-power CW output from Yb-doped fiber lasers.

7. Conclusion

In conclusion, study on high-power Yb-doped CW fiber laser shows that higher intracavity losses lead to generation of unwanted high-peak power self-pulses in laser output which may damage fiber components and inhibit high-power laser generation. Hence, in order to achieve high-power CW fiber laser output, it is necessary to carefully select doped fiber length, resonator configuration, and reduce intracavity losses by minimizing losses at splice joints in the cavity and also to effectively manage heat load of active fiber and all-fiber components.

Author details

B.N. Upadhyaya[*]

Address all correspondence to: bnand@rrcat.gov.in

Solid State Laser Division, Raja Ramanna Centre for Advanced Technology, Indore, India

References

[1] <http//www.ipgphotonics.com>.

[2] Y. Jeong, J. K. Sahu, D. N. Payne, and J. Nilsson, "Ytterbium-doped large-core fiber laser with 1.36 kW continuous-wave output power", Opt. Express 12, 6088–6092 (2004).

[3] D. C. Hanna, R. M. Percival, I. R. Perry, R. G. Smart, P. J. Suni, and A. C. Tropper, "An ytterbium-doped monomode fibre laser: broadly tunable from 1.010 μm to1.162 μm and three-level operation at 974 nm", J. Modern Optics 37, 517–525 (1990).

[4] Jay W. Dawson, M. J. Messerly, R. J. Beach, M. Y. Shverdin, E. A. Stappaerts, A. K. Sridharan, P. H. Pax, J. E. Heebner, C. W. Siders, and C. P. J. Barty, "Analysis of the scalability of diffraction-limited fiber lasers and amplifiers to high average power", Opt. Express 6, 13240–13266 (2008).

[5] A. Hideur, T. Chartier, C. Ozkul, and F. Sanchez, "Dynamics and stabilization of a high power side-pumped Yb-doped double-clad fiber laser", Opt. Commun. 186, 311–317 (2000).

[6] P. Glas, M. Naumann, A. Schirrmacher, L. Daweritz, and R. Hey, "Self pulsing versus self locking in a cw pumped neodymium doped double clad fiber laser", Opt. Commun. 161, 345–358 (1999).

[7] F. Brunet, Y. Taillon, P. Galarneau, and S. LaRochelle, " A simple model describing both self-mode locking and sustained self-pulsing in ytterbium-doped ring fiber lasers", J. Lightwave Technol. 23, 2131–2138 (2005).

[8] F. Sanchez, P. Le Boudec, P.-L. François, and G. Stephan, "Effects of ion pairs on the dynamics of erbium-doped fiber lasers", Phys. Rev. A 48, 2220–2229 (1993).

[9] A. V. Kir'yanov, Yuri O. Barmenkov, and I. L. Martinez, "Cooperative luminescence and absorption in Ytterbium-doped silica fiber and the fiber nonlinear transmission coefficient at λ=980 nm with a regard to the ytterbium ion-pairs' effect", Opt. Express 14, 3981–3992 (2006).

[10] P. Le Boudec, M. Le Flohic, P. L. Francois, F. Sanchez, and G. Stephan, "Influence of ion pairs on the dynamical behaviour of Er^{3+}-doped fibre lasers," Opt. Quantum Electron. 25, 359 (1993).

[11] S. D. Jackson, "Direct evidence for laser re-absorption as initial cause for self-pulsing in three-level fiber lasers", Electron. Lett. 38, 1640–1642 (2002).

[12] D. Marcuse, "Pulsing behavior of a three-level laser with saturable absorber", IEEE J. Quantum Electron. 29, 2390–2396 (1993).

[13] E. Lacot, F. Stoeckel, and M. Chenevier, "Dynamics of an erbium-doped fiber laser," Phys. Rev. A 49, 3997–4008 (1994).

[14] S. D. Jackson and T. A. King, "Dynamics of the output of heavily Tm-doped double-clad silica fiber lasers", J. Opt. Soc. Am. B 16, 2178–2188 (1999).

[15] B. N. Upadhyaya, U. Chakravarty, A. Kuruvilla, A. K. Nath, M. R. Shenoy, and K. Thyagarajan, "Effect of steady-state conditions on self-pulsing characteristics of Yb-doped cw fiber lasers", Opt. Commun. 281, 146–153 (2008).

[16] R. Rangel-Rojo and M. Mohebi, "Study of the onset of self-pulsing behaviour in an Er-doped fiber laser", Opt. Commun. 137, 98–102 (1997).

[17] J. Li, K. Ueda, M. Musha and A. Shirakawa, "Residual pump light as a probe of self-pulsing instability in an ytterbium-doped fiber laser", Opt. Lett. 31, 1450–1452 (2006).

[18] Y. O. Barmenkov and A. V. Kir'yanov, "Pump noise as the source of self-modulation and self-pulsing in erbium fiber laser", Opt. Express 12, 3171–3177 (2004).

[19] B. Ortac, A. Hideur, T. Chartier, M. Brunel, G. Martel, M. Salhi, and F. Sanchez, "Influence of cavity losses on stimulated Brillouin scattering in a self-pulsing side-pumped ytterbium-doped double-clad fiber laser", Opt. Commun. 215, 389–395 (2003).

[20] M. Salhi, A. Hideur, T. Chartier, M. Brunel, G. Martel, C. Ozkul, and F. Sanchez, "Evidence of Brillouin scattering in an ytterbium-doped double-clad fiber laser", Opt. Lett. 27, 1294–1296 (2002).

[21] A. A. Fotiadi, P. Mégret, and M. Blondel, "Dynamics of a self-Q-switched fiber laser with a Rayleigh-stimulated Brillouin scattering ring mirror", Opt. Lett. 29, 1078–1080 (2004).

[22] A. Martinez-Rios, I. Torres-Gomez, G. Anzueto-Sanchez, and R. Selvas-Aguilar, "Self-pulsing in a double-clad ytterbium fiber laser induced by high scattering loss", Opt. Commun. 281, 663–667 (2008).

[23] Z. J. Chen, A. P. Grudinin, J. Porta, and J. D. Minelly, "Enhanced Q switching in double-clad fiber lasers," Opt. Lett. 23 (1998) 454.

[24] A. Hideur, T. Chartier, and C. Özkul, "All-fiber tunable ytterbium-doped double-clad fiber ring laser", Opt. Lett. 26, 1054–1057 (2001).

[25] W. Guan and J. R. Marciante, "Complete elimination of self-pulsations in dual-clad ytterbium-doped fiber lasers at all pumping levels", Opt. Lett. 34, 815–817 (2009).

[26] Y. H. Tsang, T. A. King, D.K. Ko, and J. Lee, "Output dynamics and stabilization of a multi-mode double-clad Yb-doped silica fiber laser", Opt. Commun. 259, 236–241 (2006).

[27] H. Takara, S. Kawanishi, and M. Saruwatari, "Stabilization of a mode-locked Er-doped fiber laser by suppressing the relaxation oscillation frequency component," Electron. Lett. 31, 292–293 (1995).

[28] L. Luo and P. L. Chu, "Suppression of self-pulsing in an erbium-doped fiber laser", Opt. Lett. 22, 1174–1176 (1997).

[29] W. H. Loh and J. P. de Sandro, "Suppression of self-pulsing behaviour in erbium-doped fiber lasers with resonant pumping: experimental results", Opt. Lett. 21, 1475–1477 (1996).

[30] V. Mizrahi, D. J. DiGiovanni, R. M. Atkins, S. G. Grubb, Y.-K. Park, and J.-M. P. Delavaux, "Stable single-mode erbium fiber-grating laser for digital communication", J. Lightwave Technol. 11, 2021–2025 (1993).

[31] A. Suzuki, Y. Takahashi, M. Yoshida, and M. Nakazawa, "An ultra low noise and narrow linewidth $\lambda/4$-shifted DFB Er-doped fiber laser with a ring cavity configuration", IEEE Photonics Technol. Lett. 19, 1463–1465 (2007).

[32] H. Chen, G. Zhu, N. K. Dutta, K. Dreyer, "Suppression of self-pulsing behavior in erbium-doped fiber lasers with a semiconductor optical amplifier", Appl. Opt. 41, 3511–3516 (2002).

[33] B. N. Upadhyaya, U. Chakravarty, A. Kuruvilla, S. M. Oak, M. R. Shenoy, and K. Thyagarajan, "Self-pulsing characteristics of a high-power single transverse mode Yb-doped CW fiber laser", Opt. Commun. 283, pp. 2206–2213 (2010).

[34] B. N. Upadhyaya, A. Kuruvilla, U. Chakravarty, M. R. Shenoy, K. Thyagarajan, and S. M. Oak, "Effect of laser linewidth and fiber length on self-pulsing dynamics and output stabilization of single-mode Yb-doped double-clad fiber laser", Appl. Opt. 49, pp. 2316–2325 (2010).

[35] U. Chakravarty, A. Kuruvilla, H. Harikrishnan, B. N. Upadhyaya, K. S. Bindra, and S. M. Oak, "Study on self-pulsing dynamics in Yb-doped photonic crystal fiber laser", Opt Laser Technol. 51, 82–89 (2013).

[36] B. N. Upadhyaya, "High-power Yb-doped continuous-wave and pulsed fibre lasers", Pramana J. Phys. 8215–27 (Jan. 2014).

[37] B. N. Upadhyaya, P. Misra, A. Kuruvilla, R. K. Jain, R. Singh, K. S. Bindra, and S. M. Oak, "Study and development of 165 W of single transverse mode Yb-doped CW fi-

ber laser", International conference on Fiber Optics and Photonics, Photonics-2012, IITM, Chennai, India, Dec 09–12, 2012.

[38] P. Misra, R. K. Jain, A. Kuruvilla, R. Singh, B. N. Upadhyaya, K. S. Bindra, and S. M. Oak, "Development of 115 W of narrow linewidth single transverse mode all-fiber Yb-doped CW fiber laser", 22nd DAE-BRNS National Laser Symposium (NLS-22), Manipal University, Manipal, Jan 8–11, 2014, India.

[39] P. Misra, R. K. Jain, A. Kuruvilla, R. Singh, B. N. Upadhyaya, K. S. Bindra, and S. M. Oak, "Development of 215 W of narrow linewidth all-fiber Yb-doped CW fiber laser based on MOPA configuration", Proceedings of National Laser Symposium, Tirupati, Dec 3–6, 2014, India.

Chapter 2

Active Q-switched Fiber Lasers with Single and Dual-wavelength Operation

R. Iván Álvarez-Tamayo, Manuel Durán-Sánchez, Olivier Pottiez, Baldemar Ibarra-Escamilla, Evgeny A. Kuzin and M. Espinosa-Martínez

Additional information is available at the end of the chapter

http://dx.doi.org/10.5772/61571

Abstract

A brief explanation on Q-switched fiber laser operating principle for active technique in terms of operation characteristics is presented. Experimental analysis of proposed pulsed fiber lasers by the active Q-switched technique is demonstrated. Experimental setups include the use of Er/Yb doped fiber as a gain medium and an acousto-optic modulator as cavity elements. Setup variations include the use of fiber Bragg gratings for wavelength selection and tuning and Sagnac interferometer for wavelength selection in single wavelength operation and for cavity loss adjustment in dual wavelength operation. The experimental analysis of principal characteristics of single-wavelength operation of the fiber laser and cavity loss adjustment method for dual-wavelength laser operation are discussed.

Keywords: Q-switched lasers, Er/Yb double-clad fibers, fiber Bragg gratings, Sagnac interferometer

1. Introduction

Fiber lasers have been studied almost from the onset of laser demonstration. Research on the development of innovative laser systems has been of constant interest in optics and photonics, having fast growth and becoming a central research area in scientific and industrial implementations. Fiber lasers have been widely studied because their unique characteristics of high power confinement, high beam quality, low insertion loss, compactness, and ruggedness. In general, they are attractive for different application areas such as medicine, telecommunications, optical sensing, and industrial material processing.

Fiber lasers make use of stimulated emission to generate light by using an active medium for gain supply. The use of fibers doped with rare-earth elements provides a gain medium with great thermal and optical properties for fiber laser development, in contrast to solid state lasers. Erbium-doped fibers (EDF) have been widely used for fiber laser implementations; however, in the last decade, the constant search of efficiency improvement in terms of very high gain with low pumping thresholds has significantly increased the use of Ytterbium doped fibers because they offer an efficiency above 80% [1].

Moreover, high-power fiber lasers are also of high interest for different applications such as spectroscopy, pump sources, and the study of nonlinear phenomena. In contrast with solid state lasers, a fiber laser requires longer interaction lengths favoring the occurrence of nonlinear effects when high power is achieved, making them desirable for optical switching, nonlinear frequency conversion, solitons, and supercontinuum generation, among other applications.

As it is known, pump diodes provide pump power limited to a few watts. This restriction also limits the fiber laser output power when conventional-doped (single-clad) fibers are used. With the development of double-clad fibers (DCF), high-power fiber lasers experienced significant advances since DCF makes an output power increase attainable. In conjunction with clad-pumping techniques, the DCF feature of large surface area-to-gain volume ratio, in addition to high doping concentration, offer high output power with an improved spatial beam confinement, in contrast with the use of single-clad doped fibers [2].

However, achieving high power continuous-wave (CW) operation of a fiber laser without output power fluctuations is not as straightforward. Taking into account this fact, the development of fiber lasers in pulsed regime provides a feasible alternative. In comparison with CW fiber lasers, pulsed fiber lasers provide high peak power that can be used in the generated wavelength or shifted to other wavelength range by nonlinear frequency conversion.

The most important pulsed regimes are Q-switching and mode-locking. In contrast with CW operation, in pulsed regimes the output is time dependent. In pulsed lasers by the Q-switched technique, stable and regular short pulses are obtained with pulse durations in the nanoseconds range; in contrast to ultrashort pulses obtained by mode-locked techniques, in this case the pulse duration corresponds to several cavity round trips. Q-switching can be developed by passive and active techniques.

Passive Q-switching is performed by using a saturable absorber element placed inside the cavity including graphene [3–5], carbon nanotubes (CNT) [6–8], transition metal-doped crystals [9–11], and semiconductor saturable absorber mirrors (SESAM) [12, 13]. On the other hand, the active Q-switching technique is based on the use of a modulator driven by an external electrical generator. Cavity loss modulation is typically performed by electro-optic modulators (EOM) [14, 15], and acousto-optic modulators (AOM) [16–18]. The EOM and the AOM are based on completely different principles of operation. While the EOM is based on the Pockels effect, the AOM modulates the refractive index of sound waves that generate a periodic grating as it propagates through the medium. In terms of operation, the main difference is the modulation bandwidth. Typically, the modulation bandwidth of an EOM is 500 kHz to 1 MHz, while for AOM is in the range of 50 to 100 MHz.

The use of the active Q-switching technique for pulsed laser operation allows higher energy pulses and stability. These advantages are increased in lasers based on integrated optics (all-optic) or all-fiber setups.

Otherwise, from the onset of the fiber Bragg gratings (FBG), the incorporation into the design of optical fiber lasers has been almost immediate, contributing significantly to the progress in this particular area. FBGs have been widely used as narrow band reflectors for generated laser wavelength selection. FBGs have unique advantages as optical devices including easy manufacture, fiber compatibility, low cost, and wavelength selection, among others. Moreover, the FBG central wavelength can be displaced or modified by mechanical strain or temperature application [19, 20]. This feature makes them capable devices for fiber lasers wavelength tuning [21] and for all-fiber modulation techniques [22, 23].

Moreover, dual wavelength fiber lasers (DWFL) have been studied previously [24–26]. Obtaining two wavelengths by using a single laser cavity has been attractive for potential application in areas such as optical sources, optical communications, optical instrumentation and others. The phenomenon of obtaining two wavelengths simultaneously with equal powers has been studied in terms of the competition between generated laser lines to improve the stability and DWFL emission control methods.

The use of variable optical attenuators (VOA) and high birefringence fiber loop Sagnac interferometer (*high birefringence fiber optical loop mirror*, Hi-Bi FOLM) have been demonstrated as efficient methods for generating two laser lines simultaneously through the adjustment of losses within the cavity.

Furthermore, wavelength tuning in pulsed DWFL development suggests its possible application in microwave and mainly terahertz generation. For DWFL improvement, different techniques for tuning and setting the separation between generated laser lines have been developed. The main goal in DWFL wavelength tuning is to obtain wide separation and continuous wavelength tuning. A reliable approach for wavelength tuning in DWFLs is the use of FBGs where the Bragg wavelength is shifted by temperature changes [27] or by mechanical strain application [28, 29].

In this chapter, a brief explanation on Q-switched fiber laser operating principle for active technique is presented. Also, a description of operation characteristics of Q-switched lasers, mainly in active Q-switching technique, is presented. Additionally, the current state of the art (to our knowledge) regarding DWFL in actively Q-switched pulsed regime is reviewed.

Furthermore, experimental setups for a dual wavelength actively Q-switched fiber laser and an actively Q-switched fiber laser with single and dual wavelength operations are experimentally demonstrated and analyzed. The experimental results of the lasers are discussed and compared in terms of operation characteristics, including repetition rate, pulse duration, pulse energy, average power, and peak power.

2. Q-Switched fiber lasers: A review from operating principle to single and dual-wavelength operation

Q-switching is a significantly employed technique in fiber lasers improvement. Q-switching is a suitable technique to obtain powerful pulses at low repetition rates from a few kHz to 100 kHz, typically; it can obtain short pulses in nanoseconds range, corresponding to several cavity round trips. This is in contrast with the mode-locking technique in which ultrashort pulses are obtained.

In recent years, Q-switched fiber lasers have been attractive due to their applications in medicine, optical time-domain reflectometry, terahertz generation, optical instrumentation, remote sensing, and materials processing in the industry.

Q-switching is performed by cavity losses modulation. The intracavity losses are maintained on a high level until the gain medium accumulates a considerable amount of energy supplied by the pumping source. Then the losses are abruptly minimized to build up the laser radiation within the cavity. As a result, a pulse with energy in a range of micro-Joules (even milli-Joules) is generated. Thus, the variation of the intracavity losses corresponds to a laser resonator Q-factor modulation. In general, Q-switched fiber lasers can be obtained with continuous or pulsed pump power. In the case of continuous pump power application, an important condition must be considered: a longer gain medium upper-state lifetime is required to avoid energy loss by fluorescence emission to reach stored high energy as a consequence. Particularly, in fiber lasers the saturation energy has to be high to prevent excessive gain that can lead to an early inception of laser generation. The pulse energy is commonly higher than the gain medium saturation energy.

Although Q-switched lasers based on bulk optics are regularly preferred over fiber lasers because of their larger mode areas to store more energy, the incorporation of bulk components in fiber lasers leads to a detriment of simplicity, robustness, and alignment of the laser. Also, the use of bulk elements in fiber lasers produces a beam quality degradation and the addition of high cavity losses, resulting in a decrease of laser performance and efficiency. Thus, in bulk fiber lasers approaches, the use of higher pump powers is required to increase the laser output power [23].

The Q-switching technique can be performed by passive and active methods:

Passive Q-switching is performed by using a saturable absorber element placed inside the cavity, which modulates automatically the losses within the laser cavity. As already mentioned, the variety of saturable absorber elements in passively Q-switched fiber lasers usually includes the use of graphene, CNT, metal-doped crystals, and SESAM [3–13]. The pulse repetition rate is determined and varied by the applied pump power, while the pulse duration and pulse energy are affected by the cavity and the Q-switching element parameters and commonly remain fixed. Thus, the repetition frequency cannot be modified with independence of other operation parameters [30]. To reach an efficient performance, the saturable absorber recovery time, commonly, has to be higher than the pulse duration and lower than the gain medium upper-state lifetime. Laser pulses generated with passive Q-switching typically present a low repetition rate range because of the applied pump power range. The main advantages of passively Q-switched lasers are their simple designs and low cost due to the fact that the use of external modulators and their electronics are not required.

On the other hand, active Q-switching is based on the Q-factor modulation using a modulator included in the fiber laser design. The modulation element is driven by an external electrical generator. In the active Q-switching technique, the stored energy, when cavity loss is high, generates a pulse soon after an external electrical signal arrives on the modulator driving the intracavity losses to a low level. In this case, the pulse duration and the pulse energy depends on the energy stored by the gain medium. Hence, the pump power and the repetition rate variations modify the achieved pulse duration and pulse energy. For the active technique, the modulating switching time does not have to be similar to the pulse duration, the pulse duration is in any case of the order of the laser resonator round-trip time. As has been mentioned previously, active cavity losses modulation is typically performed by EOMs and AOMs [14–18]. According to the technological progress, the used modulators have been experiencing important changes. In early actively Q-switched lasers approaches, modulators were mainly using bulk components. Then, they were designed by using integrated optics coupled to optical fibers. Recently, all-fiber modulators have been included in fiber laser designs to increase the overall performance of the laser. The acousto-optic Q-switching is performed by a radio frequency (RF) power controlling a transducer. The generated acoustic wave provides a sinusoidal optical modulation of the gain medium density resulting in an intracavity loss modulation. AOMs can rapidly modulate the cavity losses that allows the Q-switched pulses generation with pulse durations of a hundred of nanoseconds. The shortest pulse durations and the highest pulse energies are obtained at the lowest allowed repetition rate, however, at the cost of obtaining low output average power. The use of the active Q-switching technique for pulsed laser operation allows higher energy pulses and stability. These advantages increase with laser designs based on integrated optics or all-fiber laser designs.

Moreover, most of the Q switched fiber laser approaches are focused on the use of single-clad fibers as a gain medium. In comparison with single-clad Er or Yb doped fibers, Er/Yb double clad co-doped fibers (EYDCF) can be used to suppress the self-pulsing of Er ion-pair [4]. This effect, in addition to cladding pump techniques, can be used to increase the pump power efficiency, minimizing the gain degradation by using EYDCF as a gain medium.

Regarding passively Q-switched fiber lasers, different approaches using EYDCF have been reported [30–33]. Laroche et al. [30] in 2002 presented a pioneer experimental setup of pulsed lasers for passive Q-switched technique using EYDCF as gain medium. They presented an open cavity laser using Cr2+:ZnSe as a saturable absorber. V. Philippov et al. [33] reported a similar configuration by using Cr2+:ZnSe and Co2+:MgAl2O4 as saturable absorbers. The maximum average power of 1.4 W was achieved in pulses with durations from 370 to 700 ns for a repetition rate between 20 kHz and 85 kHz.

In the case of actively Q-switched lasers using EYDCF, to our knowledge, a small number of approaches have been reported [34, 35]. Recently, González-García et al. [34] reported a linear cavity EYDCF laser Q-switched by an acousto-optic modulator. The pump power is introduced to the EYDCF by a free space subsystem carefully optimized by using a lenses design.

Nowadays, the development of DWFL has been of interest because of their ability to obtain two laser wavelengths simultaneously by the use of a single cavity. DWFL's advantages are low cost, simple design, fiber compatibility, and low loss insertion devices, making feasible more complex optical devices design. From its first approaches in CW fiber lasers, DWFL research has increased because of their potential applications in different areas such as optical communication systems, optical instrumentation, optical sources, and spectral analysis.

In recent years, the experience in the study of DWFL in CW regime has been incorporated into the implementation of DWFLs in pulsed regime. With this advancement, it has opened the possibility of new applications where high output power is required such as research of nonlinear phenomenon study and remote sensing. The main issue in DWFL operation is centered on the difficulty of obtaining two stable wavelengths simultaneously because doped fibers behave as a homogeneous gain medium at room temperature, causing a competition between the generated laser lines that leads to a generated laser line's instability.

Commonly used methods to balance the generated wavelengths include the use of polarization controllers (PC) and variable optical attenuators (VOA), among others; however, most of them are arbitrary methods in the absence of a measurable physical variable for wavelength competition analysis and characterization, affecting the repeatability in the laser performance. The methods are focused on adjusting laser intracavity losses to balance the simultaneously generated wavelengths. In previously reported investigations, the Sagnac interferometer with high birefringence (Hi-Bi) fiber loop capability has been theoretically and experimentally demonstrated as a trustworthy alternative for the adjustment of losses within the cavity [21, 36–38], since the Hi-Bi FOLM periodical spectrum can be finely displaced in wavelength by temperature variations applied on the Hi-Bi fiber loop [37].

Different experimental setups of DWFL by passive Q-switching technique have been reported [39–41]. However, to our knowledge, there have not been reported approaches using EYDCF as a gain medium. Concerning cavity losses adjustment for dual wavelength laser operation, the most frequent method is the use of PC in ring cavity fiber lasers. H. Ahmad et al. [40] reported a ring cavity passive Q-switched DWFL operating at 1557.8 nm and 1559 nm by using PC for dual wavelength generation. A nonlinear optical loop mirror (NOLM) with a dispersion-decreasing taper fiber (DDTF) in the fiber loop is used as a passive Q-switching element.

Regarding actively Q-switched fiber lasers, only few attempts have been reported in which dual wavelength emission is obtained. In 2013, G. Shayeganrad [42] reported a compact linear cavity actively Q-switched DWFL. The Q-switching is performed by an AOM. The gain medium is a c-cut Nd:YVO$_4$ crystal with a feature of dual-wavelength generation in Q-switched regime without adjustment elements. An undoped YVO$_4$ crystal is used for stimulated Raman scattering (SRS) effect enhancement. The SRS simultaneous wavelengths at 1066.7 and 1083 nm are shifted at 1178 and 1199.9 nm.

S.-T. Lin et al. [43] reported a selectable dual-wavelength actively Q-switched laser. By using two electro-optic periodically poled Lithium niobate (PPLN) integrated crystals, the output wavelengths between 1063 and 1342 nm are selected with voltage variations on PPLN Bragg modulators (EPBM) sections. It is worth mentioning that both cited experimental setups are designed with bulk optic elements with high pump power application, around 20 W. As has been said above, such designs require fine alignment, so typically, efficiency and instability problems are presented.

The all-fiber or optical fiber coupled laser systems promise to be an option for solving alignment issues minimizing losses within the laser cavity. Pump-to-signal efficiency can be increased and, consequently, highly increased pump power is not required to obtain more energetic pulses. However, for such designs, the output power is typically limited by the maximum signal power handled by the employed components.

Therefore, the use of double-clad doped fibers provides a stable and straightforward method to generate high energy nanosecond pulses in actively Q-switched dual-wavelength fiber lasers. From reported investigations, EYDCF offers high conversion efficiency for high-energy pulses generation [44, 45].

Regarding EYDCF use, in 2014, an actively Q-switched wavelength tunable DWFL using EYDCF as a gain medium has been reported [44]. The linear cavity laser incorporates the use of bulk components to introduce pump power in the EYDCF. The laser wavelengths are generated and simultaneously tuned by using a polarization maintaining fiber Bragg grating (PM-FBG). The maximal separation between generated wavelengths of 0.4 nm is adjusted by polarization axis adjustment performed by a PC. The simultaneous wavelength tuning range of ~11.8 nm is obtained by axial strain applied on the PM-FBG. The maximal average power of 22 mW is obtained with a repetition rate of 120 kHz with a pump power of 1.5 W.

Recently, a self-Q-switched (SQS) EYDCF laser with tunable single operation and dual wavelength operation using a Hi-Bi FOLM as a spectral filter was experimentally demonstrated [45]. The wavelength tuning in single wavelength operation and the cavity loss adjustment for dual wavelength operation is performed by temperature variations applied in the FOLM Hi-Bi fiber loop, allowing Hi-Bi FOLM spectrum wavelength shifting. Stable SQS pulses with duration of 4.1 µJ and repetition rate of 25 kHz are obtained with a pump power of 575 mW. The single wavelength tuning range over 8.4 nm is obtained with FOLM Hi-Bi fiber loop temperature variation in a range of ~7.2 °C. Separation between generated simultaneous dual wavelengths is 10.3 nm.

Then, we propose the use EYDCF fiber as gain medium for the design of actively Q-switched lasers with operation in single and dual wavelength. Also, we propose the use of FBGs and Hi-Bi FOLM as cavity elements that allow modifying the characteristics of laser operation and improve its performance by straightforward methods.

3. Actively Q-switched dual-wavelength fiber laser based on fiber Bragg gratings

In this section, an experimental analysis of a ring cavity dual-wavelength actively Q-switched fiber laser with an EYDCF as a gain medium is presented. A pair of FBGs is used for separately generated laser lines tuning by mechanical compression/stretch applied on the FBGs. Simultaneously generated dual-wavelength laser lines tuning are presented with wavelengths separation from 1 nm to the maximal separation of 4 nm (without the need of cavity loss adjustment).

The experimental setup is presented in Figure 1. The fiber ring cavity laser is based on the use of 3 m of EYDCF as a gain medium. The EYDCF is pumped with a laser source at 978 nm through a beam combiner. The pump power of 5 W is limited by the maximal AOM signal power of 1 W. An optical isolator with maximal output power of 5 W is used to ensure unidirectional operation. An optical subsystem formed with a 50/50 optical coupler with output ports connected to FBG1 and FBG2 with central wavelengths at 1543 and 1548 nm respectively, allows dual wavelength emission at FBG reflected wavelengths; it is also used for separate laser wavelength emission monitoring at outputs 1 and 2. The FBGs with approximately 99% of maximum reflectance are placed on mechanical devices for generated laser wavelength tuning by applying axial strain on the gratings. The simultaneously generated laser wavelengths are measured at 90/10 coupler output 3. A fiber-pigtailed AOM driven by a RF signal is placed for active Q-switching pulsed laser operation. The output spectra monitored at output ports (1, 2, and 3) are measured with an OSA and also the Q-switched pulses are detected and observed with a photodetector and an oscilloscope, respectively.

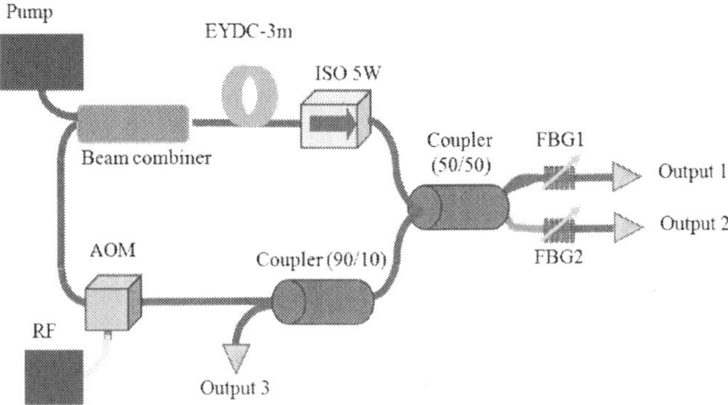

Figure 1. Experimental setup for actively Q-switched dual wavelength ring cavity fiber laser.

Figure 2 shows the experimental results for the dual-wavelength fiber laser spectrum measurements with fixed pump power of 5 W. The measurements were obtained at output 3 with an OSA with attenuation. Output power results are presented in linear scale to support the achieving of two simultaneous laser wavelengths with equal powers. Two simultaneous wavelengths are obtained without requiring cavity losses adjustment in the presented wavelength separation tuning range, however, we noticed the requirement of cavity losses adjustment for wavelength separations above 4 nm. Results for dual wavelength operation with cavity losses adjustment (wavelengths separation above 4 nm) are not presented since it was performed introducing curvature losses by fiber bending applied between 50/50 output ports and FBGs connections; an arbitrary method in which it is not possible to characterize the competition between the generated laser lines.

Figure 2(a) shows the generated laser lines spectrum measurements for dual wavelength Q-switched laser operation with different wavelength separations. The separation tuning from 1 to 4 nm is achieved by mechanical compression/stretch applied on the FBGs. The repetition rate remained fixed at 70 kHz. As it is shown, dual wavelength laser operation is generated simultaneously with approximately equal laser lines output powers without an adjustment of losses within the cavity. As it can be seen, for the repetition rate and pump power settings, a preference exists to generate the longer wavelength during the competition between the laser lines.

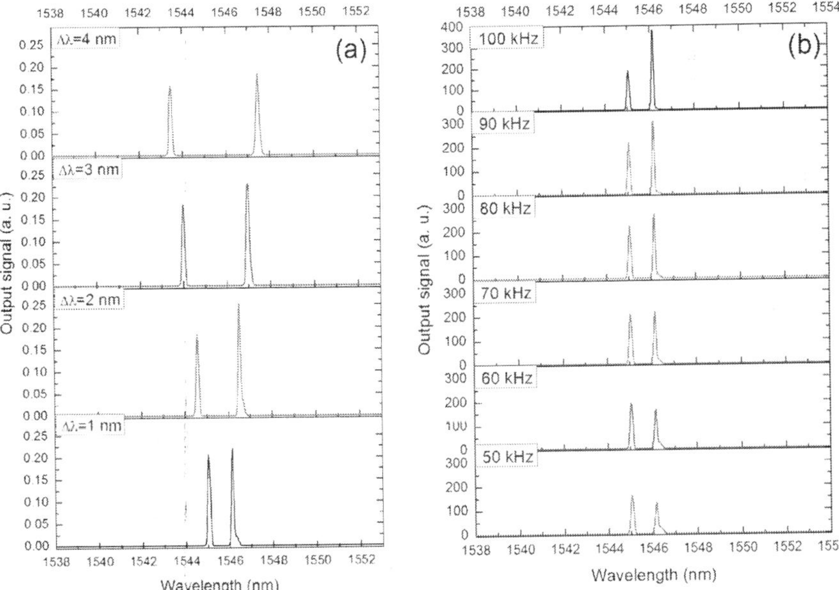

Figure 2. Spectrum measurements for dual-wavelength actively Q-switched laser operation, (a) different generated wavelengths separation tuning with fixed repetition rate, (b) generated laser lines with wavelength separation of 1 nm and repetition rate variations.

Figure 2(b) shows the generated dual wavelength laser lines with fixed wavelength separation of 1 nm (λ_1 = 1545.2 nm and λ_2 = 1546.2 nm) for repetition rate variations from 50 to 100 kHz. As it can be observed, output powers for both simultaneous wavelengths increase when the repetition rate is increased. The competition between generated laser lines presents a preference to generate the longer wavelength as the repetition rate is increased, however, dual wavelength laser operation is presented over the repetition rate range without cavity losses adjustment.

Figure 3 shows the output power ratio for the two simultaneously generated laser lines measured $P(\lambda_2)/P(\lambda_1)$, where λ_1 and λ_2 are the shorter and the longer laser wavelengths, respectively. The spectrum measurements were performed at output 3 with an OSA and output powers were individually monitored at output 1 and 2 with a photodetector and a power meter. The measurement of the power ratio between generated laser lines is a straightforward method for numerically analyzing the competition between laser line behavior. With output power ratio $0 < P(\lambda_2)/P(\lambda_1) < 1$, the shorter wavelength is generated with power above the longer wavelength. On the other hand, for $P(\lambda_2)/P(\lambda_1) > 1$, the longer wavelength presents an output power above the shorter wavelength. As it has been previously shown for the proposed experimental setup, there exists a preference to generate the longer wavelength. Furthermore, it was shown that the Q-switched dual wavelength fiber laser output powers are modified with repetition rate and tuned wavelength variations.

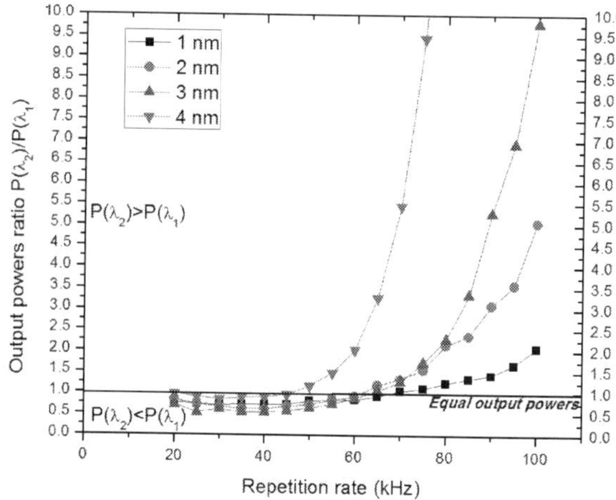

Figure 3. Dual wavelength operation generated laser wavelengths power ratio $P(\lambda_2)/P(\lambda_1)$ for repetition rate variations and different wavelengths separations.

In Figure 3, it can be clearly observed that with increasing repetition rate, the competition between laser lines has an imbalance in which the longest wavelength has a preference to be generated. Strong competition allowing dual wavelength laser operation with almost equal output powers from 20 kHz to about 60 kHz of repetition rate can be observed. With repetition rate variations from 60 kHz to 100 kHz, the longer wavelength output power increases significantly, at the expense of the shorter wavelength output power. As it can be also observed, the range of repetition values over which the longer wavelength starts growing significantly is shortened when increasing the separation between the generated laser wavelengths. As it is shown, for a wavelength separation of 1 nm, the maximum power ratio is about 2 times, with a repetition rate of 100 kHz. However, for a wavelength separation of 4 nm and a repetition rate of 70 kHz, an output powers ratio in which λ_2 is 9.5 times greater than λ_1 is observed.

Figure 4 is a group of experimental results for actively Q-switched dual wavelength laser pulses. The results also show pulse profiles for different repetition rate variations, comparison between pulses measured at different outputs, and experimental analysis of pulses time shift by repetition rate variations.

Figure 4(a) shows the optical pulse measurements for actively Q-switched dual wavelength laser operation. The wavelength separation remains fixed at 4 nm between simultaneously generated laser lines. With the use of a photodetector and an oscilloscope, the pulse traces together with the leading pulse of the signal applied to the AOM were obtained at output 3 where both generated wavelengths are simultaneously measured. The resulting pulses were obtained for different repetition rate variations from 50 to 100 kHz. For actively Q-switched operation, with the increase of repetition rate, the pulse duration typically increases as pulse amplitude decreases. As it can be seen, with a repetition rate of 50 kHz, there is a time shift of 93.7 ns between the leading edge of the electrical pulse applied to the AOM and the generated laser pulse. As we can observe, the time shift depends on the repetition rate.

The dependence of the temporal pulse shift on the repetition rate variations is shown in Figure 4(b). As it is shown, the pulse time shift increases as the repetition rate increases. Thus, it can be observed that for a repetition of 100 kHz, the pulse temporal shift between the electrical modulation signal leading edge and the generated pulse increase to ~2.3 μs.

Figure 4(c) shows the pulse traces that correspond to the same dual wavelength generation with wavelength separation of 4 nm and repetition rate of 50 kHz. Since the FBGs have a reflection close to 100% at the central wavelength, it is possible to obtain independently single laser concerning each of the generated laser wavelengths at the outputs 1 and 2 as a result of the signal transmitted by each FBG. Thereby, the pulses generated by the laser wavelength λ_1 = 1543.5 nm (blue line) obtained at output 2 and the optical pulses for wavelength λ_2 = 1547.5 nm (red line) acquired at output 1 are shown together with the optical pulse for both λ_1 and λ_2 measured at output 3. As it is shown, a slight time shift and pulse widening is observed for both wavelength pulse measurements (output 3) compared to the individual pulses observed for λ_1 and λ_2.

Figure 5 shows the output power in dual wavelength operation for generated laser wavelength separations $\Delta\lambda$ = 1 nm (λ_1 = 1545.2 nm and λ_2 = 1546.2 nm) and separation $\Delta\lambda$ = 4 nm (λ_1 = 1543.5 nm and λ_2 = 1547.5 nm) as a function of the repetition rate variations over the range from 30 kHz to 100 kHz, with the used pump power of 5 W. The difference between measured average power for both wavelength separations $P(\Delta\lambda = 1\ nm)-P(\Delta\lambda = 4\ nm)$ at the same repetition rate for both measurements is also shown. The average power was measured at output 3 with a power meter. As what typically occurs in actively Q-switched fiber lasers, it is observed that the average power increases with the repetition rate increase. As can be seen, the maximal average power is obtained with repetition rate of 100 kHz. For the dual wavelength operation with laser lines separation of 1 nm the maximal average power (red line, squared symbol) is 496 mW while it is 490 mW for a separation of 4 nm (blue line, circled symbol). The difference between average power measured for both wavelength separations tuned in a range from −10 mW to 10 mW is observed. As it can be also observed, the behavior

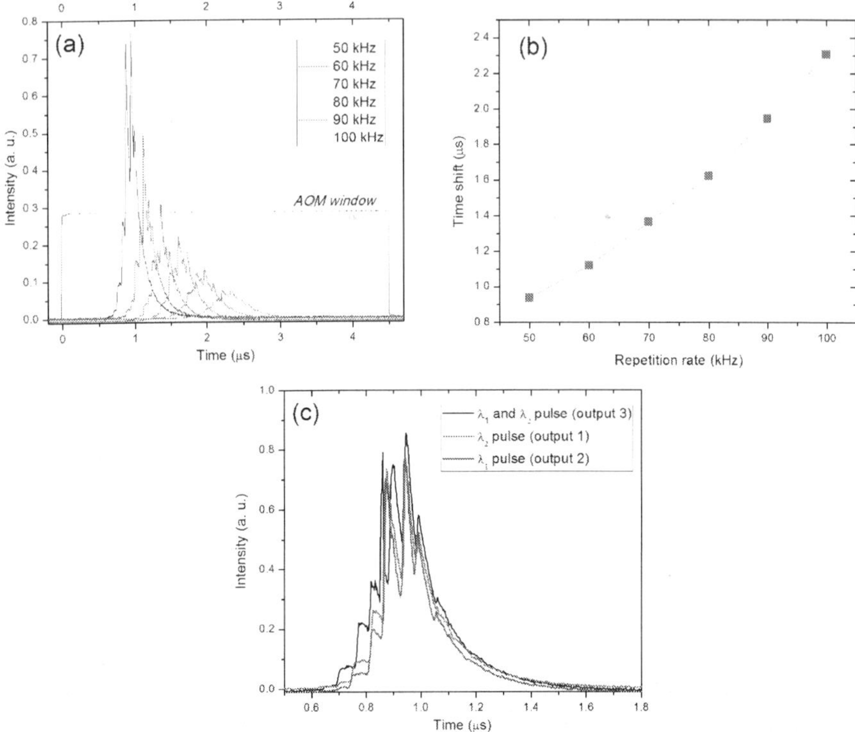

Figure 4. Optical pulses for Q-switched dual-wavelength operation with wavelength separation of 4 nm, (a)pulse profiles with repetition rate variations, (b) pulse time shifts on repetition rate variations, (c) separatepulse profiles for λ_1 and λ_2, and both λ_1 and λ_2 pulses with repetition rate of 50 kHz.

of average power on repetition rate has no significant variations with respect to the tuned wavelength separation between generated laser lines.

Figure 6 shows the measured pulse duration and the estimated pulse energy on repetition rate variations and the estimated pulse peak power for dual-wavelength laser operation. Results are obtained for wavelength separation between generated laser lines of 1 nm and 4 nm. Pulse profiles for Q-switched dual wavelength operation with both generated wavelength separations were performed with a photodetector and monitored by an oscilloscope. From pulse shape measurements, pulse duration was obtained. The pulse energy for each wavelength separations is estimated with the repetition rate and the average power results shown in Figure 5. Estimation of pulse peak power is obtained with the pulse energy and the pulse duration achieved results.

Typically for actively Q-switched lasers, with the increase of repetition rate, the obtained pulses widens increasing the pulse duration. Thus, although the pulses train average power

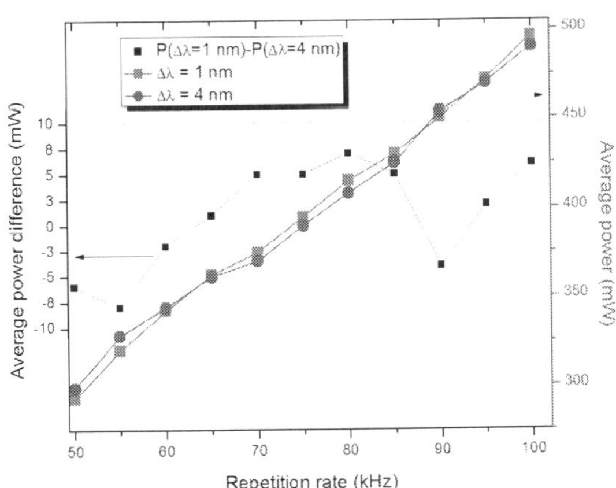

Figure 5. Average power on repetition rate variations of Q-switched dual wavelength operation with wavelength separations of 1 nm and 4 nm.

increases with the repetition rate increase (see Figure 5), the optical pulses are less energetic by the widening and the pulse peak power reduction experienced (see Figure 4(a)).

Figure 6(a) shows the pulse duration and pulse energy on repetition rate variations for dual wavelength laser operation. The pulse duration and pulse energy show a typical behavior of actively Q-switched lasers. With the repetition rate increase, pulse duration increases and pulse energy decreases are observed. For simultaneously generated laser wavelength separation of 1 nm, maximal pulse energy (blue line, circled symbol) of ~5.86 µJ and minimal pulse duration (black line, circled symbol) of 220 ns are obtained with the minimal repetition rate of 50 kHz. Similarly, for a wavelength separation of 4 nm at the same repetition rate, the maximal pulse energy (blue line, squared symbol) of 5.98 µJ and the minimal pulse duration of 295 ns is obtained.

The estimated pulse peak power on repetition rate variations is shown in Figure 6(b). As it can be observed, the pulse peak power (pulse amplitude) for the wavelength separation of 1 nm is higher compared with what is observed for the wavelength separation of 4 nm. This result is essentially attributed to a smaller increase in pulse duration for the measurements of $\Delta\lambda =$ *1 nm* as the repetition rate is increased (shown in Figure 6(a)). For $\Delta\lambda = 1$ *nm* and $\Delta\lambda = 4$ *nm*, the maximal pulse peak power of ~26.6 W and ~20.27 W, respectively, are obtained for the minimal repetition rate of 50 kHz, when the pulse widening is minimal.

In this section, an experimental analysis of an actively Q-switched ring cavity fiber laser has been presented. Through experimental and estimated results of laser spectra emission and generated laser pulses, the behavior of the dual wavelength laser operation of competitions between the simultaneously generated laser lines and the evolution of generated laser pulses has been analyzed. Actively Q-switched pulsed laser parameters as repetition rate, pulse duration, pulse energy, average power of the laser emission, and peak pulse power has been

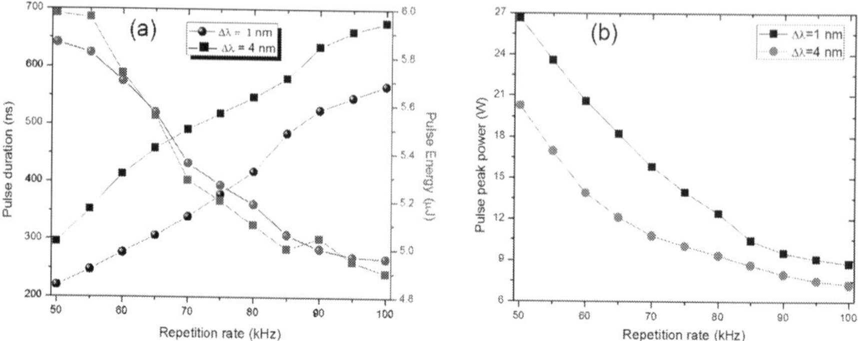

Figure 6. Q-switched dual wavelength operation with wavelength separations of 1 nm and 4 nm on repetition rate variations, (a) pulse duration and pulse energy, (b) pulses peak power.

also experimentally studied in terms of different tuned separations for two simultaneously generated wavelengths and variations on rate repetition of the electrical signal applied to the AOM. Results have been shown that are generalized to any actively Q-switched laser and particularly for lasers with dual wavelength operation. It is worth mentioning that for the proposed experimental setup, it is not necessary to implement a cavity losses adjustment method for the shown operation tuning range (wavelengths separations from 1 to 4 nm), however, a cavity losses adjusting method is required when during the competition between generated laser wavelengths there exists a wavelength preferred for the laser emission.

4. Actively Q-switched dual-wavelength fiber laser with a Sagnac interferometer for cavity losses adjustment

In this section, a linear cavity actively Q-switched fiber laser is proposed for experimental analysis. In contrast to the laser experimental setup demonstrated in the previous section (in which explaining the parameters of active Q-switched fiber lasers was intended), the experimental laser setup is a linear cavity configuration in which a method to adjust the losses within the cavity (when required) for simultaneous dual wavelength laser operation is presented. The proposed configuration includes the use of a Sagnac interferometer with high birefringence fiber in the loop (Hi-Bi FOLM) used as a spectral mirror and mainly for cavity loss adjustment during the laser lines competition in two simultaneous laser lines generation. The use of Hi-Bi FOLM as a reliable method of cavity losses adjustment for lasers operating in dual wavelength application has been extensively studied by our research group [21, 37, 38, 45]. The main objective of this section is to illustrate through a proposed experimental setup that the Hi-Bi FOLM can also be used to implement dual wavelength fiber lasers in pulsed regime for the actively Q-switched technique as well as the experimental analysis of dual wavelength laser operating parameters.

The proposed actively Q-switched fiber laser experimental setup is shown in Figure 7. The linear cavity laser is bound by two FBGs at one end and a Hi-Bi FOLM at the other end. A 3-m length of EYDCF used as a gain medium is pumped by a laser source at 978 nm through a beam combiner. The pump power was fixed to 1.5 W. An AOM driven by an RF signal generator is used for application of the active Q-switching technique. FBG1 and FBG2 with reflections of 99% at central wavelength tuned to 1542.7 nm and 1552.7 nm, respectively, are used as narrow band mirrors for generated laser wavelengths selection. With the selected FBGs central wavelengths, the separation between generated laser lines is ~10 nm. The Hi-Bi FOLM is formed by a 50/50 coupler with output ports interconnected through a Hi-Bi fiber segment of ~56 cm. The Hi-Bi FOLM is acting as a wide band mirror with a periodical spectrum. With the selected Hi-Bi fiber segment, the spectrum period is ~10.3 nm [35]. The Peltier device used for Hi-Bi fiber temperature control is used to shift the Hi-Bi FOLM spectrum in wavelength. This Hi-Bi FOLM spectrum displacement is the method for cavity losses adjustment for dual wavelength laser operation [35]. The splices between Hi-Bi fiber ends and 50/50 output ports are placed in mechanical rotation stages for Hi-Bi FOLM transmission spectrum amplitude adjustment [35]. The Hi-Bi FOLM amplitude was adjusted near maximal contrast. The unconnected 50/50 coupler port (output port) is used for Hi-Bi FOLM transmission spectrum measurement (with pump power below the laser generation threshold) and for laser spectrum measurement with an OSA. The output port is also used for pulses detection by a photodetector and observed on an oscilloscope.

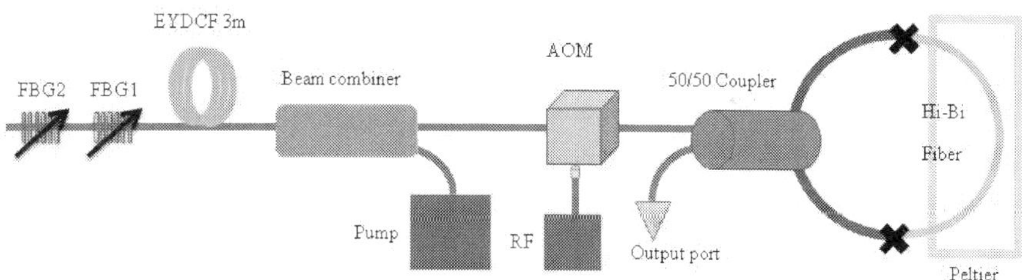

Figure 7. Experimental setup for actively Q-switched linear cavity dual wavelength fiber laser.

Figure 8 shows the cavity losses adjustment performance for single and dual wavelength laser operations. The adjustment is performed by temperature changes in the Hi-Bi FOLM fiber loop. The temperature meter and control has a resolution of 0.06 °C. The repetition rate was set to 60 kHz.

Figure 8(a) shows the three Hi-Bi FOLM transmission spectrum for the Hi-Bi fiber loop temperatures in which single wavelength operations for λ_1 and λ_2 and dual wavelength operation is generated. The Hi-Bi FOLM spectrum measurements were performed with pump power below the laser generation threshold at the output port with an OSA. As it can be seen, dual wavelength laser operation is obtained with Hi-Bi fiber loop temperature of 25.9 °C in which cavity losses are balanced. With the Hi-Bi loop temperature increase, the spectrum shifts to shorter wavelengths performing an imbalance in the competition between the laser lines, thus, the shorter wavelength (λ_1 = 1547.2 nm) laser emission is favored. On the other hand, a decrease in Hi-Bi FOLM loop temperature favors the emission of the longer wavelength ($\lambda_2 =$ 1547.2 nm). In Figure 8(b), the laser spectrum emission for dual wavelength operation and single wavelength operations for λ_1 and λ_2 are shown. The measurements were performed with pump power of 1.5 W. As it is shown, single wavelength laser operation for λ_1 and λ_2 are obtained with temperatures of 26.6 °C and 25.1 °C, respectively. Dual wavelength operation with approximately equal powers is obtained with Hi-Bi fiber loop temperature of 25.9 °C. The temperature operation range is ~1.5 °C.

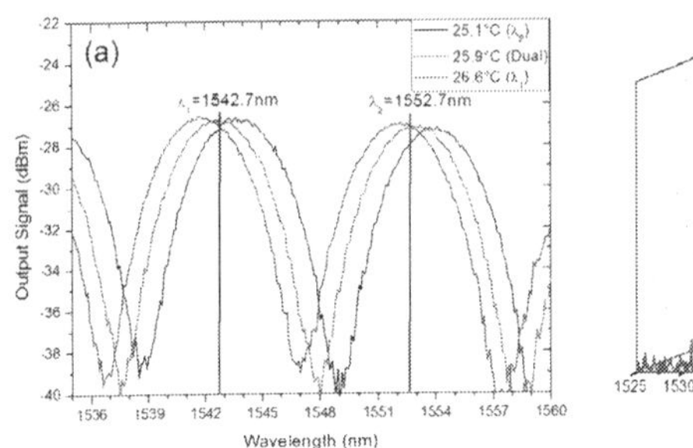
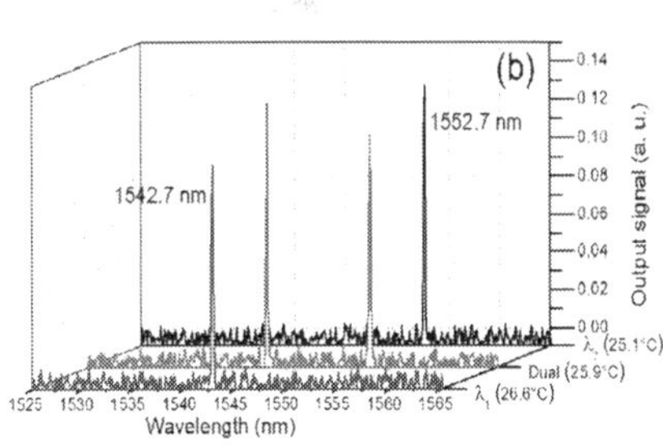

Figure 8. Cavity losses adjustment for laser operation, (a) Hi-Bi FOLM transmission spectrum wavelength displacement for single and dual wavelength operation Hi-Bi fiber temperatures, (b) generated laser spectrums for single and dual wavelength operations.

In Figure 9, pulsed regime measurements for actively Q-switched dual wavelength laser operation are presented. Pulse train profile and average power on repetition rate variations are shown.

Figure 9(a) shows the pulse train in time domain for dual wavelength laser operation with repetition rate of 60 kHz measured at the output port. The Hi-Bi fiber loop temperature was set to 25.9 °C for dual wavelength operation with equal powers as it is shown in Figure 8(b). For repetition rates below 35 kHz and above 75 kHz, unstable pulses are generated since the laser pulses displace outer the modulating AOM electrical signal time window. The inset on Figure 9(a) shows a sample pulse from the pulses train. The estimated pulse duration is ~448 ns.

In Figure 9(b), the average power on repetition rate variations for dual wavelength operation is shown. Measurements obtained with pump power of 1.5 W and repetition rate from 35 to 75 kHz, were performed at the output port with a power meter. As it can be seen, the average output power increases with the repetition rate from 58.3 to 84.9 mW.

Figure 10 shows the experimental results of pulse parameters for the actively Q-switched laser on dual wavelength operation. Measured pulse duration and estimated pulse energy and pulse peak power dependences on repetition rate variations are shown.

In Figure 10(a), results for pulse duration and pulse energy on repetition rate variations from 35 to 75 kHz are presented. As it can be observed, pulse duration and pulse energy present a behavior typically obtained in actively Q-switched lasers. Pulse duration increases as pulse

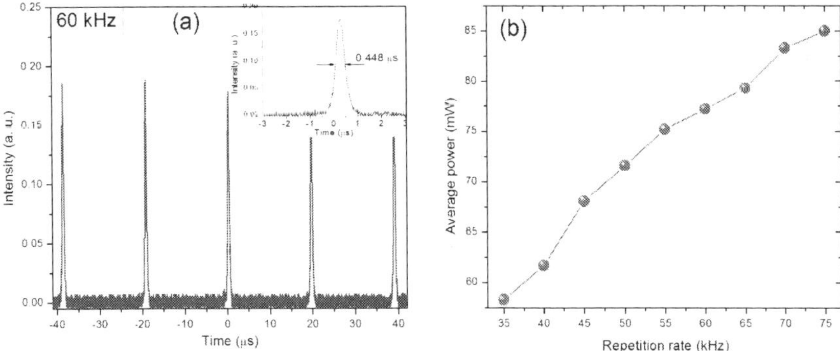

Figure 9. Actively Q-switched dual wavelength laser operation measurements, (a) pulse train with repetition rate of 60 kHz, (b) average power on repetition rate variations.

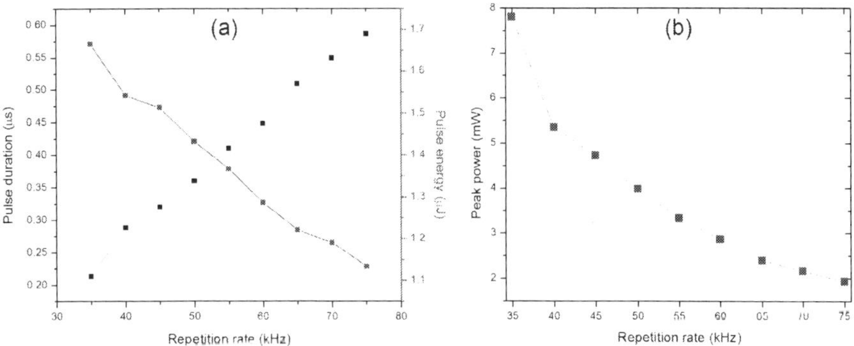

Figure 10. Actively Q-switched dual wavelength laser operation pulse parameters, (a) pulse duration and pulse energy on repetition rate variations, (b) pulse peak power on repetition rate variations.

energy decreases with the repetition rate increase. The pulse duration shows a widening in a range of 213 to 586 ns. The pulse energy decreases as the pulse widens from 1.67 to 1.13 μJ.

Figure 10(b) shows the pulse peak power dependence on repetition rate variations. As it is shown, the pulses undergo a peak power decrease as repetition rate increases. With the lower repetition rate, the pulses have less pulse duration, are more energetic, and with a higher peak power.

5. Conclusions

In this chapter, actively Q-switched fiber lasers for single and dual wavelength operation have been experimentally investigated. The investigation is based on single and dual wavelength operation of actively Q-switched fiber lasers. The documental investigation is focused on reported approaches on Q-switched fiber lasers taking into account cavity elements, configurations, experimental results, and new fiber technologies incorporation.

A review from the operating principle of pulsed lasers in the Q-switched technique to single and dual wavelength operation, mainly those lasers that use an active Q-switching method was presented. The research was led to reach the point where double clad fibers (specifically EYDCF) are used as the gain medium and the application of the active Q-switching technique by using AOM.

An analysis of the main parameters of actively Q-switched fiber lasers, including the repetition rate, pulse duration, pulse energy, average power, and peak power characteristics of the technique was experimentally discussed. This experimental study was presented in terms of a couple of proposed actively Q-switched fiber laser experimental setups.

The actively Q-switched parameters' typical behavior was mainly discussed in the first experimental setup proposed, a ring cavity dual wavelength actively Q-switched fiber laser based on the use of fiber Bragg gratings for wavelengths selection. The second experimental setup is a linear cavity actively Q-switched fiber laser with single and dual wavelength operations with a Hi-Bi FOLM. The use of the Hi-Bi FOLM as a method to adjust the losses within the cavity (when required) for simultaneous dual wavelength laser operation was discussed.

Acknowledgements

Cátedras-CONACyT project 2728. CONACyT Postdoctoral felow 160248. This work was supported by CONACyT grants 237855 and 255284.

Author details

R. Iván Álvarez-Tamayo[1*], Manuel Durán-Sánchez[1,4], Olivier Pottiez[2],
Baldemar Ibarra-Escamilla[1], Evgeny A. Kuzin[1] and M. Espinosa-Martínez[3]

*Address all correspondence to: alvarez.tamayo@hotmail.com

1 Departamento de Óptica, Instituto Nacional de Astrofísica, Óptica y Electrónica (INAOE), Puebla, México

2 Departamento de Fibras Ópticas, Centro de Investigaciones en Óptica (CIO), León,Guanajuato, México

3 Mecatrónica, Universidad Tecnológica de Puebla (UTP), Puebla, México

4 Consejo Nacional de Ciencia y Tecnología (CONACyT), México D.F., México

References

[1] Tünnermann A., Schreiber T., Limpert J. Fiber lasers and amplifiers: An ultrafast performance evolution. Applied Optics. 2010;49(25):F71–F78 / 122706. DOI: 10.1364/AO. 49.000F71.

[2] Peng B., Liu Q., Gong M., Yan P. Acousto-optic Q-switched cladding-pumped ytterbium-doped fiber laser. Chin. Opt. Lett. 2007;5(7):415–417 / 070415–03.

[3] Liu J., Xu J., Wang P. Graphene-based passively Q-switched 2 μm thulium-doped fiber laser. Optics Communications. 2012;285:5319–5322. DOI: 10.1016/j.optcom. 2012.07.063.

[4] Cao W.-J., Wang H.-Y., Luo A. -P., Luo Z.-C., Xu W.-C. Graphene-based, 50 nm wideband tunable passively Q-switched fiber laser. Laser Phys. Lett. 2012;9(1):54–58. DOI: 10.1002/lapl.201110085.

[5] Popa D., Sun. Z., Hasan T., Torrisi F., Wang F., Ferrari A. C. Graphene Q-switched, tunable fiber laser. Applied Phys. Lett. 2011;98(073106):073106-1–3. DOI: 10.1063/1.3552684.

[6] Dong B., Hao J., Hu J., Liaw C.-Y. Short linear-cavity Q-switched fiber laser with a compact short carbon nanotube based saturable absorber. Optical Fiber Technol. 2011;17:105–107. DOI: 10.1016/j.yofte.2010.12.001.

[7] Liu H. H., Chow K. K., Yamashita S., Set S.Y. Carbon-nanotube-based passively Q-switched fiber laser for high energy pulse generation. Optics & Laser Technol. 2013;45:713–716. DOI: 10.1016/j.optlasertec.2012.05.005.

[8] Zhou D -P., Wei L., Dong B., Liu W.-K. Tunable passively Q-switched Erbium-doped fiber laser with carbon nanotubes as a saturable absorber. IEEE Photon. Technol. Lett. 2010;22(1):9–11. DOI: 10.1109/LPT.2009.2035325.

[9] Laroche M., Gilles H., Girard S., Passilly N., Aït-Ameur K. Nanosecon pulse generation in passively Q-switched Yb-doped fiber laser by Cr^{4+}:YAG saturable absorber. IEEE Photon. Technol. Lett. 2006;18(6):764–766. DOI: 10.1109/LPT.2006.871678.

[10] Philippov V. N., Kir'yanov A. V., Unger S. Advanced configuration of Erbium fiber passively Q-swithed laser with Co^{2+}:ZnSe crystal as saturable absorber. IEEE Photon. Technol. Lett. 2004;16(1):57–59. DOI: 10.1109/LPT.2003.819397.

[11] Yang X.-Q., Ma Z.-G., Zheng L.-H., Shang L.-J., Su F.-F. An LD-pumped dual-wavelength actively Q-switched $Nd:Sc_{0.2}Y_{0.8}SiO_5$ laser. Optoelectronics Letters. 2015;11(2): 92–94. DOI: 10.1007/s11801-015–5003–4.

[12] Spühler G.J., Paschotta R., Kullberg M.P., Graf M., Moser M., Mix E., et al. A passively Q-switched Yb:YAG microchip laser. Applied Physics B. 2001;72(3):285–287. DOI: 10.1007/s003400100507.

[13] Nodop D., Limpert J., Hohmuth R., Richter W., Guina M., Tünnermann A. High-pulse-energy passively Q-switched quasi-monolithic microchip lasers operating in the sub- 100-ps pulse regime. Optics Letters. 2007;32(15):2115–2117. DOI: 10.1364/OL.32.002115.

[14] El-Sherif A. F., King T. A. High-energy, high-brightness Q-switched Tm^{3+}-doped fiber laser using an electro-optic modulator. Optics Communications. 2003;218(4–6): 337–344. DOI: 10.1016/S0030-4018(03)01200-8.

[15] Swiderski J., Zajac A., Konieczny P., Skorczakowski M. Numerical model of a Q-switched double-clad fiber laser. Optics Express. 2004;12(15):3554–3559. DOI: 10.1364/OPEX.12.003554.

[16] Hu T., Hudson D. D., Jackson S. D. Actively Q-switched 2.9 μm $Ho^{3+}Pr^{3+}$-doped fluoride fiber laser. Optics Letters. 2012;37(11):2145–2147. DOI: 10.1364/OL.37.002145.

[17] Wang J., Yao B. Q., Cui Z., Zhang Y. J., Ju Y. L., Du Y. Q. High efficiency actively Q-switched Ho: YVO4 laser pumped at room temperature. Laser Phys. Lett. 2014;11(8): 085003. DOI: 10.1088/1612–2011/11/8/085003

[18] Barmenkov Y. O., Kir'yanov A. V., Cruz J. L., Andres M.V. Pulsed Regimes of Erbium-Doped Fiber Laser Q-Switched Using Acousto-Optical Modulator. IEEE J. Selected topics in Quantum Electronics. 2014;20(5):0902208. DOI: 10.1109/JSTQE.2014.2304423.

[19] Du W.-C., Tao X.-M., Tam H.-Y. Fiber Bragg Grating Cavity Sensor for Simultaneous Measurement of Strain and Temperature. IEEE Photon. Technol. Lett. 1999;11(1):105–107. DOI: 10.1109/68.736409.

[20] Jung J., Nam H., Lee J. H., Park N., Lee B. Simultaneous measurement of strain and temperature by use of a single-fiber Bragg grating and an erbium-doped fiber amplifier. Applied Optics. 1999;38(13):2749–2751. DOI: 10.1364/AO.38.002749.

[21] Álvarez-Tamayo R. I., Durán-Sánchez M., Pottiez O., Kuzin E. A., Ibarra-Escamilla B. Tunable dual-wavelength fiber laser based on a polarization-maintaining fiber Bragg grating and a Hi-Bi fiber optical loop mirror. Laser Physics. 2011;21(11):1932–1935. DOI: 10.1134/S1054660X11190017.

[22] Delgado-Pinar M., Zalvidea D., Diéz A., Pérez-Millán P., Andrés M.V. Q-switching of an all-fiber laser by acousto-optic modulation of a fiber Bragg grating. Optics Express. 2006;14(3):1106–1112. DOI: 10.1364/OE.14.001106.

[23] Pérez-Millán P., Diéz A., Andrés M.V. Zalvidea D., Duchowicz R.. Q-switched all-fiber laser based on magnetostriction modulation of a Bragg grating. Optics Express. 2005;13(13):5046–5051. DOI: 10.1364/OPEX.13.005046.

[24] Mirza M. A., Stewart G. Theory and design of a simple tunable Sagnac loop filter for multiwavelength fiber lasers. Applied Optics. 2008;47(29):5242.5252. DOI: 10.1364/AO.47.005242.

[25] Sun. H. B., Liu X. M., Gong Y. K., Li X. H., Wang R. Broadly tunable dual-wavelength erbium-doped ring fiber laser based on a high-birefringence fiber loop mirror. Laser Physics. 2010;20(2):522–527. DOI: 10.1134/S1054660X10030175.

[26] Zhou K. J., Ruan Y F. Fiber ring laser employing an all-polarization-maintaining loop periodic filter. Laser Physics. 2010;20(6):1449–1452. DOI: 10.1134/ S1054660X10110393.

[27] Li S., Ngo N. Q., Tjin S. C., Binh L. N. Tunable and switchable optical bandpass filters using a single linearly chirped fiber Bragg grating. Optics Communications. 2004;239(4–6):339–344. DOI: 10.1016/j.optcom.2004.06.009.

[28] Moore P. J., Chaboyer Z. J., Das G. Tunable dual-wavelength fiber laser. Optical Fiber Technology. 2009; 15(4):377–379. DOI: 10.1016/j.yofte.2009.04.001.

[29] Moon D. S., Sun. G., Lin A., Liu X., Chung Y. Tunable dual-wavelength fiber laser based on a single fiber Bragg grating in a Sagnac loop interferometer. Optics Communications. 2008;281(9):2513–2516. DOI: 10.1016/j.optcom.2008.01.033.

[30] Laroche M., Chardon A. M., Nilsson J., Shepherd D. P., Clarkson W. A., Girard S., et al. Compact diode-pumped passively Q-switched tunable Er–Yb double-clad fiber laser. Optics Letters. 2002;27(22):1980–1982. DOI: 10.1364/OL.27.001980.

[31] Huang J. Y., Huang S. C., Chang H. L., Su K. W., Chen Y. F., Huang K. F. Passive Q switching of Er-Yb fiber laser with semiconductor saturable absorber. Optics Express. 2008;16(5):3002–3007. DOI: 10.1364/OE.16.003002.

[32] Wu D., Xiong F., Zhang C., Chen S., Xu H., Cai Z, et al. Large-energy, wavelength-tunable, all-fiber passively Q-switched Er:Yb-codoped double-clad fiber laser with mono-layer chemical vapor deposition graphene. Applied Optics. 2014;53(19):4089–4093. DOI: 10.1364/AO.53.004089.

[33] Philippov V. N., Nilsson J., Clarkson W. A., Abdolvand A., Kisel V. E., Shcherbitsky V.G., et al. Passively Q-switched Er-Yb double-clad fiber laser with Cr2+:ZnSe and Co2+:MgAl2O4 as a saturable absorber. Proc. SPIE 5335, Fiber Lasers: Technology, Systems, and Applications. 2004;5335:1–8. DOI: 10.1117/12.524767.

[34] González-García A., Ibarra-Escamilla B., Pottiez O., Kuzin E.A., Maya-Ordoñez F. M., Durán-Sánchez M. Compact wavelength-tunable actively Q-switched fiber laser in CW and pulsed operation based on a fiber Bragg grating. Laser Physics. 2015;25(045104):1–5. DOI: 10.1088/1054-660X/25/4/045104

[35] González-García A., Ibarra-Escamilla B., Pottiez O., Kuzin E.A., Maya-Ordoñez F., Durán-Sánchez M., et al. High efficiency, actively Q-switched Er/Yb fiber laser. Optics & Laser Technology. 2013;48:182–186. DOI: 10.1016/j.optlastec.2012.10.021.

[36] Durán-Sánchez M., Flores-Rosas A., Álvarez-Tamayo R.I., Kuzin E.A., Pottiez O., Bello-Jimenez M., et al. Fine adjustment of cavity loss by Sagnac loop for a dual wave-

length generation. Laser Physics. 2010;20(5):1270–1273. DOI: 10.1134/S1054660X10090446.

[37] Álvarez-Tamayo R.I., Durán-Sánchez M., Pottiez O., Kuzin E.A., Ibarra-Escamilla B., Flores-Rosas A. Theoretical and experimental analysis of tunable Sagnac high-birefringence loop filter for dual-wavelength laser application. Applied Optics. 2011;50(3):253–260. DOI: 10.1364/AO.50.000253.

[38] Álvarez-Tamayo R. I., Durán-Sánchez M., Pottiez O., Ibarra-Escamilla B., Cruz J. L., Andrés M. V., et al. A dual-wavelength tunable laser with superimposed fiber Bragg gratings. Laser Physics. 2013;23(5):2013. DOI: 10.1088/1054-660X/23/5/055104

[39] Sabran M. B. S., Jusoh Z., Babar I. M., Ahmad H., Harun S. W. Dual-wavelength passively Q-switched Erbium Ytterbium codoped fiber laser based on a nonlinear polarization rotation technique. Microwave and Optical Technol. Lett. 2015;57(3):530–533. DOI: 10.1002/mop.

[40] Ahmad H., Dernaika M., Harun S. W. All-fiber dual wavelength passive Q-switched fiber laser using a dispersion-decreasing taper fiber in a nonlinear loop mirror. Optics Express. 2014;22(19):22794-220801. DOI: 10.1364/OE.22.022794.

[41] Luo Z., Zhou M., Weng J., Huang G., Xu H., Ye C., et al. Graphene-based passively Q-switched dual-wavelength erbium-doped fiber laser. Optics Letters. 2010;35(21):3709–3711. DOI: 10.1364/OL.35.003709.

[42] Shayeganrad G. Actively Q-switched Nd:YVO$_4$ dual-wavelength stimulated Raman laser at 1178.9 nm and 1199.9 nm. Optics Communications. 2013;292:131–134. DOI: 10.1016/j.optcom.2012.11.060.

[43] Lin S.-T., Hsu S.-Y., Lin Y.-Y., Lin Y.-Y. Selectable Dual-Wavelength Actively Q-Switched Laser by Monolithic Electro-Optic Periodically Poled Lithium Niobate Bragg Modulator. IEEE Photonics Journal. 2013;5(5):1501507. DOI: 10.1109/JPHOT.2013.2280347.

[44] Durán-Sánchez M., Kuzin E.A., Pottiez O., Ibarra-Escamilla B., González-García A., Maya-Ordoñez F., et al. Tunable dual-wavelength actively Q-switched Er/Yb double-clad fiber laser. Laser Physics Letters. 2014;11(1):015102. DOI: 10.1088/1612-2011/11/1/015102

[45] Álvarez-Tamayo R. I., Durán-Sánchez M., Pottiez O., Ibarra-Escamilla B., Bello-Jiménez M., Kuzin E. A. Self-Q-switched Er–Yb double clad fiber laser with dual wavelength or tunable single wavelength operation by a Sagnac interferometer. Laser Physics. 2015;25(7):075102. DOI: 10.1088/1054-660X/25/7/075102

Passive Q-switched and Mode-locked Fiber Lasers Using Carbon-based Saturable Absorbers

Mohd Afiq Ismail, Sulaiman Wadi Harun, Harith Ahmad and Mukul Chandra Paul

Additional information is available at the end of the chapter

http://dx.doi.org/10.5772/61703

Abstract

This chapter aims to familiarize readers with general knowledge of passive Q-switched and mode-locked fiber lasers. It emphasizes on carbon-based saturable absorbers, namely graphene and carbon nanotubes (CNTs); their unique electronic band structures and optical characteristics. The methods of incorporating these carbon-based saturable absorbers into fiber laser cavity will also be discussed. Lastly, several examples of experiments where carbon-based saturable absorbers were used in generating passive Q-switched and mode-locked fiber lasers are demonstrated.

Keywords: Fiber laser, passive Q-switch, passive mode-lock, graphene, carbon nanotube

1. Introduction

Graphene and carbon nanotubes are carbon allotropes that have a lot of interesting optical properties, which are useful for fiber laser applications. For instance, both allotropes have broadband operating wavelength, fast recovery time, are easy to fabricate, and can be integrated into fiber laser cavity. As a result, they can function as saturable absorber for generating Q-switching and mode-locking pulses. There are several techniques of incorporating these carbon-based saturable absorbers into fiber laser cavity. This chapter will discuss the advantages and disadvantages of most of the techniques that have been used.

2. Introduction to passive Q-switched and mode-locked fiber laser

2.1. Q-switched fiber lasers

A laser could emit short pulses if the loss of an optical resonator is rapidly switched from a high to a low value. By controlling the Q-factor (quality factor) of a laser resonator, Q-switching allows the generation of laser pulses of short duration (from nanosecond to picosecond range) and high peak power. The Q-factor (dimensionless) is given by:

$$Q = \frac{2\pi f_0 \varepsilon}{P} \tag{1}$$

where f_o is the resonant frequency, ε is the stored energy in the cavity, and $P = -\frac{dE}{dt}$ is the power dissipated. If the Q-factor of a laser's cavity is abruptly changed from a low value to a high value, the laser will emit a pulse of light that is much more intense than the lasers' continues output. This technique is called Q-switching. There are two types of Q-switching; active and passive.

Active Q-switching uses modulation devices that change the cavity losses in accordance with an external control signal. They can be divided into three categories: mechanical, electro-optical, and acousto-optics. They inhibit laser action during pump cycle.

In passive Q-switching, the laser consists of gain medium and saturable absorber. The saturable absorber absorbs light at low intensity and transmits them at high intensity. As the gain medium is pumped, it builds up stored energy and emits photons. After many round-trips, the photon flux begins to see gain, fixed loss, and saturable loss in the absorber. If the gain medium saturates before the saturable absorber, the photon flux may build, but the laser will not emit a short and intense pulse. On the contrary, if the photon flux builds up to a level that saturates the absorber before the gain medium saturates, the laser resonator will see a rapid reduction in the intracavity loss and the laser Q-switches and therefore, will emit a short and intense pulse of light [1].

2.2. Mode-locked fiber laser

Mode-locking is a technique of generating an ultra-short pulse laser with pulse duration ranges from picoseconds (10^{-12} s) to femtoseconds (10^{-15} s). An ultra-short pulse can be generated when all the longitudinal modes have a fixed phase relationship. The fixed phase superposition between all the modes oscillating inside a laser cavity causes the cw laser to be transformed into a train of mode-locking pulse. The number of longitudinal mode that can simultaneously lase is dependent on the gain linewidth, Δv_g and the frequency separation between modes. Under sufficiently strong pumping, we can expect that the number of modes oscillating in the cavity is given by:

$$M = \frac{\Delta v_g}{c/2L} = \frac{2L}{c}\Delta v_g \tag{2}$$

where c is the speed of light and L is the length of a linear cavity. The shortest pulse duration that we can expect to obtain by a given gain line width is:

$$\tau_{min} = \tau_M = \frac{2L}{cM} = \frac{1}{v_g} \tag{3}$$

From Eq. (3), we can conclude that the shortest pulse that can be obtained is a reciprocal of gain line width (in Hz) [2]. Depending on fiber laser cavity type, the fundamental repetition rate of a mode-lock fiber laser is determined by its cavity length, as shown in the equations below:

$$Repetition\ rate\left(for\ linear\ cavity\right) = \frac{c}{2Ln} \tag{4}$$

$$Repetition\ rate\left(for\ ring\ cavity\right) = \frac{c}{Ln} \tag{5}$$

where -L, c and n denotes the length of the cavity, speed of light, and refractive index respectively. As the round-trip time, T_R, is the inverse of repetition rate, therefore,

$$T_R = \frac{Ln\left(ring\ cavity\right)\ or\ 2Ln\left(linear\ cavity\right)}{c} \tag{6}$$

Under certain conditions, the repetition rate can be some integer multiple of the fundamental repetition rate. In this case, it is called harmonic mode-locking.

Mode-locking techniques can be divided into three categories; active, passive and hybrid. Active mode-locking can be achieved by using active modulator, e.g., acousto-optic or electro-optic, Mach-Zehnder integrated-optic modulator or semiconductor electro-absorption modulator. Passive mode-locking incorporates saturable absorber (SA) into the laser cavity. An artificial saturable absorber action can also be induced artificially by using Nonlinear Polarization Rotation (NPR) technique or by using another technique called Nonlinear Amplifying Loop Mirror (NALM). In comparison, the loss modulation of an active mode-locking is significantly slower due to its sinusoidal loss modulation. As with active mode-locking, a passive mode-locked pulse is much shorter than the cavity round- trip time. Hybrid mode-locking combines active and passive mode-locking. Hybrid mode-locking uses active modulator to start mode-locking while passive mode-locking is utilized for pulse shaping.

Many nonlinear systems exhibit an instability that result in modulation of the steady state. This is due to the interplay between the nonlinear and dispersive effects. This phenomenon is referred to as the *modulation instability*. In the context of fiber optics, modulation instability requires anomalous dispersion and reveals itself as breakup of the cw or quasi-cw radiation into a train of ultrashort pulses [3-5]. The instability leads to a spontaneous temporal modulation of the cw beam and transforms it into pulse train [6].

There are many types of passive mode-locking pulses. However, for the sake of brevity, in this section, we will only discuss the soliton mode-locking pulse. It refers to a special kind of wave packets that can propagate undistorted over long distances. Soliton phenomena is formed by the interplay between the dispersive and nonlinear effects in a fiber laser cavity. Soliton mode-locking implies that the pulse shaping is solely done by soliton formation; the balance of group velocity dispersion (GVD) and self-phase modulation (SPM) at steady state. The mode-locking mechanism is not critically dependent on cavity design and no critical cavity stability regime is required. Soliton mode-locking basically works over the full cavity stability range.

In soliton mode-locking, an additional loss mechanism such as saturable absorber is essential to start the mode-locking process as well as stabilize the soliton pulse-forming process. In soliton mode-locking, the net gain window can remain open for more than 10 times longer than the ultrashort pulse, depending on the specific laser parameter [7, 8]. A stable soliton pulse is formed for all Group Delay Dispersion (GDD) values as long as the continuum loss (energy loss) is larger than the soliton loss [8] or the pulses break up into two or more pulses [9].

Soliton mode-locking can be expressed by using the following Haus master equation formalism [8, 10, 11]:

$$\sum_i \Delta A_i = \left[-iD\frac{\partial^2}{\partial t^2} + i\delta_L \left|A\left(T,t\right)\right|^2 \right] A\left(T,t\right) + \left[g - l + D_g\frac{\partial^2}{\partial t^2} - q\left(T,t\right) \right] A\left(T,t\right) = 0 \tag{7}$$

Here, $A(T, t)$ is the slowly varying field envelope, D is the intra-cavity GDD, $D_g = g/\Omega_g^2$ is the gain dispersion and Ωg is the Half-Width of Half-Maximum (HWHM) of gain bandwidth. The SPM coefficient δ is given by $\delta = (2\pi/\lambda_0 A_L)n_2\ell_L$, where n_2 is the intensity dependent refractive index of the gain medium, λ_0 is the center wavelength of the pulse, and A_L and ℓ_L is the effective mode area in the gain medium and length of light path through the gain medium within one round-trip, respectively. g is the saturated gain and l is the round-trip losses. $q(T, t)$ is the response of the saturable absorber due to an ultrashort pulse.

This soliton pulse propagates without distortion through a medium with negative GVD and positive SPM. The positive effect of SPM cancels the negative effect of dispersion. Kelly sidebands can usually be found in the optical spectrum of a soliton mode-locked fiber laser. A pronounced Kelly sidebands is an indicator that the mode-locked fiber laser is operating in its optimal pulse duration [12, 13].

3. Carbon-based saturable absorbers

Graphene and CNT have been used in Q-switching and mode-locking fiber lasers since 2003 [14-16] and 2009 [17], respectively. Compared to SESAM (Semiconductor Saturable Absorber

Mirror), CNT and graphene holds several advantages, e.g., broadband operating bandwidth, simple and low-cost fabrication process, and moderate damage threshold [18]. In this section, both electronic and optical characteristics of graphene and CNT are discussed in detail.

3.1. Graphene

3.1.1. Electronic and band structure of graphene

Graphene is the name we gave to a one-atom thick sp^2 hybridized carbon. It has a honeycomb-like structure. The sp^2 hybridization between s, p_x and p_y atomic orbitals create a strong covalent sp^2 bonds. The p_z orbital overlaps with other carbons to create a band of filled π orbitals. These bands have a filled shell and, therefore, form a deep valence band. On the other hand, the empty π^* orbitals are called the conduction band [19, 20].

Further observation of the band structure of graphene reveals three electronics properties that sparked such interest; the vanishing carrier density at Dirac point, the existence of pseudo-spin, and the relativistic nature of its carriers. The valance and conduction bands meet at high symmetry K points. Because the conduction and valence band meet at a symmetry K point, graphene is considered as zero-gap semiconductors (or zero-overlap semimetals) [21]. In intrinsic graphene, each carbon atom contributes one electron completely filling the valance band and leaving the conduction band empty. Therefore, the Fermi level, E_F, is situated precisely where the conduction and valence bands meet. These are known as the Dirac or charge neutrality points.

As mentioned before, due to this unique band structure of graphene, the following are three important features which to a large extent define the nature of electron transport of this material, namely, the zero-gap semiconductor, the existence of pseudo-spin, and the linear dispersion relation.

3.1.2. Optical properties of graphene

Graphene has three types of optical properties. They are linear optical absorption, saturable absorption, and luminescence. For generating passive Q-switch and mode-locked fiber lasers using graphene saturable absorber, our main interest is in the optical absorption and saturable absorption properties that graphene has.

Graphene saturable absorber has an ultrawide band operating wavelength as a result of its linear dispersion relation. Graphene only reflects <0.1% of the incident light in the visible region, increasing to ~2% for ten layers. Therefore, we can assume that the optical absorption of graphene layers is relative to the number of layers, for each layer absorbing $\approx 2.3\%$ over the visible spectrum [22].

The saturable absorption property of graphene is the result of Pauli blocking. Inter-band excitation by ultrafast optical pulses produces a nonequilibrium carrier population in the valence and conduction bands. In time-resolved measurements [23], two relaxation timescales are observed; a faster one of ~100 fs and a slower one, on picosecond timescale. The faster relaxation time is related to the carrier−carrier intra-band collisions and phonon emission. The slower relation time corresponds to electron inter-band relaxation and cooling of hot phonons [24, 25]. For generating mode-lock laser, a saturable absorber with relaxation time in the timescale of ~ps is necessary. In principle, single-layer graphene can provide the highest saturable absorption [26-28].

Graphene can be made luminescent by inducing a bandgap through two techniques to modify the electronic structure of graphene. One technique is by cutting it into ribbons and quantum dots [29] and the other is by chemical or physical treatments [30, 31], to reduce the connectivity of the π-electron network. A mild oxygen plasma treatment can make individual graphene flakes luminescent. The combination of photo-luminescent and conductive layers could be used in sandwich light-emitting diodes. Luminescent graphene-based material has been made to cover the infrared, visible, and blue spectral ranges [32-35].

3.2. Carbon nanotube

3.2.1. Electronic and band structure of carbon nanotube

The band structure of CNT can be assumed under a simple tight-binding model of [36]. In the model, CNT is considered as a roll of graphene layers. A very small change in diameter of the curvature of the fiber can affect the hybridization of sp^3 orbitals in which the electronic structure will also be affected. Therefore, the electronic structure depends on the geometry of the fiber and the fiber diameter.

The chiral vector, $\vec{C_h} \equiv (n, m)$, determines whether the CNT is metallic or semiconducting with bandgap. If $n-m = 3k$ (k is integer), the CNT is metallic, while when $n-m \neq 3k$, it shows that the CNT is semiconducting. At first, it was thought that the electrical and optical bandgaps of semiconducting CNTs were identical based on single particle model; however, the optical bandgap is actually smaller [37].

3.2.2. Optical properties of carbon nanotube

Just like graphene, CNT has several interesting optical properties. They are: optical absorption, saturable absorption, and electroluminescence and photoluminescence. The semiconducting CNTs have peak absorption wavelength depending on the optical bandgaps. Typical CNT with diameter d, of 7-15 nm has a bandgap energy of 1.2-0.6 eV, corresponds to the optical wavelength of 1−2 μm. Thus, the peak absorption can be tuned by choosing the appropriate diameter. However, since CNTs are a mixture of several or many kinds of semiconducting and metallic CNT as well as different diameter distribution, the absorption peak is determined by the mean tube diameter and the absorption bandwidth depends on the tube diameter distribution. Although CNT is essentially a rolled-up graphene, the absorption of CNT is nonlinear. The optical absorption in CNT is anisotropic because CNT only absorbs the light whose polarization is parallel to the axial direction of the tube; therefore, an aligned CNT sample is polarization-dependent [38]. In spite of this, since we use a randomly oriented CNT samples, the CNT is polarization-independent [39].

Single-walled carbon nanotubes (SWCNTs) have been utilized as saturable absorber (SA) for mode-locking fiber laser long before graphene. The first publication of SWCNT SA can be traced as early as 2003 [15]. CNT can saturate with high-intensity light when the states of conduction band become full and the states at valence band become empty. Furthermore, the recovery time τ is observed to be very fast. In semiconducting CNT, the recovery time of E_{11} transition is an order of 1 ps, and the transition of E_{22} is in the order of 100 fs. Slower recovery in an order of several ps is also seen in the E_{11} transition. Several mechanisms believed to be responsible for the fast relaxation have been proposed. These mechanisms multi-phonon emission [40], tube—tube interaction [41], and exciton—exciton annihilation [42, 43].

The direct bandgap that exists in CNTs suggests that they can be efficient light absorbers and emitters. Studies regarding electroluminescence properties of SWCNT—polymer composites have been performed [44, 45]. Electron and hole carriers in semiconductors can recombine by different sorts of mechanisms. In most of the cases, the energy will be released as heat. Nevertheless, a fraction of the recombination events may involve the emission of a photon. The process is called electroluminescence and is responsible for producing solid-state light sources such as light-emitting diodes (LED). The direct bandgap of a semiconducting CNT is responsible for the photoluminescence phenomena in CNT. An electron in a CNT absorbs excitation light via transition from v_2 to c_2 and creates another excitation [46]. Electron and hole rapidly relax from c_2 to c_1 and from v_2 to v_1 states, respectively. Luminescence can only be observed in isolated semiconducting CNTs because the bundled CNTs have rapid transfer process from semiconducting to metallic CNTs [47]. Additionally, a semiconducting CNT can function as a nanoscale photodetector that converts light into current or voltage [48].

4. Preparation and fabrication of carbon-based saturable absorber

4.1. Optical deposition technique

In optical deposition technique, an intense light is injected into a CNT/ graphene-dispersed solution from a fiber-optic end to attract the CNT/graphene particle onto the ferrule tip [49-54]. There are generally two types of effects that lead to self-channeling of light in fluid suspension; the optical gradient force and thermal effects relying on (weak) absorption. When the particle size is smaller than the optical wavelength, the optical gradient force is weak; therefore, a higher particle density is required to cause significant change in the refractive index to trap narrow beams. With higher particle density, multiple scattering dominates; in-turn the direction of scattered light becomes random. This effect pushes the particles toward the beam center. Self-trap through very narrow beam is difficult because it requires high particle densities and this, in turn, involves multiple scattering, which acts as effective loss. Thermal effect can lead to a significant refractive index change, but in liquid, the refractive index typically decreases with increasing temperature [55].

Thermophoresis is a thermal mechanism that is often observed in colloidal suspension, which uses a strong reaction of the suspended particles to temperature gradients. Also known as the Soret effect, thermophoresis describes the ability of a macromolecule or particle to drift along a temperature gradient [55]. Although various mechanisms are capable of depositing carbon-based saturable absorber to the fiber core, [52] has considered thermophoresis as the most likely process that is responsible for creating carbon-based saturable absorber via optical deposition. When the laser is turned on with the fiber in the solution, a strong convection current centered at the tips is observed. The induced current moves the carbon-based particles upward toward the fiber tip.

When fabricating a carbon-based saturable absorber using optical deposition, one must start by dispersing the carbon-nanotube/graphene bundles through ultrasonification process. Centrifugation follows to separate the macroscopic flakes and the agglomerated carbon nanotube/graphene. Only the homogeneous part of the solution is used for optical deposition process [49]. For depositing the carbon-based particles to the fiber core, precise optical power is essential. Lower optical power will not make the carbon-based particles adhere to the fiber core, while higher optical power will concentrate the deposition to the area around the fiber core. The difference between optimal and higher optical power is 1 dB [53]. The optimum optical power is influenced by other factors such as the size of particles, solution temperature, concentration levels of the solution, and the optical wavelength used in the deposition process [49, 53, 54]. For monitoring the optical deposition process, [50] has devised a setup that involves optical circulator and power meters. For controlling the insertion loss, the duration of the deposition has to be observed. Long deposition duration would create higher insertion loss.

Optical deposition technique is highly efficient as it only uses a small/required amount of carbon nanotube/graphene as saturable absorber. However, the disadvantages of this technique are large scattering loss [39], and the process itself is quite tedious as many factors can influence the required optical power in-order to make the carbon nanotube/graphene particles adhere to the fiber core. Furthermore, the success margin is small as only 1 dB can differentiate between optimal and high optical power.

4.2. Drop cast technique

Drop cast technique is simple and straightforward. Drip a graphene/SWCNT solution onto a fiber ferrule and let it dry to make an SA. This technique has been utilized in [56-58]. The SA insertion loss can be controlled depending on the concentration of the solution and the number of times this process is repeated. The process can also be repeated until the desired insertion loss is achieved.

Although the insertion loss is high, this can be overcome by increasing the pulse energy. As the pulse energy depends on the power and frequency, it can be altered by increasing pump power and/or lengthen the fiber laser cavity. The drawback of this technique is that in the course of increasing the pulse energy, we inevitably change the repetition rate and pulse width of the mode-locked fiber laser. Scattering loss is also an issue for this technique.

4.3. Mechanical exfoliation technique

Mechanical exfoliation uses scotch tape to repeatedly peel the graphene layers from a Highly Ordered Pyrolitic Graphite (HOPG) or graphitic flakes and transferring the layers to the surface of the fiber ferrule. A fiberscope is normally used to examine that the graphene is transferred directly onto the core of the fiber ferrule. Mechanically exfoliated graphene SA has been demonstrated by [59-62] and [63]. The advantage of this technique is that it yields the best quality graphene SA. However, the drawback of this technique is that it is time-consuming. In addition, it could be difficult to control the desired graphene layer(s) that need to be transferred onto the core of the fiber ferrule.

4.4. Thin film and polymer composite

SWCNT and graphene thin films have been reported in many literatures [15, 17, 64]. For instance, [65] sprayed a liquid with dispersed CNTs onto a fiber end surface that acts as a substrate, while [15] sandwiched a thin layer of purified SWCNTs between two quartz substrates. CNTs and graphene polymer composite have also been demonstrated in numerous publications [26, 66-69]. Many kinds of polymer materials can be used as a host to graphene and CNTs, e.g., polymethylmethacrylate (PMMA), polymide, and polycarbonate. The main advantages of using polymer composite as a host are that it reduces scattering and facilitates homogeneous dispersion of CNTs and graphene. It is thin enough to be sandwiched between fiber ferrules and has higher damage threshold compared to pure CNT/graphene layer. In spite of this, in terms of the amount used, it is less efficient than optical deposition technique and involves extra processing.

Other types of SA are also available, including tapered fiber [51, 70], D-shaped fiber [71], as well as CNT/graphene solutions embedded in photonic crystal fiber [72, 73].

5. Passive Q-switch and mode-lock experiments using carbon-based saturable absorbers

In this subsection, three pulsed Erbium-doped fiber lasers (EDFLs) are demonstrated using a comparatively simple and cost-effective carbon-based saturable absorber.

5.1. Passive Q-switch and mode-lock generation using Single-Walled Carbon Nanotubes (SWCNTs) saturable absorber via drop cast technique

Figure 1(a) shows the experimental setup for the proposed Q-switched EDFL, using a comparatively simple and cost-effective alternative technique based on SWCNTs SA. It consists of a 4 m long Erbium-doped fiber (EDF), two 1480/1550nm wavelength division multiplexers (WDM), an SWCNTs-based SA, an optical isolator, and a 20 dB output coupler in a ring configuration. The EDF is a commercial fiber with Erbium ion concentration of 2000 ppm, cut off wavelength of 920 nm, and numerical aperture of 0.24. It is backward pumped by a 1480 nm laser diode (LD) with the maximum output power of 129 mW via the WDM. Another 1480/1550 WDM is used after the gain medium to dispose excess power from the LD. An isolator is used to ensure unidirectional propagation of light inside the cavity. The homemade SWCNT SA is placed between the isolator and the WDM to act as a Q-switcher. The SWCNT SA was fabricated using the drop cast method.

The output of the laser is extracted from the cavity using the 1% output port of the optical coupler. An optical spectrum analyzer (OSA, AQ6317B) is utilized for the spectral analysis of the Q-switched EDFL, which has the spectral resolution set to 0.02 nm, whereas an oscilloscope (OSC, Tektronix, TDS 3052C) is used to monitor the pulse train of the Q-switched operation via coupling the oscilloscope with a 6 GHz bandwidth photo-detector. Total cavity length is 23 m. Except for the gain medium, the rest of the cavity uses a standard SMF-28 fiber. The total cavity length of the ring resonator is measured to be around 23 m. Furthermore, all optical components are polarization-independent.

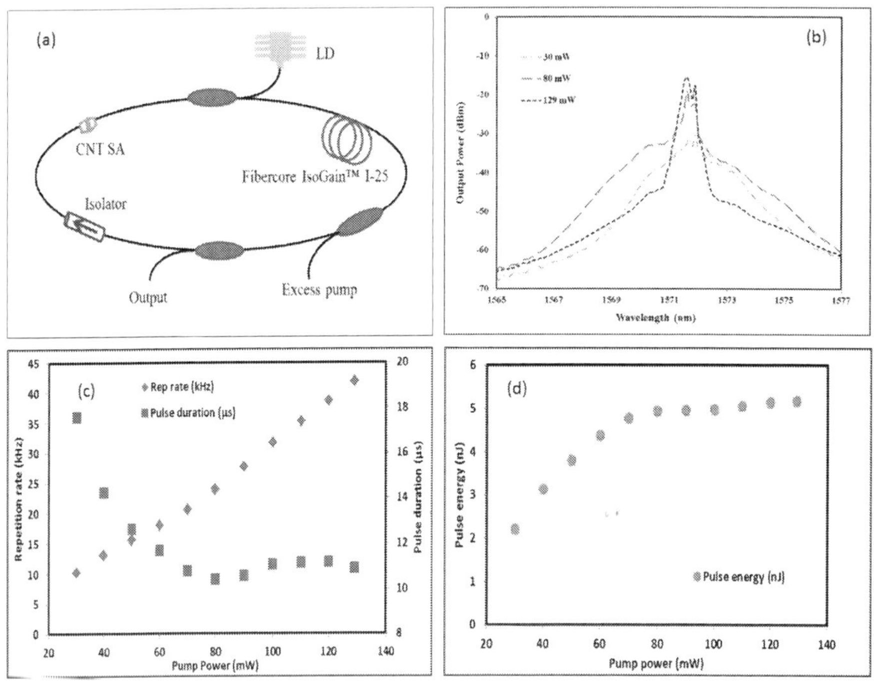

Figure 1. (a) Experimental setup for the proposed SWCNTs-based Q-switched EDFL. (b) Output spectra of the Q-switched EDFL at three different pump powers: 30 mw, 80 mW, and 129 mW. (c) Repetition rate and pulse duration relationship with pump power. (d) Pulse energy relationships with pump power.

Stable and self-starting Q-switching operation is obtained just by increasing the pump power over 30 mW. Figure 1(b) compares the output spectra of the EDFL at three different pump powers; 30 mW, 80 mW, and 129 mW. As shown in the figure, the laser operates at center wavelength of around 1571.6 nm. Spectral broadening is observed in the spectrum especially at a pump power of 80 mW, which corresponds to the minimum pulse width region. This is attributed to the Self-Phase Modulation (SPM) effect in the laser cavity [74]. The maximum

Full-Width at Half-Maximum (FWHM) of 0.6 nm is obtained when the pump power was increased to maximum (129 mW).

Figure 1(c) shows the relationship between repetition rates and pulse durations with pump power. As pump power increases from 30 mW to 129 mW, the repetition rate increases linearly from 10.25 kHz to 41.87 kHz. As pump power increases, more gain is provided to saturate the SA and thus increases repetition rate. In contrast, pulse duration decreases from 17.6 µs to 10.92 µs as the pump power increases. However, the lowest pulse duration of 10.42 µs is achieved at 80 mW pump power. After the pump power increases from 80 mW to 129 mW, the pulse durations increase slightly before decreasing back at 129 mW. Hence, the minimum attainable pulse duration is 10.24 µs, which is related to modulation depth of the saturable absorber [75, 76]. Based on the minimum attainable pulse duration, the modulation depth of the SWCNT SA is calculated to be around 3.7%. Figure 1(d) shows the relationship between pulse energy and pump power in the proposed Q-switched EDFL. As the pump power increases, the average output power also increases, which gives rise to pulse energy. It is obtained that the pulse energy can be increased from 2.23 nJ to 4.94 nJ by tuning the pump power from 30 to 80 mW, and from 4.94 nJ to 5.19 nJ when the pump power increases from 80 mW to 129 mW. The calculated average slope efficiency is 12% when the pump power increases from 30 mW to 80 mW. From 80 mW to 129 mW pump power, the calculated average slope efficiency is 16%. The pulse energy is saturated as the pump power is further increased above 80 mW.

5.2. Mode-locked erbium-doped fiber laser using Single-Walled Carbon Nanotubes (SWCNTs) saturable absorber via drop cast technique

In order to saturate the SA in a single-pass, the laser cavity is slightly changed compared to the previous subsection. The experimental setup for proposed SWCNTs-based mode-locked EDFL is shown in Figure 2(a). Compared to the previous setup, a 200 m long SMF is added in the mode-locked setup to reduce the repetition rate of the output pulse and thus increase the pulse energy in the cavity. This gives total cavity length of ~223 meter and a total Group Velocity Dispersion (GVD) of 3.6364 ps nm^{-1}. Therefore, the fiber laser is operating in the anomalous dispersion regime. For pulse duration measurement, an autocorrelator with 25 fs resolution was used. The SNR is measured using Anritsu MS2667C Radio Frequency Spectrum Analyzer (RFSA).

Soliton mode-locking operation self-starts at 56.75 mW without Q-switching instabilities. It is observed that the pulse state diminishes into continuous-wave (CW) when the pump power is below 30 mW. The resultant repetition rate is 907 kHz, which corresponds to 1.1 µs round-trip time. Figure 2(b) shows output spectrum of the proposed mode-locked EDFL. As shown in the figure, the laser operates at a central wavelength, λ_C, of 1570.5 nm with 3-dB bandwidth of 1.080 nm. Compared to the Q-switched laser, the mode-locked laser operates at a shorter wavelength due to the incorporation of 200 m long SMF in the cavity, which increases the cavity loss. The operating wavelength shifts to shorter wavelength to acquire more gain to compensate the loss.

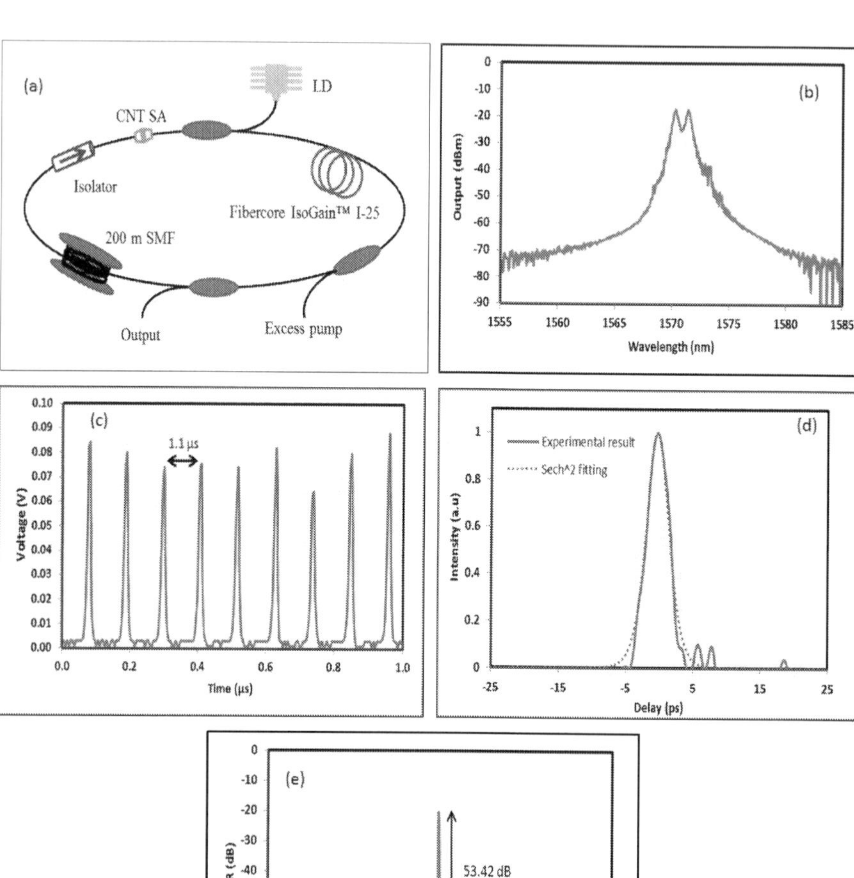

Figure 2. (a) Experimental setup for soliton mode-lock operation. (b) Output spectrum of the proposed soliton mode-locked EDFL when the pump power is fixed at 129 mW. (c) OSC trace of mode-locked fiber laser (d) Autocorrelation trace of mode-locked fiber laser at 129 mW. (e) RFSA trace of soliton mode-locked fiber laser at 129 mW.

Figure 2(c) shows the typical pulse train of the mode-locked EDFL at pump power of 129 mW. Figure 2(d) shows the corresponding autocorrelation trace of the mode-locked pulse showing the pulse duration, T_{FWHM}, of 2.52 ps. The RF spectrum of the mode-locked laser is also investigated using a RF spectrum analyzer. Figure 2(e) shows the result, which indicates a strong mode-locked pulse at frequency of 907 kHz. Figure 2(e), SNR is obtained at 53.42 dB, which is limited by the available pump power. The average output power of the soliton mode-locked fiber laser is measured to be -6.54 dBm. Based on the 3 dB bandwidth of the output spectrum, a Time Bandwidth Product (TBP) of laser is calculated to be around 0.331, which shows that the soliton pulse is slightly chirped.

Referring to Figure 2(d), the autocorrelation trace does not follow exactly the $sech^2$ fitting. When mode-locked pulse self-starts at 56.75 mW, the pulse shape does follow exactly the shape of the $sech^2$ fitting. However, in the RFSA, the SNR value is below the threshold value required to be qualified as mode-locked pulse. This is due to the 20 dB output coupler used in the experimental setup. Only 1% of the total energy that is circulating inside the cavity is extracted for measurement purpose.

Therefore, the pump power is increased to 129 mW to achieve a satisfactory SNR value. However, this, in turn, increases the pulse intensity in the autocorrelator. As a result, the pulse shape in the autocorrelator does not follow the $sech^2$ fitting.

5.3. Passive Q-switched fiber laser generation using graphene saturable absorber

Graphene was first produced by a mechanical exfoliation method in 2004 [77]. In this work, a fresh surface of a layered crystal was rubbed against another surface, which left a variety of flakes attached to it. Among the resulting flakes, a single layer flake can be found. Despite there being other methods to produce graphene, mechanical exfoliation still gives the best samples in terms of purity, defects, electron mobility, and optoelectronic properties. A single-layer graphene saturable absorber has an ultrafast relaxation time, lower scattering loss, and it performs better than multilayer graphene saturable absorber in terms of pulse-shaping ability, pulse stability, and output energy [28]. However, this method has disadvantages in terms of yield and throughput, and thus it is impractical for large-scale production. Graphene can be optically distinguished, regardless of being one-atom thick and its transmittance (T) can be expressed in terms of the fine-structure constant. Due to some properties of graphene such as linear dispersion of the Dirac electrons and Pauli blocking, it makes broadband applications and saturable absorption possible [35]. In this section, the preparation of a single-layer graphene SA (GSA) based on mechanical exfoliation technique, also known as the "scotch tape" method, is demonstrated. The position of graphene SA on the fiber core can easily be recognized by using a fiber probe.

The material is a commercially available, highly ordered pyrolytic graphite (HOPG). An HOPG flake was inserted on a strip of scotch tape and then were pressed and peeled off repeatedly in order to reduce the graphene layers to a single layer. The resultant graphene layers were then pressed against the end facet of an optical fiber ferrule in order to transfer it. The scotch tape was slowly peeled off and subsequently, a few graphene layers stick on the optical fiber ferrule. Graphical presentation of this technique is explained in detail by [60]. The result is inspected using EXFO's FIP-400 fiber inspection probe to ensure that the graphene sheets lie on top of the fiber core. The microscope image of the end surface of the ferrule after coating the graphene is illustrated in Figure 3.

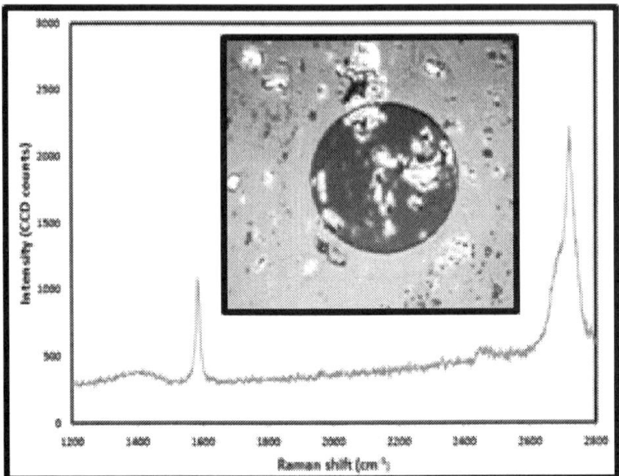

Figure 3. Raman Spectrum of the GSA. Fiberscope image of fiber ferrule with graphene (inset).

The setup of the proposed Q-switched EDFL with the newly developed GSA is similar to the previous section, except for the gain medium, SA, and 200 m long fiber. It is based on unidirectional ring cavity configuration consisting of two wavelength division multiplexer (WDM) coupler, 49 cm long bismuth-based Erbium-doped fiber (Bi-EDF) as gain medium. Since there is no polarizer in the laser cavity, the graphene is the sole responsible mechanism for creating saturable absorption.

In the experiment, the continuous wave (CW) lasing threshold was about 30 mW. When the pump power was increased to about 80 mW, the Q-switched pulses were observed by introducing physical disturbance to the cavity. Then, the pump power is further increased to the maximum pump power of 130 mW and observed the Q-switched operation. Figure 4(a) shows the output spectrum of the Q-switched EDFL at 130 mW pump power. A slight spectral broadening is also observed in the optical spectrum, which is caused by Self-phase Modulation effect (SPM). Correspondingly, the typical Q-switched pulse train is presented in Figure 4(b). As shown in the figure, the peak-to-peak pulse interval is measured to be around 43.3 µs, which can be translated into repetition rate of 26 kHz. At 130 mW pump power, the Q-switched laser has an average output power of 0.5656 mW, which corresponds to pulse energy of 24.399 nJ.

Figure 4(c) represents the pulse repetition rate and the pulse energy of the Q-switched fiber laser as a function of the pump power. The repetition rate can be tuned from 16.7 kHz to 23.1 kHz by increasing the pump power. The attainable energy is lower than the previous Q-switched laser with optical deposition technique based GSA. This is attributed to the large modulation depth of the single-layer graphene SA. A large modulation depth implies a large change in absorption for the incident light. Therefore, a lower repetition rate and pulsewidth are achieved with higher modulation depth [78].

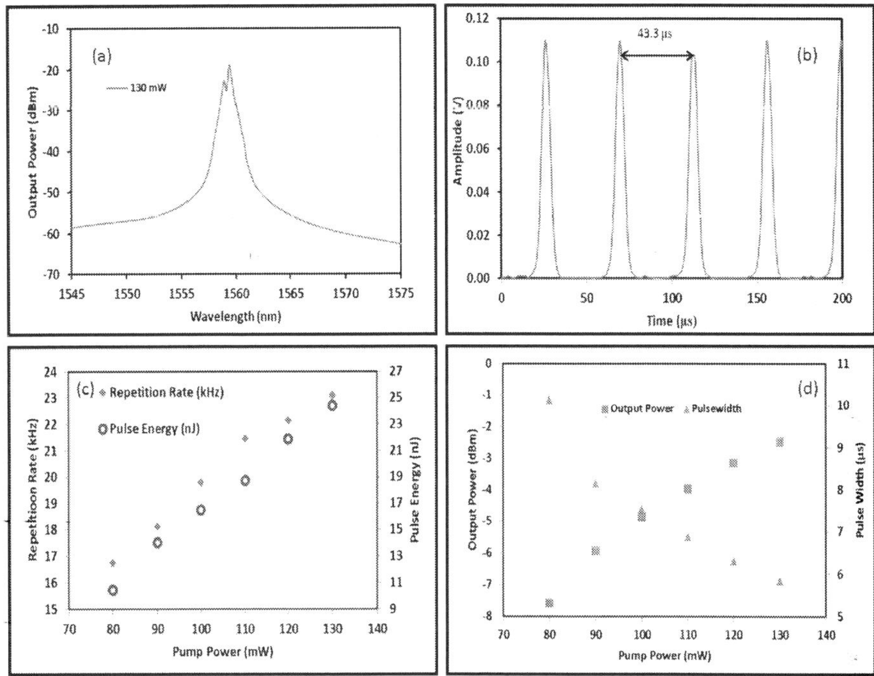

Figure 4. (a) Optical spectrum of the Q-switched laser at pump power of 130 mW. (b) Pulse train of the Q-switched laser at 130 mW pump power. (c) Repetition rate and pulse energy as a function of pump power. (d) Output power and pulse width as a function of pump power.

5.4. Passive mode-locked fiber laser using graphene saturable absorber

Figure 5(a) shows the modified configuration where a longer EDF is used in conjunction with an additional 200 m long SMF-28 to reduce the repetition rate and increase the pulse energy. The length of total cavity is about 207 m, including 1.6 m EDF, 205.4 m SMF-28 fiber from the WDM, isolator, coupler, and additional spool of SMF. The net GVD in the cavity was calculated to be 3.457 ps/nm, confirming that the laser was operating at an anomalous dispersion regime.

The mode-locking operation is not self-started in the proposed setup. Stable mode-locked pulses were observed as shown in Figure 5(b) by introducing physical disturbance to the SA after increasing the 1480 nm pump power to the maximum (130 mW). The mode-locking pulse then disappears when pump power falls below 84.5 mW. As shown in Figure 5(b), the cavity round-trip time is measured to be 1.034 µs, which corresponds to repetition rate of 967 kHz.

Figure 5(c) shows the measured optical spectrum of the soliton pulses at the launched pump power of 130 mW. Although the resultant pulse fiber laser is a soliton mode-locked fiber laser, Kelly sidebands are less prominent due to excessive nonlinearity caused by high pump power and cavity length [12]. Figure 5(d) shows the measured interference autocorrelation trace of

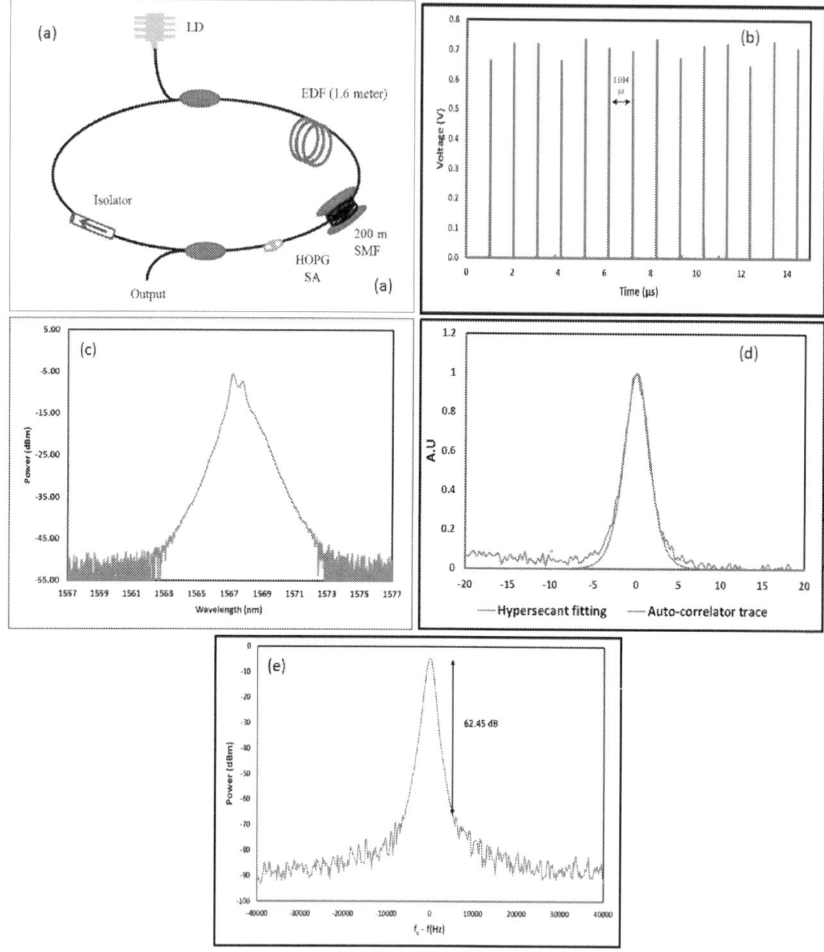

Figure 5. (a) Configuration of the mode-locked EDFL with GSA. (b) Typical mode-locking pulse train on oscilloscope. (c) OSA trace of the mode-locked EDFL. (d) Autocorrelation trace of the mode-locking pulse at launched pump power of 130 mW. (e) RF spectrum of the mode-locked pulse train.

the mode-locked pulses at a scanning range of 40 ps. As shown in Figure 5(d), the pulse was very well fitted by a sech^2 pulse profile, and the pulse duration was measured to be 3.41 ps. Consequently, the TBP was calculated to be 0.38, which is almost 1.2 times larger than the ideal TBP value (0.315). This is most probably due to the large GVD in the laser cavity and high pump power. Figure 5(e) shows the RF spectrum of the output at the launched pump power of 130 mW. The SNR of 62.45 dB indicated that the oscillator operated at stable mode-locking regime.

5.5. Passive mode-locked fiber laser using nonconductive graphene oxide paper

Figure 6 shows the result of Raman spectroscopy on graphene oxide paper. The spectroscopy was performed using a 532 nm laser with only 10% power and exposure time was set to 20 s. From the result, there are two distinctive peaks that can be observed; at 1349 cm^{-1} and 1588 cm^{-1}. These two peaks are D-band and G-band, respectively [79]. The excitation at D-band is caused by the hybridized vibrational mode related to graphene edges, and it also shows a disorder in the graphene structure. The graphite or tangential band (G-band) exists due to the energy in the sp^2 bonded carbon in planar sheets. The in-plane optical vibration of the bond resulted in Raman spectrum at the mentioned frequency [80]. A small peak at 2700 cm^{-1}, which is also known as G′ or 2D band, is barely observable because the laser power is low. The graphene layers can be indicated by the ratio of G′ and G bands. Because the intensity of G′ band is lower than the G band, it also shows that the GO paper is more than one layer.

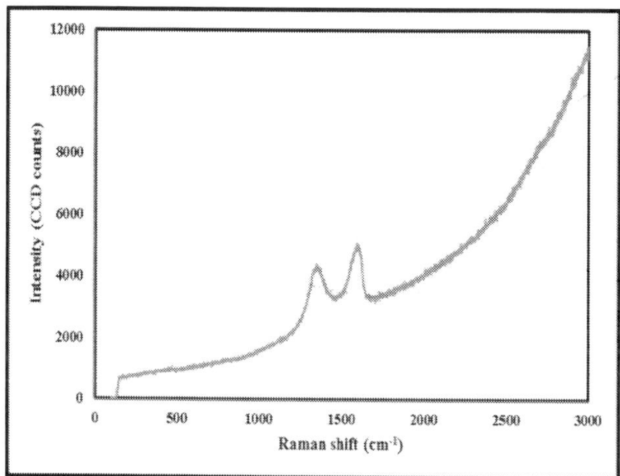

Figure 6. Raman spectroscopy result of graphene oxide paper.

The schematic of the proposed mode-locked EDFL is shown in Figure 7(a). It was constructed using a simple ring cavity, in which a 1.6 m long EDF with an Erbium ion concentration of 2000 ppm was used for the active medium and the GO paper SA was used as a mode-locker.

The SA was fabricated by cutting a small piece (2×2 mm^2) of a commercially nonconductive graphene oxide paper [81] and sandwiching it between two FC/PC fiber connectors, after depositing index-matching gel onto the fiber ends. The thickness of the GO paper is 10 μm, while the measured insertion loss of the SA to be 1.0 dB at 1550 nm. The total length of the laser cavity was measured to be approximately 12.6 meter. The resultant total GVD for this mode-locked fiber laser was calculated to be 0.1513 ps/nm. This indicated that the proposed mode-locked fiber laser operated in the anomalous dispersion regime and thus it could be classified as a soliton fiber laser.

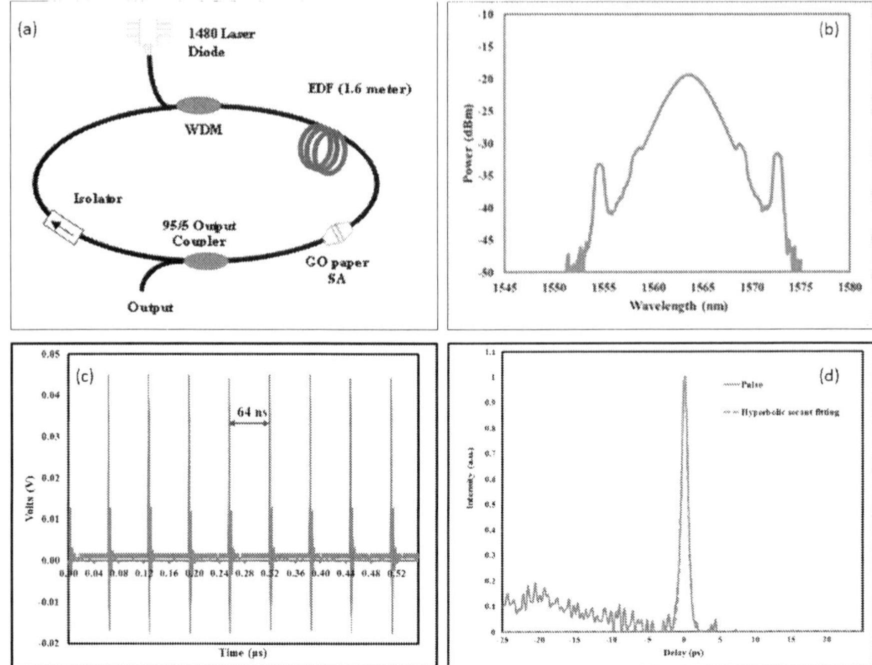

Figure 7. (a) Experimental setup. (b) Spectral characteristic of mode-locked fiber laser using GO paper. (c) Mode-locked pulse train with cavity round-trip time. (d) Autocorrelation trace.

The mode-locked fiber laser had a low self-starting threshold; approximately at 17.5 mW. Before all the modes were locked, multiple pulsing could be seen to occur at pump power as low as 10 mW. Figure 7(b) shows the spectral profile where the presence of soliton is confirmed. The presence of Kelly sidebands confirms that this mode-lock fiber laser is operating in anomalous dispersion regime. Figure 7(c) shows the pulse train of the passive mode-locked fiber. It has a cavity round-trip time of 64 ns, corresponding to a pulse repetition rate 15.6 MHz.

Figure 7(d) shows the autocorrelation trace with measured pulsewidth of 680 fs at its FWHM. The sech² fitting, which indicates the generation of the soliton pulse, is also included in the figure. The autocorrelation trace reveals that the experimental result follows the sech² fitting closely. A TBP of 0.315 is calculated from the 3-dB bandwidth of the optical spectrum and the acquired pulsewidth. This shows that the pulse is a transform-limited pulse. Since the pulsing threshold is low, the output power for this fiber laser is 0.134 mW. Consequently, the resultant pulse energy and peak power are 0.0085 nJ and 11.85 W, respectively.

As the data measurements are performed, it is observed that the pulse increasingly expanded and the spectral profile gradually changed from soliton to laser. Therefore, it is suspected that the pulse gradually is destroying the SA. Moreover, at approximately 24.4 mW, the output power was attenuated. Thus, it is concluded that a 1-layer GO paper SA is only effective in a short period of time and has low damage threshold. For the same reason, a satisfactory SNR data using RFSA is unable to be acquired. Together with low pulsing and damage thresholds, combined with the 5% of the intracavity energy taken out for performing measurements, it seems that the SNR is unable to extend more than 30 dB.

It is found that CNT and graphene saturable absorber may have inconsistent properties when prepared by different groups of researchers; despite repeating the same process. Currently, scientists are moving forward toward finding new materials that can be utilized as saturable absorber. The discovery of 2D material such as graphene has sparked interest in the potential of other 2D material. In recent developments, other 2D materials have been incorporated into fiber laser cavity to generate Q-switched and mode-locked fiber lasers. Topological insulator [19-22], transition metal dichalcogenides [23-25], and black phosphorus [26, 27] are among the recent materials being developed as saturable absorber.

6. Conclusion

We have discussed graphene and CNT saturable absorbers and their applications in generating passive Q-switched and mode-locked fiber lasers. We have demonstrated several examples of experiments where carbon-based saturable absorbers were used in generating Q-switched and mode-locked pulse in EDFL cavity. Currently, scientists are also moving forward toward new saturable absorber materials such as topological insulator and black phosphorus.

Acknowledgements

The authors acknowledge funding from the University of Malaya (Project Number: UM.C/ 625/1/HIR/MOE/ENG/09 and SF014-2014)

Author details

Mohd Afiq Ismail[1*], Sulaiman Wadi Harun[2], Harith Ahmad[1] and Mukul Chandra Paul[3]

*Address all correspondence to: afiq.ismail@siswa.um.edu.my

1 Photonics Research Centre, University of Malaya, Kuala Lumpur, Malaysia

2 Dept. of Electrical Engineering, Fac. of Engineering, University of Malaya, Kuala Lumpur, Malaysia

3 Fibre Optics Division, Central Glass and Ceramic Research Institute, Kolkata, India

References

[1] Welford D. Passively Q-switched lasers. Circuits and Devices Magazine, IEEE. 2003;19(4):31-6.

[2] Milonni P, Eberly J. Lasers physics. Hoboken. New Jersey: Wiley; 2010.

[3] Hasegawa A. Generation of a train of soliton pulses by induced modulational instability in optical fibers. Opt Lett. 1984 Jul 1;9(7):288-90. PubMed PMID: 19721573.

[4] Islam MN, Dijaili S, Gordon JP. Modulation-instability-based fiber interferometer switch near 1.5 μm. Optics letters. 1988;13(6):518-20.

[5] Boyd RW, Raymer MG, Narducci LM. Optical instabilities. Cambridge University Press, New York, NY, 1986.

[6] Agrawal GP. Nonlinear fiber optics: Springer; 2000.

[7] Jung I, Kärtner F, Brovelli L, Kamp M, Keller U. Experimental verification of soliton mode locking using only a slow saturable absorber. Optics letters. 1995;20(18):1892-4.

[8] Kartner F, Jung I, Keller U. Soliton mode-locking with saturable absorbers. Selected Topics in Quantum Electronics, IEEE Journal of. 1996;2(3):540-56.

[9] Au J, Kopf D, Morier-Genoud F, Moser M, Keller U. 60-fs pulses from a diode-pumped Nd: glass laser. Optics letters. 1997;22(5):307-9.

[10] Duling III IN. Compact sources of ultrashort pulses: Cambridge University Press; 2006.

[11] Kärtner FX, Keller U. Stabilization of solitonlike pulses with a slow saturable absorber. Optics Letters. 1995 1995/01/01;20(1):16-8.

[12] Dennis ML, Duling III IN. Experimental study of sideband generation in femtosecond fiber lasers. Quantum Electronics, IEEE Journal of. 1994;30(6):1469-77.

[13] Kelly S. Characteristic sideband instability of periodically amplified average soliton. Electronics Letters. 1992;28(8):806-7.

[14] Set S, Yaguchi H, Tanaka Y, Jablonski M, Sakakibara Y, Tokomuto M, et al., editors. A dual-regime mode-locked/Q-switched laser using a saturable absorber incorporating carbon nanotubes (SAINT). Lasers and Electro-Optics, 2003 CLEO'03 Conference on; 2003: IEEE.

[15] Set SY, Yaguchi H, Tanaka Y, Jablonski M, Sakakibara Y, Rozhin A, et al., editors. Mode-locked fiber lasers based on a saturable absorber incorporating carbon nanotubes. Optical Fiber Communication Conference; 2003: Optical Society of America.

[16] Set SY, Yaguchi H, Jablonski M, Tanaka Y, Sakakibara Y, Rozhin AG, et al., editors. A noise suppressing saturable absorber at 1550nm based on carbon nanotube technology. Optical Fiber Communication Conference; 2003: Optical Society of America.

[17] Bao Q, Zhang H, Wang Y, Ni Z, Yan Y, Shen ZX, et al. Atomic-Layer Graphene as a Saturable Absorber for Ultrafast Pulsed Lasers. Adv Funct Mater. 2009;19:3077-83.

[18] Yamashita S, editor Carbon-nanotube and graphene photonics2011. Optical Society of America.

[19] Neto AC, Guinea F, Peres N, Novoselov K, Geim A. The electronic properties of graphene. Reviews of modern physics. 2009;81(1):109.

[20] Allen MJ, Tung VC, Kaner RB. Honeycomb Carbon: A Review of Graphene. Chem Rev. 2010;110(1):132-45. English.

[21] Geim AK, Novoselov KS. The rise of graphene. Nature materials. 2007;6(3):183-91.

[22] Nair R, Blake P, Grigorenko A, Novoselov K, Booth T, Stauber T, et al. Fine structure constant defines visual transparency of graphene. Science. 2008;320(5881):1308-.

[23] Breusing M, Ropers C, Elsaesser T. Ultrafast Carrier Dynamics in Graphite. Physical Review Letters. 2009 02/27/;102(8):086809.

[24] Lazzeri M, Piscanec S, Mauri F, Ferrari AC, Robertson J. Electron Transport and Hot Phonons in Carbon Nanotubes. Physical Review Letters. 2005 11/30/;95(23):236802.

[25] Kampfrath T, Perfetti L, Schapper F, Frischkorn C, Wolf M. Strongly coupled optical phonons in the ultrafast dynamics of the electronic energy and current relaxation in graphite. Physical review letters. 2005;95(18):187403.

[26] Sun Z, Hasan T, Torrisi F, Popa D, Privitera G, Wang F, et al. Graphene mode-locked ultrafast laser. Acs Nano. 2010;4(2):803-10.

[27] González J, Guinea F, Vozmediano M. Unconventional quasiparticle lifetime in graphite. Physical review letters. 1996;77(17):3589-92.

[28] Bao Q, Zhang H, Ni Z, Wang Y, Polavarapu L, Shen Z, et al. Monolayer graphene as a saturable absorber in a mode-locked laser. Nano Research. 2011;4(3):297-307.

[29] Dössel L, Gherghel L, Feng X, Müllen K. Graphene Nanoribbons by Chemists: Nanometer-Sized, Soluble, and Defect-Free. Angewandte Chemie International Edition. 2011;50(11):2540-3.

[30] Gokus T, Nair R, Bonetti A, Bohmler M, Lombardo A, Novoselov K, et al. Making graphene luminescent by oxygen plasma treatment. ACS nano. 2009;3(12):3963-8.

[31] Stöhr RJ, Kolesov R, Pflaum J, Wrachtrup J. Fluorescence of laser-created electron-hole plasma in graphene. Physical Review B. 2010 09/14/;82(12):121408.

[32] Eda G, Lin YY, Mattevi C, Yamaguchi H, Chen HA, Chen I, et al. Blue photoluminescence from chemically derived graphene oxide. Advanced Materials. 2010;22(4): 505-9.

[33] Sun X, Liu Z, Welsher K, Robinson JT, Goodwin A, Zaric S, et al. Nano-graphene oxide for cellular imaging and drug delivery. Nano research. 2008;1(3):203-12.

[34] Lu J, Yang J-x, Wang J, Lim A, Wang S, Loh KP. One-Pot Synthesis of Fluorescent Carbon Nanoribbons, Nanoparticles, and Graphene by the Exfoliation of Graphite in Ionic Liquids. ACS Nano. 2009 2009/08/25;3(8):2367-75.

[35] Bonaccorso F, Sun Z, Hasan T, Ferrari A. Graphene photonics and optoelectronics. Nature Photonics. 2010;4(9):611-22.

[36] Saito R, Fujita M, Dresselhaus G, Dresselhaus MS. Electronic structure of graphene tubules based on C_ {60}. Physical Review B. 1992;46(3):1804.

[37] Wang F, Dukovic G, Brus LE, Heinz TF. The optical resonances in carbon nanotubes arise from excitons. Science. 2005;308(5723):838-41.

[38] Song YW, Yamashita S, Einarsson E, Maruyama S. All-fiber pulsed lasers passively mode locked by transferable vertically aligned carbon nanotube film. Optics letters. 2007;32(11):1399-401.

[39] Yamashita S. A Tutorial on Nonlinear Photonic Applications of Carbon Nanotube and Graphene. Lightwave Technology, Journal of. 2012;30(4):427-47.

[40] Ichida M, Hamanaka Y, Kataura H, Achiba Y, Nakamura A. Ultrafast relaxation dynamics of photoexcited states in semiconducting single-walled carbon nanotubes. Physica B: Condensed Matter. 2002;323(1):237-8.

[41] Tatsuura S, Furuki M, Sato Y, Iwasa I, Tian M, Mitsu H. Semiconductor carbon nanotubes as ultrafast switching materials for optical telecommunications. Advanced Materials. 2003;15(6):534-7.

[42] Ma Y-Z, Hertel T, Vardeny ZV, Fleming GR, Valkunas L. Ultrafast spectroscopy of carbon nanotubes. Carbon Nanotubes: Springer; 2008. p. 321-52.

[43] Ma Y-Z, Stenger J, Zimmermann J, Dexheimer SL, Fleming GR, Bachilo SM, et al., editors. Ultrafast carrier dynamics in single-walled carbon nanotubes probed by femtosecond spectroscopy. International Quantum Electronics Conference; 2004: Optical Society of America.

[44] Kazaoui S, Minami N, Nalini B, Kim Y, Takada N, Hara K. Near-infrared electroluminescent devices using single-wall carbon nanotubes thin flms. Applied Physics Letters. 2005;87(21):211914--3.

[45] Xu Z, Wu Y, Hu B, Ivanov IN, Geohegan DB. Carbon nanotube effects on electroluminescence and photovoltaic response in conjugated polymers. Applied Physics Letters. 2005;87(26):263118--3.

[46] Bachilo SM, Strano MS, Kittrell C, Hauge RH, Smalley RE, Weisman RB. Structure-assigned optical spectra of single-walled carbon nanotubes. Science. 2002;298(5602): 2361-6.

[47] Gutsche C. S. Reich, C. Thomsen, J. Maultzsch: Carbon nanotubes, basic concepts and physical properties. Colloid & Polymer Science. 2004;282(11):1299-.

[48] Freitag M, Martin Y, Misewich J, Martel R, Avouris P. Photoconductivity of single carbon nanotubes. Nano Letters. 2003;3(8):1067-71.

[49] Martinez A, Fuse K, Xu B, Yamashita S. Optical deposition of graphene and carbon nanotubes in a fiber ferrule for passive mode-locked lasing. Optics Express. 2010;18(22):23054-61.

[50] Kashiwagi K, Yamashita S, Set SY. In-situ monitoring of optical deposition of carbon nanotubes onto fiber end. Opt Express. 2009 03/30;17(7):5711-5.

[51] Kashiwagi K, Yamashita S. Deposition of carbon nanotubes around microfiber via evanascent light. Opt Express. 2009 09/28;17(20):18364-70.

[52] Nicholson J, Windeler R, DiGiovanni D. Optically driven deposition of single-walled carbon-nanotube saturable absorbers on optical fiber end-faces. Optics Express. 2007;15(15):9176-83.

[53] Ji H, Oxenløwe LK, Galili M, Rottwitt K, Jeppesen P, Grüner-Nielsen L, editors. Fiber optical trap deposition of carbon nanotubes on fiber end-faces in a modelocked laser. Conference on Lasers and Electro-Optics; 2008: Optical Society of America.

[54] Nicholson JW, editor Optically assisted deposition of carbon nanotube saturable absorbers. Lasers and Electro-Optics, 2007 CLEO 2007 Conference on; 2007 6-11 May 2007.

[55] Lamhot Y, Barak A, Peleg O, Segev M. Self-trapping of optical beams through thermophoresis. Physical review letters. 2010;105(16):163906.

[56] Ismail M, Ahmad F, Harun S, Arof H, Ahmad H. A Q-switched erbium-doped fiber laser with a graphene saturable absorber. Laser Physics Letters. 2013;10(2):025102.

[57] Harun S, Ismail M, Ahmad F, Ismail M, Nor R, Zulkepely N, et al. A Q-switched erbium-doped fiber laser with a carbon nanotube based saturable absorber. Chinese Physics Letters. 2012;29(11):114202.

[58] Harun SW, Ahmad H, Ismail MA, Ahmad F, editors. Q-switched and soliton pulses generation based on carbon nanotubes saturable absorber. Electronics, Communications and Photonics Conference (SIECPC), 2013 Saudi International; 2013: IEEE.

[59] Martinez A, Fuse K, Yamashita S. Mechanical exfoliation of graphene for the passive mode-locking of fiber lasers. Applied Physics Letters. 2011 09/19/;99(12):121107-3.

[60] Chang YM, Kim H, Lee JH, Song YW. Multilayered graphene efficiently formed by mechanical exfoliation for nonlinear saturable absorbers in fiber mode-locked lasers. Applied Physics Letters. 2010;97(21):211102--3.

[61] Yang Y, Chow K, editors. Optical pulse generation using multi-layered graphene saturable absorber. Information, Communications and Signal Processing (ICICS) 2013 9th International Conference on; 2013: IEEE.

[62] Chang YM, Kim H, Lee JH, Song Y-W, editors. Passive mode-locker incorporating physically exfoliated graphene for fiber ring lasers. Optical Fiber Communication Conference; 2011: Optical Society of America.

[63] Saidin N, Zen DIM, Ahmad F, Damanhuri SSA, Ahmad H, Dimyati K, et al. Q-switched thulium-doped fibre laser operating at 1900 nm using multi-layered graphene based saturable absorber. Iet Optoelectronics. 2014;8(4):155-60.

[64] Yamashita S, Set SY, Goh CS, Kikuchi K. Ultrafast saturable absorbers based on carbon nanotubes and their applications to passively mode-locked fiber lasers. Electronics and Communications in Japan (Part II: Electronics). 2007;90(2):17-24.

[65] Yamashita S, Inoue Y, Maruyama S, Murakami Y, Yaguchi H, Jablonski M, et al. Saturable absorbers incorporating carbon nanotubes directly synthesized onto substrates and fibers and their application to mode-locked fiber lasers. Optics letters. 2004;29(14):1581-3.

[66] Wang F, Torrisi F, Jiang Z, Popa D, Hasan T, Sun Z, et al., editors. Graphene passively Q-switched two-micron fiber lasers. CLEO: Science and Innovations; 2012: Optical Society of America.

[67] Popa D, Sun Z, Torrisi F, Hasan T, Wang F, Ferrari A. Sub 200 fs pulse generation from a graphene mode-locked fiber laser. Applied Physics Letters. 2010;97(20): 203106.

[68] Hasan T, Sun Z, Wang F, Bonaccorso F, Tan PH, Rozhin AG, et al. Nanotube–polymer composites for ultrafast photonics. Advanced materials. 2009;21(38-39):3874-99.

[69] Kelleher E, Travers J, Sun Z, Rozhin A, Ferrari A, Popov S, et al. Nanosecond-pulse fiber lasers mode-locked with nanotubes. Applied Physics Letters. 2009;95(11): 111108--3.

[70] Wang J, Luo Z, Zhou M, Ye C, Fu H, Cai Z, et al. Evanescent-light deposition of graphene onto tapered fibers for passive Q-switch and mode-locker. Photonics Journal, IEEE. 2012;4(5):1295-305.

[71] Song Y-W, Yamashita S, Goh CS, Set SY. Carbon nanotube mode lockers with enhanced nonlinearity via evanescent field interaction in D-shaped fibers. Opt Lett. 2007 01/15;32(2):148-50.

[72] Lin Y-H, Yang C-Y, Liou J-H, Yu C-P, Lin G-R. Using graphene nano-particle embedded in photonic crystal fiber for evanescent wave mode-locking of fiber laser. Optics Express. 2013 2013/07/15;21(14):16763-76.

[73] Zhao J, Ruan S, Yan P, Zhang H, Yu Y, Wei H, et al. Cladding-filled graphene in a photonic crystal fiber as a saturable absorber and its first application for ultrafast all-fiber laser. Optical Engineering. 2013;52(10):106105-.

[74] Digonnet MJF. Rare-Earth-Doped Fiber Lasers and Amplifiers, Revised and Expanded: CRC; 2001.

[75] Zayhowski JJ, Dill Iii C. Diode-pumped passively Q-switched picosecond microchip lasers. Opt Lett. 1994 09/15;19(18):1427-9.

[76] Zayhowski JJ, Kelley PL. Optimization of <e1>Q</e1>-switched lasers. Quantum Electronics, IEEE Journal of. 1991;27(9):2220-5.

[77] Novoselov K, Geim A, Morozov S, Jiang D, Zhang Y, Dubonos S, et al. Electric field effect in atomically thin carbon films. Science. 2004;306(5696):666-9.

[78] Spühler G, Paschotta R, Fluck R, Braun B, Moser M, Zhang G, et al. Experimentally confirmed design guidelines for passively Q-switched microchip lasers using semi-conductor saturable absorbers. JOSA B. 1999;16(3):376-88.

[79] Yang D, Velamakanni A, Bozoklu G, Park S, Stoller M, Piner RD, et al. Chemical analysis of graphene oxide films after heat and chemical treatments by X-ray photo-electron and micro-Raman spectroscopy. Carbon. 2009;47(1):145-52.

[80] Zhu Y, Murali S, Cai W, Li X, Suk JW, Potts JR, et al. Graphene and graphene oxide: synthesis, properties, and applications. Advanced materials. 2010;22(35):3906-24.

[81] Graphene Laboratories Inc. Available from: https://graphene-supermarket.com/.

Chapter 4

Double-scale Pulses Generated by Mode-locked Fibre Lasers and Their Applications

Sergey Kobtsev, Sergey Smirnov and Sergey Kukarin

Additional information is available at the end of the chapter

http://dx.doi.org/10.5772/61956

Abstract

This chapter presents a detailed analysis of properties of double-scale pulses (also called noise-like pulses and femtosecond clusters) generated in fibre lasers and gives an in-depth discussion of promising applications of such pulses.

Keywords: Fibre laser, passive mode locking, ultrashort pulse, non-linear polarisation evolution, double-scale pulse, noise-like pulse, laser radiation, mode correlations

1. Introduction

Mode-locked fibre lasers possess an intriguing capability of emitting, in certain generation regimes, a regular train of pico- or nanosecond wave packets stochastically filled with femtosecond sub-pulses. In the available literature on the subject, these wave packets are referred to as noise-like pulses [1–3], noise bursts [4], double-scale lumps [5], femtosecond clusters [6] or double-scale pulses [7]. In the scope of this chapter, we use the latter of these terms while discussing properties and applications of such pulses.

Until recently, double-scale pulse generation was only achieved in mode-locked fibre lasers, and there is still no understanding of physical mechanisms leading to formation of double-scale pulses, *i.e.*, to the co-existence of a virtually stable envelope of wave packets with power and phase fluctuations inside the wave packets. Parameters of these ultrashort pulses (such as peak power, duration, energy or instantaneous frequency) may experience significant fluctuations both during a single double-scale pulse and from one such pulse to another. Coherence time of double-scale pulses is determined by the duration of sub-pulses, which latter may be substantially shorter than that of the wave packet as a whole. As a result, such pulses have two different typical duration scales. The shape of the intensity autocorrelation

function (ACF) is a characteristic 'fingerprint' of double-scale pulses with its very distinct narrow central peak sitting on top of a broad pedestal. The width of this narrow peak corresponds to the typical sub-pulse duration and usually amounts to several hundred femtoseconds.

Double-scale pulses were mentioned for the first time at the close of the 1990s [1], when pulses with broad spectrum and noise-like behaviour of their intensity and phase were generated in a mode-locked erbium laser. For a long time, double-scale pulses were disregarded as 'not sufficiently coherent', and thus did not draw substantial attention to the respective laser generation regimes. Nevertheless, over recent years, this topic has been rapidly gaining popularity after the observation of double-scale pulses with relatively high energy in ultralong fibre lasers. Active research prompted by this discovery has shown that double-scale pulses may even be preferable in a number of applications, including non-linear frequency conversion, such as harmonic generation [8], Raman conversion [9] or super-continuum generation [10–13], as well as applications in imaging and sensing systems with high temporal and/or spatial resolution. It was furthermore demonstrated that generation of double-scale pulses in long lasers represents a remarkably multiform phenomenon encompassing many non-linear optical mechanisms, whose interaction may result in the emergence of diverse spatio-temporal coherent structures in laser radiation [14].

Here, we present a detailed analysis of double-scale pulse properties and provide an in-depth discussion of the above-mentioned and other important and promising applications of double-scale pulses generated in mode-locked fibre lasers.

2. Generation of double-scale pulses

The most common way to generate double-scale pulses is *via* fibre lasers passively locked due to the effect of non-linear polarisation evolution (NPE). It should also be noted that to date double-scale pulses have been demonstrated in other types of passively mode-locked lasers, including those using saturable absorbers made of single-walled carbon nanotubes [15], as well as topological insulators [16] and non-linear loop mirrors or amplifiers (NOLM/NALM) [11, 17, 18]. Nonetheless, NPE lasers possess the largest number of degrees of freedom in adjustment of the generation regime. This greatly facilitates realisation of various generation regimes in them, including double-scale generation ones [5, 7, 14].

A typical configuration of a fibre laser passively mode-locked due to NPE is presented in Fig. 1. In most reported research, laser-diode-pumped (LD) erbium- or ytterbium-doped optical fibres are used as the active medium. Since their generation spectrum is rather wide, NPE-mode-locked lasers are most often implemented in ring-cavity configuration, thus avoiding the need for broadband reflectors. Optical isolators are normally used to ensure unidirectional generation. Adjustment and switching of generation regimes are done with intra-cavity polarisation controllers PC1 and PC2. The generated laser radiation can be extracted from the cavity either through fibre couplers or through one of the ports of the fibre-optical polarisation beam splitter, which also carries the function of introducing polarisation-dependent optical losses, thus ensuring passive NPE mode locking.

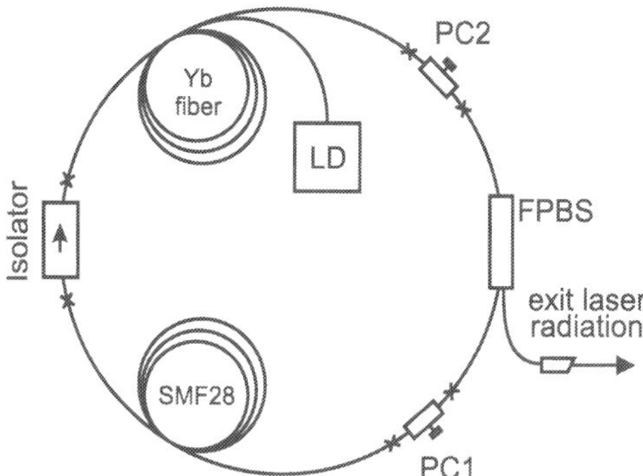

Figure 1. Typical layout of a fibre laser passively mode-locked due to the NPE effect.

In relatively short lasers (with few-meter-long cavities), both double-scale and 'usual' pulses (as well as some transitional or intermediate regimes between these two) may be generated at different settings of the intra-cavity polarisation controllers or the pump power level (see Fig. 2). In contrast, elongation of the fibre laser cavity to several hundred metres or several kilometres leads to predominant generation of double-scale pulses [19, 20]. Cavity elongation is one of the most effective ways to raise the output pulse energy in passively mode-locked lasers [21]; therefore, double-scale pulses became the focal point of research conducted by many groups around the world. Various publications studied spectral and temporal properties of double-scale pulses produced in different generation regimes [22–25].

3. Direct numerical simulation of double-scale pulses

Numerical modelling of double-scale pulse generation in fibre lasers is usually based on the generalised non-linear Schrödinger equation (GNLSE) [20] or on a system of simultaneous equations for polarisation components of the intra-cavity radiation [5]:

$$\frac{\partial A_x}{\partial z} = i\gamma \left\{ |A_x|^2 A_x + \frac{2}{3}|A_y|^2 A_x + \frac{1}{3}A_y^2 A_x^* \right\} + \frac{g_0/2}{1 + E/(P_{sat} \cdot \tau)} A_x - \frac{i}{2}\beta_2 \frac{\partial^2 A_x}{\partial t^2} \tag{1}$$

$$\frac{\partial A_y}{\partial z} = i\gamma \left\{ |A_y|^2 A_y + \frac{2}{3}|A_x|^2 A_y + \frac{1}{3}A_x^2 A_y^* \right\} + \frac{g_0/2}{1 + E/(P_{sat} \cdot \tau)} A_y - \frac{i}{2}\beta_2 \frac{\partial^2 A_y}{\partial t^2} \tag{2}$$

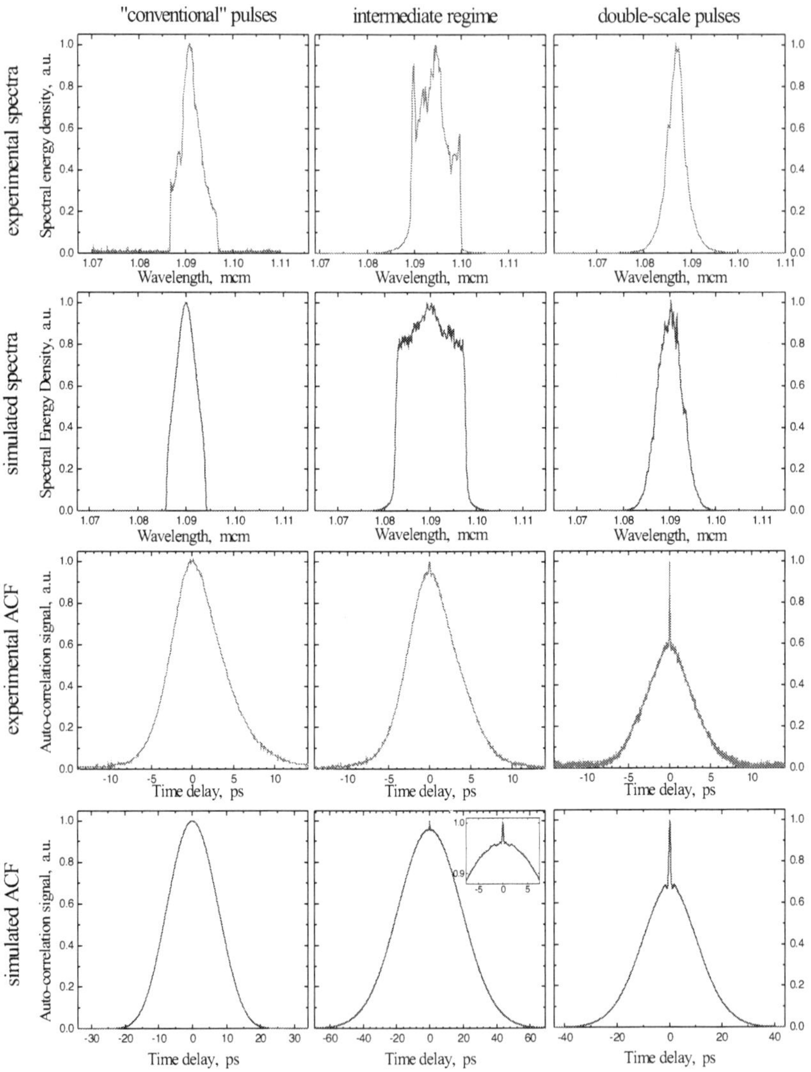

Figure 2. Experimentally measured (red) and simulated (blue) spectra and ACFs for three different lasing regimes: stable single-pulse (left column), intermediate (middle column) and noise-like generation (right column) [24].

where A_x and A_y are the polarisation components of the field envelope, z is the longitudinal coordinate along the fibre, t is the time in the retarded frame of reference, γ and β_2 are non-linear and dispersion coefficients correspondingly, g_0 and P_{sat} stand for unsaturated gain coefficient and saturation power for the active fibre.

Equations (1, 2) describe propagation of radiation along an active fibre. Taking $g_0=0$, we can use the same equations to model laser pulse propagation inside passive resonator fibre. The fibre-optical polarisation beam splitter is represented in the model by the following matrix:

$$\hat{T}_{PBS} = \begin{pmatrix} 1 & 0 \\ 0 & 0 \end{pmatrix} \tag{3}$$

Unitary 2 × 2 matrices must be used to describe polarisation controllers. In particular, a polarisation controller based on the principle of fibre compression in the direction at an angle ϕ can be represented by a matrix introducing phase delay α rotated by angle ϕ by multiplying it by the corresponding rotation matrix:

$$\hat{T}_{PC1}\left(\alpha,\phi\right) = \begin{pmatrix} \cos\phi & -\sin\phi \\ \sin\phi & \cos\phi \end{pmatrix} \begin{pmatrix} e^{i\alpha/2} & 0 \\ 0 & e^{-i\alpha/2} \end{pmatrix} \begin{pmatrix} \cos\phi & \sin\phi \\ -\sin\phi & \cos\phi \end{pmatrix} \tag{4}$$

Parameter α in Eq. (4) stands for phase delay introduced by the polarisation controller and depends on the transverse fibre deformation. Similarly, a polarisation controller utilising fibre torsion can be expressed through its eigenvector projections: $\hat{T}_{PC2}=e^{i\alpha/2}\hat{P}_+ + e^{-i\alpha/2}\hat{P}_-$, where circular polarisation state projections can be written as

$$\hat{P}_\pm = \frac{1}{2} \begin{pmatrix} 1 & \mp i \\ \pm i & 1 \end{pmatrix} \tag{5}$$

In order to model the propagation of laser pulses around the fibre cavity, Eqs. (1, 2) can be integrated numerically by the step-split Fourier method [26]. At the required points along the optical path, the polarisation transformations (3–5) are applied and losses corresponding to the intra-cavity elements are taken into account.

This modelling step is carried out repeatedly until a stationary state is reached. As a rule, anywhere between several hundred and several thousand modelled cavity round trips are needed, depending on the system parameters and the initial conditions, mostly taken as white noise or seed pulses. In certain cases, a laser may exhibit bistability: the limit cycle of the propagation equations may depend on the initial conditions. It is relevant to mention here that a similar phenomenon is observed in the experiment as hysteresis in switching between generation regimes [27]. If the limit cycle of the propagation equations corresponds to the 'conventional' pulse generation regime, the generation parameters (power, pulse duration, spectrum width, *etc.*) of the limit cycle are highly stable (of the 10^{-3} order and higher). Conversely, when the laser generates double-scale pulses, the pulse parameters in the generated train fluctuate around their average values by a few per cent or even more between two successive cavity round trips. This circumstance can be used in modelling as a basic criterion to distinguish between generation regimes. We need to point out as well that the average generation parameters must be insensitive to variation within reasonable limits of the 'non-physical' modelling parameters, such as the mesh node count, mesh width and the step of numerical integration of Eqs. (1, 2). This has to be controlled during modelling.

The approach outlined in the foregoing paragraphs is sufficiently powerful and generalised to adequately model a variety of regimes observed in fibre lasers mode-locked due to NPE (see Fig. 2 and Ref. [24]), as well as NOLM/NALM and generation of both double-scale and 'regular' pulses. Such power and generality come, however, at the expense of very significant amount of computations necessary to perform in order to compare the model with the experiment. This downside results largely from the specific nature of fibre laser configurations relying on fibre-optical polarisation controllers. These controllers introduce a phase shift, which is impossible to directly measure in regular experimental implementations. As a consequence, for a valid match between modelling and experiment, it is necessary to perform a large series of computations for a range of settings of the intra-cavity polarisation controllers corresponding to a range of parameters α and ϕ in Eqs. (4, 5). For each set of parameters, the entire numerical modelling sequence has to be carried out, including specification of the initial conditions, multiple runs of radiation propagation along the cavity and analysis of the generation regime. In full analogy to the experiment, laser generation only emerges at certain combinations of intra-cavity polarisation element settings and levels of the pumping power launched into the active fibre. Therefore, a considerable part of calculations carried for randomly chosen polarisation controller parameters does not result in a pulsed generation regime. Nonetheless, this approach gives a fair idea of the great diversity of generation regimes accessible in an NPE-mode-locked fibre laser through adjustment of its settings. For instance, in Fig. 3, we present histograms of generation parameter distributions (rms bandwidth and rms pulse duration) for 'regular' and double-scale pulses generated in numerical modelling [7] at different settings of the polarisation controllers. Evidently, there is a considerable spread in laser parameter values, up to an order of magnitude and even wider. This result effectively indicates an opportunity to modify parameters of the output laser pulses (including double-scale ones) by adjusting the intra-cavity polarisation controllers in order to achieve optimal values for specific applications.

4. Simplified phenomenological model

Research in efficiency of double-scale pulses in practical applications ideally needs a less complicated numerical model, which would enable numerical studies at much more affordable expense of computation resources. With this objective, we have developed a phenomenological model of double-scale pulses relying on superposition of uncorrelated modes. Using this model, it is comparatively easy to re-create modelling pulses with parameters known from the experiment (shape, duration and spectral width) for studying their various applications and parameter optimisation. The proposed phenomenological approach to modelling requires that the following function be calculated:

$$A(t) \sim \sqrt{P(t)} \cdot \sum_j A_j \exp\left(i\omega_j t\right). \tag{6}$$

Here, ω_j is the frequency of the j-th mode, t is the time, A_j is the complex amplitude of the j-th mode and $P(t)$ is the temporal pulse profile. The phases of complex amplitudes arg$\{A_j\}$ are taken as independent random values uniformly distributed over the range of 0–2π.

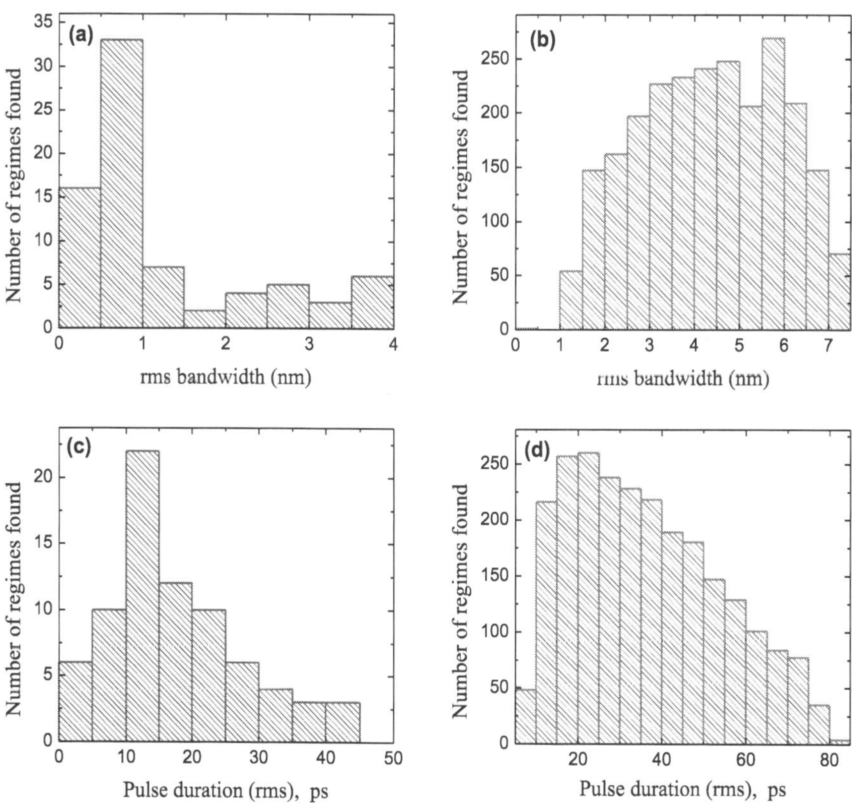

Figure 3. Rms bandwidth (a, b) and rms-pulse duration (c, d) variability in simulated 'conventional' (a, c) and double-scale (b, d) generation regimes [7].

In the simplest case, scalar values A_j in Eq. (6) can be assumed constant, while in more involved implementations of this model, their random character may also be taken into account ($|A_j|$ fluctuations). It is interesting to observe that the expression of Eq. (6) can be considered as a model of thermal (incoherent) source with a preset spectral profile. The sum part of Eq. (6) is multiplied by a specified profile $P(t)$, as a result shaping noise-like incoherent radiation into a pulse with a shape known from the experiment.

For an illustration, Fig. 4 presents the temporal distribution, spectrum and autocorrelation function of double-scale pulses generated in the simplified phenomenological model (Eq. (6)). It can be seen that the diagrams exhibit all the main features of double-scale pulses observed in both experiment and direct numerical modelling of laser generation (see Fig. 2). For instance, the temporal distribution of pulses in Fig. 4 looks like a wave packet stochastically filled with sub-pulses. Single-shot spectrum of the modelled pulses also has a noisy appearance, but the spectrum averaged over many pulses takes the form identical to that of $|A_j(\omega_j)|$ on condition that the duration of wave packets is much longer than the inverse spectrum width. The pulse ACF takes the shape of a broad pedestal with a narrow peak in its centre (Fig. 4 (c)). The relative pedestal magnitude is 0.5, and its width in the case of Gaussian pulses by a factor of $\sqrt{2}$ exceeds the duration of the pulse envelope $P(t)$ in Eq. (6)). The width of the central ACF peak, conversely, is equal to the inverse spectral width of pulses (see Fig. 4 (d)).

It should be noted here that although the proposed simplified model (Eq. (6)) is founded on the assumption of independent phases $arg\{A_j\}$ of modes (spectral radiation amplitudes), multiplication by $P(t)$ introduces correlation of neighbouring modes inside a spectral domain with the width of the order of pulse envelope $P(t)$ width. In reality, the level of an inter-mode correlation for double-scale pulses observed in the experiment and in direct numerical modelling may be higher than that of the explained simplified model. This is indicated by the ACF pedestal magnitude exceeding the 0.5 value in some generation regimes. In particular, the transient generation regime may be characterised by the maximum value of the central ACF peak much below unity (see Fig. 2). This corresponds to the strong inter-mode correlation and/or relatively small fluctuations.

The phenomenological model (Eq. (6)) analysed in the preceding discussion also improves our understanding of the origin of sub-pulses in the internal filling of double-scale pulses. According to Eq. (6), sub-pulses are closer in their nature to power oscillations resulting from interference of incoherent modes rather than to separate femtosecond pulses that are independent of each other. This circumstance leads to radical differences in behaviour of double-scale pulses undergoing temporal compression or stretching as compared with 'regular' laser pulses. This difference will be given a detailed treatment in the following section.

5. Pulse compression

One of the salient differences of double-scale pulses from 'conventional' laser pulses is related to limited possibilities of their compression. In most experimental configurations, double-scale

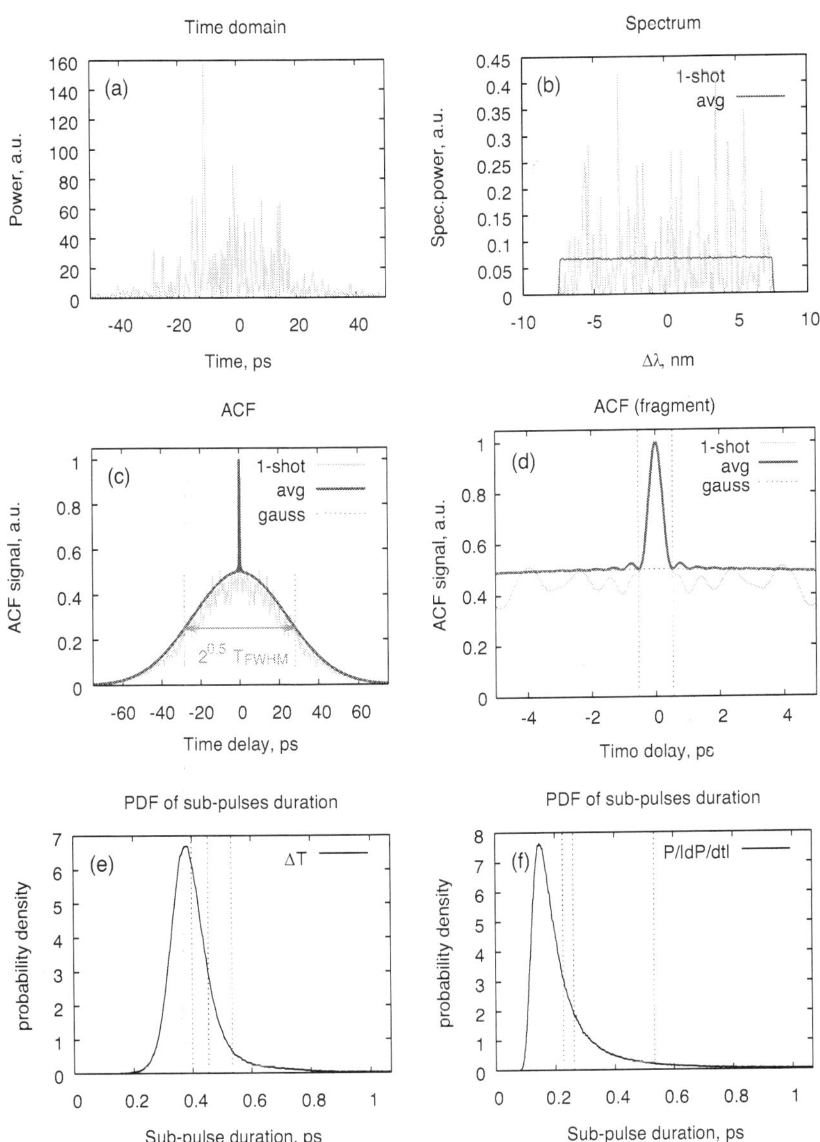

Figure 4. Time trace of power (a), spectral distribution (b) and ACF (c, d) of double-scale pulses in the simplified model. Red dotted line in (d) denotes the value of the inverse width of a Π-shaped spectrum. Grey lines correspond to single-shot traces and blue lines to those averaged over a large number of random realisations.

pulses are produced in fibre lasers with a significant positive dispersion of their cavities. If the laser generates 'conventional' pulses, the presence of uncompensated cavity dispersion leads to the formation of strongly chirped pulses. Such pulses can be subsequently compressed to a small fraction of their initial duration down to (or close to) the Fourier limit by passing them through optical compressors based on diffraction gratings, optical prisms and/or optical fibres with dispersion of different signs. Conversely, double-scale pulses present very limited possibilities of compression [24]. It has to be stressed that these limitations arise from the structure of the pulses themselves rather than from non-linear effects in fibre-optical compressors, which may also constrain compression coefficient of 'regular' laser pulses at sufficiently high-peak-power values [28].

Direct numerical modelling of laser generation based on coupled GNLSE (1, 2) explains the measured low compressibility of double-scale pulses in the experiment. As calculations demonstrated [24], unlike that of 'regular' pulses, the optical phase of double-scale pulses is not a smooth continuous function of time but rather is filled with fluctuations, as shown in Fig. 5. These phase fluctuations lead, in particular, to the lack of phase coherence across the seemingly regular array of pulses observed experimentally [29]. Fluctuations of the temporal dependence of the optical phase give rise to jitter of the instantaneous frequency of double-scale pulses. This, in turn, leads to presence at each moment in time (in each point within a double-scale pulse) of various frequencies covering the wide spectrum of the double-scale pulse, as can be readily seen in a simulated FROG diagram of Fig. 6. The fundamental principle of optical compression is to create a different temporal delay for the front and rear pulse edges, which have slightly different optical frequencies because of a pulse chirp. Passing along a phase-delay element, the components of the chirped pulse move closer to each other, producing the effect of pulse compression. Since different optical frequencies are present at each moment in double-scale pulses, frequency-dependent temporal delay of their components cannot compress such pulses as well as the 'regular' ones, as shown in Fig. 6.

Figure 6 showcases pulse spectrograms generated in direct numerical modelling of various generation regimes (indicated on top). The horizontal axis shows time in picoseconds and the vertical axis, frequency and wavelength λ (left- and right-hand axes, respectively). Radiation intensity is colour-coded on a log scale of 0 to –25 dB, as shown in the colour scale at the bottom of Fig. 6. Pulse duration in the included spectrograms corresponds to the horizontal dimension of intensity distribution, whereas the spectrum width corresponds to the vertical dimension. The distribution slant in the diagrams reflects the pulse chirp. Linear compressibility of pulses in optical compressors is predicated on the existence of chirp. The top row of diagrams demonstrates pulses exiting the laser and the bottom rows correspond to the result of various degrees of linear compression (the respective values of the optical compressor dispersion are specified to the left of the rows).

As seen in the left column of Fig. 6, the duration of 'regular' laser pulses may be reduced in a linear optical compressor almost down to the Fourier limit. The compression degree in this case is somewhat below the theoretical limit because of the non-linear chirp present at the pulse edges and clearly visible as pointed deformations of the colour patches of the left column in Fig. 6. In experiments, this limitation may be overcome by spectral filtration of such pulses:

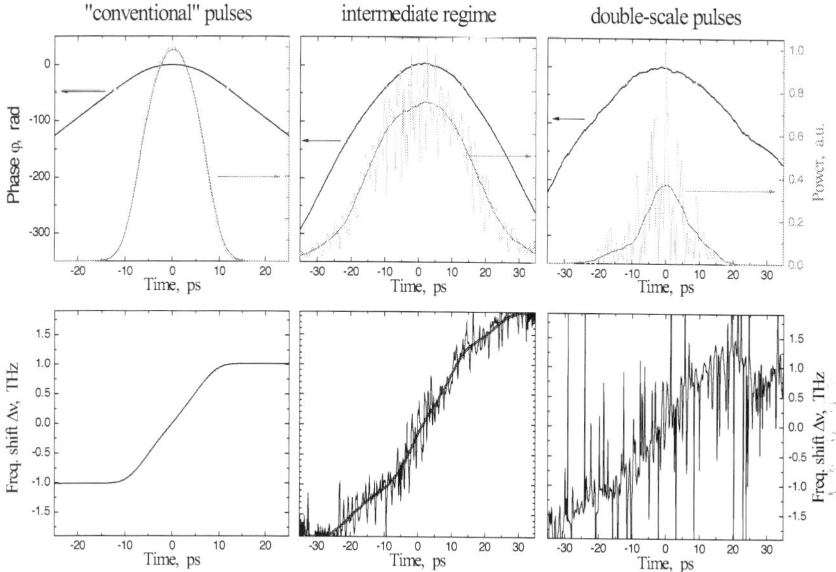

Figure 5. Simulated time dependencies of pulse power and phase (upper graphs), and frequency shifts (lower graphs) for different lasing regimes [24].

when the spectral extremities of chirped pulses are 'cut off', the temporal wings of these pulses containing non-linear chirps are also eliminated.

In contrast, double-scale pulses are not really amenable to compression (see the right column in Fig. 6). Although they have approximately three times the spectral width of a 'conventional' laser pulse (measured at –25 dB) in Fig. 6, after a pass through an optical compressor they retain a duration exceeding that of 'normal' pulses by a factor of over 5. Although these particular calculations indicate the possibility of almost twofold compression, the minimal achievable pulse duration far exceeds the Fourier limit, as a direct consequence of the pulse structure. At each moment in time (along any vertical section of the spectrogram), double-scale pulses contain practically all the components of its spectrum (see the bottom left diagram in Fig. 6).

The intermediate generation regime, with its comparatively low fluctuations of phase and other parameters of the generated pulses, takes an accordingly middle position with respect to linear compression, as indicated by the central column of Fig. 6. It is characterised by medium compressibility, which is limited by both factors mentioned earlier: the non-linear chirp (curved ends of the spectrogram) and phase fluctuations, making the spectrogram considerably wider than the spectrally limited pulses at the same or even at narrower spectrum width. This can be readily seen when comparing the central and left-hand diagrams in the bottom row of Fig. 6.

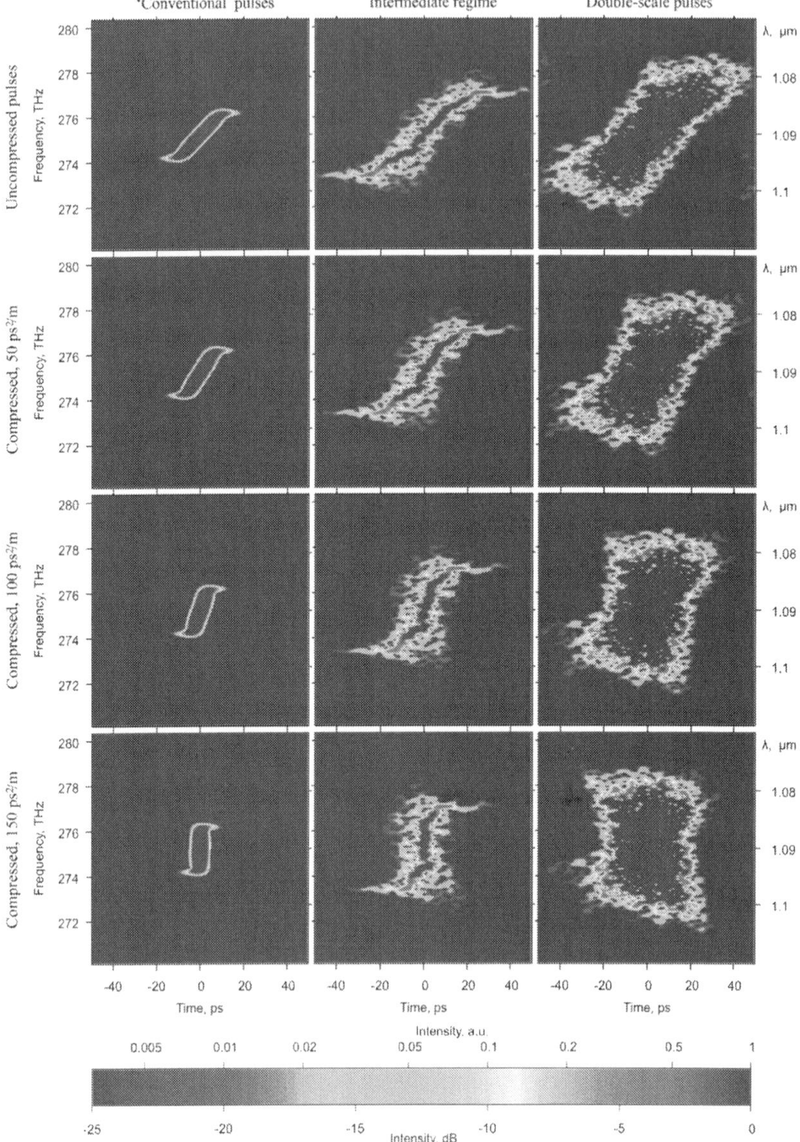

Figure 6. Simulated FROG of 'conventional' (left column), intermediate (at the middle) and double-scale (right column) pulses generated by all-normal-dispersion fibre laser passively mode-locked due to NPE and different stages of compression of these pulses.

6. Pulse stretching

It is clear that the optical phase fluctuations condition the difference of double-scale pulses from 'conventional' laser pulses not only in relation to compression but also to stretching, *i.e.*, to lengthening of the pulse duration as it propagates along a medium with chromatic dispersion (*e.g.* an optical fibre). Indeed, if we consider the diagrams of Fig. 6 in the reversed order (*i.e.* bottom to top instead of top to bottom, as before), we will observe dynamics of stretching of laser pulses with different levels of phase fluctuations in a dispersive medium. In this case, the bottom row of diagrams corresponds to the state with the smallest durations (for a given pulse type with a specific spectral width). Pulse duration increases during the process of propagation along the medium (see Fig. 6 from bottom to top).

The spectrograms presented in Fig. 6 demonstrate that double-scale pulses are significantly less stretchable than the 'regular' laser pulses of comparable or even narrower width of their optical spectrum. This circumstance makes double-scale pulses attractive in a number of applications where low temporal coherence of radiation is required, for example, to achieve high resolution in optical imaging. Unlike 'regular' ultrashort laser pulses with short coherence time, double-scale pulses may be delivered to the desired location over optical fibres practically without loss of the system's resolving power defined by the pulse coherence time.

7. Applications

Until recently, the possibility of practical application of double-scale pulses remained uncertain because stochastic filling of generated wave packets was understood to result in highly unstable output parameters. As it was experimentally shown, however, the average output power of a mode-locked fibre laser generating double-scale pulses remains relatively stable and pulse-to-pulse average power fluctuations being fairly small (a few per cent, as a rule). It was further discovered that double-scale pulses exhibit a number of remarkable properties making them radically different from 'usual' laser pulses. Double-scale pulses may carry relatively high energy (several μJ) directly at the output of a fibre master oscillator [21, 25]. Moreover, peak power of sub-pulses may far exceed the average peak power of the entire wave packet because of their short duration. It was also established that double-scale pulses feature comparatively high efficiency of non-linear interaction with the propagation medium in harmonic generation [8], Raman conversion [9] or super-continuum generation [10–13]. We analyse these and other promising applications of double-scale pulses in the following discussion.

The potential of double-scale pulses in configurations with non-linear frequency conversion is directly related to the high peak power of sub-pulse filling. The sub-pulse peak power may be several times as high as the average power of the double-scale envelope due to fluctuations resulting from stochastic nature of double-scale pulses. High peak power leads to relatively efficient non-linear optical transformation of double-scale pulses, for instance, harmonic generation, super-continuum generation, *etc*. As an illustration, let us consider the frequency-doubling process in a thin non-linear crystal that we studied both in direct numerical modelling and in experiment [8]. Second-harmonic generation (SHG) was achieved in a Yb-doped fibre laser passively mode-locked due to NPE (see Fig. 1 for the optical layout). Through adjustment of the intra-cavity polarisation controllers, this laser could be switched between generation of 'regular' and double-scale pulses as needed during the experiment. Direct numerical modelling was carried out by integrating Eqs. (1, 2) with the following parameters: non-linear coefficient $\gamma = 4.7 \times 10^{-5}$ $(\text{cm} \bullet \text{W})^{-1}$, dispersion coefficient $\beta_2 = 23$ ps^2/km, small-signal gain $g_0 = 540$ dB/km and saturation power of the active fibre $P_{sat} = 52$ mW.

Second-harmonic generation was further modelled with the following equation for the spectral wave amplitudes:

$$\frac{\partial A_{2n}}{\partial z} + \frac{1}{u}\frac{\partial A_{2n}}{\partial t} = -i\sigma_2 \sum_{j+k=n+1} A_{1j}A_{1k} \tag{7}$$

where A_{1j} and A_{2n} are spectral amplitudes of the first- and second-harmonic waves, respectively, u is the group velocity and σ_2 is the non-linear coefficient. Apart from the process of frequency doubling of the laser modes corresponding to terms with $j = k$, Eq. (7) also reflects processes of sum frequency generation (terms with $j \neq k$ under summation in Eq. (7)). Both types of processes are schematically shown in Fig. 7 (a) for $N = 3$ equidistantly spaced laser modes. Thick solid lines correspond to frequency doubling, and dotted lines show sum frequency generation. In reality, the number of laser modes N is very large; however, all of them are separated by the same distance $\Delta\omega$ from the neighbouring ones. The second-harmonic radiation, following Fig. 7 (a), will consist of $2N - 1$ equidistant modes spaced at the same spectral interval $\Delta\omega$.

Eq. (7) is integrable in the approximation of a thin non-linear crystal and in the absence of the pumping wave depletion. Assuming $A_{1i} = $ constant along z and $A_{2n} = 0$ at $z = 0$, let us integrate Eq. (7) from 0 to L and divide the full power of the second-harmonic wave $I_{2n} = \Sigma |A_{2n}|^2$ by $(\sigma_2 L P_1)^2$, where P_1 is the first harmonic sum power:

$$\zeta = \left(\sum_n \left| \sum_{j+k=n+1} A_{1j}A_{1k} \right|^2 \right) \cdot \left(\sum_j |A_{1j}|^2 \right)^{-2} \tag{8}$$

The resulting coefficient ζ is the SHG relative efficiency, which equals to the ratio of two SH powers, of which the first is obtained when the non-linear crystal is pumped by a double-scale laser pulse with given mode amplitudes A_{1j} and the second is generated with the use of single-mode monochromatic pumping of the same power $P_1 = \Sigma |A_{1j}|^2$. An important feature of the dimensionless relative efficiency ζ is that it does not depend on power and thickness of the thin non-linear crystal but is sensitive to mode correlations and fluctuations, thus allowing us to easily compare different lasing regimes from the viewpoint of efficiency of non-linear frequency. To do that, we simulated the generation of double-scale pulses using Eqs. (1, 2) and

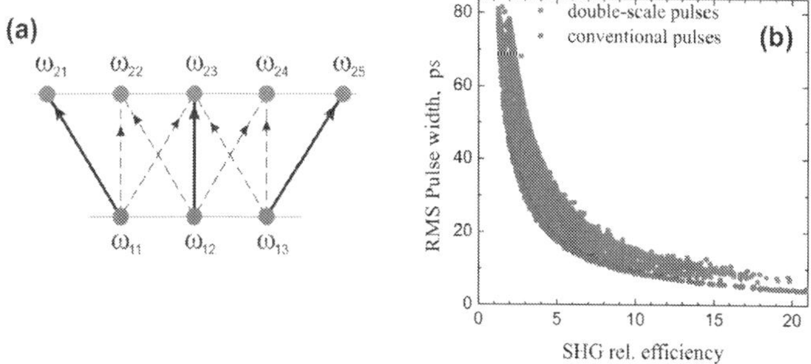

Figure 7. (a) Schematic diagram of SHG processes for $N = 3$ modes. (b) Correlation between relative SHG efficiency and rms pulse width of double-scale (red points) and conventional (green points) pulses.

averaged SHG relative efficiency (Eq. (8)) over multiple successive pulses produced by the fibre laser.

In our modelling, the parameters of the intra-cavity polarisation controllers were selected randomly. After 10,000 cavity passes, the generation mode of the modelled laser was analysed. If the laser generated double-scale or 'regular' pulses, then second-harmonic generation was modelled, which consisted in calculation of the relative efficiency ζ (Eq. (8)). Both the generation parameters and the ζ value were then saved by the program. The modelling cycle was then repeated with the intra-cavity polarisation controller settings chosen once again at random.

As a result of a large number of modelling cycles, statistical data were collected on the generation parameters and relative efficiency of frequency doubling in double-scale and 'regular' generation regimes. These data allowed the comparison of these regimes in relation to their efficiency for frequency doubling, and they also could be used to study the correlation between the frequency-doubling efficiency and the generation parameters. The most immediate link was identified between the relative efficiency of frequency doubling ζ and rms pulse duration T_{rms} with Pearson's correlation coefficient $\rho(\zeta, 1/T_{rms}) = 0.97$. The obtained results are presented in Fig. 7 (b). Each point on the shown diagram was generated as a result of direct numerical modelling of laser generation at fixed values of the intra-cavity polarisation controller settings. These values differ randomly between one point and another. The X-coordinates of the diagram points are proportional to the relative efficiency ζ of frequency doubling, whereas their Y-coordinates are proportional to the rms pulse width.

As it can be seen from Fig. 7 (b), ζ is roughly inversely proportional to T_{rms}, which means that ζ grows linearly with the pulse peak power, as it should for a second-order non-linear process. One can also note in Fig. 7 (b) that for any fixed pulse width T_{rms}, the relative SHG efficiencies are comparable for both regimes, being slightly higher for double-scale pulses. The experimental results that we measured in [8] agree with the numerical modelling, featuring higher efficiency of power transformation into the second harmonics for double-scale pulses in comparison to 'conventional' laser pulses.

We carried out similar experiments on cascaded Raman conversion of different pulse types in a long phosphosilicate fibre [9]. It was shown that feeding double-scale pulses into the fibre produces broader output spectrum as compared with using 'regular' laser pulses. Furthermore, different statistics of laser radiation in these regimes led to specific spectral features in the Raman spectrum of double-scale pulses [9]. In general, the results reported in Ref. [9] corroborate the conclusion made in the foregoing discussion about better efficiency of non-linear transformation of double-scale pulses.

Super-continuum generation [11, 30] may be also listed among promising applications related to non-linear frequency transformation. Apart from a relatively higher peak power, double-scale pulses are also less coherent. As it was demonstrated earlier, this may boost super-continuum generation and allow one to obtain wider and smoother supercontinuum spectra [11, 31, 32].

Two other prominent features of double-scale pulses, their very short coherence time and their broad spectrum [33–36], enable their application in imaging and sensor systems with high temporal and/or spatial resolution [37–39]. Importantly, double-scale pulses are not strongly affected by dispersive broadening [1], unlike regular ultrashort laser pulses with comparable coherence time. This makes it possible to transmit such pulses over optical fibres towards the target objects without loss of system's resolving power.

Inner structure of double-scale pulses filled with ultrashort sub-pulses and their significant spectrum width also make them attractive for applications such as laser-induced breakdown spectroscopy (LIBS) [6], a type of atomic emission spectroscopy. High-energy laser pulses are focused on a surface and cause ablation and plasma formation. The radiation generated by plasma is then registered in an optical system for spectral analysis. Comparison of recorded spectral lines with the known atomic optical spectra allows relatively easy and fast identification of the chemical composition of the studied sample. This method has the advantages of needing little or no sample preparation, and the possibility of depth profile generation by layer-wise ablation of material from the sample surface. This latter circumstance also solves the problem of surface contamination of the studied samples. Another important benefit of LIBS is its minimally or non-destructive nature. LIBS features sufficiently high precision and does not rely on ionising radiation, which is important for biological safety. Double-scale pulses generated in passively mode-locked fibre lasers present a virtually ideal solution for LIBS, because they deliver bursts of femtosecond sub-pulses at megahertz repetition rates [6]. High peak power of their sub-pulses drives correspondingly high efficiency of two-photon processes while maintaining low thermal stress of the studied samples.

8. Conclusions

Double-scale pulses offer a number of unique properties, including comparatively high peak power, broad spectral width and low coherence time. Therefore, such pulses hold much promise in a number of applications (*e.g.* non-linear optical frequency transformation, spectroscopy, material micro-processing and imaging techniques). They furthermore exhibit a wide variety of generation regimes and an improbable combination of stochastic and coherent laser dynamics, still far from being understood. Although this factor draws the attention of researchers to double-scale pulses, it also prevents, for the time being, broader applications of double-scale pulses in research and technology applications of fibre lasers.

Acknowledgements

This work was supported by the Grants of Ministry of Education and Science of the Russian Federation (agreement No. 14.B25.31.0003, ZN-06-14/2419, order No. 3.162.2014/K).

Author details

Sergey Kobtsev*, Sergey Smirnov and Sergey Kukarin

*Address all correspondence to: sergey.kobtsev@gmail.com

Division of Laser Physics and Innovative Technologies, Novosibirsk National Research State University, Novosibirsk, Russia

References

[1] M. Horowitz, Y. Barad, and Y. Silberberg, "Noiselike pulses with a broadband spectrum generated from an erbium-doped fiber laser," Opt. Lett. 22(11), 799–801 (1997).

[2] O. Pottiez, R. Grajales-Coutiño, B. Ibarra-Escamilla, E. A. Kuzin, and J. C. Hernández-García, "Adjustable noiselike pulses from a figure-eight fiber laser," Appl. Opt. 50(25), E24–E31 (2011).

[3] D. Lei, H. Yang, H. Dong, S. Wen, H. Xu, and J. Zhang, "Effect of birefringence on the bandwidth of noise-like pulse in an erbium-doped fiber laser," J. Mod. Opt. 56(4), 572–576 (2009).

[4] Rare-Earth-Doped Fiber Lasers and Amplifiers, Revised and Expanded. Ed. M. J. Digonnet. CRC Press, New York, 2001. M. E. Fermann, and Martin Hofer, Chapter 8 "Mode-locked fiber lasers", pp. 499-500.

[5] S. Kobtsev, S. Kukarin, S. Smirnov, S. Turitsyn, and A. Latkin, "Generation of double-scale femto/pico-second optical lumps in mode-locked fiber lasers," Opt. Express 17, 20707–20713 (2009).

[6] B. Nie, G. Parker, V. V. Lozovoy, and M. Dantus, "Energy scaling of Yb fiber oscillator producing clusters of femtosecond pulses," Opt. Eng. 53(5), 051505 (2014).

[7] S. Kobtsev, S. Smirnov, S. Kukarin, and S. Turitsyn, "Mode-locked fiber lasers with significant variability of generation regimes," Opt. Fiber Technol. 20(6), 615–620 (2014).

[8] S. Smirnov, S. Kobtsev, and S. Kukarin, "Efficiency of non-linear frequency conversion of double-scale pico-femtosecond pulses of passively mode-locked fiber laser," Opt. Express 22(1), 1058–1064 (2014).

[9] S. Kobtsev, S. Kukarin, S. Smirnov, and I. Ankudinov, "Cascaded SRS of single- and double-scale fiber laser pulses in long extra-cavity fiber," Opt. Express 22(17), 20770–20775 (2014).

[10] S. M. Kobtsev, S. V. Kukarin, and S. V. Smirnov, "All-fiber high-energy supercontinuum pulse generator," Laser Phys. 20(2), 375–378 (2010).

[11] J. C. Hernandez-Garcia, O. Pottiez, and J. M. Estudillo-Ayala, "Supercontinuum generation in a standard fiber pumped by noise-like pulses from a figure-eight fiber laser," Laser Phys. 22(1), 221–226 (2012).

[12] A. Zaytsev, C. H. Lin, Y. J. You, C. C. Chung, C. L. Wang, and C. L. Pan, "Supercontinuum generation by noise-like pulses transmitted through normally dispersive standard single-mode fibers," Opt. Express 21(13), 16056–16062 (2013).

[13] S. S. Lin, S. K. Hwang, and J. M. Liu, "Supercontinuum generation in highly nonlinear fibers using amplified noise-like optical pulses," Opt. Express 22(4), 4152–4160 (2014).

[14] D. V. Churkin, S. Sugavanam, N. Tarasov, S. Khorev, S. V. Smirnov, S. M. Kobtsev, and S. K. Turitsyn, "Stochasticity, periodicity and localized light structures in partially mode-locked fibre lasers," Nat. Commun. 6, 7004, 1–6 (2015). doi:10.1038/ncomms8004

[15] Q. Wang, T. Chen, M. Li, B. Zhang, and Y. Lu, "All-fiber ultrafast thulium-doped fiber ring laser with dissipative soliton and noise-like output in normal dispersion by single-wall carbon nanotubes," Appl. Phys. Lett. 103, 011103-1–3 (2013).

[16] Y. Chen, M. Wu, P. Tang, S. Chen, J. Du, G. Jiang, Y. Li, C. Zhao, H. Zhang, and S. Wen, "The formation of various multi-soliton patterns and noise-like pulse in a fiber laser passively mode-locked by a topological insulator based saturable absorber," Laser Phys. Lett. 11(5), 055101 (2014).

[17] Y. S. Fedotov, A. V. Ivanenko, S. M. Kobtsev, and S. V. Smirnov, "High average power mode-locked figure-eight Yb fibre master oscillator," Opt. Express 22(25), 31379–31386 (2014).

[18] O. Pottiez, R. Grajales-Coutiño, B. Ibarra-Escamilla, E. A. Kuzin, and J. C. Hernández-García, "Adjustable noiselike pulses from a figure-eight fiber laser," Appl. Opt. 50 (25), E24–E31 (2011).

[19] S. M. Kobtsev and S. V. Smirnov, "Fiber lasers mode-locked due to nonlinear polarization evolution: golden mean of cavity length," Laser Phys., 21(2), 272–276 (2011).

[20] I. A. Yarutkina, O. V. Shtyrina, M. P. Fedoruk, and S. K. Turitsyn, "Numerical modeling of fiber lasers with long and ultra-long ring cavity," Opt. Express 21(10), 12942–12950 (2013).

[21] S. Kobtsev, S. Kukarin, and Yu. Fedotov, "Ultra-low repetition rate mode-locked fiber laser with high-energy pulses," Opt. Express 16(26), 21936–21941 (2008).

[22] P. Grelu and N. Akhmediev, "Dissipative solitons for mode-locked lasers," Nat. Photonics 26, 84–92 (2012).

[23] L. Wang, X. Liu, Y. Gong, D. Mao, and L. Duan, "Observations of four types of pulses in a fiber laser with large net-normal dispersion," Opt. Express 19, 7616–7624 (2011).

[24] S. Smirnov, S. Kobtsev, S. Kukarin, and A. Ivanenko, "Three key regimes of single pulse generation per round trip of all-normal-dispersion fiber lasers mode-locked with nonlinear polarization rotation," Opt. Express 20(24), 27447–27453 (2012).

[25] S.V.Smirnov, S.M. Kobtsev, S.V.Kukarin and S.K.Turitsyn (2011). Mode-locked fibre lasers with high-energy pulses, Laser systems for applications, Dr Krzysztof Jakubczak (Ed.), ISBN: 978-953-307-429-0, InTech, DOI: 10.5772/24360. Available from: http://www.intechopen.com/books/laser-systems-for-applications/mode-locked-fibre-lasers-with-high-energy-pulses.

[26] G. P. Agrawal, Nonlinear Fiber Optics, third ed., Academic Press, San Diego, USA 2001.

[27] A. Komarov, H. Leblond, and F. Sanchez, "Multistability and hysteresis phenomena in passively mode-locked fiber lasers," Phys. Rev. A 71, 053809-1–9 (2005).

[28] S. V. Smirnov, S. M. Kobtsev, and S. V Kukarin. "Linear compression of chirped pulses in optical fibre with large step-index mode area," Opt. Express 23(4), 3914–3919 (2015).

[29] A. F. J. Runge, C. Aguergaray, N. G. R. Broderick, and M. Erkintalo, "Coherence and shot-to-shot spectral fluctuations in noise-like ultrafast fiber lasers," Opt. Letters 38 (21), 4327–4330 (2013).

[30] L. Wang, X. Liu, Y. Gong, D. Mao, and L. Duan, "Observations of four types of pulses in a fiber laser with large net-normal dispersion," Opt. Express 19 (8), 7616–7624 (2011).

[31] F. Vanholsbeeck, S. Martin-Lopez, M. González-Herráez, and S. Coen, "The role of pump incoherence in continuous-wave supercontinuum generation," Opt. Express 13 (17), 6615–6625 (2005).

[32]   S. Kobtsev and S. Smirnov, "Modelling of high-power supercontinuum generation in highly nonlinear, dispersion shifted fibers at CW pump," Opt. Express 13 (18), 6912–6918 (2005).

[33]   Y. Takushima, K. Yasunaka, Y. Ozeki, and K. Kikuchi, "87 nm bandwidth noise-like pulse generation from erbium-doped fibre laser," Electron. Letters 41, 399–400 (2005).

[34]   L. M. Zhao, D. Y. Tang, T. H. Cheng, H. Y. Tam, and C. Lu, "120 nm bandwidth noise-like pulse generation in an erbium-doped fiber laser," Opt. Commun. 281, 157–161 (2008).

[35]   L. A. Vázquez-Zuñiga and Y. Jeong, "Super-broadband noise-like pulse erbium-doped fiber ring laser with a highly nonlinear fiber for Raman gain enhancement," IEEE Photonic. Technol. Lett. 24, 1549–1551 (2012).

[36]   L. A. Vazquez-Zuniga and Y. Jeong, "Super-broadband noise-like pulse Erbium-doped fiber ring laser with a highly nonlinear fiber for Raman gain enhancement," IEEE Photonic. Technol. Lett. 24 (17), 1549–1551 (2012).

[37]   S. Keren, A. Rosenthal, and M. Horowitz. "Measuring the structure of highly reflecting fiber Bragg gratings," IEEE Photonic. Technol. Lett. 15(4), 575–577 (2003).

[38]   V. Goloborodko, S. Keren, A. Rosenthal, B. Levit, and M. Horowitz, "Measuring temperature profiles in high-power optical fiber components," Appl. Opt. 42(13), 2284–2288 (2003).

[39]   S. Keren and M. Horowitz, "Interrogation of fiber gratings by use of low-coherence spectral interferometry of noiselike pulses," Opt. Lett. 26(6), 328–330 (2001).

Chapter 5

Passively Stabilized Doubly-Resonant Brillouin Fiber Lasers

Vasily V. Spirin, Patrice Mégret and Andrei A. Fotiadi

Additional information is available at the end of the chapter

http://dx.doi.org/10.5772/61714

Abstract

We consider ultra narrow-line lasers based on doubly-resonant fiber cavities, describe experimental techniques, and present two methods for passive stabilization of single-frequency fiber Brillouin lasers. In the first approach, Brillouin fiber laser is passively stabilized at the pump resonance frequency by employing the self-injection locking phenomenon. We have demonstrated that this locking phenomenon delivers a significant narrowing of the pump laser linewidth and generates the Stokes wave with linewidth of about 0.5 kHz. In the second methodology, the fiber laser is stabilized with an adaptive dynamical grating self-organized in un-pumped Er-doped optical fiber. The laser radiates a single-frequency Stokes wave with a linewidth narrower than 100 Hz. The ring resonators of both presented lasers are simultaneously resonant for the pump and the Stokes radiations. For adjusting the double resonance at any preselected pump laser wavelength, we offer a procedure that provides a good accuracy of the final resonance peak location with ordinary measurement and cutting errors. The stable regime for both Brillouin lasers is observed during some intervals, which are interrupted by short-time jumping-intervals. The lasers' stability can be improved by utilizing polarization-maintaining (PM) fiber configuration and a cavity protection system.

Keywords: Brillouin fiber laser, self-injection locking, ring fiber resonator

1. Introduction

Stimulated Brillouin Scattering (SBS) is the nonlinear process with the lowest threshold in optical fibers [1]. SBS in optical fiber is very valuable for numerous applications such as fiber lasers, fiber amplifiers, fiber sensors, and optical processing of radio-frequency signals [2-9].

Recently developed Brillouin fiber lasers with the so-called doubly-resonant cavity (DRC) demonstrate low threshold, high spectral purity, and low intensity noise [10]. These lasers are

very promising for a variety of special uses, such as coherent interferometric sensing, optical communication, microwave photonics, and coherent radar detection [11-13].

The Stokes wave in these lasers is generated within a short fiber ring cavity, which is simultaneously resonant for the pump and the Stokes radiations. This configuration also delivers the minimal Brillouin threshold. However, stable operation of the DRC Brillouin laser is extremely sensitive to the resonance detuning between the pump frequency and the ring cavity mode. Typically, in order to obtain an established single-longitudinal mode (SLM) operation, various types of active stabilization systems that adjust the cavity length or alter the frequency of the pump laser are used [11-12].

In Sections 3 and 4, we report two schemes of passively stabilized SLM DRC Brillouin lasers.

1. In the first scheme, the single-mode doubly-resonant Brillouin fiber laser is passively stabilized at the pump resonance frequency by employing the self-injection locking phenomena. Optical self-injection locking has drawn a lot of attention for optical communication application as an efficient technique to improve the spectral and polarization properties of semiconductor lasers [14-16]. The physical phenomena observed with optical feedback in DFB varied from linewidth narrowing with weak feedback to irregular chaotic oscillations with the strong one [14,17]. In this work, we utilize self-injection locking for matching the DFB pump laser frequency with the Brillouin laser resonance cavity mode as well as for decreasing the pump laser linewidth.

2. In the second approach, the pump fiber laser is stabilized with a population inversion grating imprinted in an un-pumped Er-doped optical fiber.

In Section 2, we present a special procedure for the precise setting of the ring fiber cavity to the double resonance condition at any preselected pump wavelength. We show that experimental record of the cavity Brillouin response to a frequency scan of the pump laser allows calculating the excess ring cavity length that must be removed from the ring in order to shift the Brillouin resonance to the right position. We have demonstrated that the proposed algorithm delivers a good accuracy of the resonance peak location with ordinary measurement and cutting errors.

2. Adjustment of double resonance in Brillouin fiber lasers

Doubly-resonant Brillouin laser cavity should be simultaneously resonant for the pump and the Stokes radiations. The free spectral range (FSR) of the fiber cavity with a length of about several meters is equal to tenths of MHz. Therefore, typical pump DFB laser can be easily adjusted to a cavity mode by current or temperature setting. The problem then is how to keep this matching despite random deviations of the cavity length and/or pump laser frequency. However, even if we provide precise and stable coincidence of the pump laser frequency with one of the cavity resonant modes, it does not guarantee the resonance for the down-shifted Stokes wave for short cavities.

In this paragraph, we present an algorithm and an experimental technique for the adjustment of the double resonance of the fiber optical ring cavities at any preselected wavelength. The experimental setup is depicted in Fig. 1.

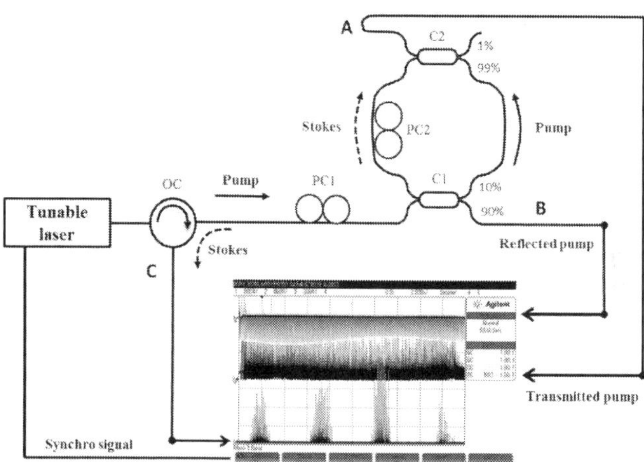

Figure 1. Double resonance peaks recorded with single wavelength scan.

A fiber ring cavity with the lengths about 4 m is pumped by a tunable laser Agilent 81940A with 100 kHz linewidth and output power up to 25 mW. The behavior of the fiber-optic ring resonator is very similar to a linear Fabry-Perot cavity, so the ports B and A are equivalent to the reflected and the transmitted ports of the linear cavity, respectively. Figure 1 also illustrates how the double resonance looks in the experiment. When the tunable laser sweeps in the interval 1540–1560 nm synchronously with oscilloscope trace, the Stokes peaks appear only for the certain pump wavelengths that correspond to the double resonance condition.

The length L of the fiber inside the ring defines the FSR of the cavity:

$$FSR = c \ / \ nL, \tag{1}$$

where c is the speed of light in a vacuum, and n is the refractive index of the SMF-28 fiber equal to 1.468 at 1550 nm [18].

The double resonance is achieved when the Brillouin shift v_B is equal to an integer multiple of the FSR (see Fig. 2):

$$v_B = m \ FSR, \tag{2}$$

where m is an integer, which here we refer to as the order of the Stokes peak.

With this condition, the cavity became resonant for both the pump and Stokes radiations, which drastically reduces the SBS threshold and leads to Stokes peaks emergence.

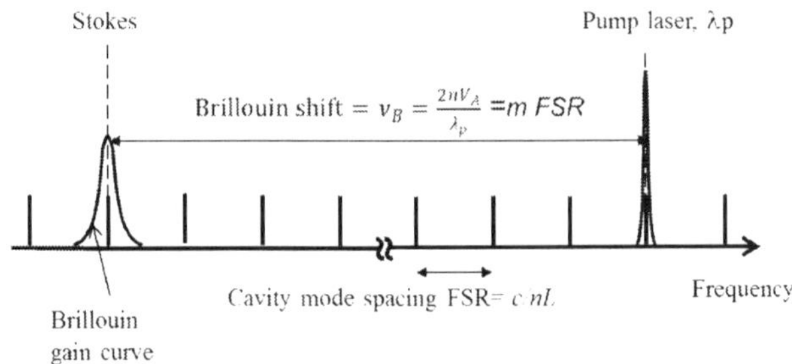

Figure 2. Doubly-resonant conditions.

In order to find the doubly-resonant states, we adjust the Brillouin shift by tuning the wavelength of the pump laser λ_p, because the Brillouin shift depends on pump wavelength as:

$$v_B\left(\lambda_p\right) = \frac{2nV_A}{\lambda_p},\qquad(3)$$

where $V_A \approx 5800$ m/s [19] is the acoustic velocity in the optical fiber.

For a precise measurement of the acoustic velocity, we have developed the experimental setup presented in Fig. 3. The radiation of the tunable laser at a fixed wavelength equal to 1549.7 nm was injected into the cavity ring through the circulator OC and the coupler C1. This wavelength was chosen to lead to Stokes emission in the ring cavity. The beat frequency between the Stokes component and the backscattered Rayleigh radiation of the pump is registered at port A with an RF spectrum analyzer HP 70908A. This beat frequency was found to be equal to 10.87689 GHz, which gives us with eq. 3 an acoustic velocity equal to 5740 m/s, so very close to the expected theoretical value.

The resonant coupling between the scanning pump laser and the cavity mode occurs during very short multiple time intervals. Indeed, the increase of the transmitted pump power at the resonance is accompanied by a decrease of the reflected power as shown in Fig. 4. Strong pulses at the Stokes wavelength are recorded at port C when the pump power is sufficient to exceed the Brillouin threshold. These pulses are red-shifted by ~0.08 nm from the pump wavelength and can only appear when the Brillouin frequency shift is an integer multiple of the FSR. Moreover, there is a perfect synchronization between these Stokes pulses and the transmitted ones. This is precisely the double resonance condition.

In order to estimate the exact position of the double resonance peaks, we use an averaging procedure. Envelopes of the Stokes pulses after averaging of 7 laser scans, recorded for initial unadjusted cavity, are shown in Fig. 5.

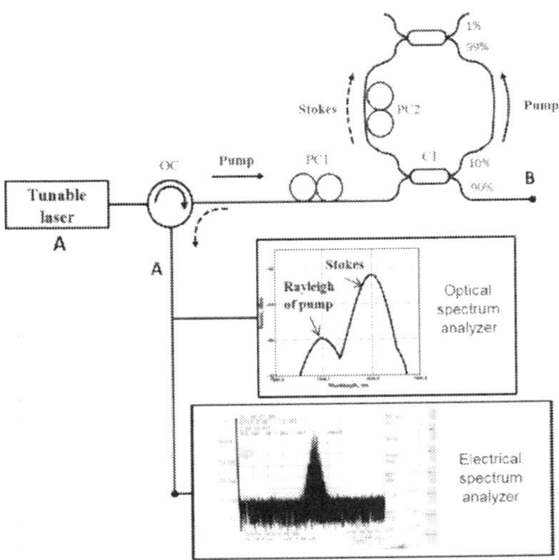

Figure 3. Experimental setup for precise acoustic velocity measurement.

The order of the envelope peak m can be found from the eqns. (1–3) as the nearest integer function [20]:

$$m = nint\left(\frac{\lambda_{m-1}}{\lambda_{m-1} - \lambda_m}\right), \tag{4}$$

where λ_m, λ_{m-1} are the central wavelengths of the neighboring m and $(m-1)$ – orders of the envelopes inside the measuring interval.

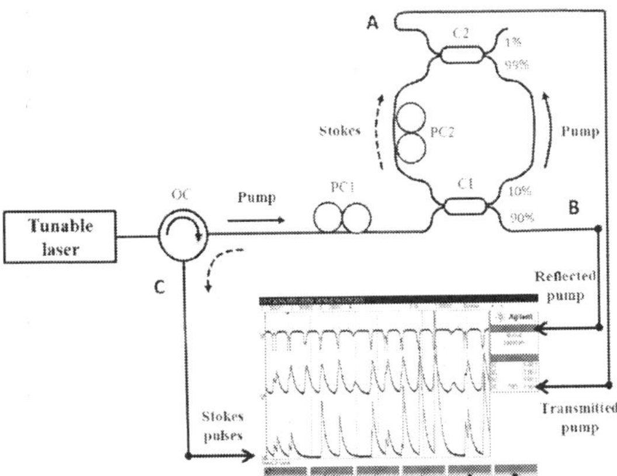

Figure 4. Signals at double resonance during the scanning.

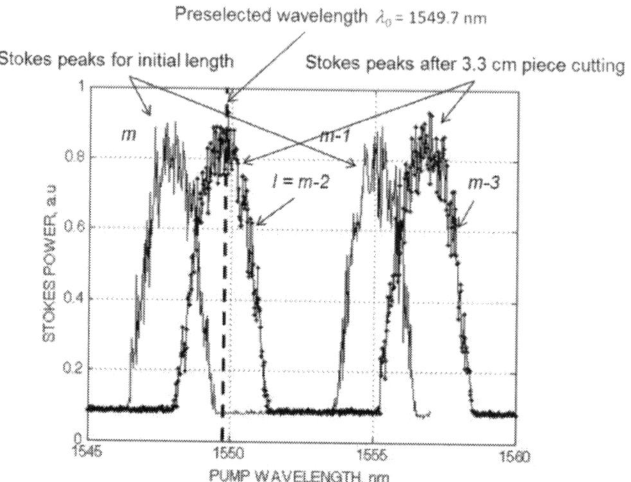

Figure 5. Stokes envelopes vs. pump wavelength after 7 scans averaging for initial and adjusted cavities.

The DFB pump laser we use in the Brillouin ring laser has a fixed wavelength equal to 1549.7 nm. As we see from Fig. 5, the envelope positions for unadjusted cavity are noticeably away from the wanted wavelength λ_0 = 1549.7 nm. The nearby peaks to λ_0 Stokes envelopes have central wavelengths equal to 1547.9 nm and 1555 nm, which corresponds to m = 219 and m-1 = 218, respectively (see Fig. 5).

From the eqns. (1–3), we find that to move the l -order envelope peak, calculated for the original cavity, to its new position with the central wavelength equal to λ_0, we should reduce the cavity length by ΔL:

$$\Delta L = \frac{c}{2n^2 V_A}\left(m\lambda_m - l\lambda_0\right),\tag{5}$$

Figure 6 shows the lengths ΔL calculated for our experimental data as a function of m-l. In theory, we can move to the new position any envelope peak that is located at the right side from λ_0 [20]. For example, we can move at the required position the nearest 218-order peak (m - l = 1), but technically it is difficult to splice the fiber when cutting less than 2-cm piece of the fiber. In the experiment, we replace the 217-order peak (m - l = 2) to the preselected wavelength 1549.7 nm, by cutting ΔL = 3.3 cm piece of the fiber. As we can see in Fig. 5, the result of the adjustment is in good agreement with the analytical estimation.

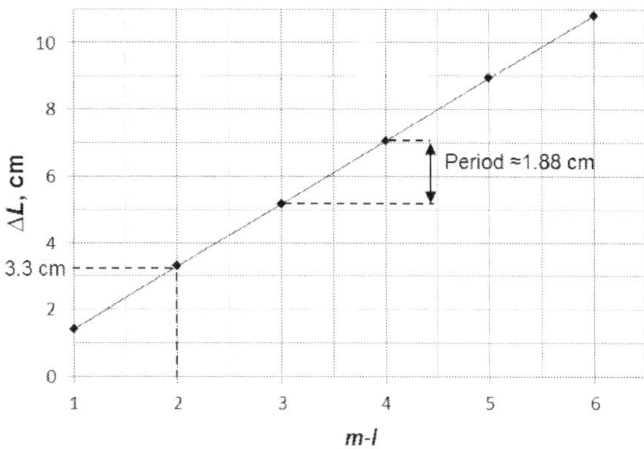

Figure 6. Cutting length ΔL vs. difference in peak orders m-l.

Let us now consider the uncertainties of the algorithm for adjusting the doubly-resonant condition in the ring Brillouin fiber laser. The cavity adjustment algorithm includes three main steps:

1. Experimental recording of the Brillouin traces with the original cavity. Determination of the Stokes envelope positions.

2. Calculation of the length ΔL, which must be taken out from the cavity in order to transfer the resonance peak to the preselected position.

3. Cutting and splicing the fiber into the cavity in order to reduce the cavity length by ΔL.

Each of these actions introduces some errors that decrease the accuracy of the technique [21].

The experimental determination of the central wavelengths of the envelopes λ_{m-1}, λ_m, is accompanied by some errors $\delta\lambda_{m-1}$, $\delta\lambda_m$, which depend on the accuracy of the equipment used to measure the wavelength. The standard uncertainty of the peak order m can be estimated from eqn. (4) with the error propagation formula as:

$$\delta m = \frac{\sqrt{2}\lambda_m}{\left(\lambda_{m-1} - \lambda_m\right)^2}\,\delta\lambda_m,\tag{6}$$

where $\delta\lambda_{m-1} = \delta\lambda_m = \delta\lambda$ are the standard errors in determination of the peak location.

The ordinary error of the peak position $\delta\lambda$ in our experiment is equal to 0.1 nm; however, with additional averaging over a number of oscilloscope traces, we get an uncertainty equal to 0.05 nm. Even higher precision can be obtained because the tunable laser Agilent 81940A is arranged for typical absolute wavelength accuracy equal to ±5 pm with repeatability ±1 pm.

Figure 7 shows the standard error of m versus the cavity length L for different accuracies of the peak position measurement. The dependence is calculated utilizing eqns. (6) and (1–3) for λ_m equal to 1550 nm. The method delivers absolutely correct value of the peak order only for the very short cavities. For the cavities with lengths 4.0–8.0 m, the uncertainty of m reaches 1–2%. Nevertheless, the relatively high error of the peak order m makes minor contribution to the final error of the adjusted resonance peak position.

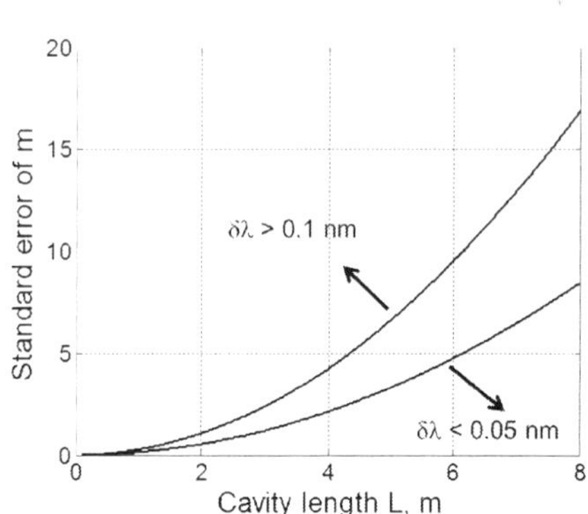

Figure 7. Standard error in calculations of Stokes peak order m.

Without loss of generality, we can assume that we transfer the (m-1) order peak toward the preselected position, which is located between the m and (m+1) order peaks. With this assumption, the maximum standard uncertainty of the required length reduction ΔL can be estimated as:

$$\delta\Delta L \cong \frac{c}{2n^2 V_A}\left(\frac{\sqrt{5}\lambda_m}{\lambda_{m-1}-\lambda_m}\right)\delta\lambda \tag{7}$$

Figure 8 shows the uncertainty $\delta\Delta L$ versus the length of the laser cavity L. The errors in the estimation of the cutting piece ΔL are less than 0.6 mm for the 0.1 nm error in the position of the resonance peak for 4 m-length fiber cavity.

Now with eqn. (5) we can estimate the inaccuracy in the position of the (m-1) order peak which we move at desired wavelength $\lambda 0$. The resulting errors in the position of the peak due to cutting and measurement errors are shown in Fig.9.

The error in the position of the adjusted peak due to measurement errors is equal to 1.6 MHz. This error is thus significantly less than the linewidth of Brillouin gain profile in the fiber, which is equal to about 30 MHz. However, the main errors of the method occur due to inaccuracies of the cut-splice procedure. Fortunately, even for the 1 m-long fiber cavity with

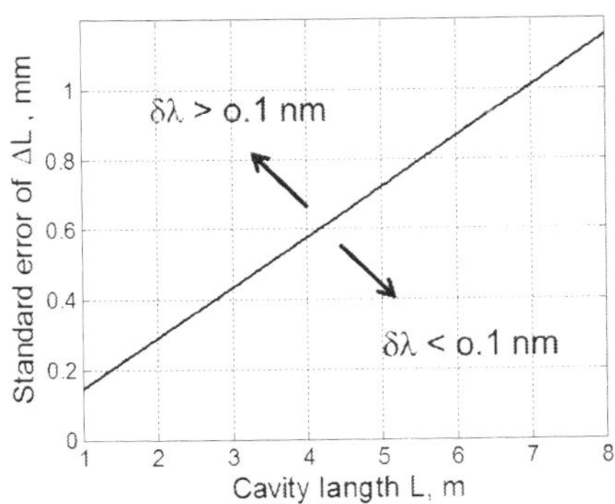

Figure 8. Standard error in estimations of cavity reduction length ΔL.

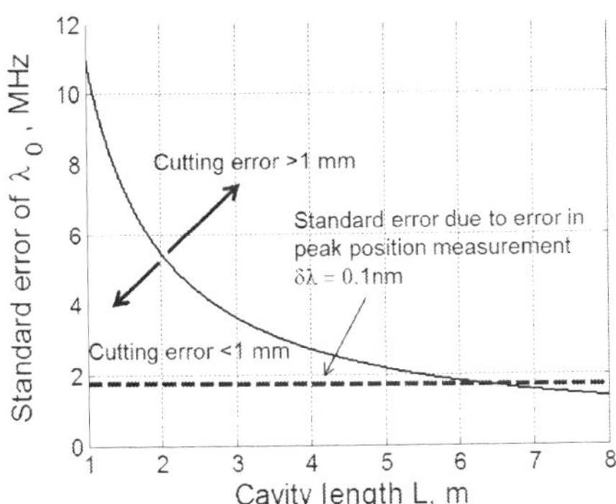

Figure 9. Standard error of the adjusted peak position.

1 mm cutting error, we get only an error of 11 MHz in the final peak position, which is quite reasonable for practical applications. Therefore, the algorithm provides good precision for the adjusted peak location with ordinary measurement and cutting accuracy.

Besides the adjustment of the double resonance, the proposed technique allow finding of the fiber cavity length:

$$L = \frac{c}{2n^2 V_A} \left(\frac{\lambda_m \lambda_{m-1}}{\lambda_{m-1} - \lambda_m} \right). \tag{8}$$

However, the error in the length estimation with eqn. (8) is relatively high. Figure 10 shows the standard errors in the estimation of the fiber cavity length, which was calculated utilizing eqn. (8) with error propagation formula. The standard uncertainty of the cavity length exceeds 30 cm for the 8-m-length cavity and 0.1-nm errors in the peak position measurement. Thus, more precise measurement of the peak location is required for the accurate determination of the cavity length with this method.

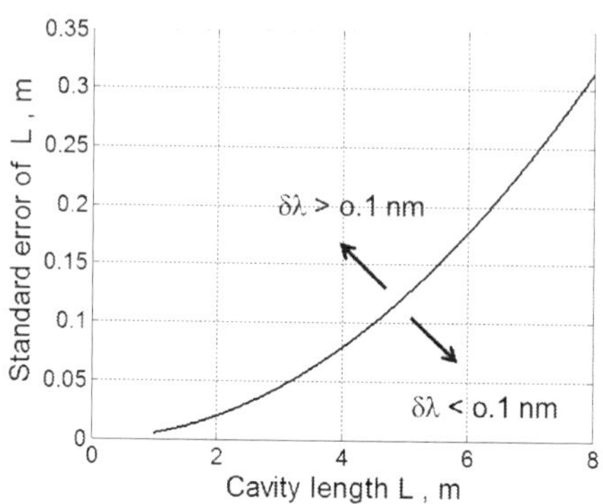

Figure 10. Error of the estimation of the cavity length.

3. Doubly-resonant Brillouin fiber laser passively stabilized with self-injection locking phenomena

In this section, we present a single-longitudinal-mode, doubly-resonant Brillouin fiber laser, which is passively stabilized at the pump resonance frequency by employing the self-injection locking phenomena.

The experimental configuration of this passively stabilized SLM Brillouin fiber laser is shown in Fig. 11. The Brillouin laser is pumped by a standard MITSUBISHI FU-68PDF-V520M27B DFB laser with a fiber output and a built-in optical isolator. The output radiation of pump laser, which operates at a wavelength equal to 1549.7 nm, is connected to the optical circulator OC1, amplified by an Er-doped fiber amplifier (EDFA) up to 30 mW and launched into the fiber-optic ring resonator (FORR) through another optical circulator OC2. The FORR comprises two couplers, polarization controller, and contains approximately 4-m-long fiber. The feed-back switch (FS) provides the possibility to switch-on or switch-off the optical feedback signal reaching the DBF laser. The FORR was preliminary adjusted for double resonance at pump wavelength equal to 1549.7 nm by utilizing the algorithm described in the previous section.

Once the pump laser gets a resonance with the ring laser cavity, the growing optical feedback forces the DFB pump laser to operate at the cavity resonance frequency. After such locking, any slow detuning of the cavity resonance caused, for example, by temperature variations, is compensated by a matching deviation of the pump laser operation wavelength.

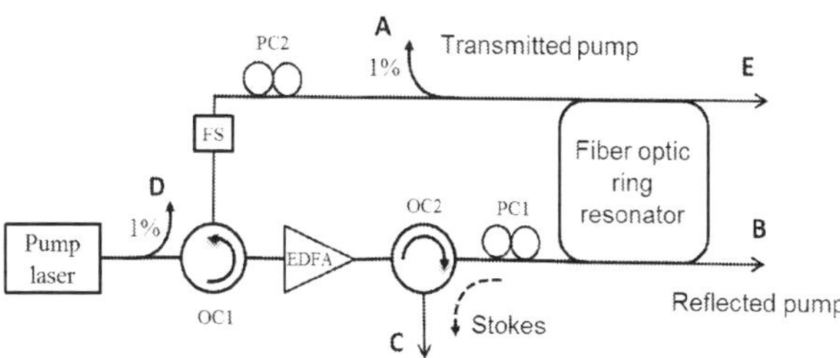

Figure 11. Configuration of the Brillouin laser with self-locked DFB pump laser; OC – optical circulator, PC – polarization controller, FS – feedback switcher.

The linewidth of pump laser was measured with the delayed self-heterodyne technique. An all-fiber spliced Mach–Zehnder interferometer with a 15-km delay fiber in one arm and 20 MHz phase modulator supplied by polarization controller in the second arm is used for this purpose. Figure 12 shows delayed self-heterodyne spectra of unlocked (Fig. 12a) and locked DFB (Fig. 12b) pump laser at port D of Fig. 11.

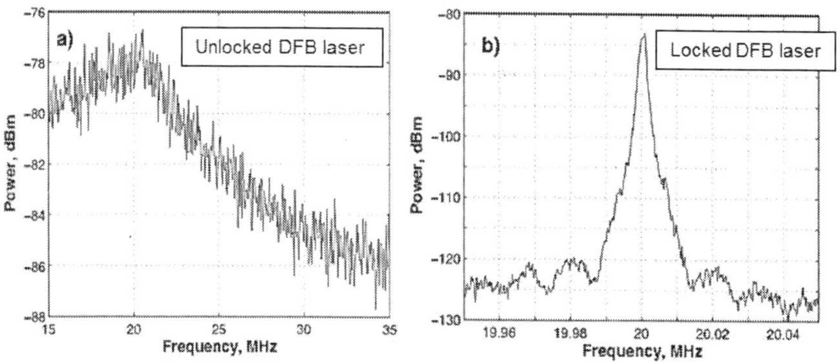

Figure 12. Delayed self-heterodyne spectra of the DFB pump lasers; a) unlocked laser, b) self-locked pump laser.

For the unlocked DFB laser, the full-width at half-maximum (FWHM) linewidth was estimated equal to 4 MHz, assuming that the line shape is Lorentzian. Meantime, the linewidth of the locked pump laser reduces by more than 1000 times (see Fig. 12b) [22].

With the resonance, the power, which is circulating inside the FORR, is increasing up to the Brillouin threshold. After the SBS threshold is reached, some of the pump energy is converted into a Stokes wave, traveling in the opposite direction inside the ring. The main part of the backward-propagating Stokes radiation is relaunched into the FORR but the other part goes out and is available at ports C and E.

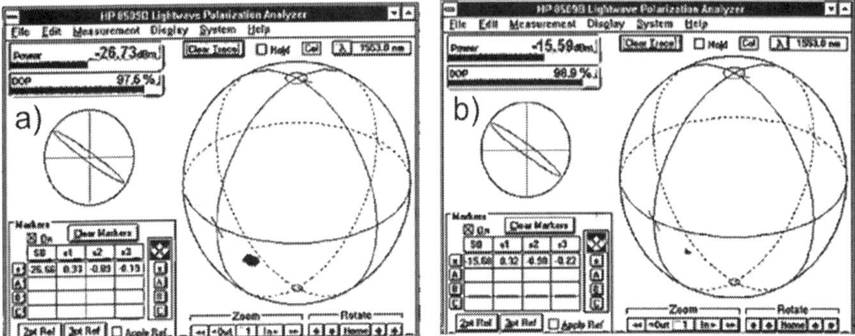

Figure 13. Polarization state during 30 s: a) unlocked, and b) locked DFB pump laser.

Figure 13 shows time-behavior of the polarization states of the unlocked and locked DFB pump laser recorded at port D during 30 s. The free running and locked laser emitted nearly completely polarized light with degree of polarization (DOP) equal to 98–99%. However, the locked laser demonstrates a slightly better temporal stability of the polarization. Whatever the feedback, the Stokes wave polarization significantly varies in time (see Fig. 14).

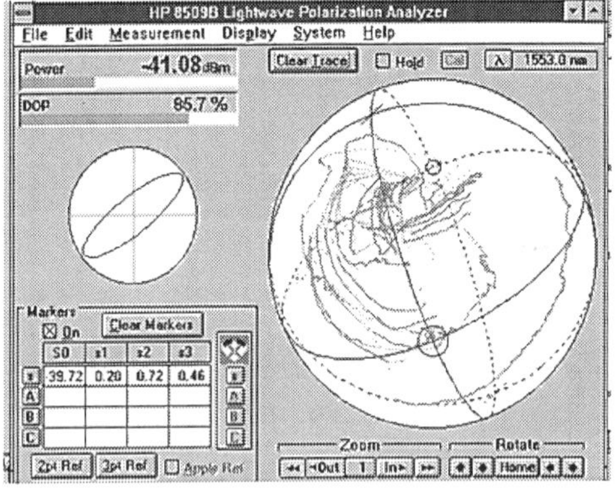

Figure 14. Polarization state of the Stokes radiation during 30 s.

The maximum output Stokes power at port C was about 5 mW providing approximately 40% Brillouin laser slope efficiency. The Stokes linewidth is expected to be narrower than the pump one [23]. Indeed, Fig. 15 shows the delayed self-heterodyne spectra of the Stokes component of the Brillouin laser from which the double linewidth $2\Delta v$ can be estimated to be equal to 10 kHz with 20 dB criterion that corresponds to 1 kHz with 3 dB criterion [24]. It should be noted that the delay length of 15 km is too short to obtain the incoherent mixing, which is required in the measurements utilizing delayed self-heterodyne technique. For the 0.5 kHz linewidth, the length of the delay fiber should exceed 100 km. However, the error due to the shorter delay fiber can lead to visible large-scale oscillations in self-heterodyne spectra and an increasing of the estimated linewidth in comparison with actual one [24]. For this reason, the actual linewidth of the Brillouin laser is better than the measured value equal to 500 Hz.

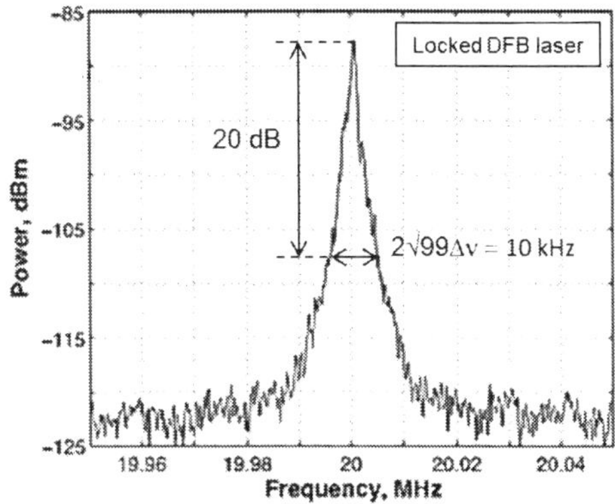

Figure 15. Delayed self-heterodyne spectra of the Stokes radiation of Brillouin laser.

Figure 16 illustrates a typical time-behavior of the Brillouin laser without locking. When optical-feedback (FS) is switched off, the injection locking of the pump laser is disabled and the stable resonance is never observed. Indeed, the pump power is almost totally reflected from FORR to port B. Without locking, the pump power inside the Brillouin laser cavity is small. So the transmitted pump power and the Stokes component are almost negligible at ports A and C, correspondingly.

On the other hand, optical feedback drastically changes the operation mode of the Brillouin laser (see Fig. 17). When FS is switched on, the Brillouin laser starts to operate mostly at the resonance mode of the FORR. The pump power mainly goes inside the FORR and the reflected power at port B becomes negligible. Analysis of the powerful single-frequency Stokes emission measured at port C shows that the observed regime consists of stable Stokes radiation for some duration, interrupted by short-time jumping intervals that are believed to be related to the hoping of the resonance mode of the fiber laser. It is clear that a more detailed analysis of this phenomenon is needed.

The variation of the transmitted pump power varies in antiphase with the reflected power, leading to a strong time correlation between the pump power and the Stokes radiation. The experimental conditions greatly modify the durations of the stable resonance-intervals and jumping-intervals. In the experiment presented here, ordinary SMF-28 fiber was used inside the FORR and the cavity was not stabilized against vibration and temperature. Under these conditions, stable Stokes radiation is usually observed during 0.5–5 s.

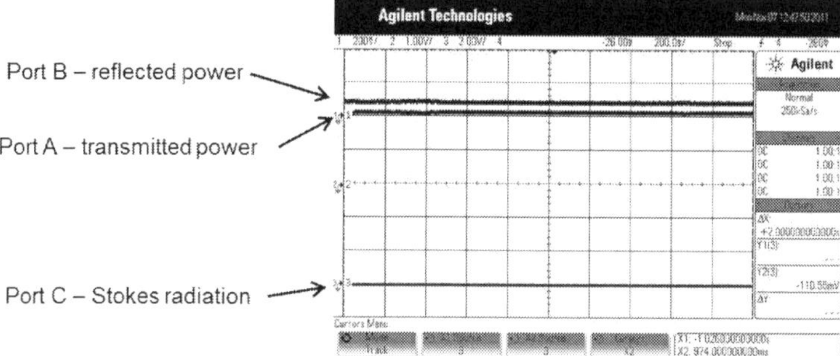

Figure 16. Oscilloscope traces when optical feedback is "switch-off."

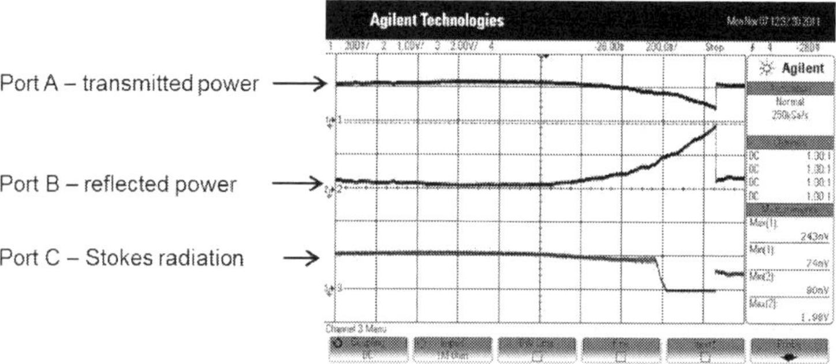

Figure 17. Typical oscilloscope traces when optical feedback is "switch-on."

We believe that stability and efficiency of proposed technique can be noticeably improved with temperature stabilization and, for example, by utilizing polarization-maintaining fiber in the Brillouin ring cavity.

4. Brillouin fiber lasers passively stabilized with adaptive grating

In this section, we present another approach for passive stabilization of SLM DRC Brillouin laser that demonstrates even narrower laser linewidth than the one that was reported in the previous section. In this method, we utilize a population inversion grating, a side effect of the population inversion mechanism, which, along with the refractive index change effect, is widely used in many devices based on rare-earth doped fibers [25–32]. Here, dynamical grating is self-organized in the piece of the un-pumped Er-doped optical fiber and able to adjust the laser mode to the cavity resonance frequency. The adaptive all-fiber method was applied in combination with two coupled FORRs that were used for preliminary pump-mode selection and Stokes generation.

The experimental laser configuration is shown in Fig.18. Optical gain is supplied by a 4.5-m segment of erbium-doped single mode fiber (SM-EDF) pumped by 1480-nm laser diode through a 1480/1550 wavelength-division multiplexer (WDM). Optical circulator (OC) and two isolators (OI) provide unidirectional pump propagation through the cavity. To get single-frequency generation, the laser cavity also comprises two coupled FORRs.

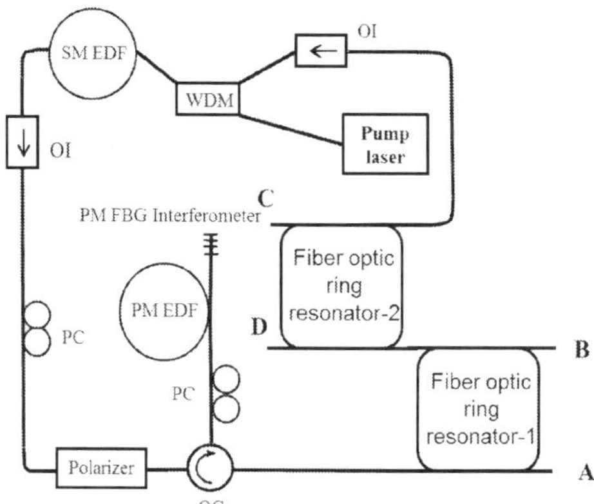

Figure 18. The laser configuration; OC- optical circulator, PC – polarization controller, OI – optical isolator, WDM – wavelength-division multiplexer, PM EDF – polarization-maintaining erbium-doped fiber, SM EDF – single-mode erbium-doped fiber.

This combined cavity has superior intermodal spacing, allowing the laser to operate only at modes that are common for both ring resonators, i.e., to suppress multilongitudinal-mode generation [33, 34]. The FORR-2 with a short fiber length is used only as a frequency-selective element only. High-fidelity FORR-1, comprising relatively long fiber, is used as a frequency-selective element and as the Brillouin laser ring cavity. The FORR-1 consists of two 95/5 couplers, polarization controller and contains 20-m length of a standard SMF-28 fiber used as an efficient media for Brillouin Stokes generation. The radiation at the Stokes wavelength is mainly emitted through port B, while port D is used to measure the parameters of the pump radiation. We install a fiber polarizer inside the main laser cavity that ensures single polarization mode of the pump radiation. A 1.7-m section of un-pumped erbium-doped polarization maintaining fiber (PM-EDF) has absorption of about 5.5 dB/m at 1530 nm and, in combination with the narrow-band fiber Bragg grating (FBG) interferometer, operates as an adaptive ultra-narrow-band reflector. The FBG interferometer is a Fabry-Perot cavity made of two uniform FBGs inscribed in the PM single mode optical fiber and separated by ~0.8 cm. It has 95% main reflectivity peak at 1547.37 nm with a FWHM equal to about 0.07 nm. In the course of this laser operation, the optical waves traveling in the un-pumped fiber section in opposite directions interfere to form a standing wave that causes inscription of a population inversion grating in the fiber. Indeed, the two-wave mixing effect [35] is the key phenomena used here to make the grating highly reflective at the inscribing light wavelength. The two-wave mixing being strongly power-dependent, the grating reflectivity peak is enhanced at the laser resonance frequency and this high reflection reduces the cavity losses, leading to a further enhancement of the resonance frequency. Any slow detuning of the frequency, for example, due to temperature variation, is followed by the matching shift of the adaptive grating reflectivity peak.

The measured Brillouin threshold for this laser was about 4 mW. Optical spectra of the Brillouin laser recorded at the ports A and B with 5 mW pump power are shown in Fig. 19.

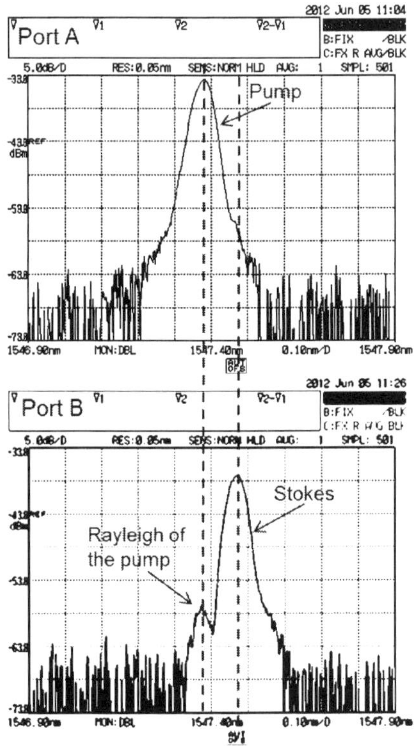

Figure 19. Brillouin laser optical spectra at ports A and B.

The Stokes power at port B was about 0.3 mW, providing approximately 30% slope efficiency of the pump-to-Stokes power conversion. The peak at the Stokes wavelength is about 20 dB higher than the peak associated with the Rayleigh scattering of the pump wave (see Fig. 19). Because of this high spectral contrast, this signal could be further amplified by an extra Er-doped amplifier without significant degradation of the signal-to-noise ratio.

The linewidths of pump and Stokes radiations were measured with the delayed self-heterodyne technique with a 25.3-km delay fiber in one arm and 15 MHz phase modulator in the second arm. In this experiment, the beat signal is detected by a photodiode and an RF spectrum analyzer with a frequency resolution of about 100 Hz. Figure 20 shows the delayed self-heterodyne spectrum of the pump radiation recorded at port D. The pump FWHM is estimated to be 300 Hz. When increasing the pump power, the Stokes radiation appears at port B. As the pump frequency is resonant for FORR -1, the power circulating inside FORR -1 is about 35 higher than the power at the resonator input. This supports effective generation of the Brillouin radiation in FORR -1 comprising longer peace of fiber rather than FORR -2. The pump-to-Stokes power conversion efficiency is higher when FORR -1 is resonant not only for the pump, but also for the Stokes wave frequency. So, theoretically, we need to adjust the double resonance at FORR-1. However for the 20-m long cavity, the FSR is significantly smaller than the Brillouin gain width, so the resonance peaks are overlapping and this provides the double resonance without any adjustment [20].

The Stokes wave linewidth is expected to be narrower than the pump one. Figure 20 shows the delayed self-heterodyne spectrum of the Stokes wave that corresponds to a linewidth at half-maximum equal to about 100 Hz.

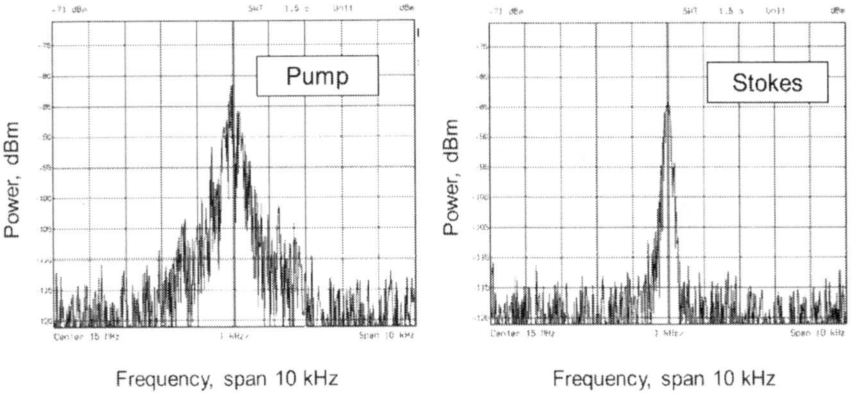

Figure 20. Delayed self-heterodyne spectra of the pump and Stokes radiations.

Typical time-behavior of this Brillouin laser is very similar to that reported before. Stable pump and Stokes power activities are observed during some time intervals that are interrupted by short-time jumping. There is a strong correlation between the time-behavior of the pump and Stokes radiations inside the cavities, where jumps of the pump power are always followed by Stokes power fluctuations. Moreover, below the Brillouin threshold, the Stokes radiation is negligible, but the laser dynamics is nearly the same as the one observed above the threshold. Environmental conditions strongly affect the stable interval durations and these stable durations could be significantly increased by an effective acoustic noise protection of the laser cavity. The stable time intervals of 0.2–1 s are typical for unprotected laser, which operates in regular lab environment [36].

5. Conclusion

We have presented two approaches for designing passively stabilized single-longitudinal-mode doubly-resonant Brillouin fiber lasers.

In the first approach, the SLM DRC Brillouin fiber laser is passively stabilized at pump resonance frequency by employing the self-injection locking phenomenon. For precise setting of the ring fiber cavity to the double resonance at any preselected pump laser wavelength, we design a special algorithm that delivers good accuracy of the resonance peak location with ordinary measurement and cutting errors. We have demonstrated that the locking phenomenon decreases the pump laser linewidth by a factor more than 1000 times and generates a Stokes wave with linewidth of about 0.5 kHz.

In the second approach, the pump fiber laser is stabilized with adaptive grating recorded in un-pumped Er-doped optical fiber. The laser is shown to emit a single-frequency Stokes wave with a linewidth narrower than 100 Hz. This result is a significant progress in the field of passively stabilized Brillouin lasers.

The stable regime for both Brillouin lasers is observed during some intervals that are interrupted by short-time jumping-intervals. We believe that the laser stability can be further improved by utilizing fully polarization-maintaining fiber spliced configuration and a proper cavity protection system.

Acknowledgements

We gratefully acknowledge financial support from the CONACYT, Mexico; the IAP program VII/35 of the Belgian Science Policy; and the Ministry of Education and Science of the Russian Federation (14.Z50.31.0015).

Author details

Vasily V. Spirin[1*], Patrice Mégret[2] and Andrei A. Fotiadi[2,3,4]

*Address all correspondence to: vaspir@cicese.mx

1 División de Física Aplicada, CICESE, Ensenada, B.C., México

2 University of Mons, Electromagnetism and Telecommunication Department, Mons, Belgium

3 Ioffe Physico-Technical Institute of the RAS, St. Petersburg, Russia

4 Ulyanovsk State University, Ulyanovsk, Russia

References

[1]   Agrawal G.P.: *Nonlinear Fiber Optics*. 3rd ed., Academic Press, San Diego, 2001, 1-466.

[2]   Cowle G.J., Stepanov D. Yu.: Hybrid Brillouin/erbium fiber laser. *Opt. Lett.* 1996; 21: 1250-1252.

[3]   Fotiadi A.A., Mégret P.: Self-Q-switched Er-Brillouin fiber source with extra cavity generation of a Raman supercontinuum in a dispersion shifted fiber. *Opt. Lett.* 2006; 31:1621-1623.

[4]   Boucon A., Fotiadi A.A., Mégret P., Maillotte H., Sylvestre T.: Low-threshold all-fiber 1000 nm supercontinuum source based on highly non-linear fiber. *Opt. Comm.* 2008; 281: 4095-4098.

[5]   Preda C.E., Fotiadi A.A., and Mégret P.: Numerical approximation for Brillouin fiber ring resonator. *Opt. Express.* 2012; 20, 5783-5788.

[6]   Grukh D.A., Kurkov A.S., Razdobreev I.M., Fotiadi A.A.: Self-Q-switched ytterbium-doped cladding-pumped fibre laser. *Quant. Electron.* 2002 : 32, 1017-1019.

[7]   Fotiadi A.A.: Random lasers: An incoherent fibre laser. *Natur. Photon.* 2010; 4: 204-205.

[8]   Spirin V.V., Kellerman J., Swart P.L., and Fotiadi A.A.: Intensity noise in SBS with injection locking generation of Stokes seed signal. *Opt. Express.* 2006; 14: 8328-8335.

[9]   Fotiadi A.A., Kiyan R., Deparis O., Megret P., and Blondel M.: Statistical properties of stimulated Brillouin scattering in single-mode optical fibers above threshold. *Opt. Lett.* 2002; 27: 83-85.

[10]   Stokes L.F., Chodorow M., and Shaw H.J.: All-single-mode fiber resonator. *Opt. Lett.* 1982; 7:, 288-290.

[11]   Norcia S., Tonda-Goldstein S., Dolfi D., Huignard J.P., and Frey R.: Efficient single mode Brillouin fiber laser for low-noise optical carrier reduction of microwave signals. *Opt. Lett.* 2003; 28: 1888-1890.

[12]   Geng J., Staines S., Wang Z., Zong J., Blake M., and Jiang S.: Highly stable low-noise Brillouin fiber laser with ultranarrow spectral linewidth. *IEEE Photon. Technol. Lett.* 2006; 18: 1813-1815.

[13]   Molin S., Baili G., Alouini M., Dolfi D., and Huignard J.-P.: Experimental investigation of relative intensity noise in Brillouin fiber ring lasers for microwave photonics applications. *Opt. Lett..* 2008; 33: 1681-1683.

[14]   Othsubo, J.: *Semiconductor Laser. Stability, Instability and Chaos.* 2nd ed., Springer-Verlag ;Berlin; Heidelberg, 2008, 1-417.

[15] Li X.H., Liu X.M., Gong Y.K., Sun H.B., Wang L.R., and Lu K.Q.: A novel erbium/ytterbium co-doped distributed feedback fiber laser with single-polarization and unidirectional output. *Laser Phys. Lett.* 2010; 7: 55-58.

[16] Kim Y.-J., Chun B.J., Kim Y., Hyun S., and Kim S.-W.: Generation of optical frequencies out of the frequency comb of a femtosecond laser for DWDM telecommunication. *Laser Phys. Lett.* 2010; 7: 522-527.

[17] Cardoza-Avendaño L., Spirin, V. V., López-Gutiérrez R.M., López-Mercado C.A., and Cruz Hernández C.: Experimental characterization of DFB and FP chaotic lasers with strong incoherent optical feedback. *Opt. Laser Technol.* 2011; 43: 949-955.

[18] Available from: http://www.corning.com/opticalfiber/index.as

[19] Thevenaz L.: Brillouin distributed time-domain sensing in optical fibers: state of the art and perspectives. *Front. Optoelectron. China.* 2010; 3: 13-21.

[20] Spirin V.V., Lypez-Mercado C.A., Kablukov S.I., Zlobina E.A., Zolotovskiy I.O., Megret P., and Fotiadi A.A.: Single cut technique for adjustment of doubly resonant Brillouin laser cavities. *Opt. Lett.* 2013; 38: 2528-2530.

[21] López-Mercado C.A., Spirin V.V., Kablukov S.I., Zlobina E.A., Zolotovskiy I.O., Megret P., and Fotiadi A.A.: Accuracy of single-cut adjustment technique for double resonant Brillouin fiber lasers. *Opt. Fiber Technol.* 2014; 20: 194-198.

[22] Spirin V.V., Lopez-Mercado C.A, Megret P., and Fotiadi A.A.: Single-mode Brillouin fiber laser passively stabilized at resonance frequency with self-injection locked pump laser. *Laser Phys. Lett.* 2012; 9: 377-380.

[23] Debut A., Randoux S., and Zemmouri J.: Linewidth narrowing in Brillouin Lasers: Theoretical analysis. *Phys. Rev.* A. 2000; 62: 023803/1- 023803/4.

[24] Baney D. M., and Sorin W.V.: High resolution optical frequency analysis. In: Derickson, D. (ed.) *Fiber Optic Test and Measurement.* Prentice-Hall PTR, 1998, 169-219.

[25] Stepanov S.I., Fotiadi A.A., and Mégret P.: Effective recording of dynamic phase gratings in Yb-doped fibers with saturable absorption at 1064nm. *Opt. Express.* 2007; 15: 8832-8837.

[26] Fotiadi A.A., Antipov O.L., and Mégret P.: Resonantly induced refractive index changes in Yb-doped fibers: the origin, properties, and application for all-fiber coherent beam combining. In: B. Pal (ed.) *Frontiers in Guided Wave Optics and Optoelectronics.* InTech, 2010; 209-234.

[27] Lobach I.A., Kablukov S.I., Podivilov E.V., Fotiadi A.A., and Babin S.A.: Fourier synthesis with single-mode pulses from a multimode laser. *Opt. Lett.* 2015; 40: 3671-3674.

[28] Fotiadi A.A., Antipov O.L., and Mégret P.: Dynamics of pump-induced refractive index changes in single-mode Yb-doped optical fibers. *Opt. Express.* 2008; 16: 12658-12663.

[29] Fotiadi A.A., Zakharov N.G., Antipov O.L., and Mégret P.: All-fiber coherent combining of Er-doped amplifiers through refractive index control in Yb-doped fibers. *Opt. Lett.* 2009; 34: 3574-3576.

[30] Fotiadi A.A., Antipov O.L., Kuznetsov M.S., Panajotov K., and Mégret P.: Rate equation for the nonlinear phase shift in Yb-doped optical fibers under resonant diode-laser pumping. *J. Holography Speckle.* 2009; 5: 299–302.

[31] Fotiadi A., Antipov O., Kuznetsov M., and Mégret P.: Refractive index changes in rare earth-doped optical fibers and their applications in all-fiber coherent beam combining. In: A. Brignon (ed.) *Coherent Laser Beam Combining.* John Wiley & Sons, 2013, chap. 7, pp. 193-230.

[32] Kuznetsov M.S., Antipov O.L., Fotiadi A.A., and Mégret P.: Electronic and thermal refractive index changes in Ytterbium-doped fiber amplifiers. *Opt. Express.* 2013. 21: 22374-22388.

[33] Saleh A.A.M., and Stone J.: Two-stage Fabry-Perot filters as demultiplexers in optical FDMA LANs. *J. Lightwave Technol.* 1989; 7: 323-335.

[34] Stadt H., and Muller J.M.: Multimirror Fabry-Perot interferometers. *JOSA A.* 1985; 2: 1363-1370.

[35] Stepanov S.: Dynamic population gratings in rare-earth doped optical fibres. *J. Phys. D: Appl. Phys.* 2008; 41: 224002/1-224002/23.

[36] Spirin V.V., Lopez-Mercado C.A., Kinet D., Megret P., Zolotovskiy I.O., and Fotiadi A.: A single-longitudinal-mode Brillouin fiber laser passively stabilized at the pump resonance frequency with a dynamic population inversion grating. Laser Phys. Lett. 2013; 10: 015102/1-015102/4.

Chapter 6

Dual-Wavelength Fiber Lasers for the Optical Generation of Microwave and Terahertz Radiation

Kavintheran Thambiratnam, Harith Ahmad and Mukul C. Paul

Additional information is available at the end of the chapter

http://dx.doi.org/10.5772/61690

Abstract

Dual-Wavelength Fiber Lasers (DWFLs), which provide a simple and cost-effective approach for the optical generation of Microwave (MHz) and Terahertz (THz) radiation. The emphasis of this review is to trace the early development of DWFLs, including the issues and limitations faced by the various gain media right to the latest advancements in this field as well as their roles in generating the desired output. This review covers both the simple approaches of narrow-band filters and comb filters for microwave radiation generation, as well as the use of DWFLs with diethylaminosulfurtetrafluoride or LiNbO$_3$ crystals for generating THz radiation.

Keywords: Fibre Laser, Microwave, Terahertz, Dual-Wavelength

1. Introduction

The generation of Microwave (MHz) and Terahertz (THz) signals has been an area of intense research efforts due to the potential application for these signals in a multitude of applications. These new applications are spread across the board, encompassing conventional applications that are typically associated with MHz and THz signals such as communications and manufacturing [1-3] to novel and niche areas such as defence and science [4, 5].

While the generation of MHz and THz signals has been traditionally confined exclusively within the electromagnetic and radio frequency domains, recent developments in the field of photonics have made it possible for the generation of MHz and THz in this domain. This allows researchers to now overcome some of the technological limits on the development of these signals in the electromagnetic and radio frequency domains, and at the same time allowing them to realize a vast new array of applications. Photonic based MHz systems could allow for

the development of applications such as photonic generation, processing, control and distribution of MHz and millimeter-wave signals, while photonic based THz signal generation could in turns realize the development of new non-invasive sensing and measurement systems as well as expanding communications bandwidths. The interest in MHz based photonics systems is in the interactions between microwave and optical signals. The generation of microwave signals is typically a complicated affair, requiring expensive electronic circuits with multiple frequency doubling stages. This creates a major issue in terms of practically, and for safety purposes most microwave generation facilities are located at remote sites, thus making the distribution of MHz signals in the electrical domain a highly inefficient exercise as these signals will experience high losses during the transmission process. This is where photonic based MHZ systems has the significant advantage, as the generation of microwave signals in the optical domain alleviates these problems by allowing low-loss and broad bandwidth transmissions over optical fibres and at the same time taking advantage of the inherent immunity of optical fibres to electromagnetic waves. In most cases, microwave signals in the optical domain can be generated by the relatively simple process of optical heterodyning [6, 7], whereby the beating of two closely spaced but distinct frequencies results in the generation of the desired MHz output at a photodetector. Using this approach, MHz waves can be generated from a central office and distributed through an optical grid, thus simplifying equipment requirements and in turn reducing the cost and complexity of the system.

At the same time, intensive research has also been undertaken to investigate the field of THz radiation. THz radiation lies in the frequency gap between the infrared and microwave regions of the emission spectra, falling between the band gaps of 100 GHz to 30 THz and although long been eminent only in the domains of astronomy and analytical science, recent advances in technology have now seen its expansion into new areas of interest. This is due to the almost unlimited potential THZ radiation has for a wide variety of applications such as communications, security and defence, biological and medical sciences and spectroscopy to name a few. THz generation by optical sources typically entails non-linear optical frequency conversion, although initial research into this area was limited due to frequency and conversion efficiencies. However, with the availability of femtosecond lasers, THz frequencies and conversion efficiencies that could not be previously attained from optical sources can now be realized. Where the typical method of generating THz pulses would be by the use of femtosecond oscillators in the form of photoconductive switches and optical rectification, these pulses can now be realized using a variety of techniques that include the adaptation of femtosecond laser systems for generating of high intensity THz pulses as well as the exploitation of beats generated by two closely spaced lasing wavelengths. THz radiation is highly important for a multitude of applications, in particular the analysis of biological materials as the energy of THz photons is typically several orders of magnitude below the energy level required to ionize the valence electrons from biological molecules. This makes THz radiation a non-ionizing radiation, and gives it a fundamental advantage over ionizing radiation as it negates any harmful effects to biological structures that would result in the formation of highly reactive free radicals, which can cause secondary or indirect damage to other biomolecules. As such, THz radiation would find substantial potential for applications that require the analysis of biological matter, encompassing a wide range of effects from research such as spectroscopic analysis to real-world applications such as security applications.

2. Optical generation of microwave and THz radiation

The generation of THz and microwave radiation by optical means can be typically accomplished by two methods. The first method entails the creation of an ultrafast photocurrent in a photoconductive switch or semiconductor using electric-field carrier acceleration, otherwise known as the photo-Dember effect. Alternatively, the desired radiation can be optically generated by non-linear optical effects such as optical rectification or optical parametric oscillations. Both these approaches are, however, rather difficult to achieve, requiring highly sensitive electronics or exotic media such as GaAs, GaSe, GaP, ZnTe, CdTe, diethylaminosulfurtetrafluoride (DAST) or $LiNbO_3$ crystals.

Recently however, a new approach has shown much potential for the optical generation of THz radiation; optical heterodyning [6, 7]. By employing the optical heterodyning approach, both microwave and THz radiation can be generated as a result of the frequency differences between two closely spaced laser modes. When a photo-mixer with a short photo-carrier lifetime is exposed to these laser modes, the photocurrent is modulated at the frequency difference of the two laser modes, thereby generating an electromagnetic wave within the desired spectral range. The generated heterodyne beat can be represented mathematically as:

$$E_1\left(t\right) = E_{01}\cos\left(\omega_1 t + \phi_1\right)$$

(1)

$$E_2\left(t\right) = E_{02}\cos\left(\omega_2 t + \phi_2\right)$$

(2)

With E_{01} and E_{02} representing the amplitude of the generated modes, while ω_1 and ω_2 representing the angular frequency of the same modes. The phases of the modes are given in (1) and (2) as ϕ_1 and ϕ_2 respectively, while t denotes the time. Superimposing Equations (1) and (2) yields Equation (3), which is as follows:

$$I_{RF} = A\cos\left[\left(\omega_1 - \omega_2\right) + \left(\phi_1 - \phi_2\right)\right]$$

(3)

whereby I_{RF} represents the frequency of the generated electromagnetic wave and A being a constant determined by E_{01} and E_{02}. Tuning the spacing between the two wavelengths will now yield the electromagnetic output in the desired spectrum, be it in the microwave or THz regions.

3. MHz generation from DWFLs

In a perfect scenario, the generation of MHz and THz radiation from DWFLs would require the interaction of two closely Single-Longitudinal Mode (SLM) beams propagating in the same laser cavity. An SLM output is desired, as the long cavities in most fiber laser configurations would typically lead to a dense group of wavelengths propagating alongside a central wavelength, in turn preventing the generation of a clean and stable output [8]. SLM outputs can be obtained from fibre lasers by employing either narrow band or comb filters. Narrow band filters encompass filters such as Fiber Bragg Gratings (FBGs) [9, 10] and arrayed wave-guide gratings (AWGs), while comb filters encompass interferometers, such as the Mach-Zehnder filter, Fabry-Perot filter and ring resonators. Both approaches have their advantages; narrow-band filters are easily integrated into laser cavities, allowing for the accurate filtering of wavelengths from a broad spectral band, while interferometers are easy to fabricate and provide uniform channel spacings. However, one would also need to consider the drawbacks of each approach; FBGs and AWGs are difficult to fabricate, whereas interferometers are usually very sensitive with substantial sizes and do not function well outside a laboratory environment.

The earliest, and arguably the easiest DWFL configuration for generating an SLM output would be the use of FBGs with two distinct but close filtering wavelengths to isolate the desired wavelengths from a broadband spectrum and allow only these two wavelengths to oscillate within the laser cavity. A typically example of this is seen in the configuration as proposed by D. S. Moon and Y. Chung [11], which uses a simple ring cavity with an Erbium Doped Fiber (EDF) acting both the broadband spectrum source and also the amplifier for the lasing wavelengths. The desired dual-wavelength output is achieved by integrating two FBGs into the laser cavity. Although the proposed design in simple, nevertheless it achieves its crucial objectives, which is to stable and uniform dual wavelength SLM output. The system as proposed in Figure 1 is highly stable, with a power fluctuation of less than 0.2 dB at room temperature and a relatively high Optical Signal to Noise Ratio (OSNR) of more than 60 dB.

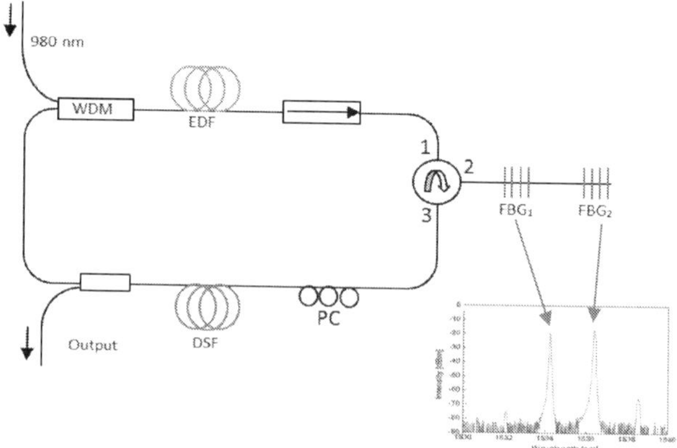

Figure 1. Experimental setup of a simple DWFL using an EDF as the gain medium. The inset of the figure shows the generated dual wavelength output obtained from the system [11]. Bear in mind though that the inset indicates where the dual-wavelength output is generated, not where the sampling is made.

A particular problem arises though in the design of Figure 1, which revolves around the use of the EDF. Specifically, the issue that arises is that of homogenous broadening and cross-gain saturation, both of which are effects strongly felt in the EDF and will lead to a condition of unstable lasing. A straight-forward solution to this is to cool the EDF to sub-zero temperatures, typically around 77K using liquid nitrogen. While this alleviates the effects of homogenous broadening and cross-gain saturation, it is for all purposes impractical and cannot be practically realized. Alternatively, exotic EDFs can also be used, such as those with specially designed structures or by taking advantage of various optical effects such as Polarization Hole Burning (PHB). Again, these may allow the desired goals to be achieved, but also increases the cost and complexity of the setup. The authors of [11] opted for a different approach, which is to suppress the line-broadening using a 2 km long Dispersion Shifted Fiber (DSF). While this approach was successful in creating the desired SLM output, careful balancing is required when using the DSF, thus adding a certain level of complexity to the system.

As a result of the effects of homogenous line broadening, in-homogenous gain media has also been investigated for the development of a stable, dual-wavelength SLM output. Among the successful demonstrations of this approach was that by X. Chen et al. [12] who were able to use two Ultranarrow Transmission Bands (UNTB) FBGs in conjunction with a Semiconductor Optical Amplifier (SOA) to generate the desired output for operation in the microwave regime. The key advantage of this setup, as shown in Figure 2 is the use of the SOA as the gain medium which is has a very much lower homogeneous line-broadening effect as compared to the EDF.

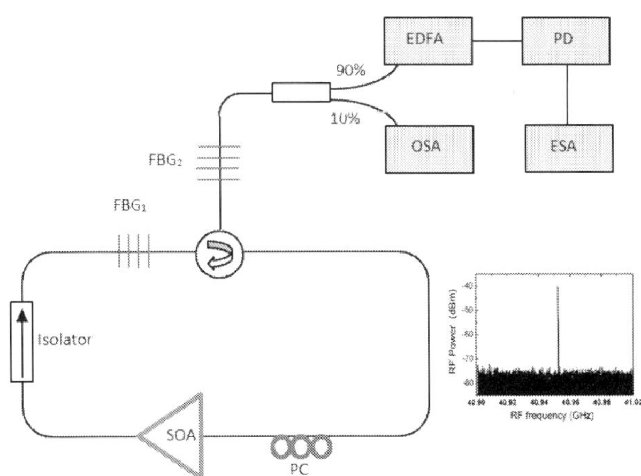

Figure 2. SLM dual wavelength fiber laser utilizing two UTNB FBGs. The mixing of the two wavelengths results in the generation of a microwave signal as seen from a radio frequency spectrum analyser, as shown in the figure [12].

As a result, the effects that typically arise in homogenously broadened gain media, such as mode-competition is negligible, and thus the two SLM wavelengths are able to lase in the same cavity with the relative phase fluctuations between the two wavelengths being low and thus allowing the system to generate a low-phase-noise microwave signal without requiring a microwave reference source. This particular work was able to generate microwave signals at 40.95, 18.68 and 6.95 GHz, corresponding to the channel spacings of the three different dual-wavelength UNTB FBGs incorporated into the laser cavity. A spectral width as small as 80 kHz was with a frequency stability better than 1 MHz when operating at room temperature in a free-running mode.

While the use of the SOA resolved the issues related to the stability of the lasing wavelengths, improved performance in terms of the channel spacing and thus, the frequencies of the generated microwave and THz radiation outputs could be achieved by using other wavelength filtering mechanisms in place of the conventional FBGs, such as Phase-Shifted FBGs (PSFBGs) [13], Inverse-Gaussian Apodized FBGs (IGAFBGs) [14] and Chirped Moire FBGs (CMFBG) [15]. The advantages of these approaches has already been proven, such as the wider range of the microwave signals generated as reported by E. Guillermo et al. [13]. E. Guillermo proposed and demonstrated a setup utilizing two cascaded PSFBGs written onto FBGs and configured to form a distributed feedback fiber laser. Using the PSFBGs, a dual-wavelength laser output is achieved with each wavelength having a Full-Width at Half Maximum (FWHM) linewidth of 0.22 pm in the single-wavelength regime. Piezoelectric actuators were utilized as the tuning mechanism, thus giving the user dynamic control over the wavelength spacing between the two lasing modes. A continuous tuning range of between 5 to 724 pm of was obtained, giving an equivalent microwave frequency range of 0.72 GHz to 92 GHz.

While the generation of the dual-wavelength SLM outputs was made possible by the FBG, real-world applications would demand a more consistent and uniform mechanism of obtaining these wavelengths. In this regard, the FBG falls short, as each time the system is deployed, fine tuning of the FBG's operating wavelengths and other such parameters needs to be made to ensure that the desired result is obtained. As such, focus on the development of the DWFL now shifted towards new wavelength filtering mechanisms that would give the same output regardless of the environment it was operating in. In this manner, a viable solution was found in the form of the Arrayed Waveguide Grating (AWG). When employed in a laser cavity, the AWG allowed for the quick and easy manipulation of the spacing between two oscillating wavelengths simply by selecting different channels on the AWG. As the channel spacings on the AWG were constant, thus a uniform output could be realized. Figure 3 shows such a system, which utilizes two AWGs to generate the desired wavelength spacings. This reliable yet simple approach was demonstrated by H. Ahmad, et. al. [16, 17] and used a 1 m long highly doped Leikki Er80-8/125 EDF as the linear gain medium. Two AWGs were used as the narrow-band filters. Similar to the aforementioned approach, the suppression of unwanted modes is carried out using a 7 cm long un-pumped Leikki Er80-8/125 EDF and sub-ring cavity. An output is highly stable output is obtained, with close channel spacings of between 0.01 to 0.03 nm and beating frequencies of between 1.4 to 3.2 GHz. The addition of a Mach-Zehnder modulator, driven at 180 kHz in the above setup adds the ability to obtain a distinct oscillating wavelength in addition to the dual-wavelength output should the need arise.

In addition to narrow-band filtering, the use of comb filters as a means to obtain a dual-wavelength output has also become widely accepted. One of the better examples of this approach is that of the reconfigurable dual-pass Mach-Zehnder Interferometer (MZI) filter, that was demonstrated by F. Wang et al. [18]. In this approach, the dual-pass MZI is incorporated into an EDF based ring fiber laser with a cavity that has a high extinction ratio, as shown in Figure 4.

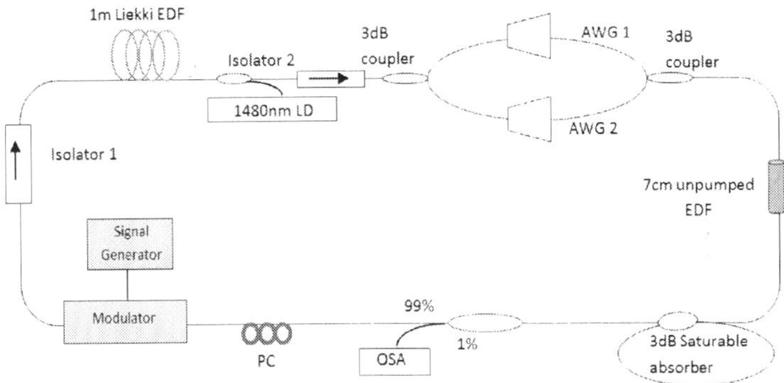

Figure 3. Experimental setup of AWG based dual wavelength SLM fiber laser (above) and the resulting microwave outputs with their corresponding dual-wavelength outputs [16].

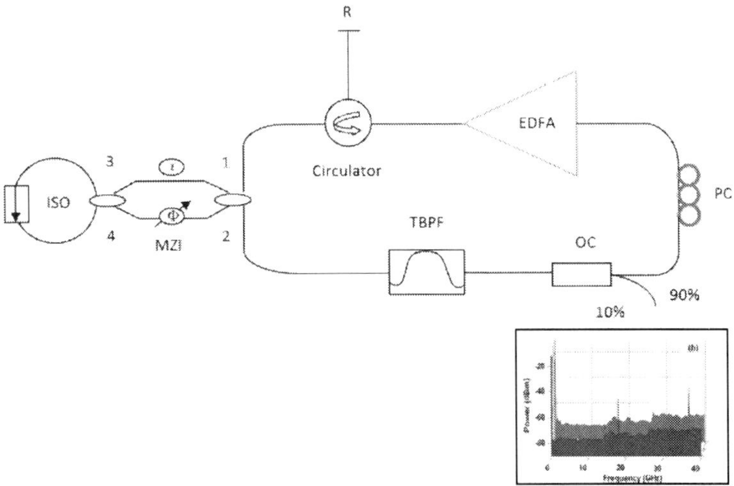

Figure 4. Experimental setup of the dual-pass MZI filter based dual-wavelength SLM fiber laser (above). The resulting microwave outputs are shown also shown in the figure (below), with the initial output shown in the forward spectrum, while the second microwave output, obtained after the channel spacing has been tuned, is shown in the spectrum at the back [18].

The proposed setup allows for the generation of a microwave output at signal frequencies of 9.76 GHz, 17.78 GHz and 35.67 GHz, although other frequencies can also be realized. The optical spectra of the system can be tuned from 1530 to 15675 nm using the Tunable Bandpass Filter (TBPF) inside the laser cavity. S. Pan and J. Yao on the other hand explored a different approach to produce a novel DWFL capable of generating an SLM output from an EDF laser. Their approach was based on a sigma architecture consisting of a ring loop and a linear standing wave arm [8] as illustrated in Figure 5. This approach is highly advantageous as it prevents gain competition among the oscillating wavelengths by placing the gain medium in the standing-wave arm and introducing the PHB effect into the system by the polarization multiplexing of the two lasing wavelengths in the ring loop of the cavity.

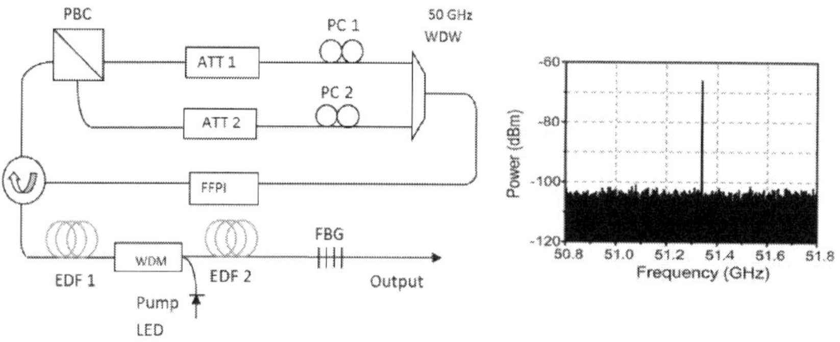

Figure 5. Experimental setup utilizing Fabry-Perot interferometer as SLM dual wavelength fiber laser. A single microwave output at approximately 51.3 GHz is obtained from the system [8].

This provides a stable output, and by integrating an ultranarrow Fabry- Perot Filter (FPF) into the cavity in the form of an un-pumped EDF, SLM operation can be realized. As a Lyot filter is formed for each wavelength, thus wavelength switching can be achieved by simply adjusting the polarization state of either wavelength. The proposed system was able to generate a microwave frequency output in a tunable range of ~10 to ~50 GHz.

Recently, a new approach towards achieving a DWFL with an SLM has been demonstrated using a ring resonator. This approach, as proposed by Y. Zhang et. al. successfully demonstrated a tunable SLM dual wavelength fiber laser based on a microwave photonic notch filter using a microfiber ring resonator [19] as shown in Figure 6.

In this experiment, the microfiber ring resonator (MRR) acts as a comb filter, while an un-pumped Polarization Maintaining EDF (PM-EDF) with an Au film coated at the end surface allows for the suppression of the mode competition and control of SLM output. A wavelength spacing of 1.02 nm, 0.82 nm, and 0.65 nm is reported by adjusting the MRR diameter, giving an equivalent microwave signal at 2.31 GHz with an OSNR of 35 dB. Similar results have also been obtained using microfiber couplers [20].

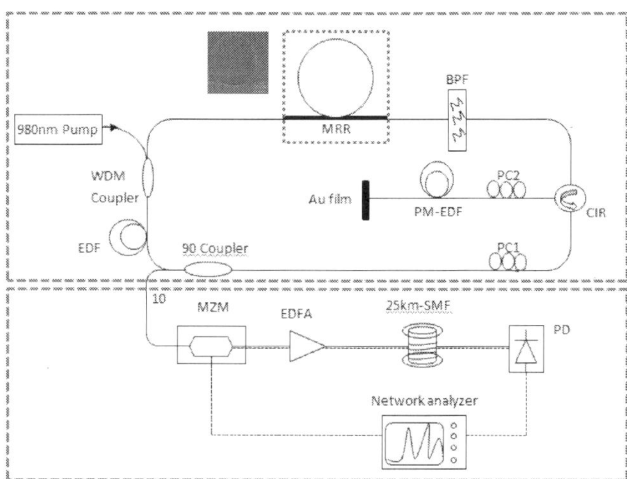

Figure 6. Experimental setup of the micro-ring resonator for dual-wavelength SLM generation. The inset shows the actual resonator ring [19].

4. Terahertz generation from dual wavelength fiber lasers

While microwave frequencies can be easy to generate, the optical generation of THz radiation can be much more complex. Unlike microwave radiation, the optical generation of THz radiation requires dual-wavelength near-Infrared (IR) laser fields to stimulate nonlinear polarizations in bulk Nonlinear Optical (NLO) crystals. Under suitable phase matching conditions, second order nonlinearity induced parametric process can lead to the generation of coherent THz radiation. However, due to the inherent characteristics of ultra-wideband THz frequency tunability (from sub-THz to tens of THz) that arise from the NLO crystals, Difference Frequency Generation (DFG) THz sources are very attractive for a wide range of applications in high-resolution spectroscopy and imaging. Traditionally, DFG THz sources were obtained using a dual wavelength approach by independent sources, although this method can be complex in terms of the necessary mechanical alignment and also the fine adjustment required in obtaining the spatial overlap of laser transverse modes [21, 22]. Alternatively, Optical Parametric Oscillators (OPO) can also be employed for this purpose, but although they can provide an output with a high power and also wide tenability, they are large, difficult to align and in most cases very expensive [23, 24]. As such, DWFLs are now being seen as a viable alternative to obtain the necessary wavelengths from a fiber laser for THz radiation generation, while at the same time leveraging on the robustness, compactness and stability of the fiber laser.

One of the most reliable approaches towards obtaining the desired frequency output is the use of a wavelength selector and laser oscillation, provided by linearly Chirped Fibre Braggs

Gratings (CFBGs) and SOAs. A system, as proposed by M. Tang et. al. [25] takes advantage of a thermo-optical phase change that allows seamless wavelength tuning; a single longitudinal mode operation is capable of achieving Continuous Wave (CW) dual-wavelength fibre laser with emissions at approximately 1060 nm using a Polarization Maintaining Ytterbium Fibre Amplifier (PM-YDFA) to amplifies the emission of the laser in order to pump a NLO DAST crystal. This setup is shown in Figure 8.

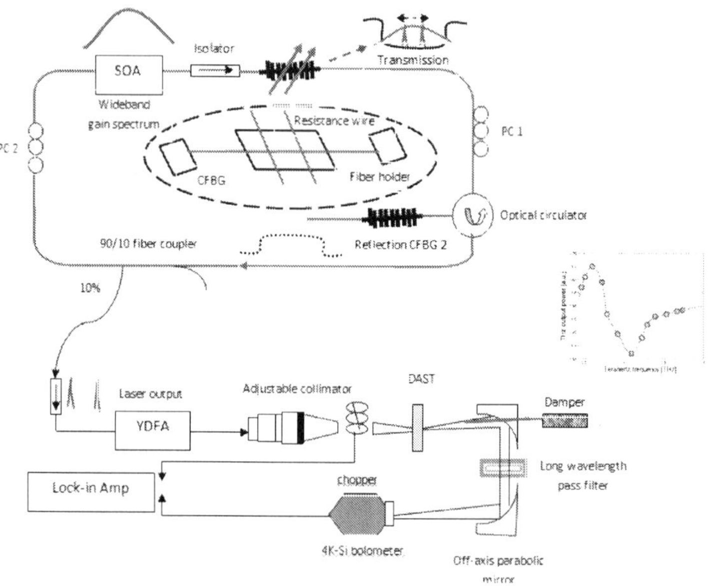

Figure 7. Setup of the DWFL for THz generation (above) and the resulting THz output that can be obtained from the system (below) [25].

The upper half of Figure 9 represents the ring configuration DWFL, which is based on a gain medium consisting of a fibre pig-tailed wideband 1060 nm SOA. This setup allows for a high saturation output power of 18 dBm and a wide optical bandwidth of 60 nm with +15 dB gain to be realized, with a relatively compact form factor as the ring cavity length is only 11.2 m long. The longitudinal mode spacing is 17.86 MHz, while unidirectional ring cavity and the stability of laser oscillations is maintained by means of optical isolator and polarization controllers at PC-1 and PC-2 respectively. The latter permits the laser polarization state to be optimized during propagation for the lasing and subsequently in amplification operations. The Two CFBGs, CFBG1 and CFBG2 are integrated into the cavity. CFBG1, which is 2.5 cm long is spliced into the cavity for transmission modes, and an identical to CFBG2 that is used for reflection alongside a 3-port optical circulator. Both CFBGs have a linear chirp rate of 4.5 nm/cm, which gives a reflection / stop band of approximately 10 nm. The inscription of the linearly chirped CFBG is achieved using hydrogen-load Nufern PS1060 photosensitive fibre with the ultraviolet scanning beam at 1060nm wavelength.

The two NiCr resistance wires shown in the figure are perpendicular to CFBG1, and lie across a heat sink while at the same time making contact with the fibre at two points. These wires have a cross-section diameter of 150 μm and are powered by a DC power supply to act as heating elements. As the refractive index of the silica glass fibre is temperature dependent, thus the points of the fibre in contact with heated elements will cause a relative phase change in a passing optical signal. This thermo-induced phase change becomes sufficiently large to break the linear chirp relationship at the CFBG, which results in a sharp transmission rise in the stop-band region of the CFBG. Two transmission peaks can occur simultaneously by means of mechanical operation on the heating elements. Using the DAST crystal, this particular DWFL setup is able to successfully generate THz radiation in the CW regime from 0.5 to 2 THz. The SOA as gain medium is able to suppress gain competition and any noise that arises from laser beating, while the CFBG wave selector efficiently controls the frequency difference of dual-wavelength single longitudinal mode lasing operations near 1060 nm. The thermo-optical induced phase changes lead to very sharp transmission peaks, and two mechanically adjustable fibre contact positions allow for two simultaneous wavelength oscillations.

The optical generation of microwave and THz radiation will undoubtedly open up a whole new aspect of research and development due to the immense potential these systems have on a variety of applications. Although this chapter has been able to give an overview of the developments of the DWFL that has led to today's technologies for microwave and THz radiation generation, it far from all-encompassing. With the continued pace of development and growing interest in this area, there is no doubt that this field will continue to grow and generate many interesting outputs in the times to come.

Author details

Kavintheran Thambiratnam[1*], Harith Ahmad[2] and Mukul C. Paul[3]

*Address all correspondence to: kavintheran@gmail.com

1 Malaysian Industry Government Group for High Technology (MiGHT), Malaysia

2 Photonics Research Centre, Malaysia

3 Central Glass and Ceramic Research Institute, India

References

[1] J. Kim, F. X. Kärtner. (2010, June). Microwave signal extraction from femtosecond mode-locked lasers with attosecond relative timing drift. *Opt. Lett.* [Online]. *35(12)*, pp. 2022-2024. Available: http://dx.doi.org/10.1364/OL.35.002022

[2] E. I. Ackerman and A. S. Daryoush. (1997, Aug). Broad-band external modulation fiber-optic links for antenna-remoting applications. *IEEE Transactions on Microwave Theory and Techniques.* [Online]. *45(8)*, pp. 1436-1442. Available: 10.1109/22.618449

[3] S. A. Pappert, C. K. Sun, R. J. Orazi, T. E. Weiner. (2000, May). Microwave fiber optic links for shipboard antenna applications. Presented at IEEE International Conference on Phased Array Systems and Technology, Proceedings. [Online]. pp. 345-348. Available: 10.1109/PAST.2000.858971

[4] A. S. Daryyoush. (2002, September). Microwave Photonics in Dual-Use Military Systems – A Personal Perspective. Presented at the RTO SET Lecture Series on "Optics Microwave Interactions", held in Jouy en Josas, France, 2-3 September 2002; Duisburg, Germany, 5-6 September 2002; Budapest, Hungary, 9-10 September 2002. [Online]. Available: http://ftp.rta.nato.int/public/PubFullText/RTO/EN/RTO-EN-028/EN-028-$I.pdf]

[5] J. Yao. (2009, Feb.). Microwave Photonics. *Journal of Lightwave Technology.* [Online]. *27(3)*, pp. 314-335. Available: 10.1109/JLT.2008.2009551

[6] J. J. Pan. (1989, Jan). Cost-Effective Microwave Fiberoptic Links Using The Heterodyne Laser. *Proc. SPIE 0995, High Frequency Analog Communications.* [Online]. *0995*, pp. 94-98, Availbale: doi:10.1117/12.960149; http://dx.doi.org/10.1117/12.960149

[7] J. Genest, M. Chamberland, P. Tremblay, M. Tetu. (1997, Jun). Microwave signals generated by optical heterodyne between injection-locked semiconductor lasers. *IEEE Journal of Quantum Electronics.* [Online]. *33(6)*, pp. 989-998. Available: 10.1109/3.585487

[8] S. Pan, and J. Yao. (2009, March). Frequency-switchable microwave generation based on a dual-wavelength single-longitudinal-mode fiber laser incorporating a high-finesse ring filter. *Optics Express.* [Online]. *17(14)*, pp. 12167-12173. Available: http://dx.doi.org/10.1364/OE.17.005414

[9] L. Bo, T. Swee Chuan, J. Meng, P. Shum. (2011, September). Dual-wavelength fiber laser using an inverse-Gaussian apodized fiber Bragg grating for tunable microwave generation. Presented at The 16th Opto-Electronics And Communications Conference, OECC 2011, Taiwan. [Online]. pp. 196-197. Available: http://dx.doi.org/10.1364/AO.50.004912

[10] [10] X. Liu, X. Yang, F. Lu, J. Ng, X. Zhou, C. Lu. (2005, Jan). Stable and uniform dual-wavelength erbium-doped fiber laser based on fiber Bragg gratings and photonic crystal fiber. *Optics Express.* [Online]. *13(1)*, pp. 142-147. Available: http://dx.doi.org/10.1364/OPEX.13.000142

[11] [11] D. S. Moon, and Y. Chung. (2013, Jan). Switchable dual-wavelength erbium-doped fiber ring laser assisted with four-wave mixing of dispersion-shifted fiber. *Optics Communications, 286(1)*, pp. 239-243.

[12] [12] C. Xiangfei, D. Zhichao, and Y. Jianping. (2006, Feb). Photonic generation of microwave signal using a dual-wavelength single-longitudinal-mode fiber ring laser. *IEEE Transactions Microwave Theory and Techniques*. [Online]. *54(2)*, pp. 804-809. Available: 10.1109/TMTT.2005.863064

[13] G. E. Villanuevaa, J. Palacía, J. L. Cruzb, M. V. Andrésb, J. Martía, P. Pérez-Millán. (2010, December). High frequency microwave signal generation using dual-wavelength emission of cascaded DFB fiber lasers with wavelength spacing tenability. *Optics Communications*. [Online]. *283(24)*, pp. 5165–5168. Available: http://www.sciencedirect.com/science/article/pii/S0030401810008357

[14] B. Lin, S. C. Tjin, M. Jiang, P. P. Shum, Y. He, Y. Ge. (2011, July). Dual-wavelength fiber laser using an inverse-Gaussian apodized fiber Bragg grating for tunable microwave generation. Presented at The 16th Opto-Electronics and Communications Conference, OECC 2011, Taiwan. [Online]. pp. 196-197. Available: http://ieeexplore.ieee.org/xpl/abstractAuthors.jsp?arnumber=6015080

[15] S. Feng, S. Lub, W. Peng, Q. Lia, C. Qia, T. Fenga, S. Jian. (2013, February). Photonic generation of microwave signal using a dual-wavelength erbium-doped fiber ring laser with CMFBG filter and saturable absorber. *Optics & Laser Technology*. [Online]. *45*, pp. 32–36. Available: http://www.sciencedirect.com/science/article/pii/S0030399212003532

[16] H. Ahmad, A. A. Latif, J. M. Taib, S. W. Harun. (2013). Tunable, low frequency microwave generation from AWG based closely-spaced dual-wavelength single-longitudinal-mode fibre laser. *J. Europ. Opt. Soc. Rap. Public.* [Online]. *8(13038)*, Available: 10.2971/jeos.2013.13038

[17] [17]H. Ahmad, M. Z. Zulkifli, A. A. Latif, S. W. Harun. (2009, December). Tunable Dual Wavelength Fiber laser Incorporating AWG and Optical Channel Selector by Controlling the Cavity Loss. *Opt. Comm.* [Online]. *282(24)*, pp. 4771-4775. Available: http://www.sciencedirect.com/science/article/pii/S003040180900844X

[18] F. Wang, X. Zhang, Y. Zhang, E. Xu. (2011, January). A Tunable and Switchable Single-longitudinal-mode Dual-wavelength Fiber Laser for Microwave Generation. *Proc. SPIE 7987, Optoelectronic Materials and Devices V.* [Online]. *7987*, Available: doi: 10.1117/12.888113

[19] [19] Z. Yu, Z. X. Liang, C. G. Jie, X. E. Ming, H. D. Xiu. (2010, July). A Microwave Photonic Notch Filter Using a Microfiber Ring Resonator. *Chin. Phys. Lett.* [Online]. *27(7)*, pp. 74207-074207. Available: http://iopscience.iop.org/0256-307X/27/7/074207

[20] [20]A. Sulaiman, S. W. Harun, M. Z. Muhammad, H. Ahmad (2013, July). Compact Dual-Wavelength Laser Generation using Highly Concentrated Erbium-Doped Fiber Loop Attached to Microfiber Coupler. *IEEE J. of Quant. Electron.* [Online]. *49(7)*, pp. 586-588. Available: http://ieeexplore.ieee.org/xpl/articleDetails.jsp?arnumber=6512030

[21] [21] A. Klehr, J. Fricke, A. Knauer, G. Erbert, M. Walther, R. Wilk, M. Mikulics, M. Koch. (2008, April). High-power monolithic two-mode DFB laser diodes for the generation of THz radiation. *IEEE J. Sel. Top. Quant. Electron.* [Online]. *14(2)*, pp. 289–294. Available: 10.1109/JSTQE.2007.913119

[22] N. J. Kim, Y. A. Leem, J. H. Shin, C. W. Lee, S. P. Han, M. Y. Jeon, D. H. Lee, D. S. Yee, S. K. Noh, and K. H. Park. (2010, September). Widely tunable dual-mode multi-section laser diode for continuous-wave THz generation. *Presented at 35th international conference on infrared, millimeter and terahertz waves (IRMMW-THz), We-C3.1, Rome.* [Online]. pp. 1-2. Available: 10.1109/ICIMW.2010.5612426

[23] [23] A. Godard, M. Raybaut, O. Lambert, J.-P. Faleni, M. Lefebvre, E. Rosencher. (2005, September). Cross-resonant optical parametric oscillators: study of and application to difference-frequency generation. *J. Opt. Soc. Am. B.* [Online]. *22(9)*, pp. 1966–1978. Available: http://www.opticsinfobase.org/josab/abstract.cfm?URI=josab-22-9-1966

[24] K. Kawase, T. Hatanaka, H. Takahashi, K. Nakamura, T. Taniuchi, H. Ito. (2000, December). Tunable terahertz-wave generation from DAST crystal by dual signal-wave parametric oscillation of periodically poled lithium niobate. *Opt. Lett.* [Online]. *25(23)*, pp. 1714–1716. Available: http://dx.doi.org/10.1364/OL.25.001714

[25] M. Tang, H. Minamide, Y. Wang, T. Notake, S. Ohno, H. Ito. (2011, January). Tunable terahertz-wave generation from DAST crystal pumped by a monolithic dual-wavelength fiber laser. *Opt. Express* [Online]. *19(2)*, pp. 779-786. Available: http://dx.doi.org/10.1364/OE.19.000779.

Chapter 7

Tunable Single-, Dual- and Multi-wavelength Fibre Lasers by Using Twin Core Fibre-based Filters

Guolu Yin, Xin Wang, Shuqin Lou and Yiping Wang

Additional information is available at the end of the chapter

http://dx.doi.org/10.5772/62097

Abstract

Tunable fibre lasers draw intensive attention because their emission wavelength can be systematically tuned within a certain spectral range, which allows using a single laser source instead of several sources. This is convenient and cost-effective for many applications in a range of fields, such as telecom, material processing, microscopy, medicine and imaging and so on. The laser wavelength can be tuned in a certain range of wavelength by inserting wavelength-selective elements into the laser's optical cavity. This chapter describes the twin core fibre (TCF)-based filters, which work as the wavelength-selective element. They are introduced into the ring cavity to implement tunable single-, dual- and multi-wavelength fibre lasers. First, we deduced the coupled-mode theory of TCF-based filter. Second, we experimentally and numerically characterized the optical properties of TCF-based filters including free spectral range, polarization dependence, strain effect and bending effect. Finally, we investigated three tunable fibre lasers which operate at single-, dual- and multi-wavelengths, respectively. The operation mechanism of the fibre lasers mainly involved the elastic-optic effect, polarization hole burning effect and non-linear optical loop mirror. We emphasized the tuning mechanism and the tuning characteristics of the tunable fibre lasers.

Keywords: Tunable fibre laser, Twin core fibre, Elastic-optic effect, Polarization hole burning effect, Non-linear optical loop mirror

1. Introduction

Tunable fibre lasers draw serious attention because their emission wavelength can be systematically tuned within a certain spectral range, which allows using a single laser source instead of several sources. This is convenient and cost-effective for many applications in a range of fields, such as telecom, material processing, microscopy, medicine and imaging and so on.

Especially in a wavelength division multiplexing system, the tunable laser is recognized as a backup light source for fixed wavelength laser replacement, with the motivating factors being cost savings and potentially higher system reliability.

Tuning the laser wavelength across a certain wavelength range can be achieved by placing wavelength-selective elements into the laser's optical cavity to provide a particular wavelength selection. Free space filters based on opto-VLSI processor [1] or two-dimensional (2D) dispersion arrangements [2] have been used as wavelength-selective elements for realizing tunable fibre lasers. These filters can provide relatively small tuning steps. However, these devices often suffer from high insertion loss which results in a low side-mode suppression ratio (SMSR). By contrast, in-line fibre filters have been widely used in the laser cavity due to the advantages of low insertion loss, compactness and integration in all fibre laser systems. Conventional Fibre Bragg gratings (FBGs) in single-mode fibres have been used for realizing tunable single-wavelength fibre lasers by highly stretching or bending the FBGs in the laser cavity [3, 4]. Various special fibre-grating structures have been designed to implement dual-wavelength fibre lasers, including a pair of identical FBGs [5], polarization maintained FBGs [6], multimode FBGs [7] and phase-shifted FBGs [8]. Besides the FBGs, a series of modal interferometers draws more attention on tunable single-wavelength fibre lasers due to their easy fabrication and low cost. These modal interferometers can be constructed by using fibre tips [9], two tapers [10, 11], and multimode fibres [12, 13]. High-birefringence Sagnac loop mirrors (HiBi-SLMs) [14, 15] and Lyot birefringence filters [16] were also employed to provide C- and L-band tunable fibre lasers. The modal interferometers and HiBi-SLMs usually lead to polarization hole burning effect which can weaken the homogeneous gain broadening of the erbium-doped fibre (EDF) and help to understand multi-wavelength fibre lasers. Generally, these lasers were switched to operate in dual-/three-wavelength oscillation by adjusting the polarization controllers (PCs). More effective mode suppression techniques have been exploited to obtain more wavelength emissions at room temperature, including phase modulation [17], four-wave mixing [18], Raman scattering [19], Brillouin scattering [20, 21] and a non-linear optical loop mirror (NOLM) [22, 23].

In this chapter, twin core fibre (TCF)-based filters acting as the wavelength-selective element are proposed to achieve tunable single-, dual- and multi-wavelength erbium-doped fibre lasers. This chapter is structured as follows: in Section 1, we will give an overview of the recent development in tunable fibre lasers; in Section 2, we will give the coupled-mode theory of TCF-based filter; in Section 3, we will characterize the optical properties of TCF-based filter, both experimentally and numerically; in Section 4, we will introduce three tunable fibre lasers which operate at single-, dual- and multi-wavelengths, respectively and the conclusion will be given in the fifth section.

2. Coupled-mode theory

The TCF used in our experiments was fabricated by means of the groove-stack-and-draw method. The TCF preform was first prepared by side grooving in a large-diameter pure silica rod; then the rectangular groove was filled with two small-diameter Ge-doped silica rods and some other small-diameter pure silica rods; finally the rods were fixed in a pure silica jacket tube. Figure 1(a), (b) and (c) illustrates the original preform, micrograph image of the TCF cross section and the magnified micrograph image of the core region, respectively. The diameters of the core and cladding are approximately 6.4 µm and 130 µm, respectively. The separation between the two core axes is 14.2 µm. The difference of the refractive index between the core and cladding is $\Delta = 0.296\%$. We intended to fabricate a TCF with two identical cores. Unfortunately, the two cores in the fabricated TCF have a tiny difference in the geometry shape. In order to evaluate the optical properties accurately, we applied a digital image processing technology to model and rebuild the geometry of the TCF [24], as shown in Figure 1(d). The rebuilt TCF will be used in the numerical analyses in Section 3.

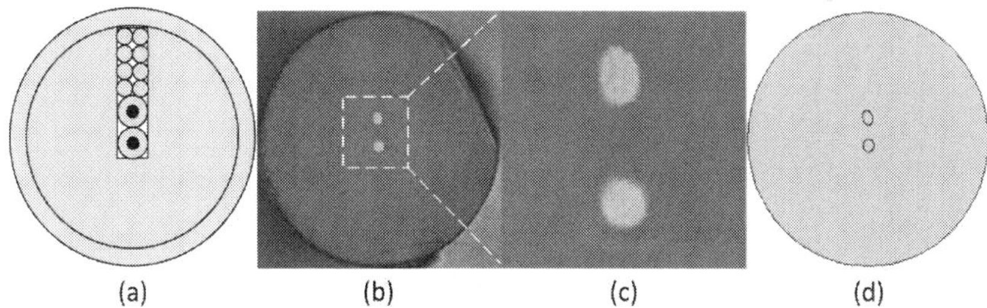

(a)   (b)   (c)   (d)

Figure 1. (a) Cross section of the original preform, (b) micrograph image of the TCF cross section, (c) magnified micrograph image of the core region in the TCF, (d) rebuilt geometry of the TCF.

In TCF, light couples back and forth between the two cores. Assuming that light launches into core 1, output amplitude of core 1 can be characterized by using the mode-coupled theory with symmetric and anti-symmetric mode [25]:

$$A^{x,y}\left(L_0\right) = \left[\cos\left(S^{x,y}L_0\right) + j\frac{\delta^{x,y}}{S^{x,y}}\sin\left(S^{x,y}L_0\right)\right] \times \exp\left(-j\delta^{x,y}L_0\right) \tag{1}$$

$$\begin{cases} S^{x,y} = \dfrac{\pi}{\lambda}\left(n_e^{x,y} - n_o^{x,y}\right) = \sqrt{\left(\delta^{x,y}\right)^2 + K^2} \\[2ex] \delta^{x,y} = \dfrac{\pi}{\lambda}\left(n_2^{x,y} - n_1^{x,y}\right) = \dfrac{\beta_2^{x,y} - \beta_1^{x,y}}{2} \end{cases} \tag{2}$$

where λ is the wavelength of light in a vacuum, L_0 is the length of the TCF, n_e and n_o are the effective refractive indices (ERIs) of the symmetric and anti-symmetric mode, respectively, n_1 and n_2 are the ERIs of the two cores when the modes in the two cores propagate independently, β_1 and β_2 are the corresponding propagation constants of core 1 and core 2, K is the coupling coefficient between the fundamental modes of two cores and superscripts x and y represent two orthogonal polarization states.

For a normalized incident light with a given state of polarization (SOP), the electric field can be given as 2D-column vector:

$$E_{in} = \begin{pmatrix} \cos\theta \\ \sin\theta \end{pmatrix} \tag{3}$$

where θ is the angle between the polarization direction of the incident light and the x-polarization direction. The transmission matrix of the TCF can be expressed as

$$T_{TCF} = \begin{pmatrix} A^x & 0 \\ 0 & A^y \end{pmatrix} \tag{4}$$

Therefore, the output field E_{out} can be analyzed with the following Jones matrix representation:

$$E_{out} = \begin{pmatrix} E_x \\ E_y \end{pmatrix} = T_{TCF} E_{in} = \begin{pmatrix} A^x & 0 \\ 0 & A^y \end{pmatrix} \begin{pmatrix} \cos\theta \\ \sin\theta \end{pmatrix} = \begin{pmatrix} A^x \cos\theta \\ A^y \sin\theta \end{pmatrix} \tag{5}$$

Substituting Eq. (1) into Eq. (5), the output power in core 1 can be expressed as [26]

$$
\begin{aligned}
P_1(L_0,\lambda) &= (E_x)^2 + (E_y)^2 \\
&= \left[1 - \frac{(S^x)^2 - (\delta^x)^2}{(S^x)^2} \sin^2(S^x L_0) \right] \cos^2\theta \\
&\quad + \left[1 - \frac{(S^y)^2 - (\delta^y)^2}{(S^y)^2} \sin^2(S^y L_0) \right] \sin^2\theta
\end{aligned}
\tag{6}
$$

If two cores have identical refractive index and diameter, the two cores have the same propagation constants which result in $\beta = \beta_1 = \beta_2$, $\delta = 0$ and $S = K$. Thus, Eq. (6) can be rewritten as

$$P_1(L_0,\lambda) = \cos^2(K L_0) \tag{7}$$

Here, the coupling coefficient K can be expressed as [27]

$$K = \frac{\left(2\Delta\right)^{1/2} U^2 K_0\left(\dfrac{Wd}{a}\right)}{aV^3 K_1^2\left(W\right)} \tag{8}$$

$$\begin{cases} U = a\sqrt{k_0^2 n_{co}^2 - \beta^2}, \ W = a\sqrt{\beta^2 - k_0^2 n_{cl}^2} \\ V = \sqrt{U^2 + W^2}, \ \Delta = \left(n_{co}^2 - n_{cl}^2\right)\big/\left(2n_{co}^2\right) \end{cases} \tag{9}$$

where $K_0 = 2\pi/\lambda$ is the wavenumber in vacuum at the wavelength λ, a is the core radius, d is the separation between the two core axes, n_{co} and n_{cl} are the refractive indices of the core and cladding, respectively, Δ is the difference between refractive indices of the core and cladding and K_0 and K_1 are the zero order and first order of modified Bessel functions of the second kind, respectively.

3. Characterization of TCF-based filter

The TCF-based filter is formed by splicing a section of TCF between two segments of SMFs as shown in Figure 2(a). The splicing between the TCF and the SMF is carried out by using an Ericsson fusion splicer (FSU 975) in manual operation mode. We used an active monitoring mode to make the SMF align to one core of the TCF. One end of the SMF was connected to a broadband light source, and one end of the TCF was connected to a power meter through a bare fibre adapter. The other ends of the SMF and TCF were placed in fibre holders of the splicer. We manually adjusted the position of the SMF and TCF by driving the motors in the splicer. When the output power reached the maximum value, we concluded that the SMF was aligned to one core of TCF. Then, we adopted a low-power arc to splice the SMF and TCF. Figure 2(b) shows the micrograph image of the splicing joint between TCF and SMF, which indicates that one core of TCF is precisely aligned to the core of SMF.

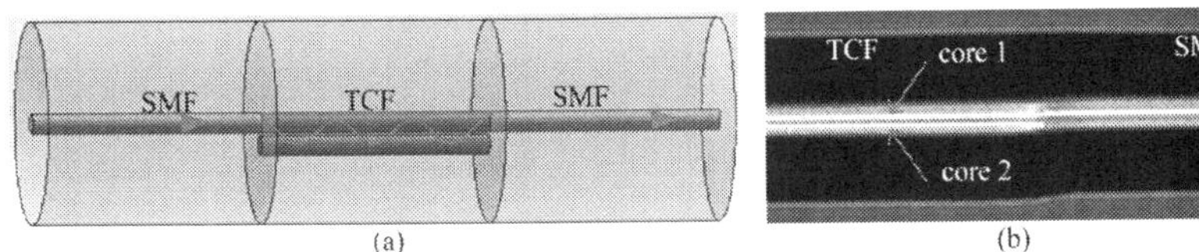

Figure 2. (a) Schematic of a TCF-based filter, (b) micrograph image of the splicing point between TCF and SMF.

Theoretically, we used the finite element method to evaluate the optical properties of TCF. The TCF structure used for numerical calculation is the rebuilt geometry as shown in Figure 1(d). The refractive index of the cladding, n_{cl}, is set as the refractive index of pure silica using

Sellmeier equation and the refractive index of the core is obtained through equation $n_{co} = (1 - 2\Delta^{-1/2}) n_{cl}$ with Δ = 0.296%. At λ = 1550 nm, n_{co} and n_{cl} are 1.448325 and 1.444023, respectively.

In the following Sections 3.1–3.4, we will investigate four optical properties of TCF-based filters, including free spectra range (FSR), polarization dependence, strain effect and bending effect. All the optical properties are closely related to the tunable fibre lasers discussed in Section 4. In short, the free spectra range mainly determines the tuning range of the fibre laser. The polarization dependence ensures stable dual-wavelength oscillations. The strain effect and bending effect help exactly reflect the tuning mechanism of the fibre lasers.

3.1. Free spectra range

In the TCF-based filter, the light from the SMF is launched into one core of the TCF at the first splicing point. Then, light couples back and forth between the two cores along the TCF. At the second splicing point, a comb-like spectrum was expected according to the coupled-mode theory in Section 2. The transmission dips and peaks occur when the following phase condition is satisfied:

$$\sqrt{\delta^2 + K^2} L_0 = \begin{cases} (m + 1/2)\pi, \text{dip} \\ m\pi, \text{peak} \end{cases} \tag{10}$$

where m is an integer. The separation of the adjacent dips/peaks can be defined as the free spectra range [28]:

$$\Delta\lambda = \frac{\pi/2}{L_0 \dfrac{\partial K}{\partial \lambda}} \tag{11}$$

which indicates that the FSR is inversely proportional to both the length of TCF and the derivative of the coupling coefficient with respect to wavelength.

We fabricated seven TCF-based filters with different length L_0 = 86.9 mm, 115 mm, 160 mm, 250 mm, 0.5 m, 1.1 m and 1.3 m, on the one hand. Light from broadband source (BBS, NKT phonics, superkTM white light laser) is launched into the TCF-based filter and measured by the optical spectrum analyser (OSA, YOKOGAWA, AQ6370C) with a resolution of 0.02 nm. Figure 3(a), (b), (d) and (e) illustrates the transmission spectra of the TCF-based filter with L_0 = 86.9, 115, 160 and 250 mm, respectively. The FSRs of the seven TCF-based filters are measured experimentally to be 49.0, 35.0, 24.2, 15.7, 7.8, 3.7 and 3.2 nm.

On the other hand, we calculated the coupling coefficient K as shown in Figure 3(c), and then calculated the theoretical FSRs through Eq. (11). The experimental FSRs (dashed line) and

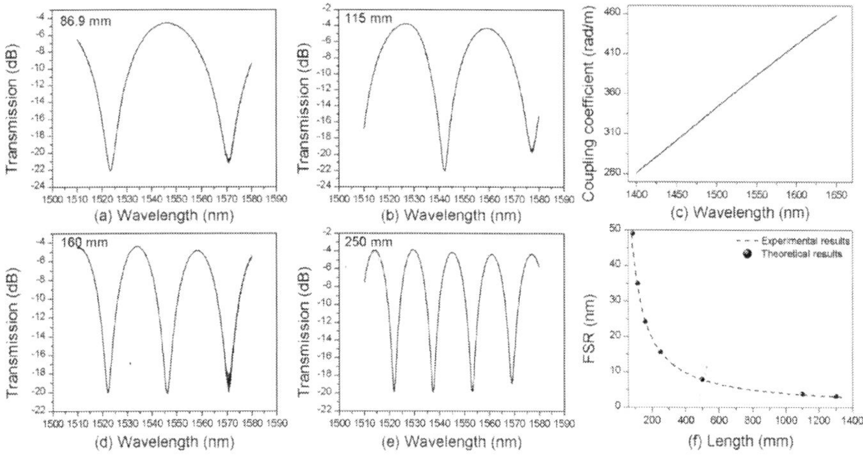

Figure 3. Transmission spectra of TCF-based filters with lengths, L_0, (a) 86.9, (b) 115, (d) 160, (e) 250 mm, (c) calculated coupling coefficient as a function of wavelength, (f) measured and calculated FSRs for various lengths of TCF.

theoretical FSRs (sphere symbols) were illustrated in Figure 3(f). It is found that the experiment results are in accord with the theoretical anticipation.

3.2. Polarization dependence

To theoretically evaluate the polarization dependence of the TCF, we numerically calculated the ERIs of n_e, n_o, n_1 and n_2 as shown in Table 1. Substituting the calculated ERIs into Eq. (2) and combining with Eq. (6), we can obtain the theoretical spectra of the TCF-based filter with different SOPs, as shown in Figure 4(a). It can be found that the position of the transmission peak varies with the SOPs. The theoretical wavelength spacing between the transmission peaks with x-polarization and y-polarization is 1.6 nm. Furthermore, the transmission peaks with SOPs at θ and $(\pi/2-\theta)$ are located symmetrically from each other with respect to the transmission peak with SOP at θ 45°.

In the experiment, we used a polarizer to obtain the linear polarized light from a broadband light source. A polarization controller was used to induce only rotation of this light. Figure 4(b) illustrates the evolution of transmission spectra of the TCF-based filter when the PCF was adjusted. It can be found that the maximum wavelength shift of the transmission peak is 1.2 nm, which is defined as the wavelength spacing between transmission peaks with x-polarization and y-polarization. Such experimental results agree well with the theoretical prediction as shown in Figure 4(a), even though the values of wavelength spacing and the FSR have tiny differences. Two causes can be considered to explain the difference. One is error of the parameters used for the theoretical calculation, including the geometric dimensions and the refractive indexes. The other is the non-uniformity of the homemade TCF in the longitudinal direction. Above all, theoretical and experimental results prove that a TCF-based filter is

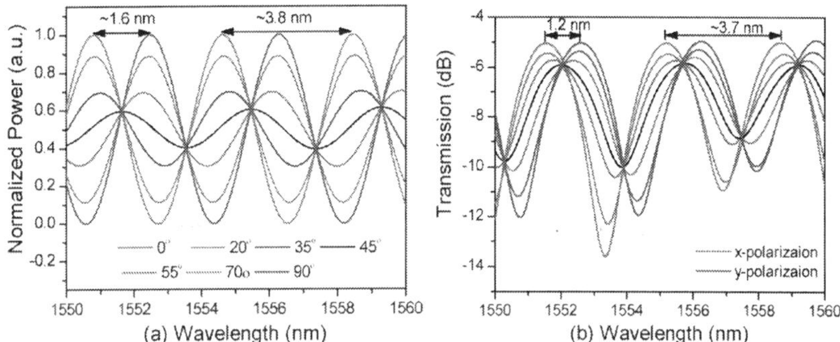

Figure 4. Transmission spectra of the TCF-based filter at different SOPs: (a) theoretical results, (b) experimental results.

polarization dependent. Such polarization dependence induces the polarization hole burning (PHB) effect in the laser cavity and ensures stable dual-wavelength oscillations with tunable wavelength spacing.

ERIs	x-polarization	y-polarization
n_e	1.444855015932922	1.444856531392316
n_o	1.444608454872977	1.444609352306018
n_1	1.444705605736814	1.444705931113986
n_2	1.444787015740477	1.444789030825202

Table 1. Effective refractive indices at $\lambda = 1550$ nm.

3.3. Strain effect

As TCF-based filter is applied to axial strain, the refractive index of both core and cladding will decrease according to the photo-elastics [29]:

$$\Delta n_{co/cl} = -\frac{1}{2}n_{co/cl}^3 \left[P_{12} - \mu\left(P_{11} + P_{12}\right)\right]\varepsilon_z \qquad (12)$$

where ε_z is the axial strain, $P_{11} = 0.12$ and $P_{12} = 0.27$ are the photo-elastic constants for the silica and $\mu = 0.17$ is the Poisson ratio. According to Eq. (12), the decreasing magnitude of the core refractive index is larger than that of the cladding refractive index, which induces a decrease in the refractive index difference between core and cladding. Such a decrease in the refractive index difference would make the electric field of the fundamental core mode extend further to the cladding, thus enhancing the coupling of the two cores, and the coupling coefficient increases as shown in Figure 5(a).

On the other hand, it is worth noting that the coupling coefficients also depend on the wavelength. As the wavelength is increased, the evanescent fields extend further away from the fibre core, the mode overlap increases and as well as the coupling coefficient. As discussed, when the TCF-based filter is stretched, the coupling coefficients would increase because of the elastic–optic effect. In order to keep the phase condition of the transmission peak unchanged in Eq. (10), the wavelength should shift to the short wavelength direction to give an opposite variation of coupling coefficient. Therefore, it is expected to find a 'blue' shift of the transmission peak when the TCF-based filter is stretched.

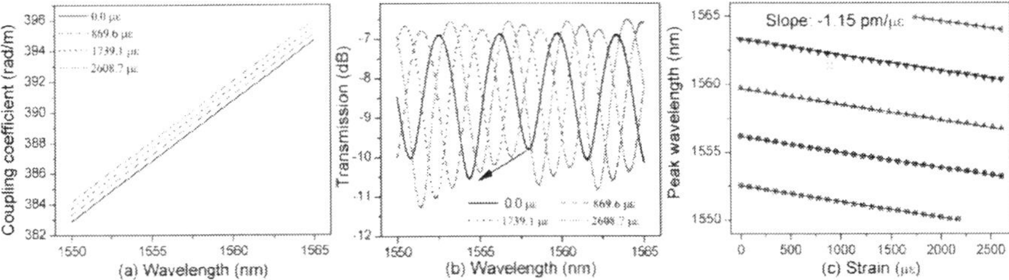

Figure 5. Theoretical and experimental results of TCF-based filter at different axial strains (a) calculated coupling coefficients, (b) transmission spectra of the TCF-based filter, (c) evolution of the peak wavelengths.

To investigate the strain effect experimentally, the TCF-based filter with a TCF length of $L_0 =$ 1.1 m was fixed by two fibre holders adhered to stages. The initial distance between the two stages is around $L = 1.15$. The fibre was stretched by manually moving the right stage away from the left one, with a 10-μm precision. To qualify the stretching, the displacement of the moving stage is defined as δz which gives rise to an axial strain $\varepsilon_z = \delta z / L$, δz was increased from 0.0 to 3.0 mm with an interval of 0.1 mm. Correspondingly, the axial strain was increased from 0.0 to 2608.7 με with a step of 86.9 με. We recorded the transmission spectra of the TCF-based filter at each axial stain. Figure 5(b) illustrates the transmission spectra when $\varepsilon_z = 0$, 869.6, 1739.1, and 2608.7 με. Figure 5(c) illustrates the evolution of peak wavelength. It can be found that the transmission spectrum shifted toward the shorter wavelength direction. This point agrees with the theoretical prediction implied by the coupling coefficient in Figure 5(a). In detail, the peak wavelength shifts linearly with an increase in axial strain. The sensitivity is around 1.15 pm/ με.

3.4. Bending effect

Figure 6(a) shows that the TCF is bent with a curvature radius of R. In order to simplify the theoretical calculation, the two cores were assumed to be located in the bend plane. The centre position between two cores is defined as the coordinate zero point. The bending direction is represented by the negative x-direction. Core 1 and core 2 are located in the negative x-direction and positive x-direction, respectively.

When the bending is applied to TCF, the part on outside of the bend (positive x-direction) is in tension, and the part on inside of the bend (negative x-direction) is in compression. According to the elastic–optic effect, the refractive indices of the outside part and inside part in the TCF increase and decrease, respectively. The refractive index profile in the x-direction can be expressed by equivalent refractive index model [30]:

$$n'(x) = n(x)\left(1+\frac{x}{R}\right) = n(x)(1+Cx) \tag{13}$$

where $C = 1/R$ is the curvature and $n(x)$ and $n'(x)$ are refractive index profiles when the TCF is straight and bent, respectively. Figure 6(b) shows the refractive index profile when $C = 0$ and $10\ \mathrm{m}^{-1}$. It can be found that the refractive indices of core 1 and core 2 decrease and increase, respectively.

When the TCF is straight, the two cores have identical refractive index, which results in the same propagation constant $\beta_1 = \beta_2$ at $C = 0\ \mathrm{m}^{-1}$. When the curvature is gradually increased, the propagation constants of core 1 and core 2 linearly increase and decrease in accordance, respectively, with the refractive index variation of the two cores, as shown in Figure 6(c). As a result, the magnitude of δ in Eq. (10) is increased, as shown in the inset of Figure 6(c).

According to Eq. (10), the phase condition of transmission dips is determined by both δ and K. As discussed, the bending TCF would induce an increase of δ, and a longer wavelength gives larger K as shown in Figure 3(c). In order to keep the phase condition of transmission dip in Eq. (10) unchanged, the wavelength of transmission dips should shift towards the shorter wavelength direction to give an opposite variation of coupling coefficient when TCF is bent. Therefore, a 'blue' wavelength shift of the transmission dip is expected when the TCF-based filter is bent.

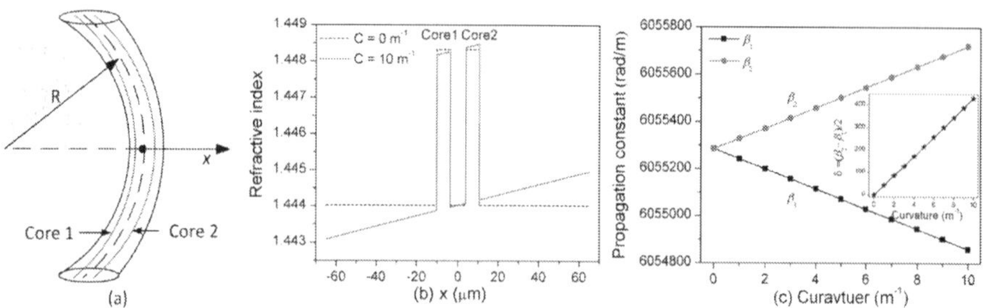

Figure 6. (a) Schematic of a bending TCF, (b) refractive index profile of TCF in x-direction, (c) propagation constants against curvature at $\lambda = 1550$ nm.

Experimentally, we chose the TCF based-filter with a TCF length of $L_0 = 86.9$ mm to evaluate the transmission spectral evolution when the TCF was bent [31]. The TCF-based filter was fixed by two stages. In order to protect the splice points between TCF and SMF, the stages were placed away from the splice points. Hence, the initial distance between the two stages was L = 103.9 mm. The curvature of the sensor head was adjusted by manually moving one stage towards another one with a 10-μm precision. The bent fibre is normally approximated as an arc of the circle. So the curvature C of the sensor is defined as $C = 1/R = 2h/[h^2+(L/2)^2]$ [32] where h is the displacement at the centre of the TCF-based filter and R is the curvature radius.

Initially, four resonance dips were found at 1461.82, 1510.38, 1558.58 and 1607.90 nm when the TCF-based filter was straight. Figure 7(a) illustrates the evolution of the transmission spectra of the TCF-based filter when its curvature was increased from 0 to 9.3 m⁻¹ gradually. It can be found that the spectra shifted towards the shorter wavelength direction, which agrees with the theoretical prospect discussed earlier. Generally, if the curvature is large enough, the resonance dip would shift to the initial position of the adjacent dip, which makes it hard to distinguish the wavelength shift. Therefore, the FSR of the filter generally determines the maximum wavelength shift that can distinguished.

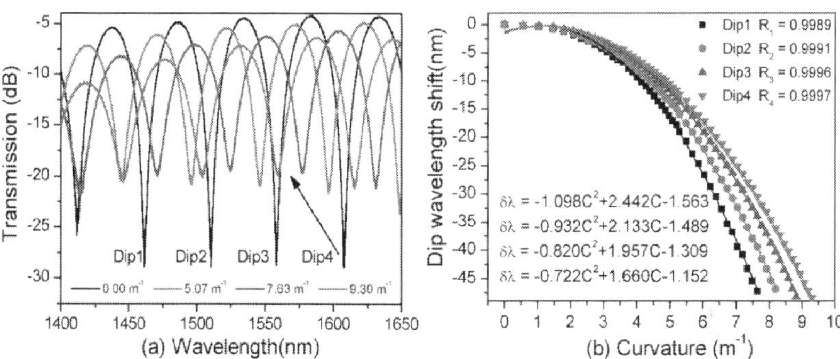

Figure 7. (a) Evolution of transmission spectra of the TCF-based filter, (b) dip wavelength shift against curvature.

Figure 7(b) illustrates the wavelength shift of the four transmission dips; the curvature was increased from 0 to 9.3 m⁻¹ gradually. The discrete experimental data can be fitted by a second-order polynomial with an R-square of up to 99.9%. The measurement ranges of the four dips are 7.63, 8.18, 8.82 and 9.30 m⁻¹. It can be found that the resonance dip with a larger resonance wavelength gives rise to a larger measurement range. In addition, the measurement range can be further increased by selecting a shorter TCF, because the FSR is inversely proportional to the TCF length according to Eq. (11).

4. Tunable fibre lasers

The tunable fibre lasers were constructed by using a ring laser cavity as shown in Figure 8(a). A 980-nm diode laser is injected into the laser cavity through a wavelength division multiplexor (WDM) to pump the erbium-doped fibre. A polarization-dependent isolator (PDI) is used to ensure that the laser oscillates in a single direction around the ring. The polarization controller is used to adjust the SOP of the light. The laser output is monitored by the optical spectrum analyser (OSA, YOKOGAWA, AQ6370C) from the 5% port of a 95:5 fibre coupler. In the following three subsections, we will investigate three in-line filters to implement the tunable single-, dual- and multi-wavelength fibre lasers.

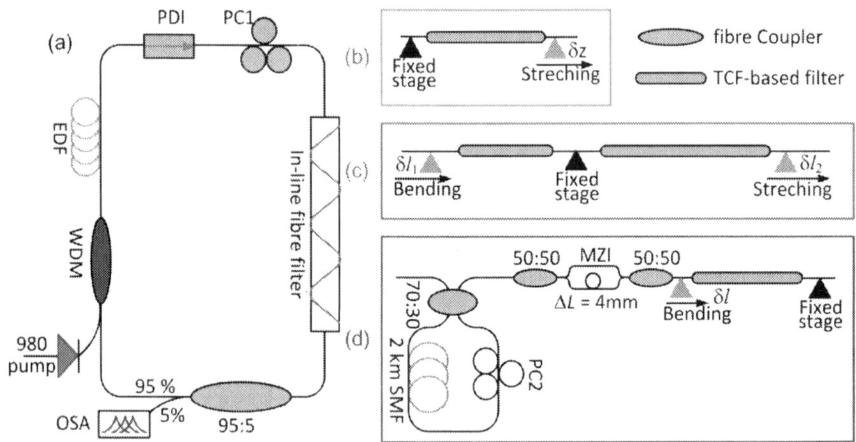

Figure 8. (a) Schematic of the tunable single-, dual- and multi-wavelength fibre lasers; (b), (c) and (d) represent three filters which are used to realize dual-, single- and multi-wavelength fibre lasers. EDF: erbium-doped fibre; WDM: wavelength division multiplexor; PC: polarization controller; PDI: polarization-dependent isolator, OSA: optical spectrum analyser.

4.1. Dual-wavelength and single-polarized fibre laser

In this work, we proposed and experimentally demonstrated a tunable dual-wavelength and single-polarized fibre laser by use of a TCF-based filter [26]. Figure 8(b) illustrates the TCF-based filter with a TCF length, L_0, of 1.1 m. In this section, we first presented the principle of the dual-wavelength fibre laser. Second, we evaluated the tunability of dual-wavelength spacing by adjusting the PC and investigated the lasing wavelength shift by stretching the TCF. Finally, we verified the single-polarized operation of the dual-wavelength fibre laser.

In the dual-wavelength fibre laser, the TCF-based filter simultaneously works as the wavelength selector and the polarization-dependent element. The TCF-based filter gives rise to a comb transmission spectrum. The SOP of the wavelengths is adjusted by using the PC, and the wavelengths with particular SOPs could become transmission peaks according to Figure 4. The polarization dependence of the TCF is helpful for inducing the PHB effect and in turn simultaneously amplifying the particular transmission peaks and achieving stable dual-wavelength oscillation.

At first, a single wavelength was obtained at λ = 1559.2 nm, then the single-wavelength was split into two wavelengths when the PC was slightly adjusted, as shown in Figure 9(a). The dual-wavelength fibre laser has a 3-dB line width of 0.047 nm and an SMSR of 40–52 dB. The maximum difference between the peak powers of the two wavelengths is around 1.2 dB. The wavelength spacing between the two wavelengths can be tuned from 0.1 to 1.2 nm by adjusting the PC.

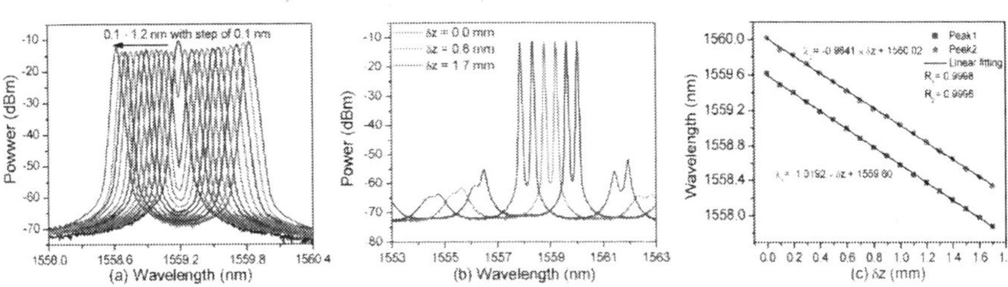

Figure 9. Experimental results of a tunable dual-wavelength fibre laser. (a) wavelength spacing is varied from 0.1 to 1.2 nm, (b) two wavelengths were simultaneously tuned by stretching the TCF, (c) two lasing wavelengths as a function of δz.

For fixed wavelength spacing, the simultaneous tuning characteristic of the dual-wavelength laser was investigated by stretching the TCF-based filter. As discussed in Section 3.3, the transmission spectrum of the TCF-based filter exhibited a 'blue shift' when it was stretched. In contrast, the value of wavelength spacing remained unchanged because the birefringence of the TCF hardly changed. Therefore, we found that the two lasing wavelengths can be simultaneously tuned towards the shorter wavelength with fixed wavelength spacing. Dual-wavelength lasing with wavelength spacing of 0.4, 0.8 and 1.2 nm was linearly tuned over 1.7, 1.0 and 0.8 nm, respectively. Here, we give only the measured results of dual-wavelength lasing with wavelength spacing of 0.4 nm as shown in Figure 9(b). When the TCF was stretched by an increment length of δz = 0.1 mm, two lasing wavelengths simultaneously moved with a step of 0.1 nm, as shown in Figure 9(c).

To study the SOPs of the dual-wavelength output, the laser output was connected to a polarization beam splitter (PBS) through a PC. The x- and y- outputs of the PBS were injected into the OSA to evaluate the dual-wavelength spectrum at two orthogonal polarizations. We adjusted the PC to make the two wavelengths polarize at x-axis and y-axis, respectively. Figure 10(a)–(d) shows the measurement results of dual-wavelength lasing with wavelength spacing of 0.2, 0.4, 0.8 and 1.2 nm. It can be found that the out spectrum at the x-axis or y-axis has only one dominating wavelength, while the other one is substantially suppressed. The cross-polarization suppression between the two wavelengths is more than 30 dB, which means the two wavelengths are orthogonal in polarization and the dual-wavelength laser operated on a single-polarization state.

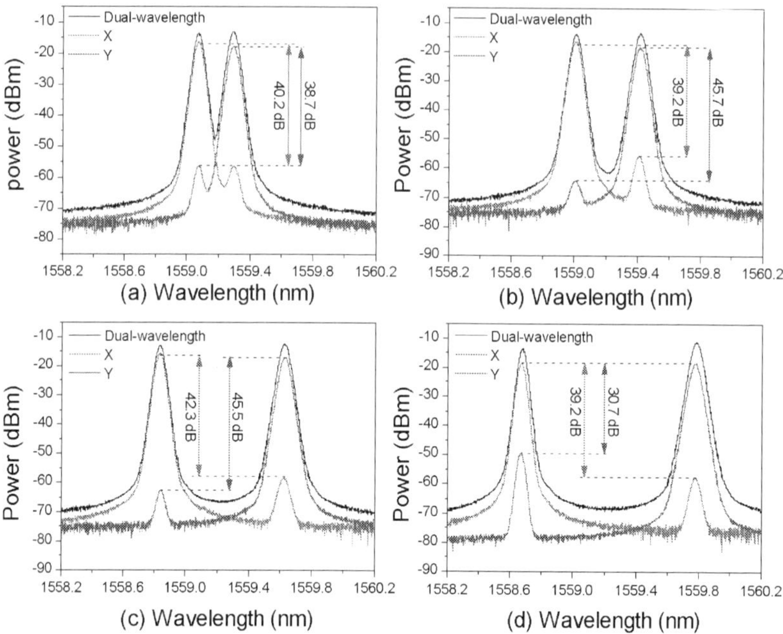

Figure 10. Output spectra of lasers with orthogonal polarization states for wavelength spacing of (a) 0.2 nm, (b) 0.4 nm, (c) 0.8 nm, (d) 1.2 nm.

4.2. Tunable single-wavelength fibre laser

In this work, we proposed and experimentally demonstrated a C-band tunable single-wavelength fibre laser by cascading two TCF-based filters [33]. Figure 8(c) illustrates the cascaded TCF-based filters with TCF lengths of 0.14 and 1.37 m. In this section, we discussed the tuning mechanism by characterizing the two TCF-based filters and investigated the tunable characteristics of the proposed tunable single-wavelength fibre laser by bending one TCF-based filter and stretching the other one.

4.2.1. Tunable mechanism

Figure 11 illustrates the tuning equipment of the cascading TCF-based filters, which consists of three transmission stages. Filter 1 was fixed by the left and middle stages with an initial distance of $L_1 = 0.20$ m, and filter 2 was fixed by the middle and right stages with an initial distance of $L_2 = 1.43$ m. The up and mid panels of Figure 12(a) illustrate the transmission spectra of the two individual TCF-based filters, which indicate that the two filters have an FSR of 28.6 and 2.9 nm. The lower panel of Figure 12(a) illustrates the spectrum of the combined structure. It can be found that the output of the cascading filters is a superposition of a slowly varying envelope with an FSR of 28.6 nm produced by filter 1 and a dense fringe with an FSR of 2.9 nm produced by filter 2.

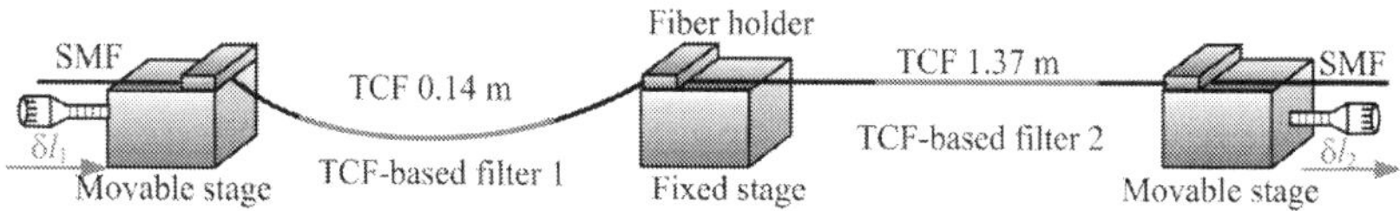

Figure 11. Schematic of the tuning technique of cascading filters.

We coarsely tuned the cascaded spectrum by bending TCF-based filter 1 and finely tuned the spectrum by stretching the TCF-based filter 2. Filter 1 is bent by moving the left stage towards the middle one. The moving distance of the left stage (δl_1) is increased from 0 to 1.9 mm with a step of ~0.25 mm. Figure 12(b) illustrates the evolution of the spectra. It can be found that the dense fringes are nearly located at the same position, and the envelope shifts towards the shorter wavelength. The envelope selects one of the dense peaks as the highest transmission peak (square symbols). The highest transmission peak can be tuned over the wavelength range of 23.2 nm with a coarse step of 2.9 nm which is determined by the FSR of TCF-based filter 2.

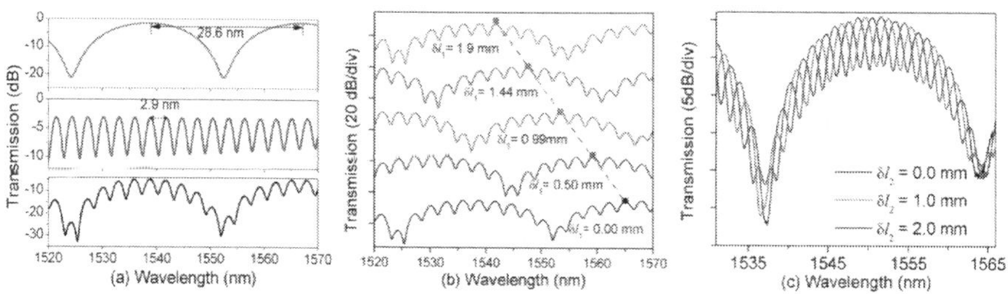

Figure 12. (a) Transmission spectra of TCF-based filter 1 (upper panel), TCF-based filter 2 (mid panel) and cascading filters (lower panel). Transmission spectra evolution of the cascading filters (b) when TCF-based filter 1 is bent and (c) when TCF-based filter 2 is stretched.

In order to realize a fine wavelength tuning, TCF-based filter 2 is stretched by moving the right stage away from the middle one. The moving distance of the right stage (δl_2) is increased from 0 to 3.0 mm with steps of 0.25 mm. Figure 12(c) illustrates the evolution of spectra when we stretched filter 2. It can be found that the envelope almost remains unchanged while the dense fringes shift towards a shorter wavelength. The highest transmission peaks can be finely tuned over the wavelength range of 2.9 nm with a certain fine step which is dependent on the strain applied on filter 2. In conclusion, the envelope and the dense fringes of the transmission spectra can be separately tuned by bending and stretching filters 1 and 2, respectively.

4.2.2. Characterization of tunable laser

The fibre laser was coarsely tuned by bending TCF-based filter 1 and finely tuned by stretching the TCF-based filter. Figure 13(a) illustrates the evolution of lasing spectra when we bent filter 1. It can be found that bending filter 1 makes the envelope shift towards shorter wavelengths and selects the transmission peak of filter 2 for lasing in a series of steps with a separation of 2.9 nm. Figure 13(b) illustrates the evolution of lasing spectra when we stretched filter 2. It was found that the lasing wavelength finely changed with a step of 0.22 nm. Figure 13(c) illustrates the whole tuning process. First, we obtained the initial single-wavelength lasing at 1565.00 nm. Second, we stretched filter 2 from $\delta l_2 = 0$ to 3.0 mm with a step of 0.25 mm and obtained fine lasing wavelengths with a step of ~0.22 nm, e.g. from 1565.04 to 1562.26 nm (black dots). Third, we loosened filter 2 to the initial position and then bent filter 1 with $\delta l_1 = 0.25$ mm (black arrow), and another coarse lasing wavelength was selected for lasing with a separation of 2.9 nm from the last one, e.g. 1562.12 nm. Fourth, the second and third steps were repeated eight times. On further bending filter 1 and stretching filter 2, the lasing wavelength jumped from 1541.8 to 1565.04 nm. Consequently, the wavelength can be linearly tuned over the wavelength range of 23.2 nm (from 1541.8 to 1564.0 nm) with a uniform step of ~0.22 nm. The total number of tunable lasing wavelength is up to 105.

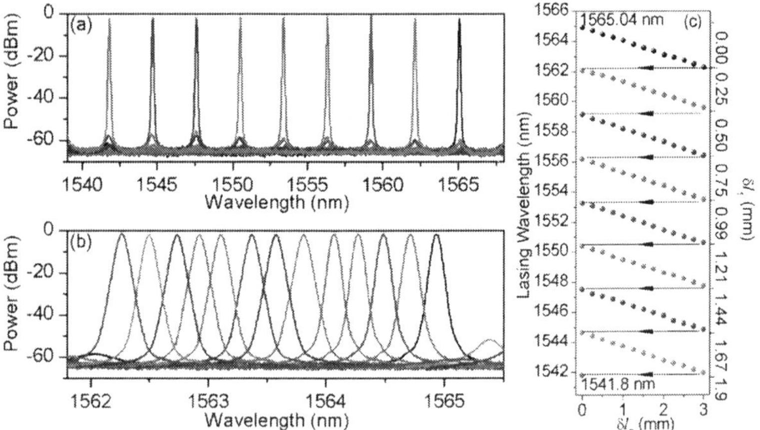

Figure 13. Output of a tunable fibre laser (a) coarse laser lines when TCF-based filter 1 is bent, (b) fine laser lines when TCF-based filter 2 is stretched, (c) all tunable laser wavelength evolution.

When we characterized the laser properties, we increased the pump power from 10 to 500 mW. The cavity loss was around 7 dB including the splicing loss (~4 dB) between the SMF and the TCF and the insertion loss of other optical devices. Such cavity loss gives rise to a threshold pump power of 20 mW. When the pump power was 500 mW, the lasing output saturated at 2.27 dB. Hence, we measured all 105 laser spectra at a pump power of 500 mW. For good visibility, only half of the lasing spectra are plotted in Figure 14. The SMSRs are 53–58 dB, and

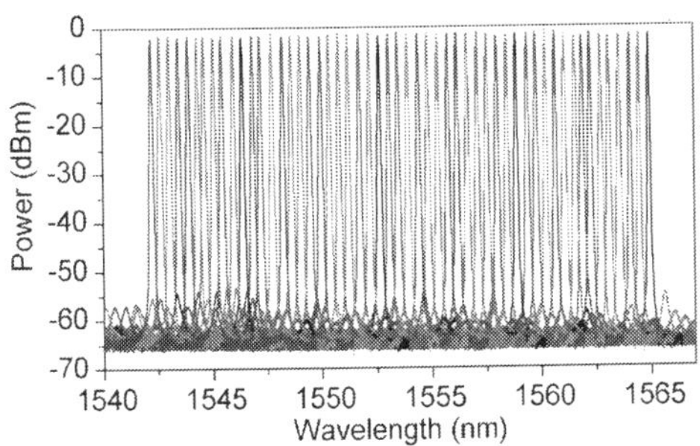

Figure 14. Output of the tunable fibre laser with a range of 23.2 nm and a uniform step of ~0.22 nm.

the linewidth is ~0.05 nm. The intensity variation over the entire tuning range was less than ±0.1 dB.

4.3. Tunable multi-wavelength fibre laser

In this work, we proposed and experimentally demonstrated a tunable multi-wavelength fibre laser based on a non-linear optical loop mirror by using a compound filter [34]. Figure 8(d) illustrates the NOLM and the compound filter. The compound filter was formed by cascading a standard Mach–Zehnder interferometer (MZI) and a TCF-based filter. In this section, we first discussed the function of the NOLM as an amplitude equalizer. Second, we discussed the tuning mechanism by characterizing the transmission spectra of the standard MZI, the TCF-based filter and the compound filter. Third, we investigated the tunable characteristics of the proposed tunable multi-wavelength fibre laser by bending the TCF. Finally, we discussed the influence of the length of the TCF on the laser characteristic.

4.3.1. Non-linear optical loop mirror

An NOLM is formed by splicing together the output ports of a 70:30 coupler. A 2-km SMF is inserted inside the loop. The transmission of the NOLM is given as [23]:

$$T = 1 - 2\rho\left(1 - \rho\right)\left\{1 + \cos\left[\alpha + \left(1 - 2\rho\right)\phi\right]\right\}$$ (14)

where ρ is the splitting ratio of the coupler, $\phi = 2\pi n_3 P L_3 / (\lambda A_{eff})$ is the non-linear phase shift induced by the self-phase modulation and cross-phase modulation, n_3 is the non-linear refractive index coefficient, L_3 is the loop length, A_{eff} is the effective fibre core area, λ is the operating wavelength, P is the input power and α is the additional linear phase difference induced by PC2 and the fibre twist.

According to Eq. (14), the transmission of NOLM is a cosine function of the total phase difference and varies with a change in α. When α is set to certain values by adjusting PC2, the NOLM can be biased at the negative feedback regime, and the transmissivity decreases with the input power. In such a case, the NOLM works as an amplitude equalizer to weaken the homogeneous gain broadening and suppress the mode competition. The multi-wavelength oscillations can be achieved at these particular PC settings.

4.3.2. Tuning mechanism

The standard MZI is established by connecting two 50:50 filters between which the fibre path difference of two arms is ΔL = 4.0 mm. The channel spacing of the standard MZI is ~ 0.4 nm, as shown in the upper panel of Figure 15(a). The mid panel of Figure 15(b) illustrates the transmission spectrum of the TCF-based filter with a TCF length of L_0 =11.5 mm and an FSR of 37 nm. The compound filter is constructed by combining the standard MZI and the TCF-based filter, and its transmission spectrum is shown in the lower panel of Figure 15(c). The fast-varying fine fringe pattern is modulated with a slow-varying one. The higher spatial frequency corresponds to the response of the standard MZI; the red line of the lower spatial frequency corresponds to the response of the TCF-based filter. In other words, the compound filter is a comb filter whose transmission peaks are modified in accordance with the shape of the spectrum of the TCF-based filter.

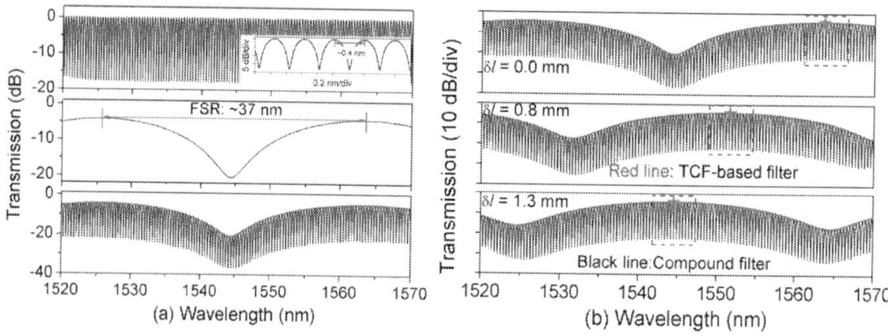

Figure 15. (a) Transmission spectra of the standard MZI (upper panel), the TCF-based filter (mid panel) and the compound filter (lower panel), (b)transmission spectra of the compound filter and the TCF-based filter when the TCF was bent with δl = 0.0, 0.8 and 1.3 mm.

The compound filter was tuned by bending the TCF. The TCF-based filter was fixed by two fibre holders adhered onto two stages with an initial distance of 25 cm. The filter was bent by moving the right stage towards the left one. To quantify the bending of the filter, we recorded the displacement of the moving stage; the initial position of the moving stage is defined as the zero point δl = 0.0 mm. Figure 15(b) illustrates the transmission spectra of the TCF-based filter (red lines) and the compound filter (black lines) when δl = 0.0, 0.8 and 1.3 mm. It can be found that the transmission peak (pentagram symbols) of TCF-based filters exhibits a 'blue shift' when δl is gradually increased. Because the transmissivity of the individual transmission peak of the standard MZI is modulated by the variation spectra of the TCF-based filter, the envelopes of the spectra of the compound filter also exhibit a 'blue shift'. In other words, the compound filter is tuned by moving its envelope towards the short-wavelength direction by bending the TCF.

The principle of the proposed tunable multi-wavelength fibre laser can be expressed as follows. The ASE spectrum of the EDF is reshaped to a comb-like transmission spectrum by the standard MZI. The intensities of the transmission peaks are modified by the TCF-based filter. Because lasing is established once the cavity loss is smaller than the EDF gain, only the peaks with high transmissivity (low loss) can oscillate. The transmission peak of TCF-based filter has the lowest insertion loss; hence, only the near region of the transmission peak would be selected for lasing simultaneously, as shown in the blue region of Figure 15(b). The wavelength range of the blue region is determined by the balance between the EDF gain profile and the profile of the total cavity loss. The NOLM can work as an amplitude equalizer to enable the multiple transmission peaks to emit as lasing wavelengths. When the TCF is bent, the transmission band of the TCF-based filter shifts towards the short-wavelength direction to cover the different channels provided by the standard MZI. Thus, the corresponding multiple lasing lines shift towards the short-wavelength direction. As a result, a tunable multi-wavelength fibre laser is realized by bending the TCF.

4.3.3. Laser tunable characterization

To study the tuning characteristic, the moving stage was adjusted with a step of 0.05 mm. Figure 16(a) illustrates that the lasing wavebands can be continuously tuned to cover a range of 24 nm from 1542 to 1566 nm. The lasing spectra were measured when the TCF was bent with δl = 0.0, 0.3, 0.8 and 1.3 mm (labelled 1–4, respectively). In the 3-dB spectral range, the lasing line amount in each lasing spectrum is around 12 peaks at least and 19 peaks at most.

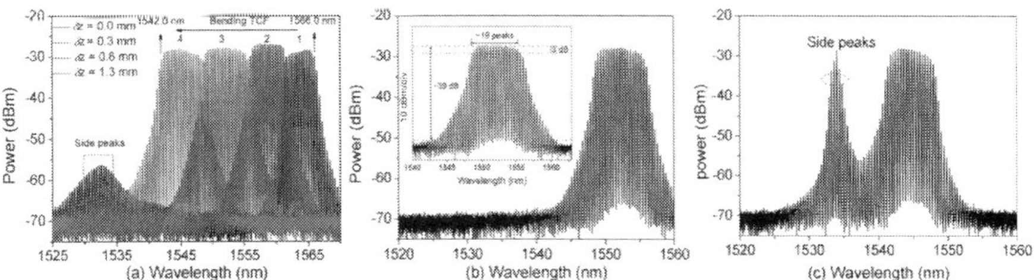

Figure 16. Lasing spectra of the multi-wavelength fibre laser (a) tunable characteristic, (b) the TCF was bent with δl = 0.8 mm and (c) δl =1.35 mm.

When δl = 0.0 mm, because the transmission peaks around 1532 nm have considerable intensities, the lasing spectrum 1 exhibits side modes around 1532 nm. Because the total fixed gain is shared by both side peaks and the lasing peaks, the lasing line amount is limited to 12 peaks in the lasing spectrum 1. When the TCF was bent, the transmission band of the TCF-based filter shifted towards the shorter wavelength direction and supressed the transmission peaks around 1532 nm, as shown in Figure 15(b). When δl is in the range from 0.4 to 0.9 mm, the transmission dip of the TCF-based filter locates around 1532 nm, resulting in transmission peaks with relatively low transmissivities. Thus, the side peaks can be efficiently suppressed. Taking δl = 0.8 mm, for example, the spectrum of the compound filter is shown in Figure 15(b), and the corresponding lasing spectrum is shown in Figure 16(b). The lasing lines are enlarged in the inset of Figure 16(b). It can be found that 19 lasing lines are located in the 3-dB spectral range. All lasing lines have an SMSR of 39 dB, a peak power of −28 dBm and a channel spacing of 0.4 nm. When we increased δl to 1.3 mm, the transmissivities of the peaks around 1532 nm increased as shown in Figure 15(b), which induced an increment of the side peaks in the laser output, as shown in the laser spectrum 4 of Figure 16(a). When we continually increased δl to 1.35 mm, the side peaks increased dramatically, as shown in Figure 16(c). As a result, we concluded that the side peaks induced by the EDF peak gain mainly limited the tuning range of the proposed laser.

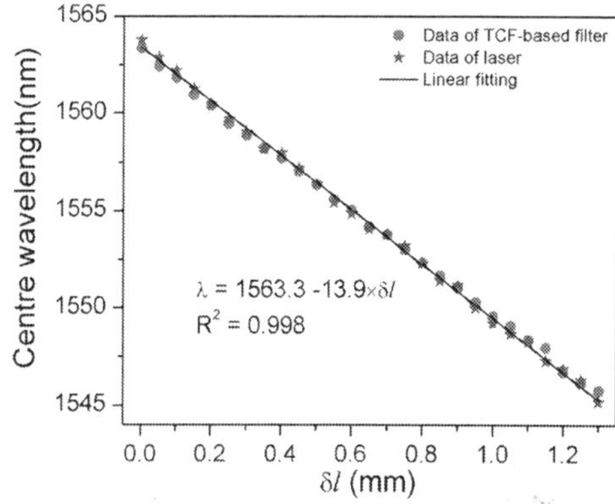

Figure 17. Centre of the lasing waveband and the transmission peak of the TCF-based filter as a function of δl.

In order to find out the relationship between the transmission spectrum of the TCF-based filter and multi-wavelength lasing spectrum, we plotted both the transmission peak of the TCF-based filter (blue pentagram symbols) and the centre of the lasing waveband (red circle symbols) as a function of δl, as shown in Figure 17. It seems that the centre of the lasing waveband tracks the transmission peak of the TCF-based filter. This point proves that the tuning mechanism of the proposed tunable fire laser takes advantage of the variation transmission band of the TCF-based filter to select different channels of the standard MZI for realizing a tunable lasing waveband.

4.3.4. Influence of the TCF length on the tuning characteristic

In order to investigate the influence of the TCF length on the tuning characteristic, we fabricated two more filters with a TCF length of L_0 = 5.0 and 22 cm. Figure 18(a) illustrates the tuning characteristic of the multi-wavelength fibre based on the TCF-based filter with a TCF length of $L_0 = 5.0$ cm. The results indicate that the lasing waveband covers a range of 15.2 nm from 1548.0 to 1563.2 nm. The lasing line amount in each lasing spectrum is around 18 peaks at the lowest and 25 peaks at the highest. Compared with the case of $L_0 = 11.5$ cm, the tuning range decreased by 8.8 nm, and the lasing line amount increased by six peaks. To explain the difference, we measured the transmission spectrum of the TCF-based filter with $L_0 = 5.0$ cm, as shown in the inset of Figure 18(a). It can be found that the FSR of the filter is 84 nm which means it is difficult for such a large FSR to suppress the side peaks around 1532 nm. Therefore, the tuning range limited by the side peaks is decreased when we increased the length of the TCF. On the contrary, the large FSR also means a flat transmission waveband, which increased the lasing line amount.

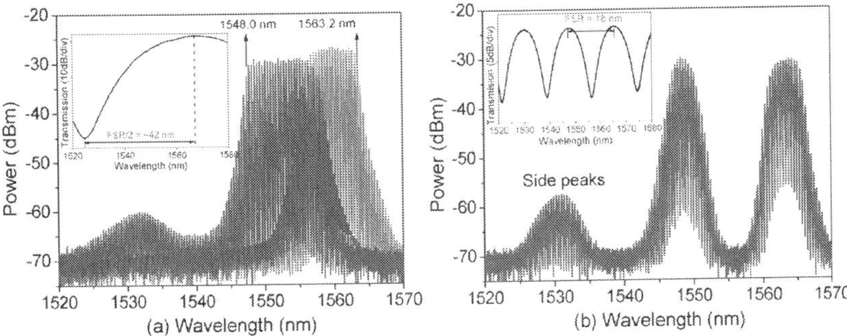

Figure 18. Lasing spectra of the proposed tunable multi-wavelength fibre laser with TCF lengths (a) $L_0 = 5$ cm and (b) $L_0 = 22$ cm. Inset: the transmission spectra of the TCF-based filter.

We carried out a reverse experiment by using a large length of TCF with $L_0 = 22$ cm. Figure 18(b) illustrated the lasing spectrum. It can be found that two lasing wavebands oscillated simultaneously. To explain the two lasing wavebands, we also measured the transmission spectrum of the TCF-based filter, as shown in the inset of Figure 18(b). It can be found that two transmission wavebands are located in the spectrum, which results in two lasing wavebands. Because the two bands shared the EDF gain, the SMSR of the lasing spectrum is drastically decreased to 21 dB. Summarizing the multi-wavelength fibre laser with TCF lengths of $L_0 = 5.0$, 11.5 and 22 cm, we can conclude that there is balance between the tuning range and the lasing line amount when we choose the length of the TCF. The shorter the TCF length, the narrower the tuning range and the larger the number of the lasing wavelengths.

5. Conclusion

In this chapter, we studied tunable single-, dual- and multi-wavelength fibre lasers by using TCF-based filters. First, we presented the coupled-mode theory of the TCF with symmetric and anti-symmetric mode and deduced the output of the TCF-based filter on the basis of the

SOP of input light. Second, we experimentally and numerically characterized the optical properties of TCF-based filter including the free spectral range, polarization dependence, strain effect and bending effect. These optical properties are closely related to the tuning mechanism and tuning characteristics of the tunable fibre lasers. Third, we implemented three tunable fibre lasers which operate at single-, dual- and multi-wavelengths, respectively. In the case of a tunable single wavelength, the wavelength can be linearly tuned over 23.2 nm (from 1541.8 to 1564.0 nm), with a uniform step of ~0.22 nm. The total number of tunable wavelength is up to 105. In the case of a tunable dual-wavelength fibre laser, the single-polarized two wavelengths can be linearly tuned by stretching the TCF. The spacing of the two wavelengths can be tuned from 0.1 to 1.2 nm by adjusting the PC located before the TCF-based filter. In the case of a tunable multi-wavelength fibre laser, 19 wavelengths were tuned over 24 nm from 1542 to 1566 nm. After discussing the influence of the TCF length on the tunable multi-wavelength fibre laser, it is concluded that there is balance between the tuning range and the wavelength amount when we choose the length of the TCF.

Acknowledgements

This work was supported by National Natural Science Foundation of China (grant no. 61405128) and China Postdoctoral Science Foundation funded project (grant nos. 2014M552227 and 2015T80913).

Author details

Guolu Yin[1], Xin Wang[2], Shuqin Lou[2] and Yiping Wang[1]

*Address all correspondence to: guoluyin@gmail.com

1 Key Laboratory of Optoelectronic Devices and Systems of Ministry of Education and Guangdong Province, College of Optoelectronic Engineering, Shenzhen University, Shenzhen, China

2 School of Electronic and Information Engineering, Beijing Jiaotong University, Beijing, China

References

[1] Xiao F, Alameh K, Lee T. Opto-VLSI-based tunable single-mode fiber laser. Optics Express. 2009; 17(21): 18676–18680. DOI: 10.1364/OE.17.018676

[2] Sinefeld D, Marom DM. Tunable fiber ring laser with an intracavity high resolution filter employing two-dimensional dispersion and LCoS modulator. Optics Letters. 2012; 37(1): 1–3. DOI: 10.1364/OL.37.000001

[3] Song YW, Havstad SA, Starodubov D, Xie Y, Willner AE, Feinberg J. 40-nm-wide tunable fiber ring laser with single-mode operation using a highly stretchable FBG. IEEE Photonics Technology Letters. 2001; 13(11): 1167–1169. DOI: 10.1109/68.959352

[4] Babin SA, Kablukov SI, Vlasov AA. Tunable fiber Bragg gratings for application in tunable fiber lasers. Laser Physics. 2007; 17(11): 1323–1326. DOI: 10.1134/s1054660x07110096

[5] He X, Fang X, Liao C, Wang D, Sun J. A tunable and switchable single-longitudinal mode dual-wavelength fiber laser with a simple linear cavity. Optics Express. 2009; 17(24): 21773–21781. DOI: 10.1364/OE.17.021773

[6] Feng SC, Xu O, Lu SH, Mao XQ, Ning TG, Jian SS. {PT ArticleTitle}Single-polarization, switchable dual-wavelength erbium-doped fiber laser with two polarization-maintaining fiber Bragg gratings. Optics Express. 2008; 16(16): 11830–11835. DOI: 10.1364/OE.16.011830

[7] Feng X, Liu Y, Yuan S, Kai G, Zhang W, Dong X. L-Band switchable dual-wavelength erbium-doped fiber laser based on a multimode fiber Bragg grating. Optics Express. 2004; 12(16): 3834–3839. DOI: 10.1364/OPEX.12.003834

[8] Villanueva GE, Perez-Millan P, Palaci J, Cruz JL, Andres MV, Marti J. Dual-wavelength DFB erbium-doped fiber laser with tunable wavelength spacing. IEEE Photonics Technology Letters. 2010; 22(4): 254–256. DOI: 10.1109/lpt.2009.2038594

[9] Wang XZ, Li Y, Bao XY. Tunable ring laser using a tapered single mode fiber tip. Applied Optics. 2009; 48(35): 6827–6831. DOI: 10.1364/AO.48.006827

[10] Wang XZ, Li Y, Bao XY. C- and L-band tunable fiber ring laser using a two-taper Mach-Zehnder interferometer filter. Optics Letters. 2010; 35(20): 3354–3356. DOI: 10.1364/OL.35.003354

[11] Yin GL, Wang XZ, Bao XY. Effect of beam waists on performance of the tunable fiber laser based on in-line two-taper Mach-Zehnder interferometer filter. Applied Optics. 2011; 50(29): 5714–5720. DOI: 10.1364/AO.50.005714

[12] Antonio-Lopez JE, Castillo-Guzman A, May-Arrioja DA, Selvas-Aguilar R, LiKamWa P. Tunable multimode-interference bandpass fiber filter. Optics Letters. 2010; 35(3): 324–326. DOI: 10.1364/OL.35.000324

[13] Castillo-Guzman A, Antonio-Lopez JE, Selvas-Aguilar R, May-Arrioja DA, Estudillo-Ayala J, LiKamWa P. Widely tunable erbium-doped fiber laser based on multimode interference effect. Optics Express. 2010; 18(2): 591–597. DOI: 10.1364/OE.18.000591

[14] Sun GY, Zhou YW, Hu YH, Chung Y. Broadly tunable fiber laser based on merged Sagnac and intermodal interferences in few-mode high-birefringence fiber loop mir-

ror. IEEE Photonics Technology Letters. 2010; 22(11): 766–768. DOI: 10.1109/lpt. 2010.2044991

[15] Zhou YW, Sun GY. Widely tunable erbium-doped fiber laser based on superimposed core-cladding-mode and Sagnac interferences. IEEE Photon Journal. 2012; 4(5): 1504–1509. DOI: 10.1109/jphot.2012.2211074

[16] Zhang J, Qiao XG, Liu F, Weng YY, Wang RH, Ma Y, Rong QZ, Hu ML, Feng ZY. A tunable erbium-doped fiber laser based on an MZ interferometer and a birefringence fiber filter. Journal of Optics. 2012; 14(1): 015402. DOI: 10.1088/2040-8978/14/1/015402

[17] Luo AP, Luo ZC, Xu WC. Tunable and switchable multiwavelength erbium-doped fiber ring laser based on a modified dual-pass Mach-Zehnder interferometer. Optics Letters. 2009; 34(14): 2135–2137. DOI: 10.1364/OL.34.002135

[18] Harun SW, Parvizi R, Shahi S. Multi-wavelength erbium-doped fiber laser assisted by four-wave mixing effect. Laser Physics Letters. 2009; 6(11): 813–815. DOI: 10.1002/ lapl.200910072

[19] Wang ZY, Cui YP, Yun BF, Lu CG. Multiwavelength generation in a Raman fiber laser with sampled Bragg grating. IEEE Photonics Technology Letters. 2005; 17(10): 2044–2046. DOI: 10.1109/LPT.2005.856430

[20] Zhang Z, Zhan L, Xia YX. Tunable self-seeded multiwavelength Brillouin-erbium fiber laser with enhanced power efficiency. Optics Express. 2007; 15(15): 9731–9736. DOI: 10.1364/OE.15.009731

[21] Peng WJ, Yan FP, Li Q, Yin GL, Feng SC, Feng T, Tan SY. Tunable self-seeded multiwavelength Brillouin–erbium fiber laser using an in-line two-taper Mach–Zehnder interferometer. Optics & Laser Technology. 2013; 45: 348–351. DOI: 10.1016/j.optlastec.2012.06.025

[22] Tran TVA, Lee K, Lee SB, Han YG. Switchable multiwavelength erbium doped fiber laser based on a nonlinear optical loop mirror incorporating multiple fiber Bragg gratings. Optics Express. 2008; 16(3): 1460–1465. DOI: 10.1364/OE.16.001460

[23] Yin GL, Lou SQ, Zou H. A multiwavelength Er-doped fiber laser using a nonlinear optical loop mirror and a twin-core fiber-based Mach-Zehnder interferometer. Laser Physics Letters. 2013; 10(4): 045103. DOI: 10.1088/1612-2011/10/4/045103

[24] Wang LW, Lou SQ, Chen WG, Li HL. A novel method of rapidly modeling optical properties of actual photonic crystal fibres. Chinese Physics B. 2010; 19(8): 084209. DOI: 10.1088/1674-1056/19/8/084209

[25] Huang WP. Coupled-mode theory for optical waveguides: an overview. Journal of the Optical Society of America A. 1994; 11(3): 963–983. DOI: 10.1364/JOSAA. 11.000963

[26] Yin GL, Lou SQ, Wang X, Han B. Dual-wavelength erbium-doped fiber laser with tunable wavelength spacing using a twin core fiber-based filter. Journal of Optics. 2014; 16(5): 055404. DOI: 10.1088/2040-8978/16/5/055404

[27] Snyder AW. Coupled-mode theory for optical fibers. Journal of the Optical Society of America. 1972; 62(11): 1267–1277. DOI: 10.1364/JOSA.62.001267

[28] Digonnet M, Shaw HJ. Wavelength multiplexing in single-mode fiber couplers. Applied Optics. 1983; 22(3): 484–491. DOI: 10.1364/AO.22.000484

[29] Zou Y, Dong X, Lin G, Adhami R. Adhami R: Wide range FBG displacement sensor based on twin-core fiber filter. IEEE/OSA Journal of Lightwave Technology. 2012; 30(3): 337–343. DOI: 10.1109/JLT.2011.2181334

[30] Nagano K, Kawakami S, Nishida S. Change of the refractive index in an optical fiber due to external forces. Applied Optics. 1978; 17(13): 2080–2085. DOI: 10.1364/AO. 17.002080

[31] Yin GL, Lou SQ, Lu WL, Wang X. A high-sensitive fiber curvature sensor using twin core fiber-based filter. Applied Physics B. 2014; 115(1): 99–104. DOI: 10.1007/ s00340-013-5578-z

[32] Shao LY, Laronche A, Smietana M, Mikulic P, Bock WJ, Albert J. Highly sensitive bend sensor with hybrid long-period and tilted fiber Bragg grating. Optics Communications. 2010; 283(13): 2690–2694. DOI: 10.1016/j.optcom.2010.03.013

[33] Yin GL, Lou SQ, Hua P, Wang X, Han B. Tunable fiber laser by cascading twin core fiber-based directional couplers. IEEE Photonics Technology Letters. 2014; 26(22): 2279–2282. DOI: 10.1109/lpt.2014.2351808

[34] Yin GL, Lou SQ, Wang X, Han B. Tunable multi-wavelength erbium-doped fiber laser by cascading a standard Mach–Zehnder interferometer and a twin-core fiber-based filter. Laser Physics Letters. 2013; 10(12): 125110. DOI: 10.1088/1612-2011/10/12/125110

Utilization of Reflective Semiconductor Optical Amplifier (RSOA) for Multiwavelength and Wavelength-Tunable Fiber Lasers

Yeh Chien-Hung and Chow Chi-Wai

Additional information is available at the end of the chapter

http://dx.doi.org/10.5772/61752

Abstract

In this chapter, there are three sections to demonstrate the reflective semiconductor optical amplifier (RSOA)-based fiber laser architectures for multiwavelength and wavelength-tunable operations. In the first section, we introduce an L-band multiwavelength laser by utilizing a C-band RSOA with a linear cavity, which is produced by a polarization controller (PC), an optical coupler (OCP), and a reflected fiber mirror (RFM). In the proposed RSOA laser scheme, two to seven wavelengths could be lased and created simultaneously in the L-band range, while the RSOA operates at various bias currents.

In the second section, we demonstrate a multiwavelength laser source by utilizing a C-band RSOA with dual-ring fiber cavity. In the measurement, the laser cavity is constructed by an RSOA, a 1×2 OCP, a 2×2 OCP, and a PC, respectively. Thus, 13–18 wavelengths around L-band could be produced simultaneously, as the bias current of C-band RSOA is driven at 30–70 mA.

Keywords: RSOA, SOA, Multiwavelength, Wavelength-tunable

1. Introduction

Recently, the wavelength-tunable and multiwavelength fiber laser schemes are very essential for several optical applications, such as the wavelength-division-multiplexed (WDM) communication systems [1], optical testing and measurement [2], fiber optics sensor [3], and radio-frequency (RF) photonics [4]. Therefore, the different laser resonators for the multiwavelength lasing have been proposed and investigated, such as utilizing the dual Sagnac loop [5], fiber ring cavity [6], compound ring scheme [7], and Fox–Smith ring cavity [8].

To achieve wavelength lasing, several gain media approaches have been proposed and utilized, such as the erbium-doped fiber amplifier (EDFA) [9], semiconductor optical amplifiers (SOA) [5,10], reflective semiconductor optical amplifiers (RSOA) [11], Raman amplifier [12], and hybrid optical amplifier [13]. However, SOA would be a better option over EDFA due to the nature of inhomogeneous broadening. As a result, the past SOA-based lasers have been also proposed and investigated to generate multiwavelength output at room temperature [14, 15].

Generally, to accomplish a wavelength-tunable output with single longitudinal mode (SLM), the several wavelength selective elements (optical filters) and laser architectures have been used inside ring cavity. The proposed fiber lasers could be designed in linear or ring architectures [16, 17]. Besides, the fiber Fabry–Perot filter (FFPF), tunable bandpass filter (TBF), and fiber Bragg grating (FBG) could be utilized inside the fiber ring cavity to select different wavelength lasing. In fact, owing to the limited gain amplification of the amplifiers, the output power would drop rapidly on both sides of their operating bandwidth [16–18].

In this chapter, there are three sections to demonstrate the RSOA-based fiber laser architectures for multiwavelength and wavelength-tunable operations. In section 2, we propose an L-band multiwavelength laser by utilizing a C-band RSOA with a linear cavity, which is produced by a polarization controller (PC), an optical coupler (OCP), and a reflected fiber mirror (RFM). In the proposed RSOA laser scheme, two to seven wavelengths could be lased and created simultaneously in L-band range, while the bias-current of RSOA is operated at various bias currents.

In section 3, we demonstrate a multiwavelength laser source by utilizing a C-band RSOA with dual-ring fiber cavity. In the measurement, the laser cavity is constructed by an RSOA, a 1×2 OCP, a 2×2 OCP, and a PC, respectively. Thus, 13–18 wavelengths around the L-band could be produced simultaneously, as the bias current of C-band RSOA is driven at 30–70 mA.

In section 4, we investigate the wavelength-tunable fiber ring laser architecture by using the RSOA and SOA. Here, the wavelength tuning range of 1538.03–1561.91 nm can be obtained. The measured output power and optical signal-to-noise ratios (OSNRs) of the proposed fiber laser are between –0.8 and –2.5 dBm and 59.1 and 61.0 dB/0.05 nm, respectively. The power and wavelength stabilities of the proposed laser are also studied. Besides, the proposed laser can be directly modulated at 2.5 Gbit/s quadrature phase shift keying-orthogonal frequency division multiplexing (QPSK-OFDM) signal. In addition, 20–50 km single-mode fiber (SMF) transmission distances are also achieved within the forward error correction (FEC) limit without dispersion compensation. We believe that it can be a cost-effective and promising candidate for next-generation wavelength-division multiplexed-passive optical network (WDM-PON).

2. Linear cavity for multiwavelength lasing

In the section, we present and demonstrate using the C-band RSOA with a linear fiber cavity to produce multiwavelength in the L-band range at room temperature [19]. Hence, two to seven lasing wavelengths could be observed by adjusting the DC bias currents of the RSOA. Fig. 1 shows the experimental setup of the proposed multiwavelength fiber laser scheme. The laser is constructed by an RSOA, a PC, a 1×2 and 50:50 OCP, and an RFM with 95% reflection in C + L bands. In the experiment, the bias current range of RSOA (*produced by CIP*) is between 30 and 80 mA. In addition, the PC can be used to maintain the polarization state and obtain a maximum output power. Moreover, the output wavelength spectrum is observed by an optical spectrum analyzer (OSA) with a resolution of 0.01 nm.

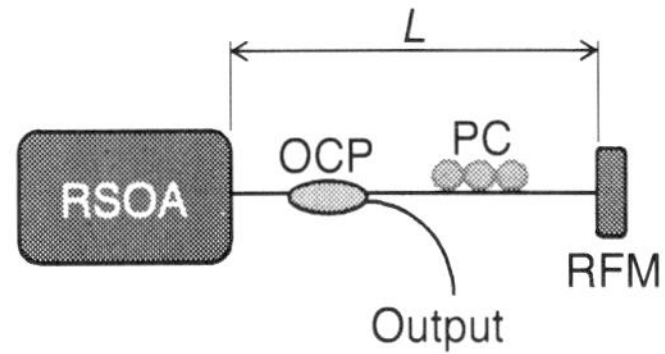

Figure 1. Experimental setup of the proposed fiber laser scheme in linear cavity scheme.

Hence, first we measure and observe the output amplified spontaneous emission (ASE) profile of RSOA in the bias currents of 30–80 mA. As illustrated in Fig. 2, when we increase the DC, the observed output power of the ASE also raises. And the center wavelength of the ASE would drift to shorter wavelength locations. In addition, the gain spectrum of the RSOA is observed at C-band in the entire operated currents.

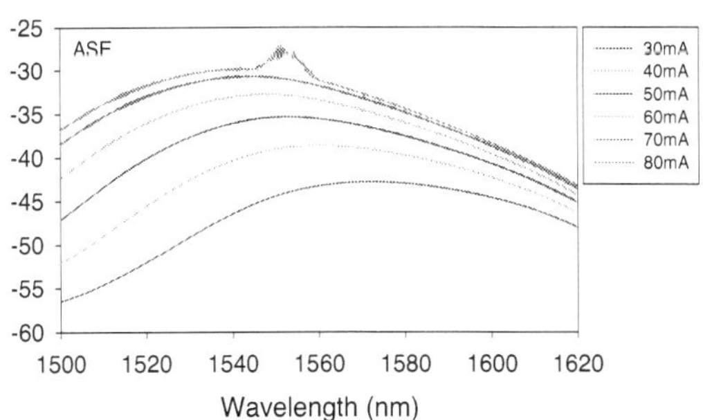

Figure 2. The output ASE spectra of RSOA at the operated currents of 30–80 mA.

In the proposed RSOA-based laser scheme, multiwavelength could be produced under the accurate polarization statement. Hence, Fig. 3 shows the observed output spectra of proposed multiwavelength RSOA laser at the bias currents of 30–80 mA. In the measurement, we obtain that seven wavelengths are generated simultaneously as the operated current of RSOA is larger than 40 mA. As a result, the threshold current of the laser is around 40 mA in the proposed laser scheme. While the operated currents are 80, 70, 60, 50, 40, and 30 mA, respectively, the

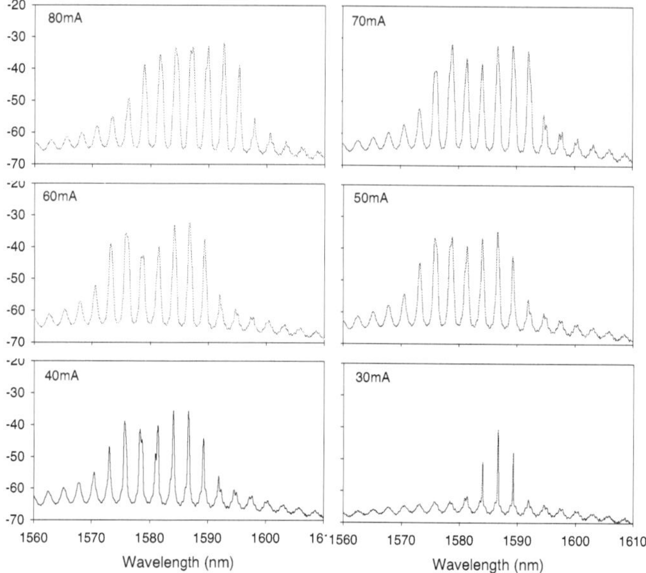

Figure 3. Output spectra of proposed fiber laser at the bias currents of 30–80 mA.

lasing wavelengths are also observed at 1592.62, 1578.88, 1586.74, 1586.68, 1586.68, and 1586.68 nm. Previous researches presented that in high input power, the SOA would show a peak gain drift toward longer wavelength range [20]. Thus, the proposed multiwavelength RSOA lasers are produced in the L-band window rather than the C-band in the lasing condition under a high feedback power. As shown in Fig. 3, the mode spacing of proposed laser is nearly 2.7 nm at various bias currents. Furthermore, when we increase the bias current of RSOA gradually, the lasing multiwavelength also drifts to longer wavelength range, as seen in Fig. 3.

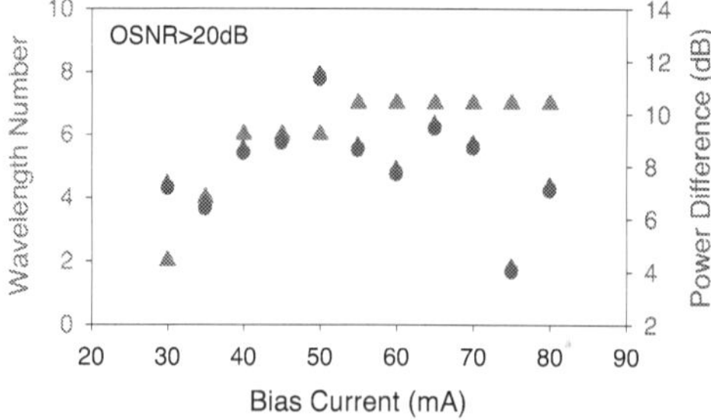

Figure 4. The number of lasing wavelength in the proposed laser, when the OSNR is larger than 20 dB.

Fig. 4 presents the number of lasing multiwavelength and the peak power difference among these wavelengths as the OSNR is larger than 20 dB. When lasing wavelengths larger than six are required, the bias current must be larger than 40 mA. Here, Fig. 4 displays the power variation (ΔP) of lasing multiwavelength at the different operated currents. The maximum and minimum power differences of 11.3 and 3.9 dB are observed, respectively, under the current of 50 and 75 mA. Besides, as seen in Fig. 5, we also execute the output power of proposed RSOA laser under a bias current range of 25–80 mA. According to the measured result, the observed output powers are between 0.001 and 0.223 mW.

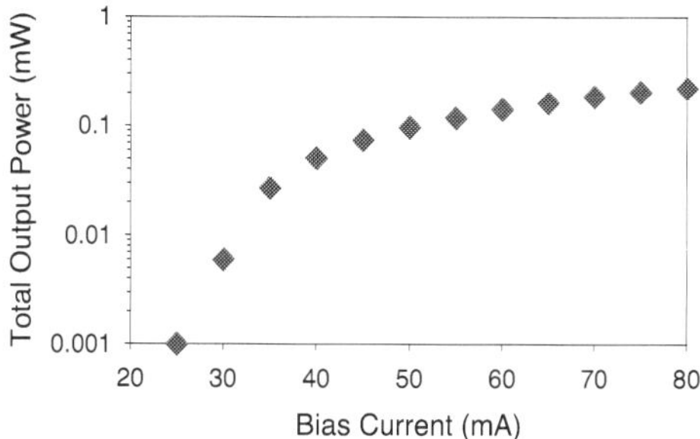

Figure 5. Total output power of proposed fiber laser at the bias currents of 25–80 mA.

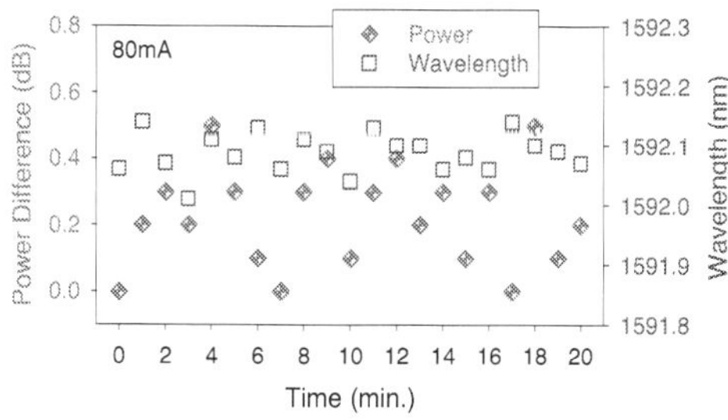

Figure 6. Wavelength and power variations of the proposed multiwavelength fiber laser in an observing time of 20 minutes.

Finally, the output stability testing of proposed multiwavelength RSOA laser is made to execute the output performance of power and wavelength. In this experiment, the RSOA is operated at 80 mA current. Besides, we select one of the lasing lightwaves at 1592.06 nm initially for measurement. Here, the observation time is set at 20 minutes. As seen in Fig. 6, the output power and wavelength variations are measured at ±0.25 dB and ±0.065 nm, respectively. Hence, the results present that the proposed multiwavelength RSOA laser has the excellent optical output stabilities for future system application. In addition, under 2 hours observing time, the measured output stabilities of the proposed laser are still maintained.

3. Dual-ring cavity for multiwavelength lasing

In this section, we introduce and demonstrate the multiwavelength fiber laser source by employing a C-band RSOA with a dual-ring fiber cavity [21]. Thus, experimental setup of the proposed multiwavelength RSOA-based laser scheme with a dual-ring cavity design is shown in Fig. 7. The proposed laser is composed of a C-band RSOA, a PC, a 1×2 OCP, a 2×2 OCP, and an optical circulator (OC). Here, two OCPs could create a dual-ring cavity of proposed fiber laser scheme. The OC is used to generate a counterclockwise direction. The PC, which is placed in the ring cavity, is used to maintain the polarization state and obtain a maximum output power.

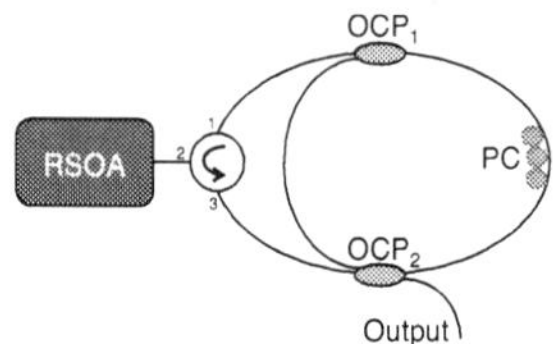

Figure 7. Experimental setup of the proposed RSOA-based fiber ring laser.

In the proposed RSOA-based dual-ring scheme laser, the generated multiwavelength would be obtained under a proper polarization tuning. Here, Fig. 8 displays the observed output spectra of proposed RSOA laser in the bias current range of 30–70 mA. As seen in Fig. 8, when the OSNR of each wavelength is larger than 20 dB and the bias current of RSOA is 30 mA, the proposed RSOA-based laser can emit ~14 wavelengths simultaneously. The multiwavelength output range could be observed at the wavelengths of 1588.48–1612.71 nm. Besides, if the bias current of RSOA is less than 30 mA, the proposed laser cannot generate any wavelength. Thus, the threshold current of the proposed fiber laser is around 30 mA. As illustrated in Fig. 8, while the bias current of RSOA is 40, 50, 60, and 70 nm, respectively, the output wavelength range of proposed RSOA laser is observed at 1588.48–1612.71 nm, 1580.74–1613.70 nm, 1584.22–1608.37 nm, and 1580.81–1614.34 nm.

In this study, the maximum peak powers of the output wavelengths are obtained at 1592.18, 1593.95, 1593.81, 1604.68, and 1610.65 nm, while the bias currents of ROSA are 30, 40, 50, 60, and 70 mA, respectively. Previous studies presented that under high input power, the SOA would show a peak gain drift toward longer wavelength range [20]. Thus, the proposed multiwavelength RSOA-lasers are produced in the L-band window rather than the C-band in the lasing condition under a high feedback power. Moreover, if the bias current of RSOA increases gradually, the output multiwavelength also drifts to a longer wavelength range. Moreover, the measured mode spacing is nearly 1.9 nm in the proposed fiber laser at different operating current ranges.

Then, Fig. 9 presents the number of lasing wavelengths under the OSNR of >20 dB. In this measurement, when the various bias currents of 30–70 mA are used, 13–18 multiwavelengths

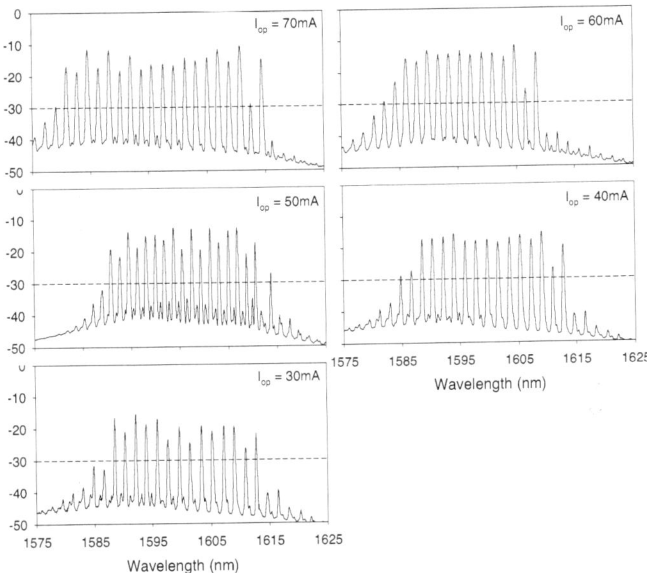

Figure 8. Output spectra of proposed multiwavelength laser under the bias currents of 30–70 mA.

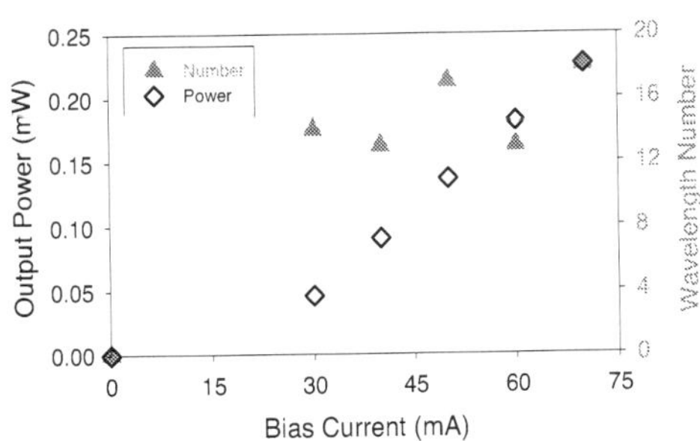

Figure 9. Measured number of lasing wavelengths and total output power of the proposed laser under different bias currents.

could be also measured. Besides, we also execute the output power of proposed RSOA-based laser under the bias current range of 30–70 mA as shown in Fig. 9. In addition, the output powers are observed at the range of 0.046–0.226 mW.

To verify the performance of output power and output wavelength, a short-term stability test of proposed multiwavelength laser is executed. In the measurement, the RSOA is set at 70 mA.

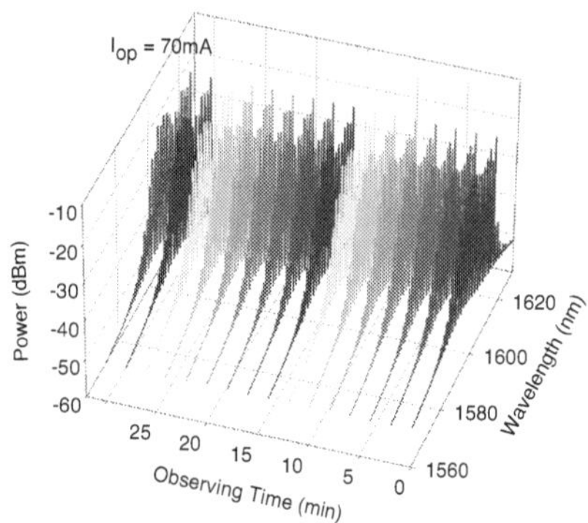

Figure 10. Output spectra of proposed RSOA-based laser under an observation of 30 minutes.

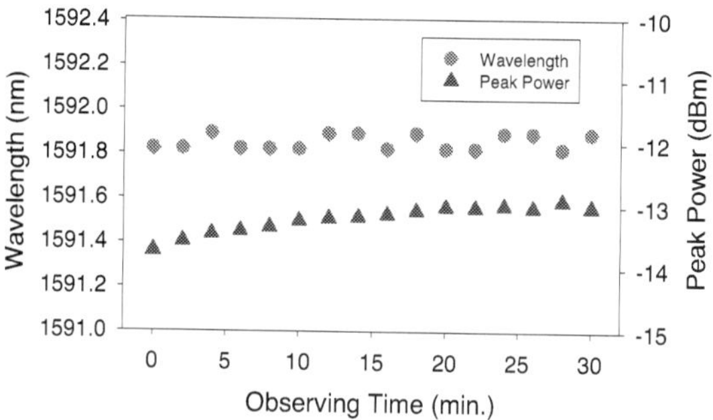

Figure 11. Output wavelength and the power fluctuations of the proposed fiber laser in an observing time of 30 minutes.

Then, Fig. 10 shows the output stability of proposed RSOA-based laser under an observation time of 30 minutes. Initially, one of lasing wavelength is chosen at 1591.82 nm with peak power −13.9 dBm for stability measurement. As illustrated in Fig. 11, the output power and wavelength fluctuations of 0.4 dB and 0.07 nm are completed, respectively, in an observing time of 30 minutes. As a result, experimental results show that the proposed fiber laser has excellent output stabilities. In addition, after 2-hour observing measurement, the measured output stabilities of the proposed laser are still maintained.

4. Wavelength-tunable laser in single mode

In the final section, a stable and continuous wavelength-tuning ring laser scheme using a C-band RSOA and a C-band SOA in the wavelength-tuning range of 1538.03–1561.91 nm is proposed and demonstrated for serving as the upstream transmitter (Tx) of colorless optical

network unit (ONU) [22]. Here, Fig. 12 presents the experimental setup of the proposed wavelength-tunable fiber ring laser scheme. The fiber laser consists of a C-band SOA, a C-band RSOA, a three-port optical circulator (OC), a 1×2 OCP, a PC, and a C-band tunable bandpass filter (TBF). The SOA is commercially available (*GIP, OAU116BB128001A*). Its gain is around 9 dB. The SOA is fixed at 120 mA pumping current and the bias current of RSOA is operated at 50, 60, 70, and 80 mA, respectively, in the experiment. The TBF is utilized inside a ring cavity for wavelength selection. The 3-dB bandwidth, tuning range, and insertion loss of TBF are 0.4 nm, 30 nm (1530–1560 nm), and ~6 dB, respectively. The PC is employed to vary the polarization status for maintaining the maximum output power. As seen in Fig. 12, a ring laser cavity is formed in which the gain media are the SOA and RSOA. Moreover, the TBF selects the desired wavelength within the gain bandwidth of the SOA. Then the signal will launch into the RSOA, which is modulated directly by data; hence, the optical signal can be modulated. Then, the modulated optical signal is observed at the output port of the laser via an OCP. The cavity length of the laser is estimated to be ~34.5 m.

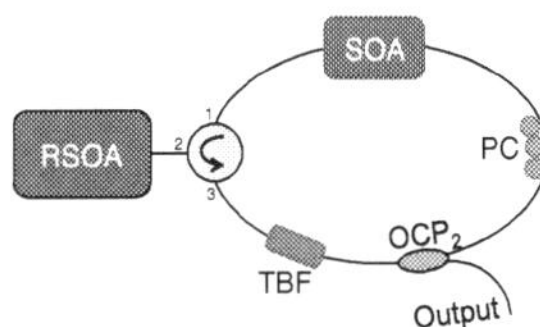

Figure 12. Experimental setup of proposed stable wavelength-tunable fiber ring laser.

Fig. 13 shows the output spectra of proposed fiber ring laser in the wavelength range of 1538.03–1561.91 nm, since the bias currents of the SOA and RSOA are 120 and 80 mA, respectively. The tunability of the fiber laser is continuous, and the wavelength tuning is performed by using the TBF. The measured peak power of lasing wavelength is between –3.9 and –2.8 dBm in this wavelength range. The observed background noise of lasing wavelength is below –61 dBm. As a result of cascading the SOA and RSOA inside the fiber cavity, the effective gain bandwidth range is enhanced and the output power of lasing wavelength in both sides of the spectrum could not drop rapidly, as shown in Fig. 13. However, other previous schemes would have a rapid drop of output power in both sides of the spectrum [23].

Fig. 14 displays the measured output power and OSNR in the lasing wavelengths from 1538.03 to 1561.91 nm. As seen in Fig. 14, the measured minimum and maximum output powers are –2.5 and –0.8 dBm at 1559.88 and 1541.03 nm, respectively. The maximum power difference (ΔP) is 1.7 dB in the lasing wavelength range. Moreover, the observed OSNR could be larger than 59.1 dB in this wavelength range. The maximum difference of OSNR measured is 1.9 dB, as illustrated in Fig. 14. Furthermore, utilizing the technique described in ref. [24], the relative intensity noise (RIN) of the proposed laser is –79.5 dB/Hz.

In the measurement, a 2.5-Gbit/s quadrature phase shift keying-orthogonal frequency division multiplexing (QPSK-OFDM) modulation format is applied on the RSOA of the ring laser for

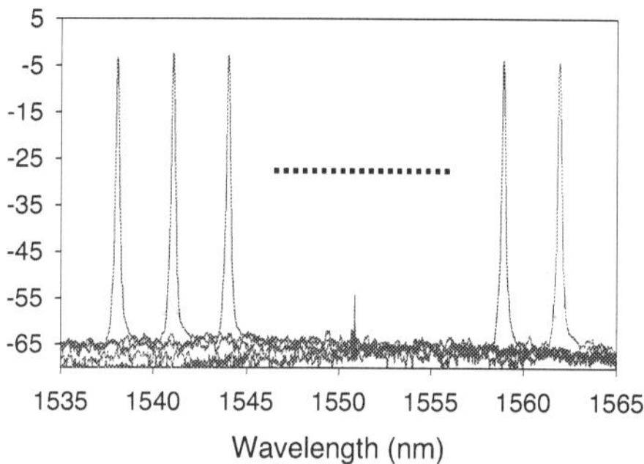

Figure 13. Output wavelength spectra of proposed fiber ring laser in the wavelength range of 1538.03–1561.91 nm, when the bias currents of SOA and RSOA are 120 and 80 mA.

direct signal modulation. Besides, the RSOA is designed to have a direct modulation bandwidth of ~1 GHz. The QPSK-OFDM signal and the DC are combined and applied to the RSOA via a bias tee (BT). Here, the electrical OFDM baseband upstream signal could be created by using an arbitrary waveform generator (AWG) and the Matlab® program. The OFDM transmitter (Tx) signal consists of the serial-to-parallel conversion, QPSK symbol encoding, inverse fast Fourier transform (IFFT), cyclic prefix (CP) insertion, and digital-to-analog conversion (DAC) for signal processing. Here, the 63 subcarriers of QPSK-OFDM format occupy ~1.25 GHz modulation bandwidth with a fast-Fourier transform (FFT) size of 128 and cyclic prefix of 1/32. Therefore, a 2.5-Gbit/s total data rate is obtained. Then, the lasing wavelength signal would be detected directly via a 9-GHz PIN receiver (Rx), and the received OFDM signal can be taken by a real-time sampling oscilloscope for signal decoding. In order to demodulate the vector signal, the use of the off-line DSP program is required. And the demodulation processing covers the synchronization, FFT, zero forcing equalization, and QPSK symbol decoding. Therefore, the bit error rate (BER) is obtained from the observed signal-to-noise ratio (SNR).

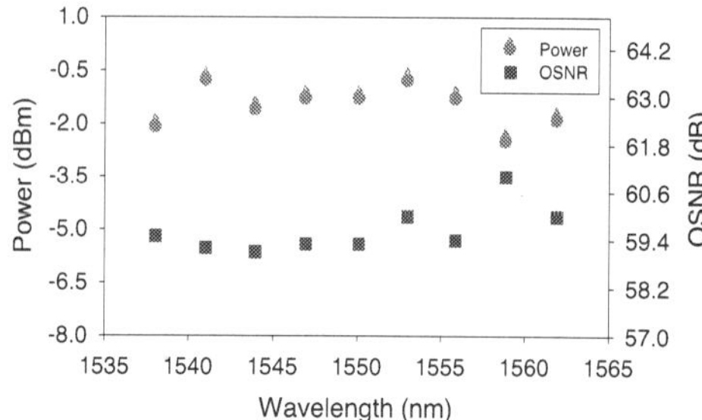

Figure 14. Measured output power and optical signal-to-noise ratio (OSNR) in the lasing wavelengths of 1538.03–1561.91 nm.

For the next experiment, the bias current of SOA is fixed at 120 mA, while the RSOA is operated at the DC bias current of 50, 60, 70, and 80 mA, respectively. The lasing wavelength is selected at 1553.0 nm for performance testing. Figs. 15(a) to 15(c) show the BER performances of proposed laser at the back-to-back (B2B) status and after 20 and 50 km fiber transmissions, respectively, when the RSOA is operated at 50, 60, 70, and 80 mA. And the insets of Figs. 15(a) to 15(c) are corresponding constellations at the BER of 3.8×10^{-3} (the FEC threshold) under the B2B and after 20 and 50 km SMF transmissions, respectively, when the SOA and RSOA are 120 and 80 mA. As seen in Figs. 15(a) to 15(b), the optical power penalties are 0.5 and 3.3 dB (RSOA@50mA); 0.6 and 1.8 dB (RSOA@60mA); 0.3 and 1.2 dB (RSOA@70mA); and 0.2 and 1 dB (RSOA@80mA), respectively, after 20 and 50 km SMF transmissions at FEC threshold. Moreover, the optical sensitivities of received powers are also observed in Fig. 15, at the B2B and after 20 and 50 km fiber transmissions, respectively, when the RSOA is operated at 50, 60, 70, and 80 mA and SOA is fixed at 120 mA. At the B2B status and after 20 km fiber transmission, by increasing the bias current of RSOA gradually, the received powers are also increased, as seen in Fig. 15. However, when the bias current of RSOA is increased after 50-km fiber transmission, the optical received sensitivities could be enhanced. That means that the received power and penalty can be improved after 50-km fiber transmission, when the bias current of RSOA is increased to 80 mA.

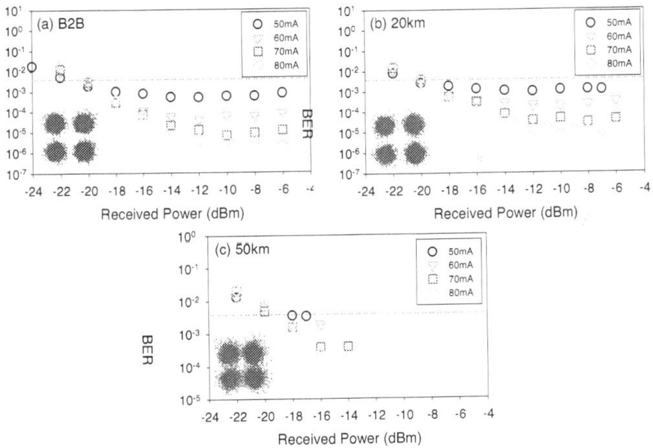

Figure 15. BER performances of proposed laser (a) at the B2B and (b) after 20- and (c) 50-km fiber transmissions, respectively, when the RSOA is operated at 50, 60, 70, and 80 mA. And the insets are corresponding constellations at the FEC threshold.

5. Conclusion

In conclusion, we have introduced three RSOA-based fiber laser architectures for multiwavelength and wavelength-tunable operations in three different laser architectures. In the first section, we have demonstrated an L-band multiwavelength laser by utilizing a C-band RSOA with a linear cavity, which is produced by a PC, an OCP, and an RFM. In the proposed RSOA laser scheme, two to seven wavelengths could be lased and created simultaneously in the L-band range, while the RSOA operates at various bias currents.

In the second section, we have proposed a multiwavelength laser source by utilizing a C-band RSOA with dual-ring fiber cavity. In the measurement, the laser cavity was constructed by an RSOA, a 1×2 OCP, a 2×2 OCP, and a PC, respectively. Thus, 13–18 wavelengths around L-band could be produced simultaneously, as the bias current of C-band RSOA was driven at 30–70 mA.

In the final section, we have investigated the wavelength-tunable fiber ring laser architecture by using the RSOA and SOA. Here, the wavelength tuning range of 1538.03–1561.91 nm could be obtained. The measured output power and OSNRs of the proposed fiber laser were between –0.8 and –2.5 dBm and 59.1 and 61.0 dB/0.05 nm, respectively. The power and wavelength stabilities of the proposed laser were also studied. Besides, the proposed laser could be directly modulated at 2.5 Gbit/s QPSK-OFDM signal. And 20–50-km SMF transmission distances were also achieved within the FEC limit without dispersion compensation. We believe that it can be a cost-effective and promising candidate for next-generation WDM-PON.

Acknowledgements

This work was supported by Ministry of Science and Technology, Taiwan, under grants MOST-103-2218-E-035-011-MY3,   MOST-104-2628-E-009-011-MY3,   and   MOST-103-2221-E-009-030-MY3.

Author details

Yeh Chien-Hung[1]* and Chow Chi-Wai[2]

*Address all correspondence to: yehch@fcu.edu.tw

1 Department of Photonics, Feng Cha University, Taichung, Taiwan

2 Department of Photonics, National Chiao Tung University, Hsinchu, Taiwan

References

[1] C. H. Yeh, C. W. Chow, Y. F. Wu, F. Y. Shih, C. H. Wang and S. Chi, "Multiwavelength erbium-doped fiber ring laser employing Fabry-Perot etalon inside cavity operating in room-temperature," Opt. Fiber Technol., vol. 15, no. 4, pp. 344–347, 2009.

[2] Z. G. Lu, F. G. Sun, G. Z. Xiao and C. P. Grover, "A tunable multiwavelength fiber ring laser for measuring polarization-mode dispersion in optical fiber," IEEE Photon. Technol. Lett., vol. 16, pp. 1280–1282, 2004.

[3] C.-H. Yeh, M.-C. Lin, B.-C. Cheng and S. Chi, "S-band long-distance fiber Bragg grating sensor system," Opt. Fiber Technol., vol. 13, pp. 170–173, 2007.

[4] C. H. Cox III, Analog Optical Links: Theory and Practice, Cambridge University Press, New York, May 2004.

[5] M. A. Ummy, N. Madamopoulos, P. Lama and R. Dorsinville, "Dual Sagnac loop mirror SOA-based widely tunable dual-output port fiber laser," Opt. Express, vol. 17, pp. 14495–14501, 2009.

[6] H. Chen, "Multiwavelength fiber ring laser lasing by use of a semiconductor optical amplifier," Opt. Lett., vol. 30, pp. 619–621, 2005.

[7] J. Zhang and J. W. Y. Lit, "Compound fiber ring resonator: theory," J. Opt. Soc. Am. A, vol. 11, pp. 1867–1873, 1994.

[8] P. Barnsley, P. Urquhart, C. Millar and M. Brierley, "Fiber Fox-Smith resonators: application to single-longitudinal-mode operation of fiber lasers," J. Opt. Soc. Am. A, vol. 5, pp. 1339–1346, 1988.

[9] C. H. Yeh, F. Y. Shih, S. Wen, C. W. Chow and S. Chi, "Using C-band erbium-doped fiber amplifier with two-ring scheme for broadly wavelength-tuning fiber ring laser," Opt. Commun., vol. 282, pp. 546–549, 2009.

[10] Y. W. Lee, J. Jung and B. Lee, "Multiwavelength-switchable SOA-fiber ring laser based on polarization maintaining fiber loop mirror and polarization beam splitter," IEEE Photon. Technol. Lett., vol. 16, pp. 54–56, 2004.

[11] Y. F. Wu, C. H. Yeh, C. W. Chow, J. Y. Sung, and J. H. Chen, "Stable and wavelength-tunable self-injected RSOA-based laser," IEEE Photon. J., vol. 7, no. 4, #1503007, 2015.

[12] C. Zhao, X. Yang, J. H. Ng, X. Dong, X. Guo, X. Wang, X. Zhou and C. Lu, "Switchable dual-wavelengtherbium-doped fiber-ring lasers using a fiber Bragg grating in high-birefringence fiber," Microw. Opt. Technol. Lett., vol. 41, pp. 73–75, 2004.

[13] C.-H. Yeh and S. Chi, "A broadband fiber ring laser technique with stable and tunable signal-frequency operation," Opt. Express, vol. 13, pp. 5240–5244, 2005.

[14] H. Ahmad, A. H. Sulaiman, S. Shahi and S. W. Harun, "SOA-based multi-wavelength laser using fiber Bragg gratings," Laser Phys., vol. 19, pp. 1002–1005, 2009.

[15] K. Vlachos, C. Bintjas, N. Pleros and H. Avramopoulos, "Ultrafast semiconductor-based fiber laser sources," IEEE J. Sel. Top. Quantum Electron., vol. 10, pp. 147–154, 2004.

[16] C. H. Yeh, C. W. Chow, Y. F. Wu, S. S. Lu and Y. H. Lin, "Stable wavelength-tuning laser in single-frequency by optical-injected Fabry-Perot laser diode and RSOA for long distance fiber propagation," Laser Phys., vol. 22, no.1, pp. 256–260, 2012.

[17] C. H. Yeh, C. W. Chow, J. H. Chen, K. H. Chen and S. S. Lu, "Broadband C- plus L-band CW wavelength-tunable fiber laser based on hybrid EDFA and SOA," Opt. Fiber Technol., vol. 19, no. 4, pp. 359–361, 2013.

[18] C.-H. Yeh, C.-C. Lee, C.-Y. Chen and S. Chi, "A stabilized and tunable erbium-doped fiber ring laser with double optical filter," IEEE Photon. Technol. Lett., vol. 16, no. 3, pp. 765–767, 2004.

[19] C. H. Yeh, C. W. Chow, and S. S. Lu, "Using a C-band reflective semiconductor optical amplifier and linear cavity laser scheme for L-band multi-wavelength lasing," Laser Phys. Lett., vol. 10, no. 4, p. 045108, 2013.

[20] M. J. Connelly, "Wideband semiconductor optical amplifier steady-state numerical model," IEEE J. Quantum Electron., vol. 37, pp. 439–447, 2001.

[21] C. H. Yeh, C. W. Chow, and S. S. Lu, "Use of reflective semiconductor optical amplifier and dual-ring architecture design for stable multi-wavelength fiber laser," Laser Phys., vol. 24, no. 5, p. 055101, 2014.

[22] C. H. Yeh, J. Y. Sung, L. G. Yang, C. W. Chow, and J. H. Chen, "Stable and wavelength-tunable RSOA- and SOA-based fiber ring laser," Opt. Fiber Technol., vol. 20, no. 3, pp. 250–253, 2014.

[23] C.-H. Yeh, C.-C. Lee, C.-Y. Chen and S. Chi, "A stabilized and tunable erbium-doped fiber ring laser with double optical filter," IEEE Photon. Technol. Lett., vol. 16, no. 3, pp. 765–767, 2004.

[24] S. E. Hashemi, "Relative intensity noise (RIN) in high-speed VCSELs for short reach communication," thesis, Chalmers University of Technology, 2012.

Chapter 9

Dissipative Solitons in Fibre Lasers

Vladimir L. Kalashnikov and Sergey V. Sergeyev

Additional information is available at the end of the chapter

http://dx.doi.org/10.5772/61713

Abstract

Interdisciplinary concept of dissipative soliton is unfolded in connection with ultrafast fibre lasers. The different mode-locking techniques as well as experimental realizations of dissipative soliton fibre lasers are surveyed briefly with an emphasis on their energy scalability. Basic topics of the dissipative soliton theory are elucidated in connection with concepts of energy scalability and stability. It is shown that the parametric space of dissipative soliton has reduced dimension and comparatively simple structure that simplifies the analysis and optimization of ultrafast fibre lasers. The main destabilization scenarios are described and the limits of energy scalability are connected with impact of optical turbulence and stimulated Raman scattering. The fast and slow dynamics of vector dissipative solitons are exposed.

Keywords: Ultrafast fibre laser, mode-locking, dissipative soliton, non-linear dynamics, vector solitons, optical turbulence, stimulated Raman scattering

1. Introduction

Over the last decades, ultrafast fibre laser technologies have demonstrated a remarkable progress. By definition [1–4], these technologies concern generation, manipulation and application of optical pulses from a fibre laser or a laser-amplifier system with (i) peak power P_0 exceeding substantially an average laser power P_{ave} and (ii) pulse widths τ which are much lesser than a laser round-trip period T. Such a definition can be re-interpreted in terms of a laser *mode-locking*, which means a phase-locked interference of the M laser eigenmodes producing an equidistant train of ultrashort pulses. Then, $\tau \propto 1/M\delta\omega$ (M is a number of locked eigenmodes and $\delta\omega$ is an inter-mode frequency interval defined by $T \propto 1/\delta\omega$) and $P_0 \propto MP_{ave}$ [5,6]. It means that a mode-locked laser generates a comb of equidistant optical frequencies comprising the broad spectral range $\Delta \propto M\delta\omega$. It is clear that the substantial enhancement of P_0 (by the factor of $M \propto 1/\tau\delta\omega$ ~$10^5 \div 10^6$, i.e. upto over-MW level [7,8]) and τ–reduction (~T/M,

i.e. down to sub-100 fs level [9]) promise an outlook for different applications [10] including non-linear and ultrasensitive laser spectroscopy [11–14], biomedical applications [15–19], micromachining [20,21], high-speed communication systems [22], metrology [23] and many others [24,25]. The extraordinary peak powers in combination with the drastic pulse width decrease bring the high-field physics on tabletops of a mid-level university lab [26–29]. Moreover, the over-MHz pulse repetition rates $\delta\omega$ provide the signal rate improvement factor of $10^3 \div 10^4$ in comparison with that of classical chirped-pulse amplifiers [26]. As a result, the signal-to-noise ratio enhances substantially, as well.

Another aspect of the ultrafast laser applications is connected with studying non-linear phenomena [30]. Ultrafast lasers became an effective platform for investigation of general non-linear processes such as instabilities and rogue waves [31,32], self-similarity [33] and turbulence [34]. A coherent self-organization in such non-linear systems [35,36] is the keystone of this review, and it will be considered below in detail. But here, we have to point at the multidisciplinary context of our topic. The ultrafast fibre lasers can be treated as an ideal playground for exploring of non-linear system phenomenology as a whole [37]. Such a playground spans gravity and cosmology [38], condensed-matter physics and quantum field theory [39–41], biology, neurosciences and informatics [42,43]. The advance of ultrafast laser technology is that the theoretical insights promise to become directly testable, controllable and, on the other part, the theory can be urged by new precise measurable experimental challenges.

To date, the solid-state lasers allowed generating shortest pulses with highest peak powers directly from an oscillator with high repetition rates ($\delta\omega > 1$ MHz) [44–49]. The main advantages of solid-state laser systems are (i) broad gain bands (i.e. very large M) allowing generation of extremely short pulses (τ approaches one optical cycle for such media as Ti:Sp or Cr:chalcogenides) and (ii) covering the spectral range from visible (Ti:Sp) through infra-red (Ti:Sp, Cr:forsterite and Cr:YAG) to mid-infrared (Cr:chalcogenides) wavelengths, as well as (iii) possibility of independent and precise dispersion [50] and non-linearity [46,48] control. Nevertheless, fibre lasers have unprecedented prospects [51] due to (i) possibility of mean power scaling provided by large gain, (ii) high quality of laser mode, (iii) reduced thermo- and environment-sensitivity, (iv) compactness and integrity of laser setup. Additionally, one has to point at broader gain bands of fibre media in comparison with the energy-scalable, thin-disk, solid-state oscillators operating within analogous wavelength ranges [9,52][1] and possibility to break into deep-UV and mid-IR optical spectral ranges [55,56].

In this review, we will concern the concepts of a mode-locking and a dissipative soliton in a nutshell.

2. Mode-Locking

The concept of mode-locking is universal and closely connected with a principle of synchronisation of coupled oscillators [57–64]. A laser is, in fact, the interferometer which possesses a

---

1 The width of a gain band is not a decisive factor per se because both pulse width and its spectrum are affected by various factors including higher-order dispersions, non-linearity, etc. [53,54].

set of eigenmodes (longitudinal modes) separated by $\delta\omega = 2\pi/T$. Simultaneously, it is an *active* resonator, which means an amplification ΔA of mode amplitude A during the resonator round-trip in the vicinity of the maximum gain frequency ω_0 as $\Delta A(\omega) \approx (g(\omega_0) - \ell) A(\omega) - \alpha(\omega - \omega_0)^2 A(\omega)$, where $g(\omega_0)$ is a gain at the frequency ω_0, ℓ is a net-loss coefficient and $\alpha \propto g(\omega_0)/\delta\Omega^2$ takes into account a frequency-dependence of gain coefficient in the vicinity of ω_0 defined by a gain bandwidth $\delta\Omega$. In such an oversimplified model, only *one* mode with the maximum net-gain $\sigma \equiv g(\omega \to \omega_0) - \ell = 0$ is generated because the gain coefficient is energy-dependent, that is, it decreases with A (i.e. a gain is *saturable* that results in a mode competition or *mode selection*, Figure 1).

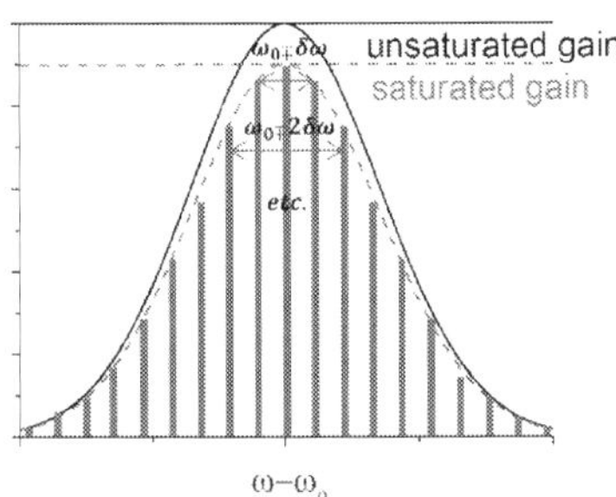

Figure 1. Comb of frequencies generated by a laser with the repetition rate $\delta\omega$. The gain band is centred at ω_0. Mode-locking consists in excitation and synchronisation of $\omega_0 \pm n\delta\omega$ sidebands.

In the case of *active mode-locking*, a periodic external modulation with the frequency of $\delta\omega$ excites the $\pm\delta\omega$ sidebands for each mode in the comb so that the modes $A(\omega)$, $A(\omega \pm \delta\omega)$ become coupled. In the framework of our oversimplified model, a steady-state regime $\Delta A = 0$ is described by the equation in time domain [59–61]:

$$(g - \ell) A(t) + \alpha \frac{\partial^2 A(t)}{\partial t^2} - vt^2 A(t) = 0, \tag{1}$$

which is the classical equation for an oscillator in the potential defined by $v \propto \delta\omega^2$. This equation has a trivial solution in the form of a Gaussian pulse [59–63]: $A(t) = A_0 \exp(-2\ln(2)\, t^2/\tau^2)$, where the pulse amplitude A_0 is defined by the condition of energy balance of $\sigma \propto v\tau^2$ (the saturable gain coefficient is energy, i.e. A_0^2-dependent) and the pulse width $\tau \propto \sqrt[4]{\alpha/v}$. In the general case, the excitation of $A(t)$ in the form of Hermitian–Gaussian solutions of Equation (1) is possible. Since the pulse width is defined by v so that $\tau \propto 1/\sqrt{\delta\Omega\delta\omega}$, the minimum pulse widths of $\sim 1/\delta\Omega$ are hardly reachable due to limitation imposed by $\delta\omega$-value. The situation can be changed in the presence of the self-phase modulation (SPM) [65] and the dynamic gain saturation. Then $\tau \propto 1/\delta\Omega\sqrt{\sigma}$ [66].

Using the non-linear processes such as SPM, loss and gain saturation allows generating the ultrashort pulses due to mechanism of the so-called *passive mode-locking* [60]. Periodical perturbations caused by transitions through non-linear laser elements such as saturable absorber or gain medium enrich the spectrum with new components $\omega_0 \pm m\delta\omega$ ($m = 1, 2,..., M$), which becomes locked through non-linear interaction [64]:

$$\frac{\partial A(\omega_1, z)}{\partial z} = -\frac{i\pi\omega_1}{2cn(\omega_1)} \int_{-\infty}^{\infty} \int_{-\infty}^{\infty} \int_{-\infty}^{\infty} d\omega_2 d\omega_3 d\omega_4\, \chi^3(\omega_1, \omega_2, \omega_3, \omega_4) \times$$
$$\times A(\omega_2, z) A(\omega_3, z) A(\omega_4, z) e^{i\Delta kz} \delta(\omega_1 - \omega_2 - \omega_3 - \omega_4), \tag{2}$$

where a four-wave non-linear process defined by non-linear susceptibility χ^3 mixes the frequencies ω_1, ω_2, ω_3 and ω_4 during the propagation through a non-linear medium along the z-coordinate ($\Delta k = k(\omega_1) - k(\omega_2) - k(\omega_3) - k(\omega_4)$ is the difference of wavenumbers, c is the speed of light, n is the frequency-dependent refractive index).

Both active and passive mode-locking concepts can be easily united from the point of view of spice-time duality [64,67–69]. For instance, let's consider *heat diffusion* equation:

$$\frac{\partial u}{\partial t} = \sigma u + \zeta \frac{\partial^2 u}{\partial x^2} - vx^2 u, \tag{3}$$

where a heat radiated at $x = 0$ by the point source σ diffuses along x-axis and is absorbed by cooler with the parabolic 'cooling potential'. The replacements $t \rightarrow z$ and $x \rightarrow t$ result in an equation for 'diffusion' of light describing an *active amplitude mode-locking* (see Eq. (1)):

$$\frac{\partial A}{\partial z} = \sigma A + \alpha \frac{\partial^2 A}{\partial t^2} - vt^2 A. \tag{4}$$

Eq. (4) is clearly understandable in the Fourier domain: $\frac{\partial \tilde{A}}{\partial z} = \sigma \tilde{A} + v \frac{\partial^2 \tilde{A}}{\partial \omega^2} - \alpha\omega^2 \tilde{A}$, where an external (i.e. active) modulation defined by the v-coefficient 'diffuses' (i.e. broadens) a field spectrum $\tilde{A}$ and such diffusion is compensated by spectral dissipation defined by the α-coefficient. In the time-domain, the spectrum broadening corresponds to a light pulse shortening due to parabolic potential action which is balanced by spectral dissipation causing a pulse widening.

The space-time duality can be extended further with the help of *diffraction-dispersion* duality:

$$\frac{\partial A}{\partial z} = -\frac{i}{2k} \frac{\partial^2 A}{\partial x^2} \quad \leftrightarrow \quad \frac{\partial A}{\partial z} = \frac{i}{2}\beta_2 \frac{\partial^2 A}{\partial t^2}, \tag{5}$$

where k and β_2 are the wave number and group-delay dispersion coefficients, respectively. Both processes describe the beam/pulse spreading with propagation which is accompanied by phase ϕ profile distortion, i.e. by appearance of the *chirp* $Q \propto d^2\phi/dx^2$ (or $d^2\phi/dt^2$). The *active phase modulation* from this point of view

$$\frac{\partial A}{\partial z} = \sigma A + \alpha \frac{\partial^2 A}{\partial t^2} - i v t^2 A \tag{6}$$

(compare with (4)) looking as $\frac{\partial \tilde{A}}{\partial z} = \sigma \tilde{A} + iv \frac{\partial^2 \tilde{A}}{\partial \omega^2} - \alpha \omega^2 \tilde{A}$ in the Fourier domain describes a 'diffraction' (dispersion) in the frequency domain inspired by phase modulator which is balanced by spectral dissipation. The main difference from (4) is that the phase modulation in (6) distorts the phase and thereby produces chirp like the action of thin lens but in the time domain. In other words, the phase modulation in (6) pushes the spectral components out of the point of stationary phase $\partial\phi/\partial t = 0$, adding the frequency shift $\propto vt$ (Doppler shift) which enhances the spectral dissipation on the pulse wings and thereby forms a pulse like the active amplitude modulator. But the phase profile $\phi(t)$ is parabolic in this case.

The transition to a *passive mode-locking* looks straightforward, but one has to be careful in this case. The spice-time duality suggests a simple way to realize the temporal focusing like that in space domain: combination of phase modulation ('time lens') from Eq. (6) with dispersion ('time diffraction') from Eq. (5) allows compressing a pulse. Therefore, a replacement of time focusing by a time self-focusing (SPM) would provide a laser pulse self-trapping like the effect of laser beam self-trapping:

$$\frac{\partial A}{\partial z} = \frac{i}{2}\beta_2 \frac{\partial^2 A}{\partial t^2} - i\gamma |A|^2 A, \tag{7}$$

which is the famous *non-linear Schrödinger equation* describing propagation of optical solitons in a fiber ($\beta_2 < 0$ corresponds to an anomalous dispersion, γ is a SPM-coefficient) [35,70,71].

It is appropriate to mention here that the space-time duality $x \to t$ allows extending the physical context of consideration beyond scopes of optics. For instance, A, $E = \int dx\, |A|^2$, and ϕ can be related for a mean-field amplitude, number of particle (mass of condensate) and momentum (wave number) for a Bose–Einstein condensate [39]. Then, it is clear that the dispersion and SPM terms in Eq. (7) describe the kinetic energy and four-particle interaction potential for gas of bosons. Such an interpretation opens a road to a quantum theory of solitons [72–74].

Following the same procedure for Eq. (4), describing the active amplitude mode-locking results in the simplest version of equation for a passive mode-locking, so-called *cubic non-linear Ginzburg–Landau equation* [35,36,75]:

$$\frac{\partial A}{\partial z} = \sigma A + \alpha \frac{\partial^2 A}{\partial t^2} + \kappa |A|^2 A. \tag{8}$$

This equation describes a combined action of saturated net-gain (σ), spectral dissipation (α) and non-linear gain (κ). The last term results from loss saturation in a non-linear absorber with the response time much lesser than the pulse width. As will be shown below, such an assumption is valid for a broad class of fibre mode-locking mechanisms. Physics of passive mode-locking resembles that of active one: self-focusing in time domain causes a spectrum broadening which is balanced by spectral dissipation. Loss and energy-dependent gain are required for developing and stabilizing the mode-locking (all these factors are included in σ-term which is < 0 for a steady-state pulse). Eqs. (7) and (8) have a similar solution $A(t) \propto \mathrm{sech}(t/\tau)$ but the mathematical structures of these equations differ substantially that created discrepancies between concepts of the 'true' [76] and *dissipative solitons* (DSs, see next section) [36]. Combining Eqs. (7) and (8) gives the famous complex cubic non-linear Ginzburg–Landau equation (cubic CNGLE) [35–37,42]:

$$\frac{\partial A}{\partial z} = i\left(\frac{\beta_2}{2} \frac{\partial^2 A}{\partial t^2} - \gamma |A|^2 A \right) + \sigma A + \alpha \frac{\partial^2 A}{\partial t^2} + \kappa |A|^2 A, \tag{9}$$

which is a playground for study of DSs. Equation (9) allows a number of further generalizations such as: (i) description of non-distributed evolution due to dependence of the equation coefficients on z [77]; (ii) generalization of non-linearity type aimed first of all to adequate description of different mode-locking mechanisms (see below); (iii) taking into account the higher-order dispersions, i.e. ω-dependence of β_2 [68]; (iv) taking into account the vector nature of light, i.e. transition to a system of coupled two-component CNGLEs [68,78–81], etc.

Now let us consider the mode-locking mechanisms for fibre lasers in more detail. Active mode-locking can be utilized for DS generation from a fibre laser [82–84], but the widespread mechanism is based on the *non-linear polarization rotation* (NPR) which uses the effect of intensity-dependent polarization mode coupling in a fibre [85–88]. There is voluminous literature concerning the experimental realization of NPR mode-locking in fibre lasers; therefore, our selection of references is rather subjective and concerns the DS context [95–111].

It is known [70] that an ideal single-mode fibre supports two degenerate orthogonally polarized modes. However, a real fibre has inherent birefringence caused by core asymmetry or mechanical stress (Figure 2).

Since SPM as well as cross-phase modulation (XPM) contribute to refractivity index with the strength defined by field intensity, such a contribution will change the state of polarization (SOP, Figure 3) [60,70,89] that can be described by coupled equations for two orthogonal (x and y) polarization components [70]:

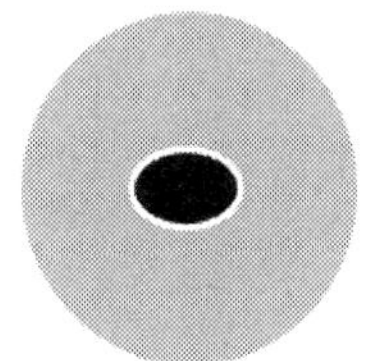

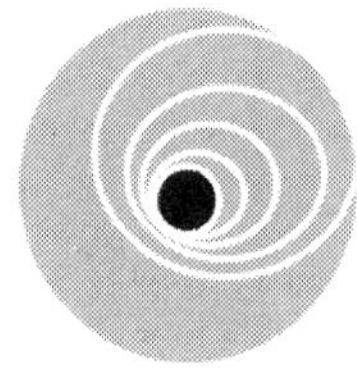

Figure 2. Fibre cross-sections with cylindrical symmetry broken by manufacturing process that leads to fibre birefringence.

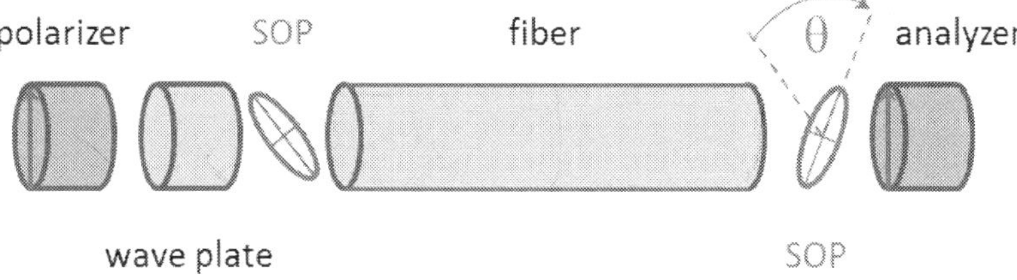

Figure 3. Block scheme of a self-amplitude modulator (SAM) based on NPR. Both linear birefringence and NPR contribute to the change of state of polarization (SOP, rotation angle θ) that leads to intensity-dependent transmission of this scheme.

$$\frac{\partial A_x}{\partial z} + \beta_{1x}\frac{\partial A_x}{\partial t} - \frac{i\beta_2}{2}\frac{\partial^2 A_x}{\partial t^2} - \sigma A_x - \alpha\frac{\partial^2 A_x}{\partial t^2} =$$

$$= -i\gamma\left(|A_x|^2 + \frac{2}{3}|A_y|^2\right)A_x - \frac{i\gamma}{3}A_x^* A_y^2 e^{-2i\Delta\beta z},$$

$$\frac{\partial A_y}{\partial z} + \beta_{1y}\frac{\partial A_y}{\partial t} - \frac{i\beta_2}{2}\frac{\partial^2 A_y}{\partial t^2} - \sigma A_y - \alpha\frac{\partial^2 A_y}{\partial t^2} =$$

$$= -i\gamma\left(|A_y|^2 + \frac{2}{3}|A_x|^2\right)A_y - \frac{i\gamma}{3}A_y^* A_x^2 e^{2i\Delta\beta z},$$

$$(10)$$

where the dissipative factors from Eq. (9) are taken into account and $\Delta\beta = 2\pi / L_b$ describes a 'strength' of linear birefringence (L_b is a beat-length). As was shown in [81,91–94,179], the multiscale averaging technique allows reducing Eq. (10) to the modified scalar non-linear Ginzburg–Landau equation (*so-called sinusoidal Ginzburg–Landau equation*) in which the self-amplitude modulation term (SAM, last term in Eq. (9)) is replaced by $\propto \log Q(|A|^2)A$, where Q is a complex function defined by birefringence and settings of laser wave plates and polarizer.

Such an approach opens a way to multi-parametrical optimization of fibre lasers mode-locked by NPR.

Despite its relative simplicity in principle as well as possibility of all-fibre-integrity of a laser, NPR in the form presented in Figure 3 is too sensitive to laser setup, uncontrollable perturbations and requires a precise manual tuning. The modified SAM setup, which can utilize both NPR and scalar SPM, is shown in Figure 4. It is the so-called *non-linear optical loop mirror* or *figure eight laser* (Figure 4) [112,113,265]. In principle, this setup is an all-fibre realization of additive-mode-locking [82,114] with inherently adjusted linear optical propagation lengths for counter-propagating beams. The main control parameter here is the beam splitting ratio ρ controlling the mutual intensities of counter-propagating beams.

The unique property of this SAM setup is its ability to utilize different types of non-linearities for mode-locking (e.g. see [115–118]). Different modifications of this mode-locking mechanism have been used in DS fibre lasers [119–125]. Nevertheless, a fibre loop defining SAM remains environment- and tuning-sensitive.

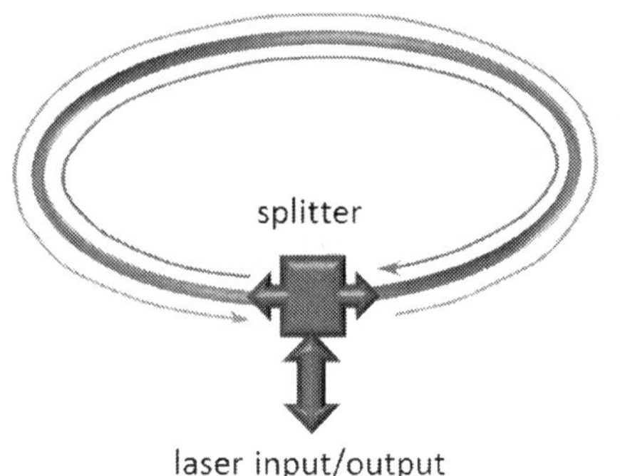

Figure 4. Block-scheme of SAM based on a non-linear optical loop mirror. Splitter splits input laser beam into two counter-propagating ones (red and green arrows) with some splitting ratio ρ. The beams interfere after round-trip and partially return into a laser. The result of interference is intensity-dependent due to NPR or/and SPM within a loop.

There is a class of alternative approaches utilizing non-fibre well-controllable non-linearities for mode-locking by the cost of broken fibre-integrity of a laser. Such an alternative was provided by development of high-non-linear *semiconductor saturable absorber mirror* (SESAM) [126–135]. The point is to put a semiconductor layer into a composed multi-layer mirror with the well-controllable spectral characteristics as well as with the adjustable intensity concertation of penetrating field within a semiconductor layer. In fact, it is an advanced non-linear Fabri–Perot interferometer with the reflectivity coefficient depending on the incident intensity (or energy) [136]. Interaction of light with a semiconductor layer can be characterized roughly as excitation of carriers from a valence band of semiconductor to its conduction band. Excited carriers thermalize inside a conduction band with the character time ~100 femtoseconds. This time defines a fastest response of SESAM to a laser radiation. Then, the thermalized carriers can relax into valence band or intra-band trapping states with the characteristic times from picoseconds to nanoseconds. Thus, SAM due to SESAM is *slow* in comparison with that due to NPR or SPM because the response times of the lasts are defined by intra-atom polarization dynamics, i.e. these times belong to femtosecond diapason. Additionally, the *spectral diapason* of SESAM response is substantially squeezed in comparison with that of pure electronic non-linearities due to resonant character of SESAM non-linearity. This can trouble the mode-locking within a spectral range exceeding the SESAM bandwidth. But the reverse side of the SESAM-band squeezing is that a non-linear response of SESAM becomes resonantly enhanced. This means that SESAM can provide more easily starting, stable and controllable mode-locking. The key characteristics of SESAM are [127] *loss saturation fluence* $E_s = hv/2\sigma_a$ (hv is a photon energy, σ_a is an absorption cross section), *modulation depth* $\mu_0 = \sigma_a N$ (N is a density of states in semiconductor), *relaxation (recovery) time T*, *unsaturable loss*, *saturable loss bandwidth* and level of *two-photon absorption*.

Akin mode-locking methods providing full fibre-integrity, broadband absorption, sub-picosecond response time and avoiding a complex multi-layer mirror weaving use nanotube and graphene saturable absorbers [30,137–143] and other low-dimensional structures [144].

From the theoretical point of view, the response of saturable absorber (SESAM or other quantum-size structures) to a laser field can be very complicate. In principle, one has to take into account finite loss bandwidth, its dispersion, dependence of refractive index on carrier's (or exciton's) density (so-called linewidth enhancement), complex kinetics of excitation and relaxation, etc. However, the praxis demonstrated that a simple model of two-level absorber is well working [145]:

$$\frac{d\mu}{dt} = -\frac{\mu - \mu_0}{T_r} - \mu\frac{|A|^2}{E_s} \tag{11}$$

with some possible modifications (e.g. see [146]). Since DSs, as a rule, have over-picosecond widths (see next section), one may use an adiabatic approximation for (11) so that the expression for SAM coefficient in the last term in Eq. (9) has to be replaced:

$$\kappa \leftrightarrow (-1) \times \mu = \frac{\mu_0}{1 + |A|^2/P_s}, \tag{12}$$

where $P_s = E_s/T_r$ is a loss saturation power.

One may propose a hypothesis that an analogue of Kerr-lens mode-locking, which is a basic mechanism for generation of femtosecond pulses from solid-state lasers [60,85,147], can be realized in a fibre laser as well. Such an insight is based on possible enhancement of the laser beam spatial-trapping induced by non-linearity in a medium with spatially inhomogeneous gain/loss or refractivity [148–152]. The model for analysis of such phenomena can be based on extension of dimensionality of Eq. (9), with taking into account the diffraction and transverse inhomogeneity of gain, loss or/and refractive index (the last can work as SAM due to the waveguide leaking loss) [153]:

$$\frac{\partial A}{\partial z} = i\left(\frac{\beta_2}{2}\frac{\partial^2 A}{\partial t^2} - \frac{1}{2k}\frac{\partial^2 A}{\partial x^2} - \gamma|A|^2 A\right) + \sigma A + \alpha\frac{\partial^2 A}{\partial t^2} - \kappa x^2,$$

(13)

where cylindrical symmetry is assumed, x is a radial coordinate and κ is a coefficient (complex in general case) which describes a transverse inhomogeneity of a fibre. Figure 5 shows the net-gain profiles (a) and the intra-laser pulse energies (b) as function of an effective aperture size obtained on the basis of variational approach for Eq. (13) [153]. The results demonstrate a principal feasibility of the Kerr-lens mode-locking regime for a DS fibre laser.

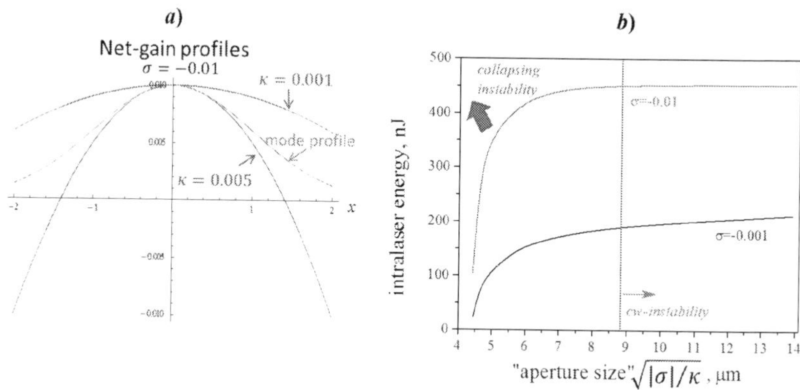

Figure 5. (a) Transverse net-gain profiles for different transverse parabolic distributions of net-gain coefficient which can be realized by inhomogeneous doping of fiber or by impact of waveguide leaking loss. (b) The dependence of in-tra-laser DS energy for a DS Yb-fibre laser with the 14 nm filter bandwidth and the average GDD of 330 fs²/cm [153]. DS collapses for large energies and cannot start for large aperture sizes (here SAM has inverse sign and the continu-ous-wave generation prevails).

All these mode-locking techniques are realizable for both soliton proper and DS fibre lasers (excluding the Kerr-lens mode-locking which requires sufficiently high pulse energies provided by only a DS laser). Now let's consider the DSs fibre lasers proper.

3. DS concept: Theory and experiment

A 'classical' soliton can be formally defined as a solution of non-linear evolution equation belonging to discrete spectrum of the inverse scattering transform [71,76,154]. The non-linear equations, which can be solved by inverse scattering transform, are 'exactly integrable'. This means that they are akin to linear equations in some sense. In particular, they obey the superposition principle and, as a result, can be canonically quantized [155,156]. One has to note that integrability of a non-linear evolution equation and non-dissipative (non-Hamiltonian) character of the latter are not equivalent because there are both non-integrable Hamiltonian systems and integrable dissipative ones [36]. The point is that the DS concept is not connected with 'integrability'; therefore, DSs are not 'true' solitons in a mathematical sense. However, many properties of DSs, in particular, their stable localization, robustness in the processes of scattering and interaction, well-organized internal structure, etc., resemble the properties of 'true' solitons. Formally, one may define *DS* as a *localized and stable structure emergent in a non-linear dissipative system far from the thermodynamic equilibrium* [36]. DSs are abundant in the different natural systems ranging from optics and condensed-matter physics to biology and medicine. In this sense, one may paraphrase that DSs "are around us. In the true sense of the word they are absolutely everywhere" [157]. Therefore, the concept of DS became well established in the last decade [36,37,42,158].

Stability of a DS under condition of strong non-equilibrium can be achieved only due to well-organized energy exchange with environment and subsequent energy redistribution within a DS. It results in energy flux inside a DS and, thereby, in DS phase inhomogeneity [36]. For a simplest case of Eq. (9), which has a DS solution in the form of $A(t) = A_0 \mathrm{sech}(t/\tau)^{1+i\psi}$ ($\psi \propto Q\tau^2 = \tau^2 \partial^2 \phi / \partial t^2$ is a dimensionless chirp parameter) [85,159,160], the DS energy generation [36]

$$ F = 2\sigma |A|^2 + 2\kappa |A|^4 - 2\alpha \left| \frac{\partial A}{\partial t} \right|^2 + \alpha \frac{\partial^2 |A|^2}{\partial t^2} \tag{14} $$

as well as the spectrum $|\tilde{A}(\omega)|^2$ are shown in Figure 6 in dependence of β_2 (the data are based on an approach of [160]). One can see that the spectrum broadening transfers the action of spectral dissipation on a pulse 'in whole' into well-structured energy exchange: inflow at pulse centrum and outflow on its wings. The key characteristic of a dissipation inhomogeneity is a *chirp*, i.e. an inhomogeneity of phase. In absence of the chirp, the spectral dissipation acts on a pulse in whole that, in particular, induces a multi-pulse instability [161]. However, a power-dependent chirp causes inhomogeneity of energy transfer (Figure 7). Energy flows in the region closer to central wavelength where the gain is maximal. This region is located in the vicinity of pulse maximum. Energy flows out from the spectrum wings which are located on the wings of pulse, that is, the pulse localization is supported by spectral dissipation through non-linear mechanism of chirping [42,160,162]. One has to note that a direction of energy fluxes inside a DS depends on parameters and can be inversely related to the direction shown in Figures 6,7 (i.e. energy can flow from wings to centre). The corresponding structure was named *dissipative anti-soliton* [210].

Thus, an additional mechanism of SAM (in addition to mechanisms considered in the previous section) appears, which provides unique robustness of DSs (i.e. DS exists within a broad range of laser parameters [163,164]).

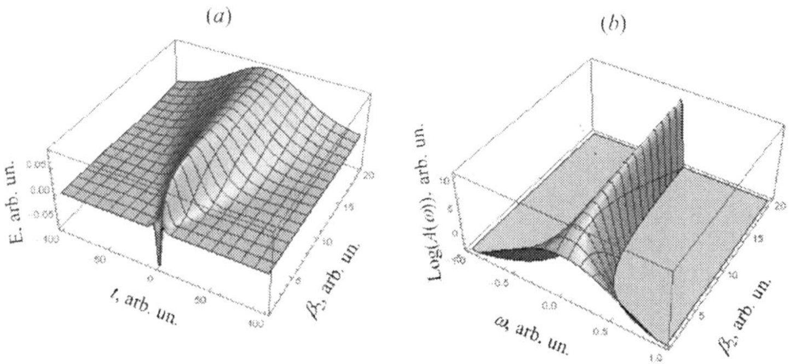

Figure 6. (a) Profile of energy generation and (b) logarithm of spectral power in dependence on GDD (normal dispersion range) for a DS of [160].

Below, we will consider a chirp as the essential characteristic of DS [210]. One of the reasons is that the chirp allows DS to accumulate energy $E \propto \psi$, which means that DS is *energy-scalable* [37,164–167]. The last statement does not mean that a chirp-free pulse is not energy-scalable. However, the energy-scalability of such pulses can be provided by only fine-tuned and separated control of SPM and GDD that can be achieved in solid-state oscillators [26,168] or in large mode area (LMA) fibre lasers [169]. For fibre lasers such an approach entails issues of full-fibre integrability, higher-order mode control [170,171][2] and thermo-effects impact [172].

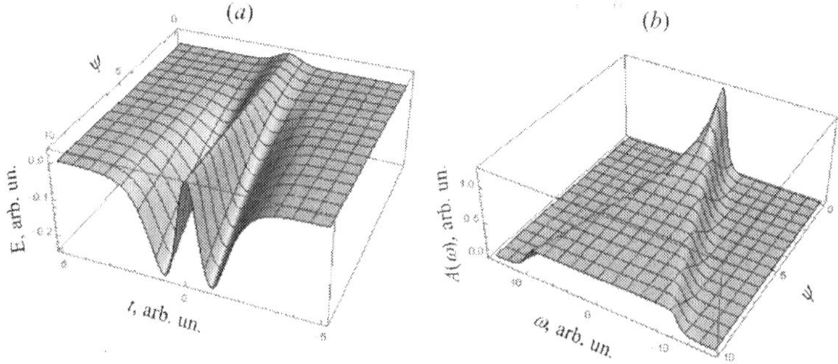

Figure 7. (a) Profile of energy generation and (b) power in dependence on chirp for a DS of Eq. (9): $\sigma = -0.01$, $\kappa = 0.025$, $\alpha = 0.05$, $A_o = 1$, $\tau = 1$.

---

2 However, namely LMA and photonic-crystal fibres could realize a Kerr-lens mode-locking in a fibre laser [152,153].

In the terms of space-time duality (see above), the mechanism of formation of time window, within which a DS is localized, resembles a phenomenon of total internal reflection from some 'borders' created by phase discontinuity. Such borders are formally defined by the equivalence of the wave number of out-/in-going radiation $k(\omega)=\beta_2\omega^2/2$ (wave number of dispersive linear wave) and the DS wave number $q=\gamma P_0 : k(\pm\Delta)=q$, where $P_0 = |A_{max}|^2$ is a DS peak power and a DS spectral width is $\Delta=\sqrt{2\gamma P_0/\beta_2}$ [173]. Since a system is dissipative, the above phase equilibrium has to be supplemented by loss compensation condition: spectral loss $\propto \alpha\Delta^2$ has to be compensated by non-linear gain κP_0. Combination of above criteria gives a definition of the parametric limits for DS [44,173]:

$$\frac{2\alpha\gamma}{\beta_2\kappa} \leq \begin{cases} 2/3, E \to \infty \\ 2, E \to 0 \end{cases}, \tag{15}$$

where E is a DS energy. Eq. (15) is valid for the *cubic-quintic CNGLE* in which SAM has a form of $(\kappa|A|^2-\kappa\zeta|A|^4)A$, that is, a non-linear gain is saturable (despite the unsaturable SAM in Eq. (9)). A SAM saturability is necessary for DS stabilization [174]. Equality in Eq. (15) corresponds to the DS stability border where $\sigma = 0$ (see Eq. (9)). The asymptotic $E \to \infty$ corresponds to a *perfectly energy-scalable* DS or to a phenomenon of *dissipative soliton reso-nance* (DSR) [37,44,165–167,175–183], which is sufficiently robust, exists in different SAM environments and even within the anomalous GDD range [178,182]. Important property of DSR is that the DS energy E can be scaled without loss of stability by plain scaling of laser average power or/and its length L [44,180,184]. The chirp scales with length as well. As a result, the DS peak power and spectrum width tend to a constant for fixed parameters of Eq. (9) (i.e. fixed α, β_2, σ and κ) and the energy scaling is provided by DS stretching in time domain.

This ideology of energy scaling by the pulse stretching goes back to the so-called wave-breaking-free or stretched pulse fibre lasers where the propagation within the anomalous-dispersion fibre sectors alternates with the propagation under normal GDD action [96,101,185–187]. As a result of pulse stretching, the non-linear effects in such systems are reduced, which allows increasing an energy and suppressing a noise. As an alternative approach, one can exclude an anomalous GDD at all and to realize a so-called *similariton* regime, when a pulse accumulates an extremely large chirp and, thereby, an energy [33,103,188]. However, a self-similar regime is not soliton-like one in nature; therefore, we will focus on the all-normal-dispersion fibre lasers (ANDi) which produces DSs possessing a high stability within a broad range of laser parameters [97,101,185,186,189,190]. In Figure 8, the energy-scalable DS lasers are sub-divided into three main types: (1) *all-fibre*, (2) *fibre with a free-space sector* and (2) LMA including rod and photonic-crystal fibre PCF.

The advantage of the *first type* of lasers is their integrity, which does not require an operational alignment, includes potentially compressing and delivering sections, and provides environ-ment insensitivity and easy integrability with fibre-amplifier cascades [200]. The last advant-age is especially attractive because it allows a direct seeding of DS into chirp-pulse amplifier without preliminary pulse stretching. The *second type* of the DS fibre lasers can be considered as a testbed for development of the first type. No wonder that the results achieved here are more impressive (Figure 8). At last, the *third type* of the DS fibre lasers is most akin to the thin-disk solid-state ones with simultaneous advantage of the broad gainbands. Such lasers provide the DS energy scalability by scaling of laser beam area in combination with the scaling of laser period and average power. Nevertheless, one has to keep in mind that both LMA and PCF technologies have some disadvantages (see above) which make them similar to solid-state lasers.

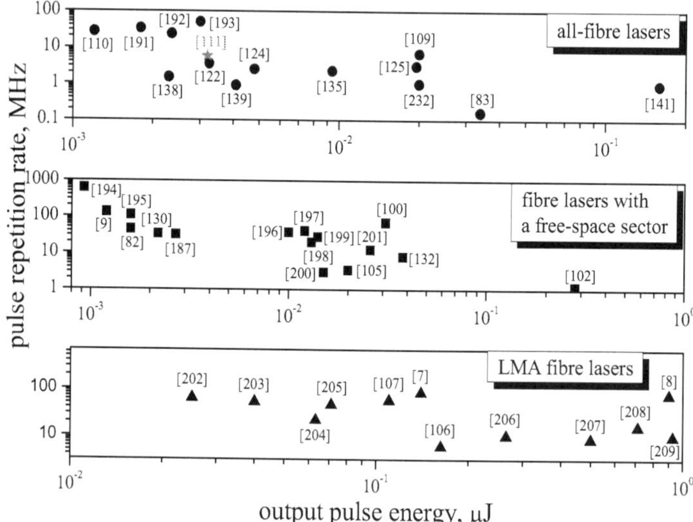

Figure 8. Experimental realizations of DSs in fibre lasers (Refs. [7-9,82,83,100,102,105-107,109-111,122,124,125,130,132,135,138,139,140,141,187,191-209,232]; Ref. [111] corresponds to a DS in the anomalous GDD region).

The diversity of the results obtained (Figure 8) needs a comprehension from a *unified viewpoint*; therefore, let's survey briefly some theoretical aspects relevant to the DS fibre lasers. There is vast literature regarding the theory of DSs. Some preliminary systematization can be found in [44]. However, it is necessary at first to declare the stumbling block of this theory: *absence of a unified viewpoint*. There exist unbroken walls between the circles of scientific community exploiting and exploring the DS concept: walls between the solid-state and fibre laser representations of the theory, condensed-matter one, numerical and analytical approaches, etc. Briefly and conditionally, the relevant theoretical approaches can be divided into (1) numerical, (2) exact analytical and (3) approximated analytical. The last includes the models based on (3.1) perturbative, (3.2) adiabatic models (AM) as well as those based on (3.3) phase-space truncation (i.e. variational approximation (VA) and method of moments (MM)).

As was emphasized repeatedly, both linear and non-linear dissipations are crucial for the DS formation. The simplest and most studied models for such a type of phenomena are based on the different versions of CNGLE (e.g. Eq. (9)).

Extensive *numerical study* of DSs of the cubic-quintic CNGLE has been carried out by N. N. Akhmediev with co-authors [35–37,42,157,166,175,176,178,182]. The simulations have allowed finding the DS stability regions for some two-dimensional projections of CNGLE parametrical space. The summarizing description of the results obtained is presented in [44]. Most impressive results are: (i) parametric space of DS has a reduced dimensionality resulting, in particular, in the appearance of DSR; (ii) DSR remains in a model with lumped evolution that is typical for the most of fibre lasers; (iii) DSR and, correspondingly, DS exist within the anomalous GDD region as well. However, the main shortcomings of the numerical approaches are: the parametrical space under consideration is not physically relevant, and the true dimensionality of DS parametric space is not identified. It is clear that the only advanced and self-consistent *analytical* theory of DS would provide, in particular, a true representation of DS parametric space and DSR conditions.

As was mentioned above, the evolution equations describing DSs are not-integrable. The efforts based on the algebraic techniques [62,213,214] and aimed to finding the generalized DS solutions of CNGLE were not successful to date. Nevertheless, few exact partial DS-solutions are known. For instance, *sole known exact analytical DS-solution* of cubic-quintic CNGLE is [110,166,176,182,189,211,212]:

$$A = \sqrt{\frac{A_0}{\cosh\left(t/\tau\right)+B}}\, \exp\left[\frac{i\psi}{2}\ln\left(\cosh\left(t/\tau\right)+B\right)+i\phi z\right],$$ (16)

where A_0, B, τ, ψ and ϕ are real constants [189]. This solution belongs to a fixed-point solution class, which means that it exists only if some constraints are imposed on the cubic-quintic CNGLE parameters. Solution (16) provides with important insights into properties of DSs. In particular, the systematical classification of DS spectra (truncated concave, convex, Lorentzian and structured spectra) and DS temporal profiles (from *sech*-shaped $A(t)$ to tabletop one) is possible in the framework of analytical approach. The transition to a DSR-regime reveals itself in the 'time-spectral' duality shown in Figure 9 [141,189]. The sense of this 'duality' will be explained below from the point of view of adiabatic theory of DSs.

The crucial shortcoming of the approach based on few exact DS solutions of evolution equations is that the strict restrictions are imposed on the equation parameters. As a result, the DS cannot be traced within a broad multidimensional parametric range and the picture obtained is rather sporadic and is of interest only in the close relation with the numerical results and experiment. Some additional information can be obtained on the basis of perturbation theory which provides with a quite accurate approximation for a low-energy DS [215–217].

Most powerful approaches to the theory of DSs have been developed in the framework of *approximated techniques* (for review see [44]): AM [165,167,217–221], VA [77,177,222–224,225]

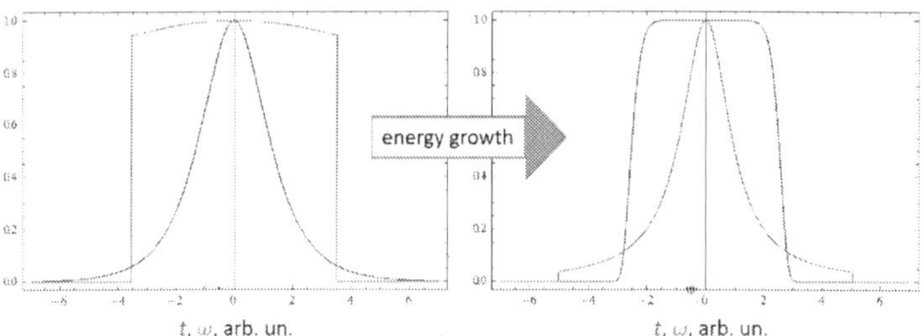

Figure 9. 'Time-spectral duality' for an energy-scalable DS: bell-like time-profile (blue curve on the left picture) transforms into tabletop one with truncated edges (right) and, inversely, truncated tabletop spectrum (red curve on the left picture) transforms into bell-like one (right) with the DS energy growth.

and MM [175,210,222,226]. The most impressive results obtained are: (i) physically relevant representation of DS parametric space was revealed (it is a so-called *master diagram*, see [44] for review and Figure 10); (ii) such a representation allows understanding the structural properties of DS and its energy-scaling laws (i.e. DSR conditions) for different mode-locking techniques; (iii) DS dynamics and an issue of optimal arrangement of laser elements providing the maximum DS stability and energy have been explored [224,225,227,228]; (iv) vectorial extension of VA concerning a vector DS (VDS) was endeavoured [229].

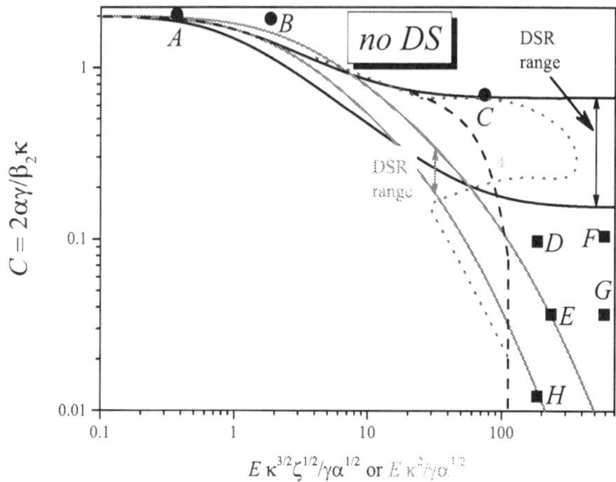

Figure 10. DS master diagrams for the cubic-quintic CNGLE (black) and the CNGLE with a SAM defined by Eq. (12) (red) (in the last case $\kappa = 1/\gamma P_s$). DSR ranges correspond to a so-called *positive branch* of DS [44,167], which has a highest DS energy-scalability and stability [221]. The dashed curve corresponds to a DS stability border obtained from numerical simulations of cubic-quintic CNGLE taking into account a quantum noise [243]. Dot blue curve shows the DS border under effect of SRS [260]. The parametrical space shown is the *physically relevant* parametrical space of DS. Points correspond to different scenarios of DS destabilization (see text).

Both AM and VA demonstrate two-dimensional representation of DS parametric space in the form of master diagram. Dimensionality can grow with complication of CNGLE non-linearity when SPM becomes saturable so that the cubic non-linear term in Eq. (9) has to be replaced by $[(\kappa - \kappa\zeta \mid A \mid^2) - i(\gamma + \chi \mid A \mid^2)] \mid A \mid^2 A$ [217]. This effect can appear in a fibre laser with NPR (e.g. see [179] where such a completely cubic-quintic CNGLE is connected with the NPR mode-locking technique). In this case, DS soliton exists in both normal and anomalous GDD regions [175,182,233].

The master diagram is a manifold of isogains (i.e. curves with $\sigma = const \leq 0$). Figure 10 demonstrates the zero-isogains ($\sigma = 0$) corresponding to the DS stability limit (upper curves) as well as the borders between 'energy-scalable' and 'energy-non-scalabe' branches of DS (lower curves, see [44,167] for a formal definition[3]). Figure 10 demonstrates that the saturation of SAM (so-called reverse saturable absorber provided, for instance, by NPR or graphene [141]; black curves) enhances the DS stability in comparison with an unsaturable SAM (red curves). Since $\lim\limits_{C \to const} E = \infty$ ($const \in$ the DSR range and its maximum value are of 2/3) for a saturable SAM (cubic-quintic CNGLE, black curves in Figure 10). Such a property of isogain curves corresponds to the DSR phenomenon. Since the stability threshold is defined by the condition of $C \equiv 2\alpha\gamma / \beta_2 \kappa \leq 2/3$, one may conclude that the broadening of spectral filter band (or gainband) enhances stability against multi-pulsing ($\alpha \propto 1/\delta\Omega^2$, see above) [107,108]. Simultaneously, SPM has to be balanced by GDD ($C \propto \gamma / \beta_2$) that, in combination with normalization of energy, gives the energy-scaling law $E \propto L$ along a DSR curve. VA predicts [230]:

$$E \propto \beta_2 / \sqrt{\alpha} \qquad (17)$$

which agrees with experimental observations of linear growth of DS energy with bandwidth [107,108] as well with a rule $E \propto L$ since $\beta_2 \propto L$ in the frameworks of distributed CNGLE.

In the case of unsaturable SAM corresponding to SESAM, some nanotube and graphene absorbers, Kerr-lensing, etc. (see Eq. (12)), the energy scaling requires scaling of the control parameter C. In this case, the asymptotic *energy scaling law* for $E\kappa^2 / \gamma\sqrt{\alpha} \gg 1$ becomes [44]:

$$E \propto \beta_2 / C\sqrt{\alpha}. \qquad (18)$$

The spectral properties of DS are described clearly in the frameworks of AM [44,167]. In the simplest case of cubic-quintic CNGLE, the DS spectrum $\tilde{A}(\omega)$ in the limit of $\mid \psi \mid \gg 1$ is a Lorentzian profile which has a characteristic width Ω_L and is truncated at frequencies $\pm\Delta$ [44,167,218]:

$$\tilde{A}(\omega) \propto \frac{H(\Delta^2 - \omega^2)}{\Omega_L^2 + \omega^2}, \qquad (19)$$

---

3 Energy-non-scalable branch has two distinguishing characteristics: it turns into solution of Eq. (9) with ζ, $\chi \to 0$ ('Schrödinger limit' [218]) and is unstable in absence of dynamic gain saturation, i.e. if σ is not energy-dependent [221].

where H is a Heaviside function. The DS energy is

$$E \propto \arctan\left(\Delta/\Omega_L\right)/\Omega_L,\tag{20}$$

and

$$\gamma\left|A(t)\right|^2 = q - \frac{\beta_2}{2}\left(\frac{d\phi(t)}{dt}\right)^2,$$

$$q = \gamma P_0 = \frac{\beta_2}{2}\Delta^2 = \frac{3\gamma}{2\varsigma}\left(1 - \frac{C}{2}\right),\tag{21}$$

$$\frac{\beta_2}{2}\Omega_L^2 = \frac{\gamma}{\varsigma}\left(1 + C\right) - \frac{5}{3}\gamma P_0,$$

Here, we trace the zero-isogain $\sigma = 0$. The DS time-profile is defined by an implicit expression:

$$t = \tau\left[\operatorname{arctanh}\left(\frac{d\phi/dt}{\Delta}\right) + \frac{\Delta}{\Omega_L}\arctan\left(\frac{d\phi/dt}{\Delta}\right)\right],\tag{22}$$

with the DS width of $\tau \propto \Delta^{-1}\left(\Delta^2 + \Omega_L^2\right)^{-1}$. Now, there are the following limiting cases:

$$\lim_{C \to 2}\begin{cases} E \to 0 \\ P_0 \to 0 \\ \Delta \to 0 \\ \Omega_L \to const \\ \tau \to 0 \\ \gamma\left|A\right|^2 = q - \beta\Delta^2\tanh^2\left(t/\tau\right) \end{cases}.\tag{23}$$

It is clear that in this 'low-energy' sector the DS time-profile is bell-like and its spectrum has tabletop form ($\Omega_L \gg \Delta$). In the DSR limit, one has:

$$\lim_{C \to 2}\begin{cases} E \to \infty \\ P_0 \to 1/\varsigma \\ \Delta \to \sqrt{\gamma/\beta\varsigma} \\ \Omega_L \to 0 \\ \tau \to const \\ \gamma\left|A\right|^2 = q - \beta\Omega_L^2\tan^2\left(t\Omega_L/\tau\Delta\right) \end{cases},\tag{24}$$

that is, a DS in the DSR sector has a flattop temporal profile and a Lorenzian spectrum ($\Omega_L \ll \Delta$). Eqs. (24) demonstrate that asymptotical growth of DS energy leads to a spectral condensation ($\Omega_L \to 0$) without a parallel temporal thermalization ($\tau \nrightarrow \infty$), which means an inevitable destabilization of a plain energy-scalability [180]. This conclusion does not mean a participial impossibility of DS energy-scaling in the frameworks of cubic-quintic CNGLE model. For instance, a saturable SPM allows DSs with tabletop profiles and $\mid d\phi / dt \mid \to \infty$ on the pulse edges. Such a DS possesses enhanced energy scalability and was observed experimentally [231].

4. DS spectrum and stability

As was explained, the dual balances in frequency domain:

$$\begin{cases} \beta_2 \Delta^2 \leftrightarrow \gamma P_0 \\ \alpha \Delta^2 \leftrightarrow \kappa P_0 \end{cases} \tag{25}$$

are formative for DS existence and stabilization. No wonder that the spectrum of DS is benchmark of its inherent properties.

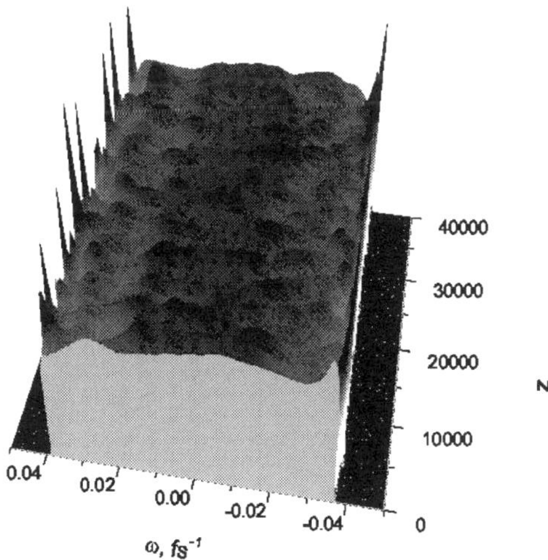

Figure 11. Perturbed DS spectrum [241].

Prior to consider the aspects of interweaving of spectral and stability properties of DSs, one has to point to a possibility of multi-wavelength multi-pulsing DSs provided by DS robustness.

As was demonstrated in [234] theoretically, the multi-DSs compounds in a mode-locked laser can be stabilized at multiple frequencies. Experimentally, such multi-frequency DS compounds can be realized by birefringence filters with a periodical (interference-like) dependence of transmission on wavelength under conditions of sufficiently broad gainband and powerful pump [235–239][4].

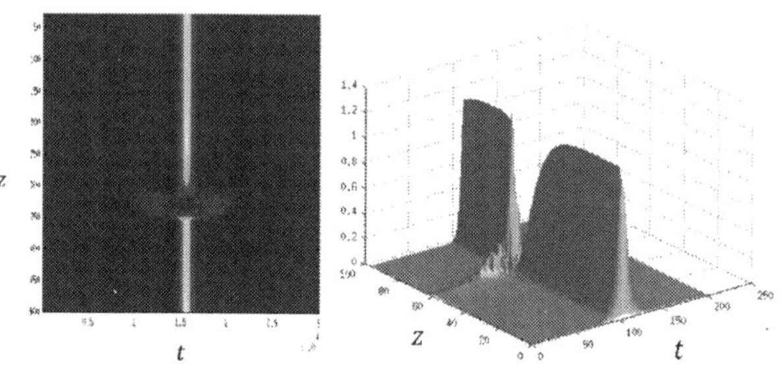

Figure 12. Exploding DS corresponding to parameters of point *A* in Figure 10. Left: contour-plot of instant power, right: 3D-graph of instant power in dependence on local time *t* and propagation distance *z* (arbitrary units) [243].

As was demonstrated in previous section, DS has non-trivial internal structure due to energy fluxes inside it. The elements of this structure (*internal modes*) can be excited that causes pulsating or chaotic dynamics of DS with preservation of its temporal and spectral localization [241]. The spectral envelope acquires a shape of 'glass with boiling water' (Figure 11). Appearance of such perturbations is understandable in frameworks of the DS perturbation theory in spectral domain [242]. One has to note that such perturbations take a place inside the DS stability region where $\sigma < 0$ (below the corresponding upper curves in Figure 10). Above the stability boarder ('no DS' region in Figure 10), there are three main destabilization scenarios [243]. For small energies and in the vicinity of stability border (point *A* in Figure 10), DS is *exploding* (Figure 12), which means its aperiodic disappearance with excitation of continuous waves and subsequent DS recreation [37,244–248]. With the energy growth (point *B* in Figure 10), the *rogue* DSs develop (Figure 13) [37,249–253]. Such a regime can be interpreted as DS structural chaotization, that is, generation of multiple DSs with strong interactions causing extreme dynamics.

For sufficiently large energies in the vicinity of stability border (point *C* in Figure 10), the typical destabilization scenario is the generation of multiple DSs (Figure 14). The source of this destabilization is the growth of spectral dissipation caused by DS spectral broadening with approaching to stability border so that the DS splitting becomes more energy advantageous [161]. Moreover, the DS splitting can be enhanced by its phase inhomogeneity because the gain (energy in-flow) is maximum at the points of stationary phase $d\phi / dt = 0$ [254]. Such a splitting

---

4 A multi-porting configuration of a DS laser supports even simultaneous generation of conventional and dissipative wavelength-separated solitons [240].

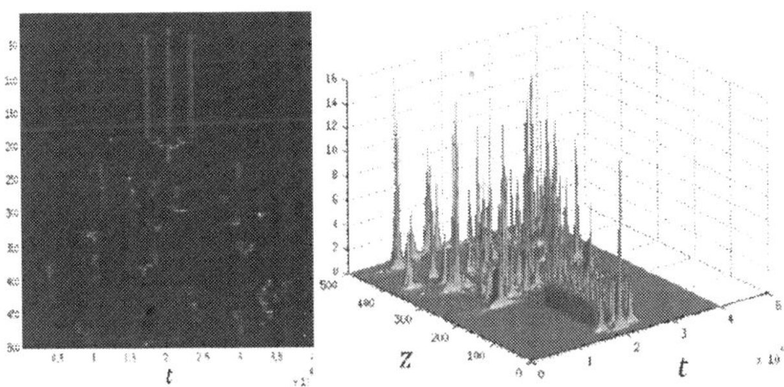

Figure 13. Rogue DSs corresponding to parameters of point B in Figure 10. Left: contour-plot of instant power, right: 3D-graph of instant power in dependence on local time t and propagation distance z (arbitrary units) [243].

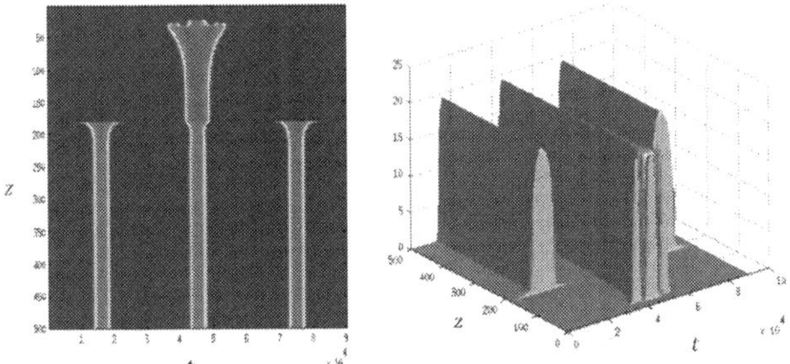

Figure 14. DS molecule corresponding to parameters of point C in Figure 10. Left: contour-plot of instant power, right: 3D-graph of instant power in dependence on local time t and propagation distance z (arbitrary units) [243].

can result in an extreme dynamics like solitonic turbulence (Figure 13) [255] or regular multi-DS complexes (Figure 14, so-called DS molecules) which can have non-extreme internal dynamics (soliton gas) and interact with a background (soliton liquid) [256-259]. If the dynamic gain saturation (see [44]) contributes, the multi-DS complexes can evolve slowly into a set of equidistant DSs with repetition rate multiple of laser one (harmonic mode-locking) [158,159].

The numerical simulations of cubic-quintic CNGLE with taking into account a quantum noise validated the fact of inconsistency of spectral condensation and absence of temporal thermo-lization that breaks the DS energy scalability (see previous section) [243]. As a result, the DS stability region breaks abruptly with energy growth (dashed curve in Figure 10) and multitude of turbulent scenarios of DSs evolution develops (Figure 15) [34,243].

Serious limitations on power and energy scalability of DSs in fibre lasers arise from stimulated Raman scattering (SRS) [2,109]. The stability border of DS under action of SRS is shown in Figure 10 by dot blue curve (DS is stable on the left of this curve) [260]. As was found, SRS enhances the tendency to multi-pulsing with energy growth caused by enhancement of

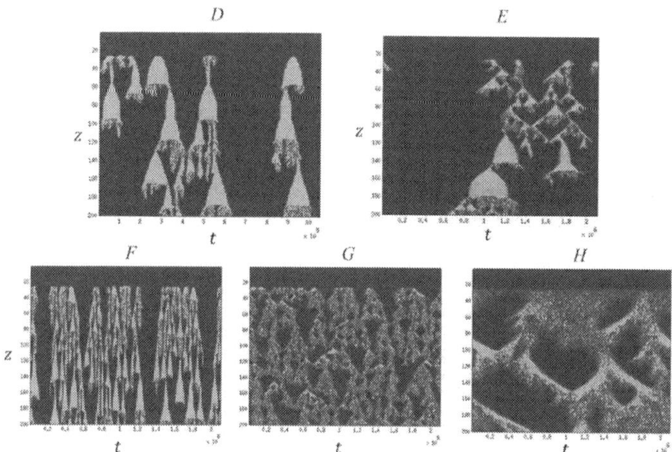

Figure 15. DS turbulent regimes (contour-plot of instant power, arbitrary units) corresponding to the parameters of points *D*, *E*, *F*, *G* and *H* in Figure 10 [243].

spectral dissipation due to SRS [260]. Simultaneously, generation of anti-Stokes radiation causes chaotization of DS dynamics and irregular modulation of DS temporal and spectral profiles [261] (Figure 16). DS profile remains localized, but it is strongly cut by colliding dark and grey soliton-like structures [34].

As was shown, the DS dynamics can be regularized by formation of *dissipative Raman soliton* (DRS). DRS can exist in the form of DS which is Stokes-shifted due to self-Raman scattering (Figure 17, *a*) [260] or as bound DS–DRS complex (Figure 17, *b*) [262]. In the last case, stabilization is achieved by feedback, i.e. reinjection of Stokes signal through a delay line [263]. In the absence of a feedback, the Raman pulse is noisy [264] (see dash line spectrum in Figure 16, *b*).

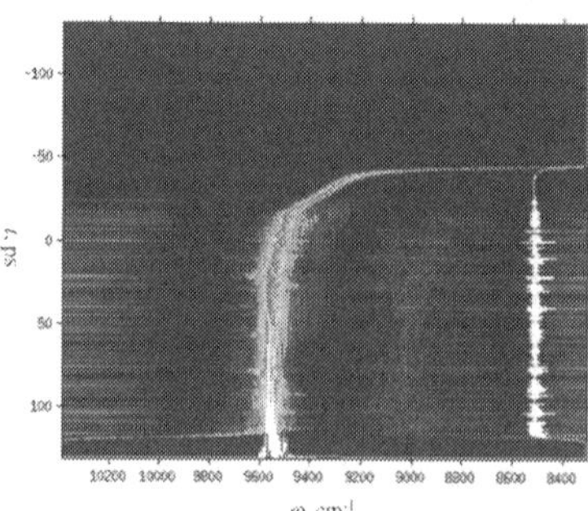

Figure 16. Wigner function of turbulent DS in the presence of SRS.

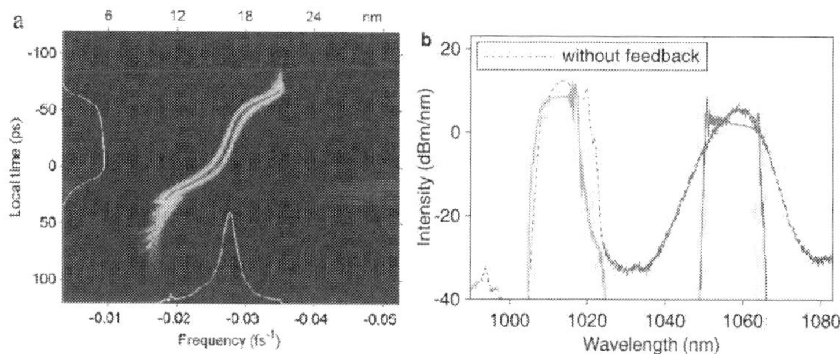

Figure 17. (a) Wigner function of single DRS [260] and (b) spectrum of bound DS–DRS complex [262].

5. Vector DSs

As was pointed above, SOP can play leading role in a fibre laser dynamics. In particular, it can contribute to mode-locking or/and spectral filtering. However, diapason of polarization phenomena in a DS fibre laser spreads essentially broader. As was found, intrinsic fibre birefringence (Figure 2) can lead to DS splitting into two independent SOPs [78]. This phenomenon is used to realize the NPR mode-locking mechanism where a DS SOP evolves (or remains locked) as a whole during propagation [265-269]. The polarization dynamics can be fast ($\leq T$) or slow ($>>T$) and vary from regular (with possible period multiplication or harmonic mode-locking) to chaotic one [270,271]. There are evidences of ultrafast SOP evolution when SOP changes across a DS profile [272].

The specific multiple pulse instability of vector dissipative solitons (VDSs) leads to generation of the bound states of DSs with different SOPs (*vector soliton molecules*) which are locked by a non-linear coupling [273,274] or by a group-velocity locking produced by spectrum shift between DSs with different SOP [275]. As was shown experimentally (Figure 18) [276], the dynamics of VDS molecules can be highly non-trivial and demonstrate both fast and slow periodic switching between fixed SOPs as well as SOP procession, which is especially interesting for fibre laser telecommunications based on polarization multiplexing.

The important breakthrough in the recent theory of VDSs is the demonstration of insufficiency of approaches based on the coupled CNGLEs (like (10)) for adequate description of DS polarization dynamics. It was demonstrated that an active medium polarizability contributes to DS dynamics substantially [277]. As was shown, the SOP-sensitive interaction between DS and a slowly relaxing active medium with taking into account the birefringence of fibre laser elements and light-induced anisotropy caused by elliptically polarised pump field change the SOP at a long time scale that results in fast and slowly evolving SOPs of VDSs (Figure 19).

The non-trivial contribution of active medium kinetics and polarizability with taking into account the pump SOP and SPM demonstrates a complex dynamics including spiral attractors

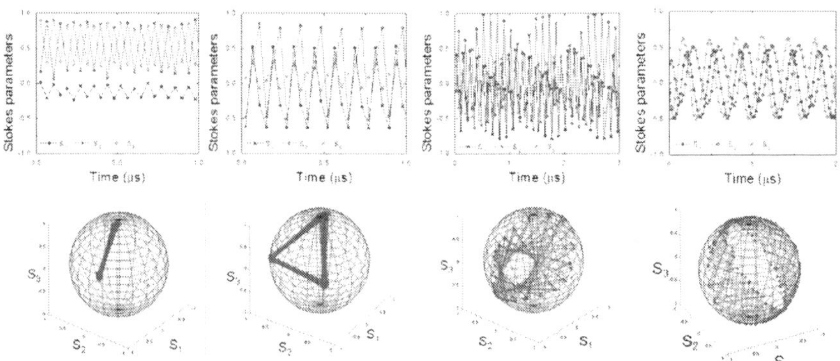

Figure 18. Polarization dynamics of VDS molecules in an Er-fibre laser mode-locked by carbon nanotubes [276]. Top row demonstrates the evolution of Stokes parameters. The bottom row reproduces this evolution on the Poincaré sphere (each point corresponds to SOP after one laser round-trip).

and dynamic chaos (Figure 20) [278]. One may assume that such a non-trivial polarization dynamics is of great importance for DS energy scaling, in particular, due to vector nature of SRS [279]. These topics remain unexplored to date.

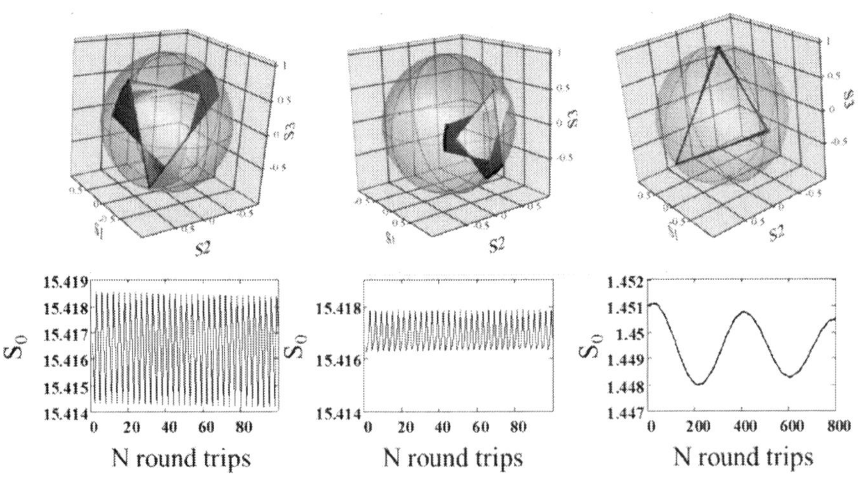

Figure 19. Polarization dynamics of VDS with taking into account an Er-fibre polarizability [277]. Both fast and slow SOP dynamics exist in dependence on fiber laser birefringence strength.

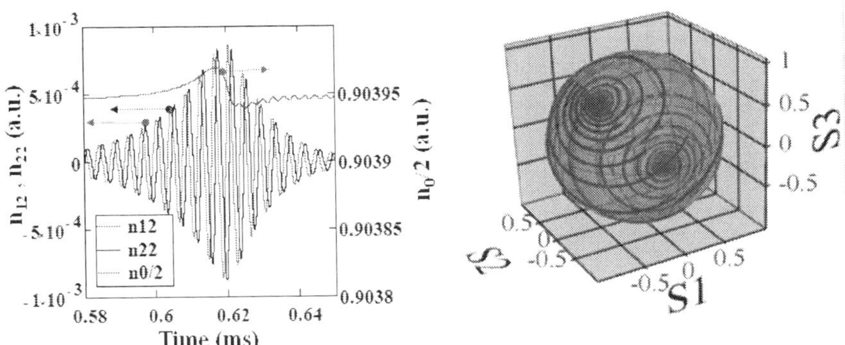

Figure 20. Dynamics of active fibre inversion components, which are polarization-sensitive (n_{12} and n_{22}) and non-sensitive (n_0), with the corresponding slow SOP evolution of VDS [278].

6. Conclusion

The recent progress in development of ultrafast fibre lasers and advances in exploring of DS are interrelated. DSs allowed scoring a great success in ultrashort pulse energy scalability that is defined by unprecedented stability and robustness of DS. At this moment, it is possible to achieve over-MW peak powers for sub-100 fs pulses directly from a fibre laser at over-MHz repetition rates. New spectral diapasons became reachable owing to development of mid-IR active fibres and using the frequency-conversion directly in a laser. Development of new mode-locking techniques, especially based on using of SESAMs, graphene and another quantum-sized structure allowed improving a laser stability, integrity and environment insensitivity. A great advance has been achieved in the theory of DSs. New powerful analytical techniques based on extensive numerical simulations and experimental advances extended understanding of the DS fundamental properties and revealed new prospects in improvement of characteristics of ultrafast fibre lasers. Based on achieved results, one may outline some unresolved problems. As was found, there are stability limits for a DS energy scaling imposed by optical turbulence and SRS. Deeper insight into the nature of these phenomena could allow to overcome these limits without substantial complication of laser setup. Simultaneously, control of intra-laser spectral conversion is a direct way to broadening of spectral range. Then, the dynamics and properties of VDSs remain scantily explored. Recent studies demonstrated a multitude of polarization phenomena, which cannot be grasped in frameworks of existing models. In particular, polarizability and kinetics of an active fibre in combination with birefringence of a laser in a whole can contribute non-trivially to a laser dynamics. As an additional aspect of further development, one may point at the development of new mode-locking techniques, which could improve DS stability and integrity of a fibre laser, decrease pulse width and extend a diapason of pulse repetition rates. At last, one has to remember that a fibre laser is an ideal playground for study of complex non-linear phenomena and, undoubtedly, new bridges between different fields of science will be built with a further progress of ultrafast fibre lasers.

Acknowledgements

This work was supported by FP7-PEOPLE-2012-IAPP (project GRIFFON, No. 324391).

Author details

Vladimir L. Kalashnikov* and Sergey V. Sergeyev

*Address all correspondence to: v.kalashnikov@aston.ac.uk

AIPT, Aston University, B 7ET, Birmingham, UK

References

[1]   Fermann ME, Hartl I. Ultrafast fiber laser technology. IEEE J Sel Topics in Quantum Electron 2009;15(1):191–206. DOI: 10.1109/JSTQE.2008.2010246.

[2]   Fermann ME, Galvanauskas A, Sucha G, Harter D. Fiber-lasers for ultrafast optics. Appl Physics B 1997;65:259–75.

[3]   Fermann ME, Hartl I. Ultrafast fiber lasers. Nature Photonics 2013;7:868–74. DOI: 10.1038/NPHOTON.2013.280.

[4]   Limpert J, Röser F, Schreiber Th, Tünnermann A. High-power ultrafast fiber laser systems. IEEE J Sel Top Quantum Electron 2006;12(2):233–44. DOI: 10.1109/JSTQE. 2006.872729.

[5]   Lamb WE. Theory of an optical laser. Phys Rev 1964;134:A1429–50.

[6]   Siegman AE. Lasers. Sausalito: University Science Book; 1986. 1283 p.

[7]   Lefrançois S, Kieu K, Deng Y, Kafka JD, Wise FW. Scaling of dissipative soliton fiber lasers to megawatt peak powers by use of large-area photonic crystal fiber. Opt Lett 2010;35(10):1569–71.

[8]   Baumgartl M, Lecaplain C, Hideur A, Limpert J, Tünnermann C. 66 W average power from a microjoule-class sub-100 fs fiber oscillator. Opt Lett 2012;37(10):1640–2. DOI: 10.1364/OL.37.001640.

[9]   Chong A, Renninger WH, Wise FW. Route to the minimum pulse duration in normal-dispersion fiber lasers. Opt Lett 2008;33(22):2638–40. DOI: 10.1364/OL.33.002638.

[10]   Sucha G. Overview of industrial and medical applications of ultrashort pulse lasers. In: Fermann ME, Galvanauskas A, Sucha G. (Eds.) Ultrafast Lasers: Technology and Applications. New York: Marcel Dekker, Inc.; 2003. pp. 323–358.

[11]   Xu C, Wise FW. Recent advances in fibre lasers for nonlinear microscopy. Nature Photonics 2013;7:875–82. DOI: 10.1038/NPHOTON.2013.284.

[12]   Müller M, Squier J. Nonlinear microscopy with ultrashort pulse lasers. In: Fermann ME, Galvanauskas A, Sucha G. (Eds.) Ultrafast Lasers: Technology and Applications. New York: Marcel Dekker, Inc.; 2003, pp. 661–97.

[13]   Kalashnikov VL, Sorokin E. Soliton absorption spectroscopy. Phys Rev A 2010;81:033840. DOI: 10.1103/PhysRevA.81.033840.

[14]   Kalashnikov VL, Sorokin E, Sorokina IT. Chirped dissipative soliton absorption spectroscopy. Opt Express 2011;19(18):17480–92.

[15]   Kurtz RM, Sarayba MA, Juhasz T. Ultrafast lasers in ophthalmology. In: Fermann ME, Galvanauskas A, Sucha G. (Eds.) Ultrafast Lasers: Technology and Applications. New York: Marcel Dekker, Inc.; 2003, pp. 745–65.

[16]   Clowes J. Next generation light sources for biomedical applications. Optic Photonic 2011;3(1):36–8.

[17]   Fujimoto JG, Brezinski M, Drexler W, Hartl I, Kärtner F, Li X, Morgner U. Optical coherence tomography. In: Fermann ME, Galvanauskas A, Sucha G. (Eds.) Ultrafast Lasers: Technology and Applications. New York: Marcel Dekker, Inc.; 2003, pp. 699–743.

[18]   Drexler W, Fujimoto JG. (Eds.) Optical Coherence Tomography. Berlin: Springer-Verlag; 2008. 1346 p.

[19]   Lanin AA, Fedotov IV, Sidorov-Biryukov DA, Doronina-Amitonova LV, Ivashkina OI, Zots MA, Sun C-K, Ilday FÖ, Fedotov AB, Anokhin KV, Zheltikov AM. Air-guided photonic-crystal-fiber pulse-compression delivery of multimegawatt femtosecond laser output for nonlinear-optical imaging and neurosurgery. Appl Phys Lett 2012;100:101104. DOI: 10.1063/1.3681777.

[20]   Osellame R, Cerullo G, Ramponi R. (Eds.) Femtosecond Laser Micromachining. Heidelberg: Springer; 2012. 483 p.

[21] Gattass RR, Mazur E. Femtosecond laser micromachining in transparent materials. Nat Photon. 2008;2:219–25. DOI: 10.1038/nphoton.2008.47.

[22] Nakazawa M. Ultrahigh bit rate communication system. In: Fermann ME, Galvanauskas A, Sucha G. (Eds.) Ultrafast Lasers: Technology and Applications. New York: Marcel Dekker, Inc.; 2003. pp. 611–660.

[23] Udem Th, Holzwarth R, Hänsch TW. Optical frequency metrology. Nature 2002;416:233–7. DOI: 10.1038/416233a.

[24] Liu Y, Tschuch S, Rudenko A, Dürr M, Siegel M, Morgner U, Moshammer R, Ullrich J. Strong-field double ionization of Ar below the recollision threshold. Phzs Rev Lett 2008;101:053001. DOI: 10.1103/PhysRevLett.101.053001.

[25] Sciaini G, Miller RJD. Femtosecond electron diffraction: heralding the era of atomically resolved dynamics. Rep Prog Phys 2011;74:096101. DOI: 10.1088/0034-4885/74/9/096101.

[26] Südmeyer T, Marchese SV, Hashimoto S, Baer CRE, Gingras G, Witzel B, Keller U. Femtosecond laser oscillators for high-field science. Nat Photon 2008;2:599–604. DOI: 10.1038/nphoton.2008.194.

[27] Krausz F, Ivanov M. Attosecond physics. Rev Mod Phys 2009;81(1):163–234. DOI: 10.1103/RevModPhys.81.163

[28] Pfeifer T, Spielmann C, Gerber G. Femtosecond x-ray science. Rep Prog Phys 2006;69(2):443–505. DOI: 10.1088/0034-4885/69/2/R04.

[29] Mourou GA, Tajima T, Bulanov SV. Optics in the relativistic regime. Rev Mod Phys 2006;78(2):309–71. DOI: 10.1103/RevModPhys.78.309.

[30] Martinez A, Sun Z. Nanotube and graphene saturable absorbers for fibre lasers. Nat Photon 2013;7:842–5.

[31] Lecaplain C, Grelu Ph, Soto-Crespo JM, Akhmediev N. Dissipative rogue waves generated by chaotic bunching in a mode-locked laser. Phys Rev Lett 2012;108:233901.

[32] Dudley JM, Dias F, Erkintalo M, Genty G. Instabilities, breathers and rogue waves in optics. Nat Photon 2014;8:755–64.

[33] Dudley JM, Finot Ch, Richardson DJ, Millot G. Self-similarity in ultrafast nonlinear optics. Nat Phys 2007;3:597–603.

[34] Turitsyna EG, Smirnov SV, Sugavanam S, Tarasov N, Shu X, Babin SA, Podivilov EV, Churkin DV, Falkovich G, Turitsyn SK. The laminar–turbulent transition in a fibre laser. Nat Photon 2013;7:783–6. DOI: 10.1038/NPHOTON.2013.246.

[35] Akhmediev NN, Ankiewicz A. Solitons: Nonlinear Pulses and Beams. London: Chapman & Hall; 1997.

[36] Akhmediev NN, Ankiewicz A. (Eds.) Dissipative Solitons. Berlin: Springer-Verlag; 2005.

[37] Grelu Ph, Akhmediev N. Dissipative solitons for mode-locked lasers. Nat Photon 2012;6:84–92. DOI: 10.1038/nphoton.2011.345

[38] Faccio D, Belgiorno F, Cacciatori S, Gorini V, Liberati S, Moschella U. (Eds.) Analogue Gravity Phenomenology. Analogue Spacetimes and Horizons, from Theory to Experiment. Heidelberg: Springer; 2013. p. 438. DOI: 10.1007/978-3-319-00266-8.

[39] Kevrekidis PG, Frantzeskakis DJ, Carretero-González R. (Eds.) Emergent Nonlinear Phenomena in Bose-Einstein Condensates. Berlin: Springer-Verlag; 2008.

[40] Yang Y. Solitons in Field Theory and Nonlinear Analysis. New York: Springer; 2001.

[41] Abdulaev FK, Konotop VV. (Eds.) Nonlinear Waves: Classical and Quantum Aspects. Dordrecht: Kluwer Academic Pub.; 2004.

[42] Akhmediev NN, Ankiewicz A. (Eds.) Dissipative Solitons: From Optics to Biology and Medicine. Berlin: Springer-Verlag; 2008.

[43] Naruse M, (Ed.) Nanophotonic Information Physics. Berlin: Springer-Verlag; 2014.

[44] Kalashnikov VL. Chirped-pulse oscillators: route to the energy-scalable femtosecond pulses. In: Al-Khursan A. (Ed.) Solid State Laser. InTech, 2008; pp. 145–184. DOI: 10.5772/37415.

[45] Baer CRE, Heckl OH, Saraceno CJ, Schriber C, Kränkel C, Südmeyer T, Keller U. Frontiers in passively mode-locked high-power thin-disk laser oscillators. Opt Express 2012;20:7054–65. DOI: 10.1364/OE.20.007054.

[46] Saraceno CJ, Emaury F, Heckl OH, Baer CRE, Hoffmann M, Schriber C, Golling M, Südmeyer Th, Keller U. 275 W average output power from a femtosecond thin disk oscillator operated in a vacuum environment. Opt Express 2012;20:23535–41. DOI: 10.1364/OE.20.023535.

[47] Zhang J, Brons J, Lilienfein N, Fedulova E, Pervak V, Bauer D, Sutter D, Wei Zh, Apolonski A, Pronin O, Krausz F. 260-megahertz, megawatt-level thin-disk oscillator. Opt Lett 2015;40:1627–30. DOI: 10.1364/OL.40.001627.

[48] Brons J, Pervak V, Fedulova E, Bauer D, Sutter D, Kalashnikov V, Apolonskiy A, Pronin O, Krausz F. Energy scaling of Kerr-lens mode-locked thin-disk oscillators. Opt Lett 2014;39:6442–5. DOI: 10.1364/OL.39.006442.

[49] Naumov S, Fernandez A, Graf R, Dombi P, Krausz F, Apolonski A. Approaching the microjoule frontier with femtosecond laser oscillators. New J Phys 2005;7:216. DOI: 10.1088/1367-2630/7/1/216.

[50] Fedulova E, Fritsch K, Brons J, Pronin O, Amotchkina T, Trubetskov M, Krausz F, Pervak V. Highly-dispersive mirrors reach new levels of dispersion. Opt Express 2015;23:13788–93. DOI: 10.1364/OE.23.013788.

[51] Richardson DJ, Nilsson J, Clarkson WA. High power fiber lasers: current status and future prospects. J Opt Soc Am B 2010;27:B63–B92.

[52] Sraceno CJ, Heckl OH, Baer CRE, Schriber C, Golling M, Beil K, Kränkel ST, Huber G, Keller U. Sub-100 femtosecond pulses from a SESAM modelocked thin disk laser. Appl Phys B 2012;106:559–62. DOI: 10.1007/s00340-012-4900-5.

[53] Zhang J, Brons J, Seidel M, Pervak V, Kalashnikov V, Wei Z, Apolonski A, Krausz F, Pronin O. 49-fs Yb:YAG thin-disk oscillator with distributed Kerr-lens mode-locking. In: CLEO/Europe-EQEC Scientific Programme; 21–25 June; Munich, Germany. 2015. p. 166 (PD-A.1 WED).

[54] Sorokin E, Kalashnikov VL, Naumov S, Teipel J, Warken F, Giessen H, Sorokina IT. Intra- and extra-cavity spectral broadening and continuum generation at 1.5 \mu m using compact low energy femtosecond Cr:YAG laser. Applied Phys B 2003;77(2–3): 197–204.

[55] Joly NY, Nold J, Chang W, Hölzer P, Nazarkin A, Wong GKL, Biancalana F, Russel P. St J. Bright spatially coherent wavelength-tunable deep-UV laser source using an Ar-filled photonic crystal fiber. Phys Rev Lett 2011;106:203901.

[56] Jackson SD. Towards high-power mid-infrared emission from a fiber laser. Nat Photon 2012;28:423–31. DOI: 10.1038/NPHOTON.2012.149.

[57] Acerbron JA, Bonilla LL, Vicente CJP, Ritort F, Spigler R. The Kuramoto model: a simple paradigm for synchronization phenomena. Rev Mod Phys 2005;77:137–85.

[58] Pikovsky A, Rosenblum M, Kurths J. Synchronization: a universal concept in nonlinear sciences. Cambridge: Cambridge University Press; 2001.

[59] Kuizenga DJ, Siegman AE. FM and AM mode locking of the homogeneous laser - Part I: Theory. IEEE J Quantum Electron 1970;6(11):694–708. DOI: 10.1109/JQE. 1970.1076343.

[60] Haus HA. Mode-locking of lasers. IEEE J Sel Topic Quantum Electron 2000;6(6):1173–85.

[61] Haus HA. Short pulse generation. In: Duling IN, III (Ed.) Compact Sources of Ultrashort Pulses. Cambridge: Cambridge University Press; 1995. pp. 1–56.

[62] Kalashnikov VL. Mathematical Ultrashort-Pulse Laser Physics [Internet]. 15/09/2000 [Updated: 29/03/2002]. Available from: http://lanl.arxiv.org/abs/physics/0009056v3.

[63] Kuizenga DI, Siegman AE. Modulator frequency detuning effects in the FM mode-locked laser. IEEE J Quantum Electron 1970;QE-6:803–8.

[64] Akhmanov SA, Vysloukh VA, Chirkin AS. Optics of Femtosecond Laser Pulses. New York: AIP; 1992.

[65] Haus HA, Silberberg Y. Laser mode locking with addition of nonlinear index. IEEE J Quantum Electron 1986;QE-22(2):325–31.

[66] Kalashnikov VL, Poloyko IG, Mikhaylov VP. Generation of ultrashort pulses in lasers with external frequency modulation. Quantum Electron 1998;28(3):264–8.

[67] Kolner BH. Space-time duality and the theory of temporal imaging. IEEE J Quantum Electron 1994;QE-30(8):1951–63.

[68] van Howe J, Xu Ch. Ultrafast optical signal processing based upon space-time dualities. IEEE J Lightwave Technology 2006;24(7):2649–62. DOI: 10.1109/JLT.2006.875229.

[69] Salem R, Foster MA, Gaeta AL. Application of space-time duality to ultrahigh-speed optical signal processing. Adv Optics Photon 2013;5:274–317. DOI: 10.1364/AOP. 5.000274.

[70] Agrawal GP. Nonlinear Fiber Optics. Third Edition ed. San Diego: AP; 2001. 466 p.

[71] Newell AC. Solitons in Mathematics and Physics. Philadelphia: SIAM; 1985. 246 p.

[72] Lai Y, Haus HA. Quantum theory of solitons in optical fibers. I. Time-dependent Hartree approximation. Phys Rev A 1989;40:844–53.

[73] Lai Y, Haus HA. Quantum theory of solitons in optical fibers. II. Exact solution. Phys Rev A 1989;40:854–66.

[74] Yoon B, Negele JW. Time-dependent approximation for a one-dimensional system of bosons with attractive δ-function interactions. Phys Rev A 1977;16:1451.

[75] Haus HA. Theory of mode-locking with a fast saturable absorber. J Appl Phys 1975; 46:3049–58.

[76] Ablowitz MJ, Clarkson PA. Solitons, Nonlinear Evolution Equations and Inverse Scattering. Cambridge: Cambridge University Press; 1991. 516 p.

[77] Turitsyn SK, Bale BG, Fedoruk MP. Dispersion-managed solitons in fibre systems and lasers. Phys Rep 2012;521:135–203. DOI: 10.1016/j.physrep.2012.09.004.

[78] Menyuk CR. Nonlinear pulse propagation in birefringent optical fibers. IEEE J Quantum Electron 1987;23(2):174–6.

[79] Menyuk GR. Pulse propagation in an elliptically birefringent Kerr medium. IEEE J Quantum Electron 1989;25(12):2674–82.

[80] Hasegawa A. Optical Solitons in Fibers. Berlin: Springer-Verlag; 1990. 79 p.

[81] Ding E, Renninger WH, Wise FW, Grelu Ph, Shlizerman E, Kutz JN. High-energy passive mode-locking of fiber lasers. Int J Optics 2012;2012(ID354156):1–17. DOI: 10.1155/2012/354156.

[82] Wang R, Dai Y, Yan L, Wu J, Xu K, Li Y, Lin J. Dissipative soliton in actively mode-locked fiber laser. Optics Express 2012;20(6):6406–11.

[83] Koliada NA, Nyushkov BN, Ivanenko AV, Kobtsev SM, Harper P, Turitsyn SK, Denisov VI, Pivtsov VS. Generation of dissipative solitons in an actively mode-locked ultralong fibre laser. Quantum Electron 2013;43(2):95–8. DOI: 10.1070/QE2013v043n02ABEH015041.

[84] Wang R, Dai Y, Yin F, Xu K, Li J, Lin J. Linear dissipative soliton in an anomalous-dispersion fiber laser. Optics Express 2014;22(24):29314–20. DOI: 10.1364/OE.22.029314.

[85] Haus HA, Fujimoto JG, Ippen EP. Analytic theory of additive pulse and Kerr lens mode locking. IEEE J. Quantum Electron 1992;28(10):2086–96.

[86] Fermann ME, Andrejco MJ, Silberberg Y, Stock ML. Passive mode locking by using nonlinear polarization evolution in a polarization-maintaining erbium-doped fiber. Optics Lett 1993;18(11):894–6. DOI: 10.1364/OL.18.000894.

[87] Hofer M, Ober MH, Haberl F, Fermann ME. Characterization of ultrashort pulse formation in passively mode-locked fiber lasers. IEEE J Quantum Electron 1992;28(3)DOI:720–8.

[88] Fermann ME. Nonlinear polarization evolution in passively modelocked fiber lasers. In: Duling IN, III (Ed.) Compact sources of ultrashort pulses. Cambridge: Cambridge University Press; 1995. pp. 179–207.

[89] Winful HG. Polarization instabilities in birefringent nonlinear media: application to fiber-optic devices. Optics Lett 1986;11(1):33–5.

[90] Kalashnikov VL, Kalosha VP, Mikhailov VP. Self-mode locking of continuous-wave solid-state lasers with a nonlinear Kerr polarization modulator. J Opt Soc Am B 1993;10:1443–6.

[91] Ding E, Kutz JN. Operating regimes, split-step modeling, and the Haus master mode-locking model. J Opt Soc Am B 2009;26(12):2290–300.

[92] Ding E, Shlizerman E, Kutz JN. Generalized master equation for high-energy passive mode-locking: the sinusoidal Ginzburg-Landau equation. IEEE J Quantum Electron 2011;47(5):705–14.

[93] Komarov A, Leblond H, Sanchez F. Multistability and hysteresis phenomena in passively mode-locked fiber lasers. Phys Rev A 2005;71(5):053809.

[94] Komarov A, Leblond H, Sanchez F. Quantic complex Ginzburg-Landau model for ring fiber lasers. Phys Rev E 2005;72(2):025604.

[95] Matsas VJ, Newson TP, Richardson DJ, Payne DN. Selfstarting passively mode-locked fibre ring soliton laser exploiting nonlinear polarization rotation. Electron Lett 1992;28(15):1391–3.

[96] Tamura K, Ippen EP, Haus HA, Nelson IE. 77-fs pulse generation from a stretched-pulse mode-locked all-fiber ring laser. Optics Lett 1993;18(13):1080–2.

[97] Chong A, Buckley J, Renninger W, Wise F. All-normal-dispersion femtosecond fiber laser. Optics Express 2006;14(21):10096–100.

[98] Zhao LM, Tang DY, Wu J. Gain-guided soliton in a positive group-dispersion fiber laser. Optics Lett 2006;31(12):1788–90.

[99] Cabasse A, Ortac B, Martel G, Hideur A, Limpert J. Dissipative solitons in a passively mode-locked Er-doped fiber with strong normal dispersion. Optics Express 2008;16(23):19323–9.

[100] Kieu K, Renninger WH, Chong A, Wise FW. Sub-100 fs pulses at watt-level powers from a dissipative-soliton fiber laser. Optics Lett 2009;34(5):593–5.

[101] Ruehl A, Wandt D, Morgner U, Kracht D. Normal dispersive ultrafast fiber oscillators. IEEE J Sel Topic Quantum Electron 2009;15(1):170–81.

[102] Wu X, Tang DY, Zhang H, Zhao LM. Dissipative soliton resonance in an all-normal-dispersion erbium-doped fiber laser. Optics Express 2009;17(7):5580–4.

[103] Renninger WH, Chong A, Wise F. Self-similar evolution in an all-normal-dispersion laser. Phys Rev A 2010;82:021805(R).

[104] Zhao L, Tang D, Wu X, Zhang H. Dissipative soliton generation in Yb-fiber laser with an invisible intracavity bandpass filter. Optics Lett 2010;35(16):2756–8.

[105] Chichkov NB, Hausmann K, Wandt D, Morgner U, Neumann J, Kracht D. High-power dissipative solitons from an all-normal dispersion erbium fiber oscillator. Optics Lett 2010;35(16):2807–9.

[106] Baumgartl M, Ortac B, Lecaplain C, Hideur A, Limpert J, Tuenermann A. Sub-80 fs dissipative soliton large-mode-area fiber laser. Optics Lett 2010;35(13):2311–3.

[107] Lecaplain C, Baumgartl M, Schreiber T, Hideur A. On the mode-locking mechanism of a dissipative-soliton fiber laser. Optics Express 2011;19(27):26742–51.

[108] Zhang Z, Dai G. All-normal-dispersion dissipative soliton Ytterbium fiber laser without dispersion compensation and additional filter. IEEE Photon J 2011;3(6):1023–9. DOI: 10.1109/JPHOT.2011.2170057.

[109] Kharenko DS, Podivilov EV, Apolonski AA, Babin SA. 20 nJ 200 fs all-fiber-highly chirped dissipative soliton oscillator. Optics Lett 2012;37(19):4104–6.

[110] Li X, Wang Y, Zhao W, Liu X, Wang Y, Tsang YH, Zhang W, Hu X, Yang Zh, Gao C, Li Ch, Shen D. All-fiber dissipative solitons evolution in a compact passively Yb-doped mode-locked fiber laser. J Lightwave Tech 2012;30(15):2502–7.

[111] Duan L, Liu X, Mao D, Wang L, Wang G. Experimental observation of dissipative soliton resonance in an anomalous-dispersion fiber laser. Optics Express 2012;20(1): 265–70.

[112] Doran NJ, Wood D. Nonlinear-optical loop mirror. Optics Lett 1988;13(1):56–8.

[113] Ippen EP, Haus HA, Liu LY. Additive pulse modelocking. J Opt Soc Am B 1989;6:1736–45.

[114] Duling IN III, Dennis ML. Modelocking of all-fiiber lasers. In: Duling IN III, (Eds.) Compact Sources of Ultrashort Pulses. Cambridge: Cambridge University Press; 1995, pp. 140–178.

[115] Mark J, Liu LY, Hall KL, Haus HA, Ippen EP. Femtosecond pulse generation in a laser with a nonlinear external resonator. Optics Lett 1989;14(1):48–50.

[116] Kalashnikov VL, Kalosha VP, Mikhailov VP, Poloyko IG. Multi-frequency continuous wave solid-state laser. Optics Commun 1995;116(4–6):383–8.

[117] Kalashnikov VL, Kalosha VP, Mikhailov VP, Poloyko IG, Demchuk MI. Efficient self-mode locking of continuous-wave solid-state lasers with resonant nonlinearity in an additional cavity. Optics Commun 1994;109:119–25.

[118] Kalashnikov VL, Kalosha VP, Mikhailov VP, Poloyko IG, Demchuk MI. Self-mode-locking of cw solid-state lasers with a nonlinear antiresonant ring. Quantum Electron 1994;24(1):35–9.

[119] Richardson DJ, Laming RI, Payne DN, Matsas V, Phillips MW. Selfstarting, passively modelocked Erbium fibre laser based on amplifying Sagnac switch. Electronics Lett 1991;27(6):542–3.

[120] Nicholson JW, Andrejco M. A polarization maintaining, dispersion managed, femtosecond figure-eight fiber laser. Optics Express 2006;14(18):8160–7.

[121] Yun L, Liu X, Mao D. Observation of dual-wavelength dissipative solitons in a figure-eight erbium-doped fiber laser. Optics Express 2012;20(19):20992–7.

[122] Wang S-K, Ning Q-Y, Luo A-P, Lin A-B, Luo Z-C, Xu W-C. Dissipative soliton resonance in a passively mode-locked figure-eight fiber laser. Optics Express 2013;21(2): 2402–7.

[123] Zhao LM, Bartnik AC, Tai QQ, Wise FW. Generation of 8 nJ pulses from a dissipative-soliton fiber laser with a nonlinear optical loop mirror. Optics Lett 2013;38(11): 1942–4.

[124] Lin H, Guo Ch, Ruan Sh, Yang J. Dissipative soliton resonance in an all-normal-dispersion Yb-doped figure-eight fibre laser with tunable output. Laser Phys Lett 2014;11:085102.

[125] Xu Y, Song Y, Du G, Yan P, Guo Ch, Zheng G, Ruan Sh. Dissipative soliton resonance in a wavelength-tunable Thulium-doped fiber laser with net-normal dispersion. IEEE Photonics J 2015;7(3):1502007. DOI: 10.1109/JPHOT.2015.2424855.

[126] Keller U. Recent developments in compact ultrafast lasers. Nature 2003;424(14):831–8.

[127] Keller U. Semiconductor Nonlinearities for Solid-State Laser Modelocking and Q-switching. In: Garmire E, Kost A. (Eds.) Nonlinear Optics in Semiconductors II (Semiconductors and Semimetals, Vol. 59). San Diego: AP; 1999. pp. 211–86.

[128] Keller U, Weingarten KJ, Kaertner FX, Kopf D, Braun B, Jung ID, Fluck R, Hoenninger C, Matuschek N, Aus der Au J. Semiconductor saturable absorber mirrors (SESAM's) for femtosecond to nanosecond pulse generation in solid-state lasers. IEEE J Sel Top Quantum Electron 1996;2(3):435–53.

[129] Kaertner F. Lecture Notes: Introduction to Ultrafast Optics, Chapter 8 [Internet]. 2005. Available from: http://ocw.mit.edu/courses/electrical-engineering-and-computer-science/6-977-ultrafast-optics-spring-2005/lecture-notes/chapter8.pdf.

[130] Chong A, Renninger WH, Wise FW. Environmentally stable all-normal-dispersion femtosecond fiber laser. Optics Lett 2008;33(10):1071–3. DOI: 10.1364/OL.33.001071.

[131] Cabasse A, Martel G, Oudar JL. High power dissipative soliton in an Erbium-doped fiber laser mode-locked with a high modulation depth saturable absorber mirror. Optics Express 2009;17(12):9537–42.

[132] Tang M, Wang H, Becheker R, Oudar J-L, Gaponov D, Godin T, Hideur A. High-energy dissipative solitons generation from a large normal dispersion Er-fiber laser. Optics Lett 2015;40(7):1414–7.

[133] Gumenyuk R, Vartianen I, Tuovinen H, Okhotnikov OG. Dissipative dispersion-managed soliton 2mm thulium/holmium fiber laser. Optics Lett 2011;36(5):609–11.

[134] Lecourt J-B, Duterte C, Narbonneau F, Kinet D, Hernandez Y, Giannone D. All-normal dispersion, all-fibered PM laser mode-locked by SESAM. Optics Express 2012;20(11):11918–23. DOI: 10.1364/OE.20.011918.

[135] Jiang K, Ouyang C, Wu K, Wong JH. High-energy dissipative soliton with MHz repetition rate from an all-fiber passively mode-locked laser. Optics Commun 2012;285(9):2422–5. DOI: 10.1016/j.optcom.2012.01.033.

[136] Poloyko IG, Kalashnikov VL. Semiconductor saturable absorber mirrors as mode-locking device for femtosecond lasers: nonlinear Fabri-Perot resonator approach. Optics Commun 1999;168:167–75.

[137] Sun Z, Hasan T, Ferrari AC. Ultrafast lasers mode-locked by nanotubes and graphene. Physica E 2012;44:1082–91. DOI: 10.1016/j.physe.2012.01.012.

[138] Zhang H, Tang D, Knize RJ, Zhao L, Bao Q, Loh KP. Graphene mode locked, wavelength tunable, dissipative soliton fiber laser. Appl Phys Lett 2010;96:111112.

[139] Zhao LM, Tang DY, Zhang H, Wu X, Bao Q, Loh KP. Dissipative soliton operation of an Ytterbium-doped fiber laser mode locked with atomic multilayer graphene. Optics Lett 2010;35(21):3622–4.

[140] Cui YD, Liu XM, Zeng C. Conventional and dissipative solitons in a CFBG-based fiber laser mode-locked with a graphene-nanotube mixture. Laser Phys Lett 2014;11:055106.

[141] Cheng Zh, Li H, Shi H, Ren J, Yang Q-H, Wang P. Dissipative soliton resonance and reverse saturable absorption in graphene oxide mode-locked all-normal-dispersion Yb-doped fiber laser. Optics Express 2015;23(6):7000–6. DOI: 10.1364/OE.23.007000.

[142] Liu X, Cui Y, Han D, Yao X, Sun Zh. Distributed ultrafast laser. Sci Rep2014;5:9101. DOI: 10.1038/srep9101.

[143] Im JH, Choi SY, Rotermund F, Yeom D-I. All-fiber Er-doped dissipative soliton laser based on evanescent field interaction with carbon nanotube saturable absorber. Optics Express 2010;18(21):22141–6.

[144] Du J, Wang Q, Jiang G, Xu Ch, Zhao Ch, Xiang Y, Chen Y, Wen Sh, Zhang H. Ytterbium-doped fiber passively mode locked by few-layer molybdenum disulfide (MoS_2) saturable absorber functioned with enhanced field interaction. Sci Rep 2014;4:6346. DOI: 10.1038/srep06346.

[145] Paschotta R, Keller U. Passive mode locking with slow saturable absorbers. Appl Phys B 2001;73:653–62. DOI: 10.1007/s003400100726.

[146] Haus HA, Silberberg Y. Theory of mode locking of a laser diode with a multiple-quantum-well structure. J Opt Soc Am B 1985;2(7):1237–43.

[147] Kaertner F. Lecture Notes: Introduction to Ultrafast Optics. Chapter 7 [Internet]. 2005. Available from: http://ocw.mit.edu/courses/electrical-engineering-and-computer-science/6-977-ultrafast-optics-spring-2005/lecture-notes/chapter7.pdf.

[148] Lam C-K, Malomed BA, Chow KW, Wai PKA. Spatial solitons supported by localized gain in nonlinear optical waveguides. Eur Phys J Special Topics 2009;173:233–43. DOI: 10.1140/epjst/e2009-01076-8.

[149] Sakaguchi H, Malomed BA. Stable two-dimensional solitons supported by radially inhomogeneous self-focusing nonlinearity. Optics Lett 2012;37(6):1035–7. DOI: 10.1364/OL.37.001035.

[150] Borovkova OV, Kartashov YV, Vysloukh VA, Lobanov VE, Malomed BA, Torner L. Solitons supported by spatially inhomogeneous nonlinear losses. Optics Express 2012;20(3):2657–67. DOI: 10.1364/OE.20.002657.

[151] Kartashov YV, Konotop VV, Vysloukh VA. Two-dimensional dissipative solitons supported by localized gain. Optics Lett 2011;36(1):82–4. DOI: 10.1364/OL.36.000082.

[152] Kalosha VP, Chen L, Bao X. Feasibility of Kerr-lens mode locking in fiber lasers. In: Vallée R, Piché M, Mascher P, Cheben P, Côté D, LaRochelle S, Schriemer HP, Albert J, Ozaki T. (Eds.) Proc SPIE 7099, Photonics North 2008; 2–4 June 2008; Montréal, Canada. Bellingham: SPIE; 2008. p. 70990S. DOI: 10.1117/12.807415.

[153] Kalashnikov V, Apolonski A. Simulation of a Kerr fiber laser. In: Proc. Meeting on Russian Fiber Lasers; 27–30 March; Novosibirsk, Russia. 2012. pp. 113–114.

[154] Shaw JK. Mathematical Principles of Optical Fiber communications. Philadelphia: SIAM; 2004. p. 93.

[155] Kaup DJ. Exact quantization of the nonlinear Schroedinger equation. J Math Phys 1975;16:2036–41.

[156] Thacker HB, Wilkinson D. Inverse scattering transform as an operator method in quantum field theory. Phys Rev D 1979;19:3660–5.

[157] Akhmediev NN, Ankiewicz A. Solitons around us: integrable, Hamiltonian and dissipative systems. In: Porsezian K, Kuriakose VC. (Eds.) Optical Solitons: Theoretical and Experimental Challanges. Berlin: Springer-Verlag; 2002. pp. 105–126.

[158] Kuszelewicz R, Barbay S, Tissoni G, Almuneau G. Editorial on dissipative optical solitons. Eur Phys JD 2010;59:1–2. DOI: 10.1140/epjd/e2010-00167-7.

[159] Martinez OE, Fork RL, Gordon JP. Theory of passively mode-locked lasers for the case of a nonlinear complex-propagation coefficient. J Opt Soc Am B 1985;2(5):753–60. DOI: 10.1364/JOSAB.2.000753.

[160] Haus HA, Fujimoto JG, Ippen EP. Structures for additive pulse mode locking. J Opt Soc Am B 1991;8(10):2068–76.

[161] Kalashnikov VL, Sorokin E, Sorokina IT. Multipulse operation and limits of the Kerr-lens mode locking stability. IEEE J Quantum Electron 2003;39(2):323–36.

[162] Proctor B, Westwig E, Wise F. Characterization of a Kerr-lens mode-locked Ti:sapphire laser with positive group-velocity dispersion. Optics Lett 1993;18(19):1654–6. DOI: 10.1364/OL.18.001654.

[163] Chen S, Liu Y, Mysyrowicz A. Unusual stability of one-parameter family of dissipative solitons due to spectral filtering and nonlinearity saturation. Phys Rev A 2010;81:061806(R).

[164] Kalashnikov VL. Chirped-pulse oscillators: route to the energy-scalable femtosecond pulses. In: Al-Khursan AH. (Ed.) Solid State Laser. InTech; 2012. pp. 145–184. DOI: 10.5772/37415.

[165] Kalashnikov VL, Apolonski A. Chirped-pulse oscillators: a unified standpoint. Phys Rev A 2009;79:043829.

[166] Akhmediev N, Soto-Crespo JM, Grelu Ph. Roadmap to ultra-short record high-energy pulses out of laser oscillators. Phys Lett A 2008;372:3124–8. DOI: 10.1016/j.physleta.2008.01.027.

[167] Kalashnikov VL. Chirped dissipative solitons. In: Babichev LF, Kuvshinov VI. (Eds.) Nonlinear Dynamics and Applications. Minsk: 2010. pp. 58–67.

[168] Brons J, Pervak V, Fedulova E, Bauer D, Sutter D, Kalashnikov V, Apolonskiy A, Pronin O, Krausz F. Energy scaling of Kerr-lens mode-locked thin-disk oscillators. Optics Lett 2014;39(22):6442–5. DOI: 10.1364/OL.39.006442.

[169] Hu M-L, Wang Ch-L, Tian Zh, Xing Q-R, Chai L, Wang Ch-Y. Environmentally stable, high pulse energy Yb-doped large-mode-area photonic crystal fiber laser operating in the soliton-like regime. IEEE Photon Technol Lett 2008;20(13):1088–90. DOI: 10.1109/LPT.2008.924300.

[170] Ramachandran S, Fini JM, Mermelstein M, Nicholson JW, Ghalmi S, Yan MF. Ultra-large effective-area, higher-order mode fibers: a new strategy for high-power lasers. Laser Photon Rev 2008;2:429–48. DOI: 10.1002/lpor.200810016.

[171] Vukovic N, Healy N, Peacock AC. Guiding properties of large mode area silicon microstructured fibers: a route to effective single mode operation. J Opt Soc Am B 2011;28(6):1529–33. DOI: 10.1364/JOSAB.28.001529.

[172] Jansen F, Stutzki F, Otto H-J, Eidam T, Liem A, Jauregui C, Limpert J, Tünnermann A. Thermally induced waveguide changes in active fibers. Optics Express 2012;20(4): 3997–4008. DOI: 10.1364/OE.20.003997.

[173] Kalashnikov VL, Sorokin E. Dissipative Raman soliton. Optics Express 2014;22(24): 30118–26. DOI: 10.1364/OE.22.030118.

[174] Chernykh AI, Turitsyn SK. Soliton and collapse regimes of pulse generation in passively mode-locking laser systems. Optics Lett 1995;20(4):398–400. DOI: 10.1364/OL.20.000398.

[175] Chang W, Ankiewicz A, Soto-Crespo JM, Akhmediev N. Disssipative soliton resonances. Phys Rev A 2008;78:023830. DOI: 10.1103/PhysRevA.78.023830.

[176] Chang W, Ankiewicz A, Soto-Crespo JM, Akhmediev N. Dissipative soliton resonances in laser models with parameter management. J Opt Soc Am B 2008;25(12):1972–7.

[177] Kalashnikov VL, Apolonski A. Energy scalability of mode-locked oscillators: a completely analytical approach to analysis. Optics Express 2010;18(25):25757–70.

[178] Grelu Ph, Chang W, Ankiewicz A, Soto-Crespo JM, Akhmediev N. Dissipative soliton resonance as a guideline for high-energy pulse laser oscillators. J Opt Soc Am B 2010;27(11):2336–41.

[179] Ding E, Grelu Ph, Kutz JN. Dissipative soliton resonance in a passively mode-locked fiber laser. Optics Lett 2011;36(7):1146–8.

[180] Kharenko DS, Shtyrina OV, Yarutkina IA, Podivilov EP, Fedoruk MP, Babin SA. Generation and scaling of highly-chirped dissipative solitons in an Yb-doped fiber laser. Laser Phys Lett 2012;9(9):662–8. DOI: 10.7452/Japl.201210060.

[181] Cheng Zh, Li H, Wang P. Simulation of generation of dissipative soliton, dissipative soliton resonance and noise-like pulse in Yb-doped mode-locked fiber lasers. Optics Express 2015;23(5):5972–81. DOI: 10.1364/OE.23.005972.

[182] Chang W, Soto-Crespo JM, Ankiewicz A, Akhmediev N. Dissipative soliton resonances in the anomalous dispersion regime. Phys Rev A 2009;79:033840. DOI: 10.1103/PhysRevA.79.033840.

[183] Zh-Ch, Ning Q-Y, Mo H-L, Cui H, Liu J, Wu L-J, Luo A-P, Xu W-Ch. Vector dissipative soliton resonance in a fiber laser. Optics Express 2013;21(8):109910204. DOI: 10.1364/OE.21.010199.

[184] Smirnov SV, Kobtsev SM, Kukarin SV, Turitsyn SK. Mode-locked fibre lasers with high-energy pulses. In: Jakubczak K. (Ed.) Laser Systems for Applications. InTech; 2011. pp. 39–58.

[185] Wise FW, Chong A, Renninger WH. High-energy femtosecond fiber lasers based on pulse propagation at normal dispersion. Laser Photon Rev 2008;2(1–2):58–73. DOI: 10.1002/lpor.200710041.

[186] Renninger WH, Chong A, Wise FW. Pulse shaping and evolution in normal-dispersion mode-locked fiber lasers. IEEE J Sel Top Quantum Electron 2012;18(1):389–98. DOI: 10.1109/JSTQE.2011.2157462.

[187] Nelson LE, Fleischer SB, Lenz G, Ippen EP. Efficient frequency doubling of a femtosecond fiber laser. Optics Lett 1996;21(21):1759–61. DOI: 10.1364/OL.21.001759.

[188] Oktem B, Ülgüdür C, Ilday FÖ. Soliton–similariton fibre laser. Nat Photon 2010;4:307–11. DOI: 10.1038/nphoton.2010.33.

[189] Renninger WH, Chong A, Wise FW. Dissipative solitons in normal-dispersion fiber lasers. Phys Rev A 2008;77:023814. DOI: 10.1103/PhysRevA.77.023814.

[190] Chong A, Renninger WH, Wise FW. Properties of normal-dispersion femtosecond fiber lasers. J Opt Soc Am B 2008;25(2):140–8.

[191] Schultz M, Karow H, Prochnow O, Wandt D, Morgner U, Kracht D. Optics Express 2008;16(24):19562–7. DOI: 10.1364/OE.16.019562.

[192] Im JH, Choi SY, Rotermund F, Yeom D-I. All-fiber Er-doped dissipative soliton laser based on evanescent field interaction with carbon nanotube saturable absorber. Optics Express 2010;18(21):22141–6. DOI: 10.1364/OE.18.022141.

[193] Kieu K, Wise FW. All-fiber normal-dispersion femtosecond laser. Optics Express 2008;16(15):11453–8. DOI: 10.1364/OE.16.011453.

[194] Yang H, Wang A, Zhang Zh. Efficient femtosecond pulse generation in an all-normal-dispersion Yb:fiber ring laser at 605 MHz repetition rate. Optics Lett 2012;37(5): 954–6. DOI: 10.1364/OL.37.000954.

[195] Chichkov NB, Hausmann K, Wandt D, Morgner U, Neumann J, Kracht D. 50 fs pulses from an all-normal dispersion erbium fiber oscillator. Optics Lett 2010;35(18):3081–3. DOI: 10.1364/OL.35.003081.

[196] Ruehl A, Kuhn V, Wandt D, Kracht D. Normal dispersion erbium-doped fiber laser with pulse energies above 10 nJ. Optics Express 2008;16(5):3130–5. DOI: 10.1364/OE.16.003130.

[197] Lhermite J, Machinet G, Lecaplain C, Boullet J, Traynor N, Hideur A, Cormier E. High-energy femtosecond fiber laser at 976 nm. Optics Express 2010;35(20):3459–61. DOI: 10.1364/OL.35.003459.

[198] Buckley J, Chong A, Zhou Sh, Renninger W, Wise FW. Stabilization of high-energy femtosecond ytterbium fiber lasers by use of a frequency filter. J Opt Soc Am B 2007;24(8):1803–6. DOI: 10.1364/JOSAB.24.001803.

[199] Buckley JR, Wise FW, Ilday FÖ, Sosnowski T. Femtosecond fiber lasers with pulse energies above 10 nJ. Optics Lett 2005;30(14):1888–90. DOI: 10.1364/OL.30.001888.

[200] Renninger WH, Chong A, Wise FW. Giant-chirp oscillators for short-pulse fiber amplifiers. Optics Lett 2008;33(24):3025–7. DOI: 10.1364/OL.33.003025.

[201] Chong A, Renninger WH, Wise FW. All-normal-dispersion femtosecond fiber laser with pulse energy above 20 nJ. Optics Lett 2007;32(16):2408–10. DOI: 10.1364/OL.32.002408.

[202] Ortaç B, Lecaplain C, Hideur A, Schreiber T, Limpert J, Tünnermann A. Passively mode-locked single-polarization microstructure fiber laser. Optics Express 2008;16(3):2122–8. DOI: 10.1364/OE.16.002122.

[203] Lefrancois S, Sosnowski ThS, Liu Ch-H, Galvanauskas A, Wise FW. Energy scaling of mode-locked fiber lasers with chirally-coupled core fiber. Optics Express 2011;19(4): 3464–70. DOI: 10.1364/OE.19.003464.

[204] Lecaplain C, Ortaç B, Hideur A. High-energy femtosecond pulses from a dissipative soliton fiber laser. Optics Lett 2009;34(23):3731–3. DOI: 10.1364/OL.34.003731.

[205] Lecaplain C, Chédot C, Hideur A, Ortaç B, Limpert J. High-power all-normal-dispersion femtosecond pulse generation from a Yb-doped large-mode-area microstructure fiber laser. Optics Lett 2007;32(18):2738–40. DOI: 10.1364/OL.32.002738.

[206] Ortaç B, Schmidt O, Schreiber T, Limpert J, Tünnermann A, Hideur A. High-energy femtosecond Yb-doped dispersion compensation free fiber laser. Optics Express 2007;15(17):10725–32. DOI: 10.1364/OE.15.010725.

[207] Lhermite J, Lecaplain C, Machinet G, Royon R, Hideur A, Cormier E. Mode-locked 0.5 µJ fiber laser at 976 nm. Optics Lett 2011;36(19):3819–21. DOI: 10.1364/OL. 36.003819.

[208] Lecaplain C, Ortaç B, Machinet G, Boullet J, Baumgartl M, Schreiber T, Cormier E, Hideur A. High-energy femtosecond photonic crystal fiber laser. Optics Lett 2010;35(19):3156–8. DOI: 10.1364/OL.35.003156.

[209] Ortaç B, Baumgartl M, Limpert J, Tünnermann A. Approaching microjoule-level pulse energy with mode-locked femtosecond fiber lasers. Optics Lett 2009;34(10): 1585–7. DOI: 10.1364/OL.34.001585.

[210] Ankiewicz A, Devine N, Akhmediev N, Soto-Crespo JM. Dissipative solitons and antisolitons. Phys Lett A 2007;370:454–8. DOI: 10.1016/j.physleta.2007.06.001.

[211] van Saarlos W, Hohenberg PC. Fronts, pulses, sources and sinks in generalized complex Ginzburg-Landau equations. Physica D 1992;56:303–67.

[212] Soto-Crespo JM, Akhmediev NN, Afanasjev VV, Wabnitz S. Pulse solutions of the cubic-quintic complex Ginzburg-Landau equation in the case of normal dispersion. Phys Rev E 55;4:4783–96.

[213] Conte R. (Ed.) The Painlevé Property: One Century Later. New York: Springer-Verlag; 1999. 810 p.

[214] Greco AM. (Ed.) Direct and Inverse Methods in Nonlinear Evolution Equations. Berlin: Springer; 2003. 282 p.

[215] Kivshar YS, Malomed BA. Dynamics of solitons in nearly integrable systems. Rev Mod Phys 1989;61(4):763. DOI: http://dx.doi.org/10.1103/RevModPhys.61.763.

[216] Malomed BA, Nepomnyashchy AA. Kinks and solitons in the generalized Ginzburg-Landau equation. Phys Rev A 1990;42:6009.

[217] Kalashnikov VL. Chirped dissipative solitons of the complex cubic-quintic nonlinear Ginzburg-Landau equation. Phys Rev E 2009;80:046606.

[218] Podivilov E, Kalashnikov VL. Heavily-chirped solitary pulses in the normal dispersion region: new solutions of the cubic-quintic complex Ginzburg-Landau equation. JETP Lett 2005;82(8):467–71.

[219] Kalashnikov VL, Podivilov E, Chernykh A, Apolonski A. Chirped-pulse oscillators: theory and experiment. Appl Phys B 2006;83(4):503–10.

[220] Ablowitz MJ, Horikis ThP. Solitons in normally dispersive mode-locked lasers. Phys Rev A 2009;79:063845.

[221] Kharenko DS, Shtyrina OV, Yarutkina IA, Podivilov EV, Fedoruk MP, Babin SA. Highly chirped dissipative solitons as a one-parameter family of stable solutions of the cubic-quintic Ginzburg-Landau equation. J Opt Soc Am B 2011;28(10):2314–9.

[222] Malomed BA. Variational methods in nonlinear fiber optics and related fields. In: Wolf E. (Ed.) Progress in Optics, Vol. 43. North-Holland: Elsevier; 2002. pp. 71–193.

[223] Ankiewicz A, Akhmediev N, Devine N. Dissipative solitons with a Lagrangian approach. Optical Fiber Technol 2007;13(2):91–7.

[224] Bale BG, Boscolo S, Kutz JN, Turitsyn SK. Intracavity dynamics in high-power mode-locked fiber lasers. Phys Rev A 2010;81:033828.

[225] Bale BG, Kutz JN. Variational method for mode-locked lasers. J Opt Soc Am B 2008;25(7):1193–202. DOI: 10.1364/JOSAB.25.001193.

[226] Tsoy EN, Ankiewicz A, Akhmediev N. Dynamical models for dissipative localized waves of the complex Ginzburg-Landau equation. Phys Rev E 2006;73:036621.

[227] Bale BG, Kutz JN, Chong A, Renninger WH, Wise FW. Spectral filtering for high-energy mode-locking in normal dispersion fiber lasers. J Opt Soc Am B 2008;25(10): 1763–70. DOI: 10.1364/JOSAB.25.001763.

[228] Bale BG, Boscolo S, Turitsyn SK. Dissipative dispersion-managed solitons in mode-locked lasers. Optics Lett 2009;34(21):3286–8. DOI: 10.1364/OL.34.003286.

[229] Ding E, Kutz JN. Stability analysis of the mode-locking dynamics in a laser cavity with a passive polarizer. J Opt Soc Am B 2009;26(7):1400–11. DOI: 10.1364/JOSAB. 26.001400.

[230] Kalashnikov VL. Dissipative soliton energy scaling. Phys. Rev. A. Forthcoming.

[231] Liu X. Pulse evolution without wave breaking in a strongly dissipative dispersive laser system. Phys Rev A 2010;81(5):053819.

[232] Shen X, Li W, Zeng H. Polarized dissipative solitons in all-polarization-maintained fiber laser with long-term stable self-started mode-locking. Appl Phys Lett 2014;105:101109.

[233] Chen Sh, Liu Y, Mysyrowicz A. Unusual stability of one-parameter family of dissipative solitons due to spectral filtering and nonlinearity saturation. Phys Rev Lett 1997;79:4047.

[234] Farnum ED, Kutz JN. Multifrequency mode-locked lasers. J Opt Soc Am B 2008;25(6): 1002–10.

[235] Zhang H, Tang DY, Wu X, Zhao LM. Multi-wavelength dissipative soliton operation of an erbium-doped fiber laser. Optics Express 2009;17(15):12692–7.

[236] Yun L, Liu X, Mao D. Observation of dual-wavelength dissipative solitons in a figure-eight erbium-doped fiber laser. Optics Express 2012;20(19):20992–7.

[237] Zhang ZX, Xu Z, Zhang L. Tunable and switching dual-wavelength dissipative soliton generation in an all-normal-dispersion Yb-doped fiber laser with birefringence fiber filter. Optics Express 2012;20(24):26736–42.

[238] Xu ZW, Zhang ZX. All-normal-dispersion multi-wavelength dissipative soliton Yb-doped fiber laser. Laser Phys Lett 2013;10:085105.

[239] Huang S, Wang Y, Yan P, Zhao J, Li H, Lin R. Tunable and switchable multi-wavelength dissipative soliton generation in a graphene oxide mode-locked Yb-doped fiber laser. Optics Express 2014;22(10):11417–26.

[240] Mao D, Liu X, Han D, Lu H. Compact all-fiber laser delivering conventional and dissipative solitons. Optics Lett 2013;38(16):3190–3.

[241] Kalashnikov VL, Chernykh A. Spectral anomalies and stability of chirped-pulse oscillators. Phys Rev A 2007;75:033820.

[242] Kalashnikov VL. Dissipative solitons: perturbations and chaos formation. In: Skiadas CH, Dimotikalis I, Skiadas C. (Eds.) Chaos Theory: Modeling, Simulation and Applications. Singapore: Worlds Scientific Publishing; 2011. pp. 199–206.

[243] Kalashnikov VL. Dissipative solitons in presence of quantum noise. Chaotic Model Simulat 2014;(1):29–37.

[244] Soto-Crespo JM, Akhmediev N, Ankiewicz A. Pulsating, creeping, and erupting solitons in dissipative systems. Phys Rev Lett 2000;85(14):2937–40.

[245] Soto-Crespo JM, Akhmediev N. Exploding soliton and front solutions of the complex cubic–quintic Ginzburg–Landau equation. Math Computers Simulat 2005;69(5–6): 526–36. DOI: 10.1016/j.matcom.2005.03.006.

[246] Cundiff ST, Soto-Crespo JM, Akhmediev N. Experimental evidence for soliton explosions. Phys Rev Lett 2002;88(7):073903.

[247] Crtes C, Descalzi O, Brand HR. Exploding dissipative solitons in the cubic-quintic complex Ginzburg-Landau equation in one and two spatial dimensions. Eur Phys J Special Topics 2014;223:2145–59. DOI: 10.1140/epjst/e2014-02255-2.

[248] Crtes C, Descalzi O, Brand HR. Noise can induce explosions for dissipative solitons. Phys Rev E 2012;85(015205(R)). DOI: 10.1103/PhysRevE.85.015205.

[249] Arecchi FT, Bortolozzo U, Montina A, Residori S. Granularity and inhomogenety are the joint generators of optical rogue waves. Phys Rev Lett 2011;106:153901.

[250] Onorato M, Residori S, Bortolozzo U, Montina A, Arecchi FT. Rogue waves and their generating mechanisms in different physical contexts. Phys Rep 2013;528:47–89.

[251] Solli DR, Ropers C, Koonath P, Jalali B. Optical rogue waves. Nature 2007;450:1054–7.

[252] Soto-Crespo JM, Grelu Ph, Akhmediev N. Dissipative rogue waves: Extreme pulses generated by passively mode-locked lasers. Phys Rev E 2011;84(1):016604.

[253] Zavyalov A, Egorov O, Iliew R, Lederer F. Rogue waves in mode-locked fiber lasers. Phys Rev A 2012;85:013828.

[254] Komarov A, Sanchez F. Structural dissipative solitons in passive mode-locked fiber lasers. Phys Rev E 2008;77:066201.

[255] Zakharov V, Dias F, Pushkarev A. One-dimensional wave turbulence. Phys Rep 2004;398:1–65.

[256] Yun L, Han D. Bound state of dissipative solitons in a nanotube-mode-locked fiber laser. Optics Commun 2014;313:70–3.

[257] Amrani F, Haboucha A, Salhi M, Leblond H, Komarov A, Sanchez F. Dissipative solitons compounds in a fiber laser. Analogy with the states of the matter. Appl Phys B 2010;99:107–14. DOI: 10.1007/s00340-009-3774-7.

[258] Kalashnikov VL. Dissipative solitons: structural chaos and chaos of destruction. Chaotic Model Simulat 2011;(1):51–9.

[259] Zhang L, Pan Zh, Zhuo Zh, Wang Y. Three multiple-pulse operation states of an all-normal-dispersion dissipative soliton fiber laser. Int J Optics 2014;(169379). DOI: 10.1155/2014/169379.

[260] Kalashnikov VL, Sorokin E. Dissipative Raman solitons. Optics Express 2014;22(24): 30118–126. DOI: 10.1364/OE.22.030118.

[261] Kalashnikov VL. Chaotic dissipative Raman solitons. Chaotic Model Simulat 2014; (4):403–10.

[262] Babin SA, Podivilov EV, Kharenko DS, Bednyakova AE, Fedoruk MP, Kalashnikov VL, Apolonski A. Multicolour nonlinearly bound chirped dissipative solitons. Nature Commun 2014;5:4653. DOI: 10.1038/ncomms5653.

[263] Bednyakova AE, Babin SA, Kharenko DS, Podivilov EV, Fedoruk MP, Kalashnikov VL, Apolonski A. Evolution of dissipative solitons in a fiber laser oscillator in the presence of strong Raman scattering. Optics Express 2013;21(18):29556–64. DOI: 10.1364/OE.21.02056.

[264] Kharenko DS, Bednyakova AE, Podivilov EV, Fedoruk MP, Apolonski A, Babin SA. Feedback-controlled Raman dissipative solitons in a fiber laser. Optics Express 2015;23(2):1857–62. DOI: 10.1364/OE.23.001857.

[265] Haus JW, Shaulov G, Kuzin EA, Sanchez-Mondragon J. Vector soliton fiber lasers. Optics Lett 1999;24(6):376–8.

[266] Barad Y, Silberberg Y. Polarization evolution and polarization instability of solitons in a birefringent optical fiber. Phys Rev Lett 1997;78(17):3290–3.

[267] Lei T, Tu Ch, Lu F, Deng Y, Li E. Numerical study on self-similar pulses in mode-locking fiber laser by coupled Ginzburg-Landau equation model. Optics Express 2009;17(2):585–91.

[268] Cundiff ST, Collings BC, Akhmediev NN, Soto-Crespo JM, Bergman K, Knox WH. Observation of polarization-locked vector solitons in an optical fiber. Phys Rev Lett 1999;82(20):3988–91.

[269] Akhmediev N, Buryak A, Soto-Crespo JM. Elliptically polarised solitons in birefringent optical fibers. Optics Commun 1994;112(5–6):278–82. DOI: 10.1016/0030-4018(94)90631-9.

[270] Wu J, Tang DY, Zhao LM, Chan CC. Soliton polarization dynamics in fiber lasers passively mode-locked by the nonlinear polarization rotation technique. Phys Rev E 2006;74:046605. DOI: 10.1103/PhysRevE.74.046605.

[271] Zhang H, Tang DY, Zhao LM, Wu X, Tam HY. Dissipative vector solitons in a dispersion-managed cavity fiber laser with net positive cavity dispersion. Optics Express 2009;17(2):455–60.

[272] Kong L, Xiao X, Yang Ch. Polarization dynamics in dissipative soliton fiber lasers mode-locked by nonlinear polarization rotation. Optics Express 2011;19(19):18339–44.

[273] Mesentsev VK, Turitsyn SK. Stability of vector solitons in optical fibers. Optics Lett 1992;17(21):1497–9.

[274] Zhang H, Tang D, Zhao L, Bao Q, Loh KP. Vector dissipative solitons in graphene mode locked fiber lasers. Optics Commun 2010;283:3334–8. DOI: 10.1016/j.optcom. 2010.04.064.

[275] Luo Zh-Ch, Ning Q-Y, Mo H-L, Cui H, Liu L, Wu L-J, Luo A-P, Xu W-Ch. Vector dissipative soliton resonance in a fiber laser. Optics Express 2013;21(8):10199–204. DOI: 10.1364/OE.21.010199.

[276] Tsatourian V, Sergeyev SV, Mou Ch, Rozhin A, Mikhailov V, Rabin B, Westbrook PS, Turitsyn SK. Polarization dynamics of vector soliton molecules in mode locked fibre laser. Sci Rep 2013;3:3154. DOI: 10.1038/srep03154.

[277] Sergeyev SV. Fast and slowly evolving vector solitons in mode-locked fibre lasers. Philos Transac Royal Soc A 2014;372:20140006. DOI: 10.1098/rsta.2014.0006.

[278] Sergeyev SV, Mou Ch, Turitsyna EG, Rozhin A, Turitsyn SK, Blow K. Spiral attractor created by vector solitons. Light Sci Applic 2014;3:e131. DOI: 10.1038/lsa.2014.12.

[279] Lin Q, Agrawal GP. Vector theory of stimulated Raman scattering and its application to fiber-based Raman amplifiers. J Opt Soc Am B 2003;20(8):1616–31.

Chapter 10

Fiber-laser-generated Noise-like Pulses and Their Applications

Ci-Ling Pan, Alexey Zaytsev, Yi-Jing You and Chih-Hsuan Lin

Additional information is available at the end of the chapter

http://dx.doi.org/10.5772/61856

Abstract

We describe generation and amplification of medium- and high-energy noise-like pulses (NLPs) using Yb-doped optical fibers. We also demonstrate supercontinuum (SC) generation techniques in which NLPs serve as the pump. SC pumped by NLPs has been employed successfully in optical coherence tomography systems.

Keywords: Yb-doped fiber laser, laser amplifier, ultrafast laser, noise-like pulse, supercontinuum generation, nonlinear optics, optical coherence, tomography

1. Introduction

In the past two decades, ultrafast fiber laser has been a field that witnessed rapid development of the technology and emergent applications. Ultrafast pulses at high average power are useful in numerous application areas including micromachining and material processing, nonlinear frequency conversion, biomedical application, and fundamental science. It is generally recognized that fiber laser is an attractive technology for compact, robust, low cost, and reliable ultrafast pulse sources. Compared with conventional bulk solid-state lasers, main advantages of fiber-based laser systems include good beam quality, ease of delivery, and excellent heat-dissipation properties. The last feature is due to the large surface area-active volume ratio of optical fibers.

Rare-earth-doped fiber lasers also exhibit much higher efficiency than crystalline solid-state lasers. Since the first rare-earth-doped fiber laser was invented over 50 years ago [1], many experimental and theoretical results on silica- and fluoride-doped fiber amplifiers and lasers with various rare-earth dopants have been reported [2–4]. The commercially available erbium-doped amplifiers have already established their key role for optical communication networks

at 1.55-μm. On the other hand, high average power ultrafast laser sources emitting in the 1-μm wavelength range, i.e., ytterbium-doped fiber laser, have also attracted much attention.

Ultrafast pulsed fiber lasers have been observed to operate in different mode-locking regimes, such as soliton, similariton, and dissipative soliton. A special regime of repetitively pulsed fiber laser, i.e., lasers generating the so-called noise-like pulses or NLPs [4–8], was also reported. NLPs are relatively long (sub-ns) wave packets, which exhibit a fine inner structure of sub-ps pulses with randomly varying amplitude and duration. Depending on settings of polarizing components such as the wave plates, fiber lasers can generate both regular mode-locked pulses (MLPs) and NLPs. The first demonstration of fiber lasers generating NLPs was that of a ring-cavity Er: doped fiber oscillator [9]. Later on, different cavity configurations and active media were employed to achieve NLPs with controllable characteristics.

In this chapter, we describe our recent work on generation and amplification of medium- and high-energy NLPs with Yb-doped fibers. We also demonstrate supercontinuum (SC) generation techniques where NLPs serve as the pump. Theoretical aspects as well as discussions about physical mechanisms, which make NLPs distinguishable from regular MLPs, are also included. SC pumped by NLPs has been employed successfully in optical coherence tomography (OCT) systems. The advantages of the present approach using NLPs will be discussed herein.

We first describe, analyze, and compare two popular cavity configurations for generating NLPs: a dispersion-mapped cavity and an all-normal dispersion (ANDi). Simulation results based on coupled nonlinear Schrödinger equations are supported by experiments. Nonlinear polarization evolution (NPE) is the main mechanism for pulsed operation in these lasers. We show that both regular Gaussian pulses and NLPs can be achieved in the same cavity by choosing proper cavity components and adjustment.

The second part of the chapter is about SC generation. Here, we analyze the possibilities of efficient SC generation using NLPs propagating in standard silica fibers. It is shown that unique features of NLPs make them very useful for such a purpose. That is, the central wavelength of the pump and zero-dispersion wavelength (ZDW) of SC generation media are not critical. We show that, even if the pump wavelength is deep in the normal dispersion regime (for example, ~1 μm where ZDW = 1.33 μm), SC can be efficiently generated. Simulations and experimental results of SC generation by NLPs using different single-mode fibers (SMFs) are presented. We discuss the optimal selection of fiber types and other characteristics to generate flat SC in spectral region above 1 μm. The pros and cons of using specialty fibers such as photonic crystal fibers (PCFs) pumped by NLPs will also be elaborated.

In the third part of the chapter, we consider the application of SC generated by NLPs for selected applications, i.e., OCT. The SC spectrum is flat with a bandwidth of 365 nm centered at 1320 nm. The light source is successfully employed in a time-domain OCT (TD-OCT) scheme, achieving an axial resolution of 2.3 μm. High-resolution fiber-based spectral-domain OCT (SD-OCT) imaging of bio-tissue (onion skin), comparable to that obtained using a commercial swept source, is also demonstrated.

2. NLP generation with fiber lasers

NLPs were first demonstrated by Horowitz *et al.* from an Er-doped mode-locked fiber laser in 1997 [9]. A NLP is defined as some kind of complicated waveform of relatively long (sub-nanosecond) duration. It also exhibits a fine inner structure of much narrow sub-pulses (a few hundred femtoseconds in width) with randomly varying intensity and duration. The inner structure varies from one waveform to another in the pulse train. On the other hand, the repetition rate and average duration of the aforementioned waveforms forming a train of pulses are relatively stable. It was found and demonstrated that such kinds of pulses have some interesting common features: (1) a very large optical bandwidth (usually of several tens of nanometers); (2) specific waveforms that exhibit a double-scaled average autocorrelation trace with a narrow peak riding on a wide pedestal; and (3) low temporal coherence. Moreover, it was found that NLPs are essentially undistorted even after the pulses have propagated through a lengthy dispersive medium, e.g., an optical fiber over a long distance.

Thus far, characteristics of fiber-based NLPs have been predominantly assessed only indirectly by means of numerical simulation or average measurements. There have been many works published on NLPs generated by fiber laser, but still there is a lack of simple qualitative explanation of physical mechanisms involved in the formation of NLPs till now. In a paper by Horowitz *et al.*, it was supposed that NLPs are caused by a polarization-dependent delay effect due to the birefringence in the laser cavity [9]. Smirnov *et al.* [10], however, demonstrated that NLP still can occur in weakly birefringent fiber laser systems. In this paper [10], the authors introduced three regimes of single pulse generation in an ANDi fiber laser. The same regimes can also be observed in our dispersion-mapped fiber laser. On the other hand, Tang *et al.* concluded that the key physical mechanism of noise-like generation is caused by the combined effect of the soliton collapse and cavity positive feedback for dispersion managed fiber lasers [11]. This contradicted with the report that NLPs are generated in an ANDi fiber laser [12]. Aguergaray *et al.*, meanwhile, proposed that NLPs are caused by the Raman-driven destabilization of mode-locked long-cavity fiber lasers. The destabilization is accompanied by the emergence of a strong frequency-downshifted stokes signal [13]. Therefore, there are several mechanisms that can lead to the generation of NLPs. In our dispersion-mapped and ANDi Yb-doped fiber lasers, we did observe Raman peaks in the output spectra [14]. Hence, Aguergaray's model is probably more suitable for our laser systems.

2.1. NLPs generated by a dispersion-mapped fiber laser

The configuration of this laser is shown in Figure 1. It is similar to the design of stretched pulse mode-locked lasers [15]. The laser cavity comprises a segment of single-mode silica fiber followed by a length of Yb-doped fiber (Yb1200-10/125DC, Liekki). A set of 915 nm laser diodes with total power up to 24 W and a power combiner are used for pumping. Nonlinear polarization rotation (NPE) is implemented by a polarization beam splitter (PBS) and wave plates. A grating pair in the near-Littrow configuration is placed after the PBS to provide negative dispersion in the cavity. The grating arrangement functions as a dispersive delay line (DDL). The net group velocity dispersion (GVD) of the cavity is kept slightly positive. An optical isolator is placed in the air space to ensure unidirectional operation of the laser so that the mode-locking process is self-starting. By adjusting the wave plates, stable mode-locked or NLPs are readily observed. The output can be taken directly from the NPE ejection port. Additional information about the laser can be found in Ref. [14].

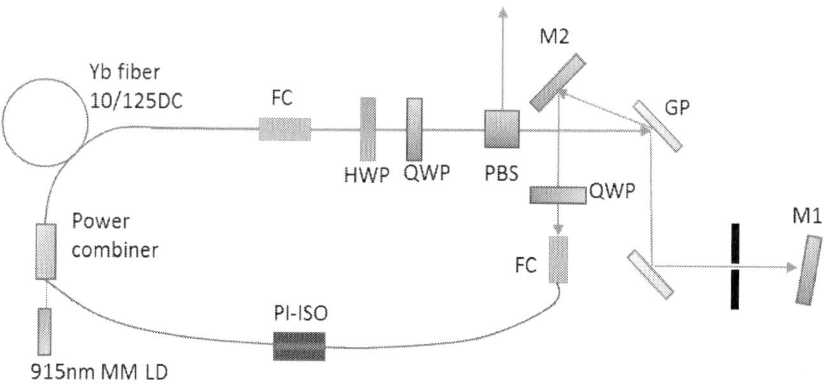

Figure 1. Schematic of the laser setup: FC, fiber coupler; HWP, half-waveplate; QWP, quarter-waveplate, GP, grating pair; PI-ISO, polarization-insensitive isolator; M1 and M2, mirrors; MM LD, multimode laser diodes.

After the single-pulse mode-locked operation was obtained in this laser, we tuned one of the wave plates while holding all the other experimental conditions unchanged. It is possible to shift the laser operation from the conventional single-pulse mode-locked operation into the NLP emission regime. This behavior was also reported by previous workers [4, 6]. We observed self-starting noise-like laser operation for a relatively large range of pumping powers. The output powers of NLPs can be varied from <0.1 W to >1.6 W. The highest pulse energy obtained was 45 nJ, with a pumping power of ~13 W. After increasing the pumping power to more than 14 W, one of the fiber couplers was damaged.

Our dispersion-mapped noise-like laser is fairly robust. We attribute this improved stability to the laser design, where NPE is accompanied by self-amplitude modulation induced by spectral filter. The role of negative GVD induced by gratings pair is also important. For the dispersion-mapped fiber laser shown in Figure 1, with a repetition rate of 31.5 MHz [14], the GVD of grating pair was set at −0.11 ps^2. This value is estimated to be slightly smaller than that of the total net positive GVD due to the fiber (~0.15 ps^2) in the cavity.

By translating the iris transversely across the laser axis (Figure 1), it is possible to modulate the central wavelength of generated pulses. We also found that the diameter of the iris determining the filter bandwidth also affects the output spectral bandwidth and duration of generated NLPs. For both narrowband and broadband NLPs, the tuning range for both cases reached 12 nm [14].

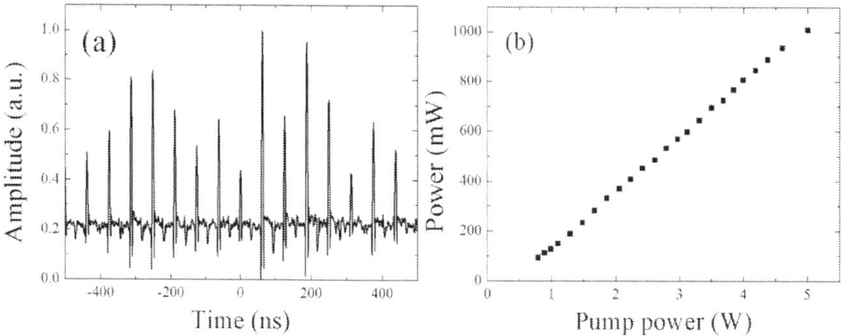

Figure 2. Oscilloscope trace of the train of noise-likes pulses (a); output power of the NLP laser vs. pumping power (b) [14].

We also set up a dispersion-mapped fiber laser in which the GVD of grating pair was set at −0.219 ps^2, while the fiber GVD was estimated to be 0.192 ps^2. Hence, the net GVD of fiber cavity became slightly negative in this configuration with a repetition rate of 15.7 MHz. The active fiber length has been increased from 1.2 m in our previous work to 7 m. Figure 2(a) shows the oscilloscope trace of the laser output as measured by a fast InGaAs detector. The behavior is typical for NLP emission state. The output powers of NLPs can be varied from <0.1 W to beyond 1 W (Figure 2(b)). The highest pulse energy obtained was ~6 nJ, with a pumping power of ~5 W. The tuning wavelength for NLP is from 1057 to 1090 nm, i.e., the tuning range is as high as 33 nm, more than twice of the previous result [14]. The spectral bandwidth of the laser output can also be varied from 6 to 54 nm. This is illustrated in Figure 3. The tuning characteristic could be important for some applications. For example, in the case of further amplification of generated pulses by fiber amplifiers, it is necessary to match the bandwidth of the oscillator with the gain spectrum of the amplifier.

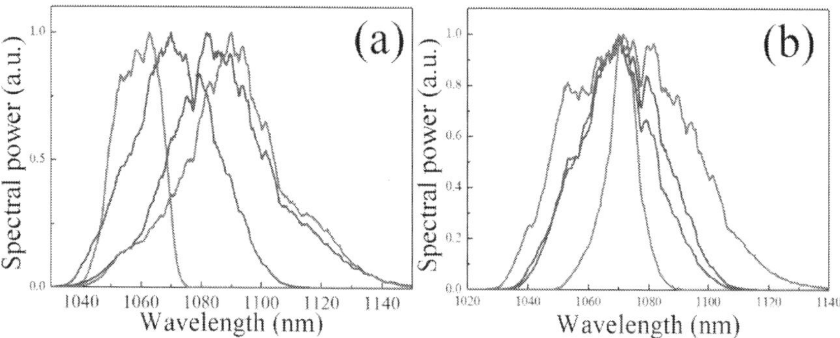

Figure 3. Optical spectra of the measured NLPs corresponding to different positions of the iris (a) and width (b).

In order to understand NLP formation, we have simulated the buildup dynamics of the laser cavity in the figure with a repetition frequency of 31.5 MHz [14]. This is performed by recognizing the cavity as consisting of several connected fiber components. Pulse propagation in each fiber section was described by the corresponding nonlinear Schrödinger coupled-mode equations [16].

$$\frac{\partial A_x}{\partial z} = i\gamma\left\{\left|A_x\right|^2 A_x + \frac{2}{3}\left|A_y\right|^2 A_x + \frac{1}{3}A_y^2 A_x^*\right\} + g\left(E_{\text{pulse}}\right)A_x - \frac{i}{2}\beta_2\frac{\partial^2 A_x}{\partial t^2}$$

$$\frac{\partial A_y}{\partial z} = i\gamma\left\{\left|A_y\right|^2 A_y + \frac{2}{3}\left|A_x\right|^2 A_y + \frac{1}{3}A_x^2 A_y^*\right\} + g\left(E_{\text{pulse}}\right)A_y - \frac{i}{2}\beta_2\frac{\partial^2 A_y}{\partial t^2}$$

(1)

where A_x and A_y are the field envelope components; γ is the nonlinear coefficient; β_2 is the GVD; $g(E_{pulse})$ is the gain function for active fiber pieces or zero for passive fiber.

Equation (1) also includes the amplification of NLPs in active optical fibers and filtering effect due to limited gain bandwidth of the aforementioned active media. If the parameter set corresponded to the condition of minimal cavity loss for specific modes, then a stable pulse train of regular (close to Gaussian) shape will be circulated in the oscillator shortly after laser onset. A negative second-order dispersion term of ~0.02 ps^2 was assumed, neglecting high-order dispersion terms. We used Jones matrices approach to represent different polarization states as the pulses propagate through the wave plates and the PBS in our calculation. Every fiber section (active and passive) was modeled using the split-step Fourier method [16]. It was found that stable NLP waveforms began to circulate in the laser cavity after just a few round-trips (Figure 4) after pumping starts. We note further that a myriad of nonlinear effects, e.g., self-phase modulation, cross-phase modulation, four-wave mixing, etc., contribute to the generation of NLPs.

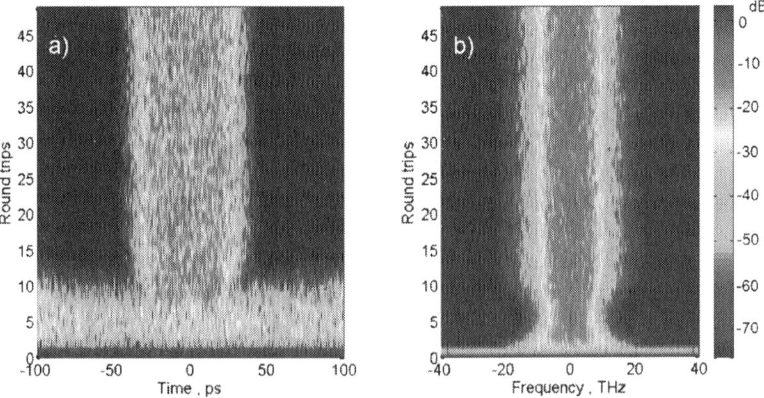

Figure 4. Building up dynamics in time (a) and spectral (b) domains of a fiber ring oscillator generating NLPs [17].

The simulation results for the steady state are in good agreement with experimental observations. The estimated iris filter bandwidth, corresponded iris diameter, output bandwidths, and bunch duration for the laser cavity with a repetition frequency of 31.5 MHz are listed in Table 1.

Iris diameter (mm)	Iris bandwidth (THz)	Measured output bandwidth (THz)	Calculated output bandwidth (THz)	Measured autocorrelation trace half-width (ps)	Calculated autocorrelation trace half-width (ps)
1	5	2.1	5	5	7
2	10	5.5	9	20	12
3	20	14	15	90	20

Table 1. Summary of measured and simulated results for NLPs [14].

By inserting an additional 200-m length of SMF inside the cavity (Figure 1), the repetition rate of the noise-like pulsed fiber laser was found to be reduced to ~928 kHz. The output power of the laser exceeded 1 W. Thus, the energy of the NLPs directly from the laser oscillator can be as high as 1.25 µJ (Figure 5).

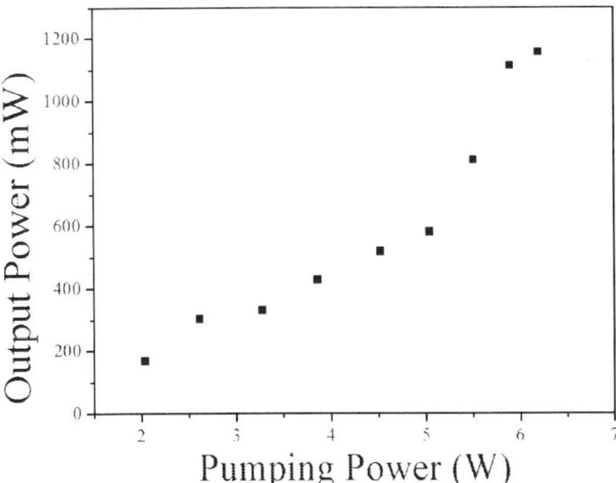

Figure 5. The pump-power dependence output of a high-energy fiber laser oscillator generating NLPs.

2.2. ANDi fiber laser

Although stable noise-like operation with controllable characteristics can be readily achieved in a dispersion-mapped fiber laser, such scheme is not practical when the required repetition rate becomes lower than 10 MHz. This is due to increased dispersion of fibers in the longer cavity and the corresponding DDL needed. Therefore, we also investigated an ANDi fiber oscillator for generation of NLPs.

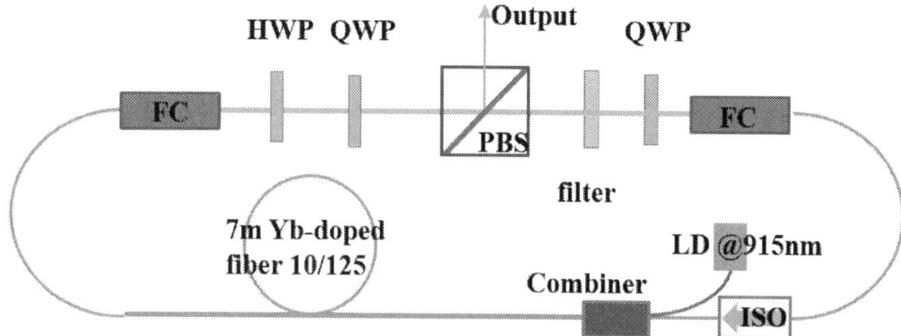

Figure 6. Schematic of ANDi fiber laser. FC, fiber coupler; HWP, half-wave plate; QWP, quarter-wave plate, PBS, polarization beam splitter; LD, laser diode; ISO, isolator.

The ANDi laser is perhaps one of the simplest designs of mode-locked Yb-doped fiber oscillators. There is no need for dispersion-compensation components such as the grating pair [18]. Typically, the dispersion in the cavity of an ANDi laser is managed by a narrow-bandwidth interference filter. We found that after proper alignment of the waveplates, an ANDi laser can also generate NLPs. In comparison with the dispersion-mapped laser, the achievable pulse energy at the output is larger. An average power of 850 mW (pumped at 2 W) at a repetition rate of ~15 MHz, which corresponds to pulse energy of 55 nJ, was obtained for a cavity length of 13 m. The bandwidth of generated NLP was as high as 21.8 nm. When the cavity length was increased to 62.5 m or a laser repetition rate of 3.2 MHz, the spectral bandwidth approaches a value at least fourfold higher than the regular mode-locked pulse. The central wavelength of the laser is 1096 nm and the spectral bandwidth can reach 83 nm. The laser can also deliver an average output power of 400 mW (pumped at 1 W) or a pulse energy of 125 nJ.

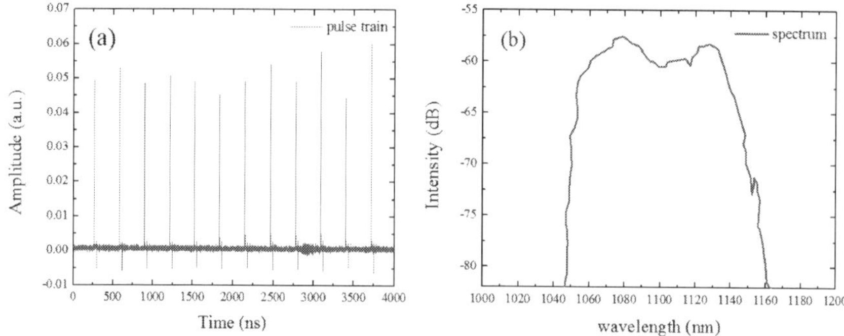

Figure 7. Characteristic of ANDi laser. (a) Pulse train and (b) optical spectrum.

In summary, using cladding-pumped Yb-doped active fibers and high-power multimode laser diodes, we demonstrated that NLPs with relatively high energies can be generated with either dispersion-mapped or ANDi cavities. The latter exhibits some advantages over the former, e.g., ease of alignment, lower threshold, and higher pulse energy. The achieved average power of NLPs can be as high as ~1 W and, limited only by damage threshold of available fiber couplers. On the other hand, it is possible to control temporal and spectral characteristics of NLPs generated by the dispersion-mapped laser over a broad range. This is not possible for the ANDi laser.

3. SC generation by NLPs

In the past decade, tremendous progress has been made in SC generation (SCG) [19]. Nowadays, SC source has been widely used in many applications such as OCT, frequency metrology, and wavelength division multiplexing systems [20–22]. Using rare-earth-doped fiber lasers, especially Yb-doped fiber lasers operating at a high power with a short pulse width as the pump laser, the spectral bandwidth and the output power of SC sources have increased dramatically [23–25].

The nonlinear mediums used for SCG have also evolved from bulk media [25, 26] to conventional optical fibers [27], and then PCFs [28–32] for visible light, near-infrared, and mid-infrared SCG. The SC generated with PCFs is popular due to the controllable ZDWs and high nonlinearity [19]. Moreover, very powerful light sources at ~800 nm and ~1 μm can be used for efficient SC pumping. However, PCFs are still expensive and not widely available. Furthermore, especially for high-power operations, splicing between PCF and pigtailed fiber pump source is a significant challenge because of mode field mismatch and the microhole collapse effect [33]. Standard SMFs for optical communication, on the other hand, are inexpensive and easy to integrate. Pumping standard SMF at ~800 nm or ~1 μm, however, is not so efficient because these popular pump wavelengths are located in the normal dispersion region of the SMF. It requires much more pump power to excite SC compared with SCG in the anomalous regime. Further, in the case of CW or picosecond duration pump sources, the SC spectrum so generated exhibits very strong oscillations [34]. Recently, it has been demonstrated that such spectral oscillations can be explained by cascaded Raman scattering [34–36]. Such modulations in spectrum are not desirable for applications such as OCT.

On the other hand, more broadband pumps have been shown to result in SC with a smoother spectrum [37]. As reported by Horowitz *et al.* [9], NLPs can be successfully used for such a purpose. Besides a very large optical bandwidth, NLPs are capable of propagating without distortion through a lengthy dispersive medium over a long distance [9]. Later, Hernandez-Garcia *et al.* [38] demonstrated flat broadband SCG by pumping SMF in the anomalous dispersion region with a train of NLP at the central wavelength of 1.5 μm. The energy threshold was as low as ~12 nJ. Recently [39], we have reported similar desirable characteristics (a low threshold and a flat spectrum) by pumping SMF in the normal dispersion regime with NLPs. In contrast to the anomalous case, the physical mechanism of such SCG is cascaded Raman scattering that results in significant spectral broadening in the longer wavelength regions of the SC whereas the Kerr effect is responsible for smoothing the spectrum of generated SC.

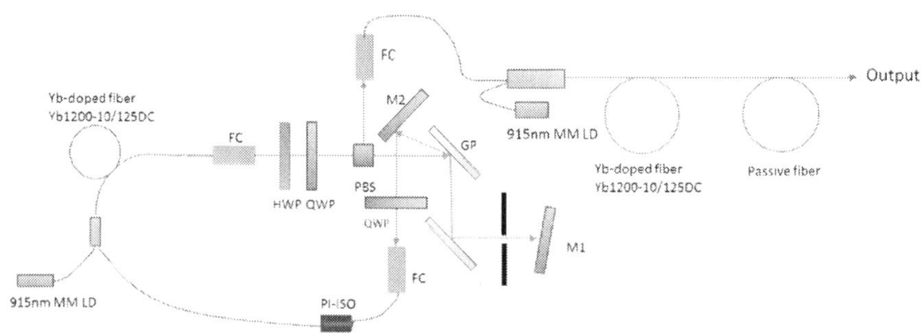

Figure 8. Schematic of the experimental setup: FC, fiber coupler; HWP, half-wave plate; QWP, quarter-wave plate; GP, grating pair; PI-ISO, polarization-insensitive isolator; ISO, Faraday isolator; M1 and M2, mirrors; MM LD, multi-mode laser diodes.

The experimental setup for SCG is shown in Figure 8. The pump source is a dispersion-mapped Yb-doped fiber laser based on the ring cavity design that can generate tunable NLPs with energies as high as ~45 nJ with a central wavelength of ~1 μm [14]. This laser can also be configured to generate stable mode-locked femtosecond pulses. The use of negative dispersion delay line and the spatial spectral filter (Figure 8) was found to be important for such a high-power noise-like operation. In the NLP regime and when pumped at ~7 W, the oscillator irradiates a pulse train with a repetition rate of ~31.5 MHz, an average power of ~600 mW, and an average noise-like bunch duration of ~45 ps. Typical spectrum and intensity autocorrelation trace of the NLP laser are shown in Figure 9. The spectral bandwidth of the NLPs was ~10 nm full width at half maximum (FWHM). We also simulated the buildup dynamics of our laser cavity by recognizing it as consisting of several connected fiber components. Assuming a thermal Gaussian noise as a source of oscillations, we found the conditions where repeated NLPs begin to circulate in the laser cavity after hundreds of round-trips. Such simulations yield results (spectra, autocorrelation traces) that are close to experimental measurements [39]. An example of simulated NLP characteristics is shown in Figure 10.

The NLPs from the seed laser described above were boosted up to an average power of ~2 W in a single 1.2 m-length Yb-doped amplifier stage and used as a pump to excite the SC. Several different SMFs were investigated for SCG. They were spliced to the output fiber end of the amplifier. All fibers are single mode at the pump wavelength (~1 μm). They were selected to examine the effects of nonlinearity on SCG. The parameters of the fiber investigated are summarized in Table 2. The dispersion coefficients β_n at 1060 nm central wavelength and the ZDW for F1060c were calculated from the Sellmeier equation for 5% GeO_2 doped silica [40]. For SMF980A and UHNA3 fibers, the parameters β_n and ZDW were calculated from dispersion curves provided by fiber manufactures.

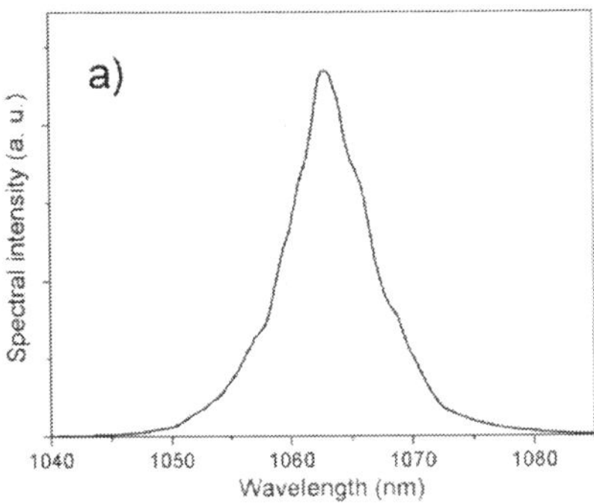
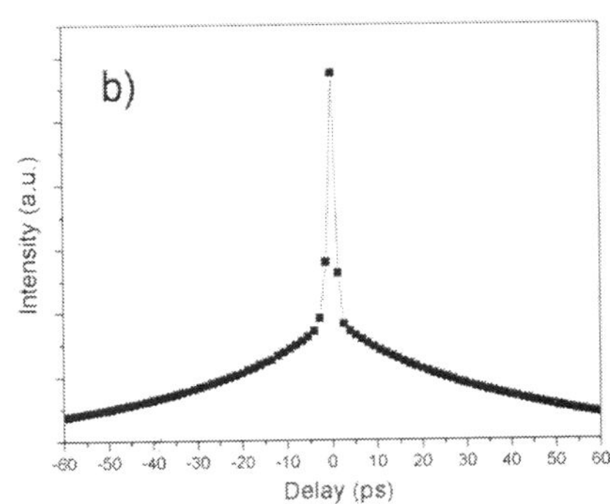

Figure 9. The experimentally measured spectrum (a) and intensity autocorrelation trace (b) of NLPs irradiated by the oscillator in Figure 1.

Fiber Type	F1060c (Lightwave Tech)	SMF980A (POFC, Taiwan)	UHNA3 (Nufern)
Mode-field diameter at 1060 nm	8.4 μm	4.5 μm	2.6 μm
Numerical aperture	0.11	0.21	0.35
Core GeO$_2$ concentration[1]	~3 mol.%	~10 mol.%	~20 mol.%
Zero-dispersion wavelength	1330 nm	1570 nm	1900 nm
Nonlinear coefficient at 1060 nm[2]	3.2 W^{-1} km^{-1}	13.7 W^{-1} km^{-1}	52.3 W^{-1} km^{-1}
Dispersion coefficients at 1060 nm[3]	$\beta_2 = 19.0\times10^{-3}$ ps^2/m, $\beta_3 = 4.5\times10^{-5}$ ps^3/m, $\beta_4 = -5.991\times10^{-8}$ ps^4/m, $\beta_5 = 1.879\times10^{-10}$ ps^5/m, $\beta_6 = -6.890\times10^{-13}$ ps^6/m	$\beta_2 = 28.1\times10^{-3}$ ps^2/m, $\beta_3 = 2.154\times10^{-5}$ ps^3/m, $\beta_4 = 8.187\times10^{-8}$ ps^4/m, $\beta_5 = 7.351\times10^{-11}$ ps^5/m, $\beta_6 = -3.965\times10^{-13}$ ps^6/m	$\beta_2 = 58.9\times10^{-3}$ ps^2/m, $\beta_3 = 2.32\times10^{-5}$ ps^3/m, $\beta_4 = 1.04\times10^{-7}$ ps^4/m, $\beta_5 = 2.078\times10^{-10}$ ps^5/m, $\beta_6 = -5.241\times10^{-13}$ ps^6/m
Spool length	50 m	50 m	10 m
Splicing loss	~0.5 dB	~1.5 dB	~3 dB

[1]Estimated data;

[2]Estimated from Eq. (2) assuming $n_2=3\times10^{-20}$m^2/W for F1060c; $n_2=3.7\times10^{-20}$m^2/W for SMF980A; $n_2=4.7\times10^{-20}$m^2/W for UHNA3, respectively;

[3]For F1060c, fibers were calculated from Sellmeier equation for 5% GeO$_2$ doped silica [41]; for SMF980A and UHNA3, fibers were calculated from dispersion curves provided by fiber manufactures.

Table 2. Fiber parameter and characteristics.

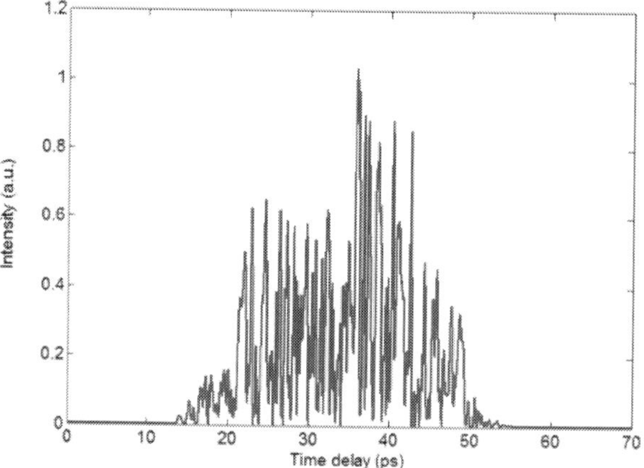

Figure 10. An example of simulated NLP waveforms.

3.1. Theoretical analysis and numerical simulation results for SCG

SC generation in an SMF by pumping with NLPs is analyzed by solving the general nonlinear Schrödinger equation in the frequency domain [19] as follows:

$$\frac{\partial A'(z,\omega)}{\partial z} = i\frac{\gamma\omega}{\omega_0}\exp\left\{-L(\omega)z\right\}F\left\{A(z,T)\int_{-\infty}^{\infty} R(T')\left|A(z,T-T')\right|^2 dT'\right\}, \tag{2}$$

where A and A' represent the electric field envelopes and γ is the nonlinear coefficient [16].

$$\gamma = \frac{2\pi n_2}{\lambda_0 A_{eff}} = \frac{8n_2}{\lambda_0 d^2}, \tag{3}$$

wherein n_2 is the nonlinear refractive index and d is the mode-field diameter. F{} denotes the Fourier transform and $R(t)$ is the full nonlinear response function, which includes instantaneous (Kerr) and delayed (Raman) terms and can be modeled by a single damped harmonic oscillator [16].

$$R(t) = (1-f_R)\delta(t) + f_R \frac{\tau_1^2 + \tau_2^2}{\tau_1\tau_2^2}\exp\left(-\frac{t}{\tau_2}\right)\sin\left(\frac{t}{\tau_1}\right)\Theta(t), \tag{4}$$

where f_R = 0.18 is the fractional contribution of the delayed Raman response, τ_1 is the time constant related to the phonon frequency, and τ_2 is the time constant related to the damping of the phonons. $\Theta(t)$ is the Heaviside step function and $\delta(t)$ is the Dirac delta function. From the Raman gain spectra, it is well known [35, 41, 42] that silica glass and Germanate glass have very similar phonon frequencies but very different phonon damping times. These time constants are listed in Table 3.

Absorption in fiber was neglected. In Eq. (2), $L(\omega)$ is the linear operator representing dispersion of the SMF, given by $L(\omega) = i(\beta(\omega)-\beta(\omega_0)-\beta_1(\omega_0)(\omega-\omega_0))$, where β_1 is the first-order dispersion coefficient associated with the Taylor series expansion of the propagation constant β about ω_0. Up to six high-order dispersion terms were taken into account for simulations (Table 2).The non-linear coefficient γ was calculated using Eq. (3) at a central wavelength of ~1.06 μm and using mode-field diameters of fibers according to Table 1. The nonlinear refractive index for fibers were assumed to be 3×10^{-20} m²/W, 3.7×10^{-20} m²/W, and 4.7×10^{-20} m²/W for F1060c, SMF980A, and UHNA3, respectively, according to Eq. (6) in Ref. [43]. Finally, the fourth-order Runge-Kutta algorithm was used to solve Eq. (2). The simulation results for the three different fiber samples (according to Table 2) are presented in Figures 11–19. For each fiber sample, we simulated the evolution of three different types of pulses of equal energy propagating through the fiber. The pulse energy in every case corresponded to the experimental conditions including splicing loss (Table 2). The first input waveform used in these calculations was a simulated NLP (Figure 10), of which characteristics (duration and bandwidth) corresponded to that of our NLP laser (Figure 9). The second input pulse waveform was that of a mode-locked Gaussian pulse with a duration of 45 ps, which is similar to the duration of our NLPs (Figures 9 and 10). The third input waveform was that of a mode-locked Gaussian pulse with a duration of 200 fs. Its spectral bandwidth (~5 THz) is similar to the bandwidth of our NLP (Figure 9).

Fiber type	F1060c (Lightwave Tech), SMF980A (POFC, Taiwan)	UHNA3 (Nufern)
τ_1	12.2 fs [42]	12 fs [35, 41]
τ_2	32 fs [42]	83 fs [35, 41]

Table 3. Single damped harmonic oscillator fitting constants.

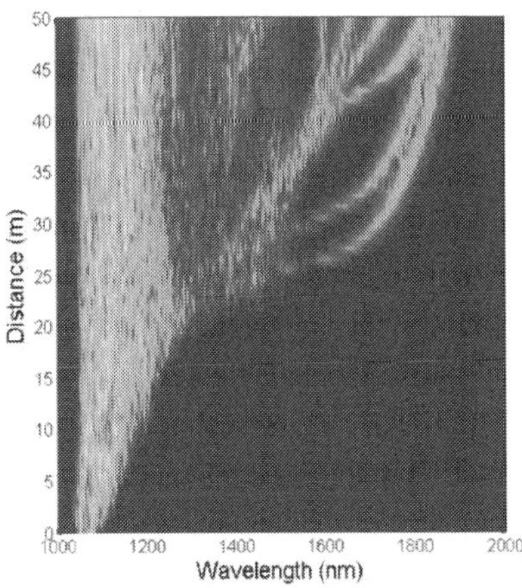

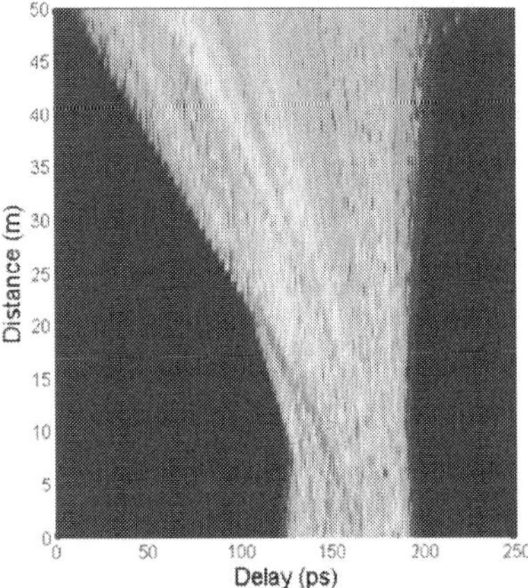

Figure 11. SC evolution in a 50-m piece of F1060c fiber pumped by NLP with 90-nJ pulse energy.

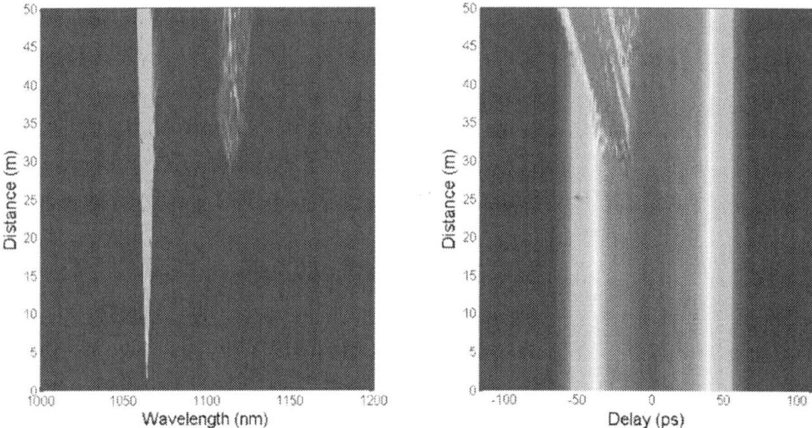

Figure 12. SC evolution in a 50-m piece of F1060c fiber pumped by Gaussian pulse with 45-ps duration and 2-kW peak power.

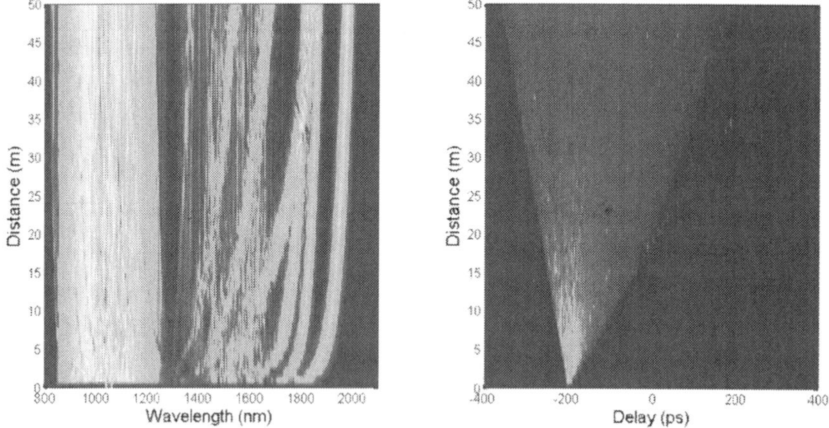

Figure 13. SC evolution in a 50-m piece of F1060c fiber pumped by Gaussian pulse with 200-fs duration and 350-kW peak power.

Examining the simulation results (Figures 11–19), we find that SC evolution in the case of NLP pumping is different from those of mode-locked picosecond or femtosecond Gaussian pulses for the three fiber samples we studied. Note that picosecond Gaussian pulses (~ 45 ps) can only generate from one to three clear Raman peaks in its spectrum (Figures 12, 15, and 18). That situation is explained by cascaded Raman scattering and is typical for narrowband pumping

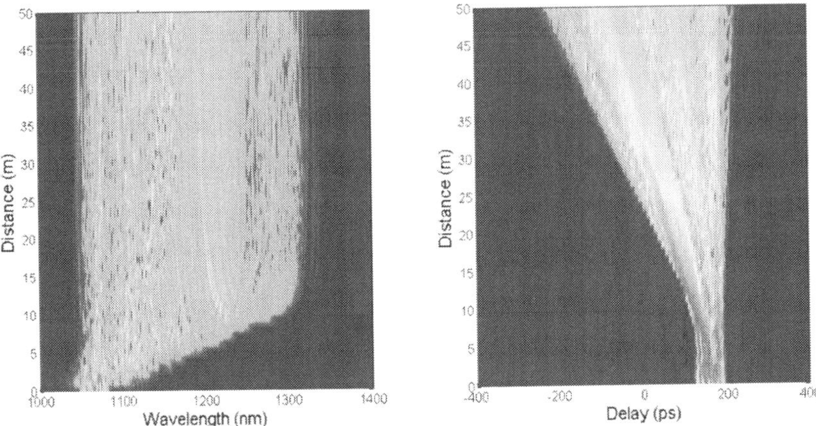

Figure 14. SC evolution in a 50-m piece of SMF980A fiber pumped by NLP with 45-nJ pulse energy.

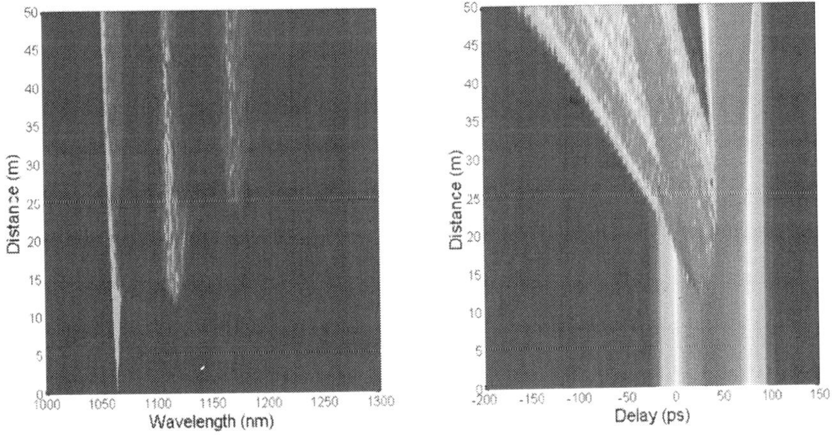

Figure 15. SC evolution in a 50-m piece of SMF980A fiber pumped by Gaussian pulse with 45-ps duration and 1-kW peak power.

with moderate powers in the normal-dispersion region [32–35, 44]. As for the shorter mode-locked Gaussian pump pulse (200 fs), it can generate broadband SC (Figures 13, 16, and 19) of which the behavior is dependent on relative positions of pump central wavelength and the ZDW. In the case when spectral difference between them is small, e.g., in the case of the F1060c fiber, initial spectral broadening caused by self-phase and cross-phase modulations is suffi-cient for the spectrum to approach the ZDW (Figure 13). Then, further spectral evolution is described by soliton fission, which is typical for SCG by ultra-short pulse [19]. We should

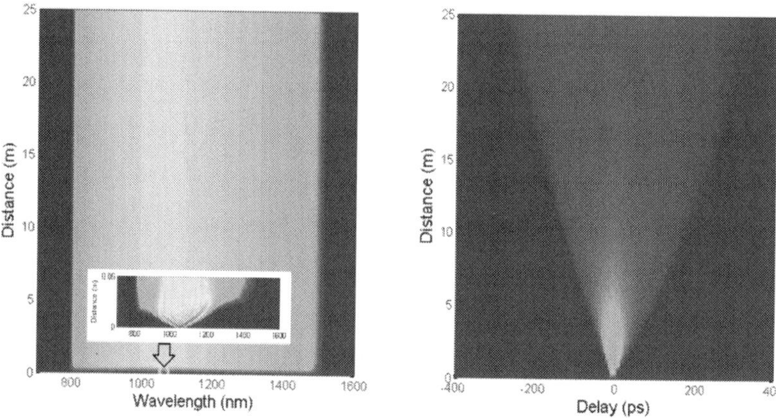

Figure 16. SC evolution in a 50-m piece of SMF980A fiber pumped by Gaussian pulse with 200-fs duration and 200-kW peak power.

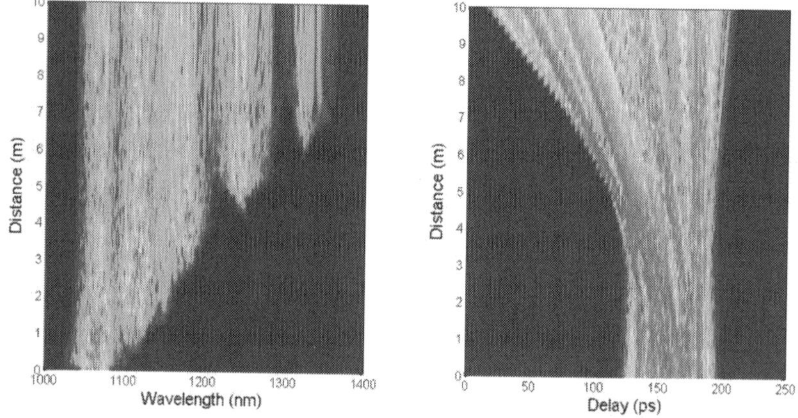

Figure 17. SC evolution in a 10-m piece of UHNA3 fiber pumped by NLP with 21-nJ pulse energy.

mention that efficient femtosecond pulse pumping requires significant peak power (hundreds of kWs) especially in the cases when ZDW is shifted to the longer wavelength region. Our simulations show that, for SMF980A and UHNA3 fibers, where the ZDWs are equal 1570 and 1900 nm, respectively, we may not be able to extend the SC to the anomalous dispersion region (Figures 16–19). Also, 200-fs-wide Gaussian pump pulses excite SC in first tens of centimeters of the SMF (see inset in Figures 16–19). Afterwards, its spectrum does not change.

It is interesting to note that NLPs, in comparison to mode-locked Gaussian pulses of similar waveform durations, exhibit the spectral broadening mainly in the longer wavelength (relative

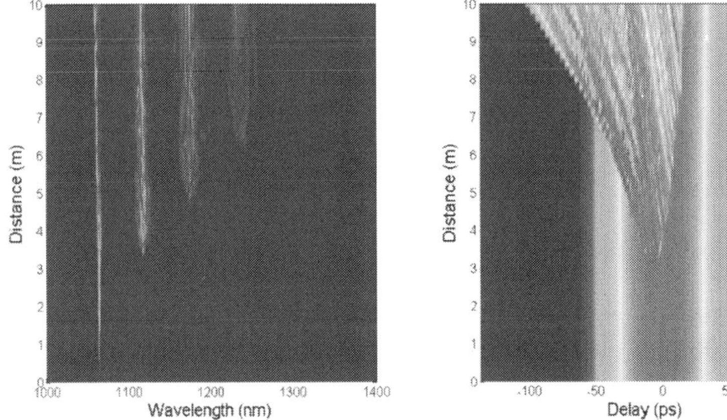

Figure 18. SC evolution in a 10-m piece of UHNA3 fiber pumped by Gaussian pulse with 45-ps duration and 450-W peak power.

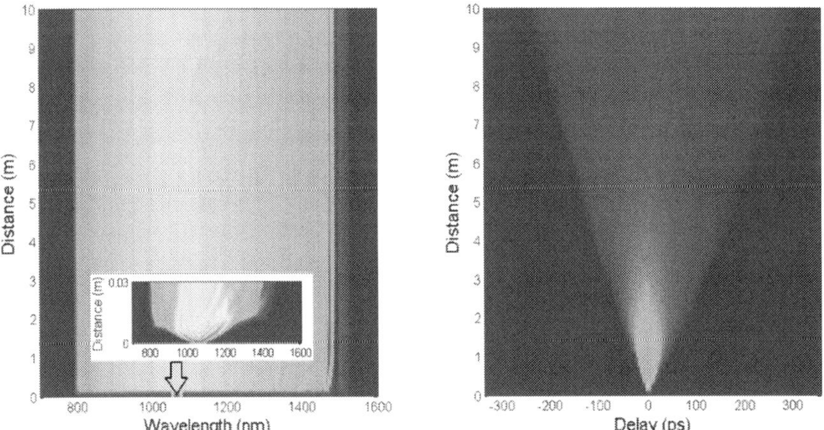

Figure 19. SC evolution in a 10-m piece of UHNA3 fiber pumped by Gaussian pulse with 200-fs duration and 100-kW peak power.

to the pump wavelength) region (Figures 11, 14, and 17). In the case of the UHNA3 fiber, the distinctive Raman orders can be found in SCG evolution spectrum (Figure 17). The broader Stokes lines and much smoother SC spectrum can be explained by broadband pumping [37], which is the feature of NLP [19]. For the two other fibers (F1060c and SMF980A), simulations suggest that even smoother SC spectra are expected to be generated. This may result from reduced Germanium concentration of their fiber core compared with that of the UHNA3 fiber.

That results in reduced phonon damping times (Table 3) and the suppression of the contribution of the Raman term in the full nonlinear response function (Eq. (4)). It is notable that pulse energy of our NLP, ~90 nJ, is sufficient to generate SC that approaches the ZDW in the case of the F1060c fiber (Figure 11). In comparison, a mode-locked pulse of a similar duration and energy can generate only one Stokes line (Figure 12). To reach a similar SC bandwidth, however, the evolution of NLP-excited SC takes longer distance in the fiber.

Recently, it has been theoretically predicted that Raman-induced spectral shift, which may happen in the normal dispersion region, is responsible for asymmetry of spectral broadening [45]. The distinctive Raman orders in Figures 12, 15, and 18 for MLPs suggested that multiple Raman processes, i.e., cascaded Raman scattering, could be responsible for initial spectral broadening for NLP pumped SMF. Therefore, we have tentatively attributed our calculated behavior of the broadband SC emission to Raman amplification of noise in the SMF [19]. Later on, after the SC spectrum approaching ZDW, further SCG process becomes soliton fission like.

First, we spliced F1060c fiber directly to the output of amplifier. Since the mode-field diameters of spliced fibers were not too much different, the splice loss was low (<0.5 dB). We measured a series of SCG spectra with the F1060c pumped at different powers (Figure 20(a)). The simulated SCG spectrum at an input power of 1500 mW in Figure 20(b) agrees quite well with our experimental observations.

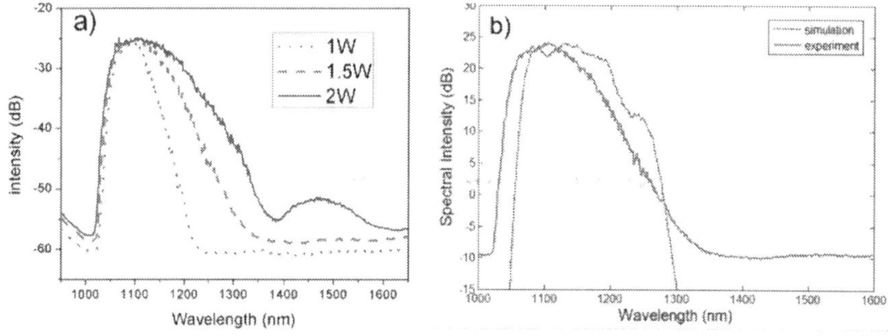

Figure 20. (a) The experimental spectra of SCG in F1060c fiber at different input powers. (b) The comparison between experimental and simulated spectra of SCG in F1060c fiber at 1500-mW input power.

Then, we examined SCG in the SMF980A fiber, which has smaller core diameter (Table 2). Owing to a large difference in mode areas, we used a short piece (~20 cm) of Newport F-SV fiber in between the connecting fibers. Hence, the total splice loss between the active fiber in the amplifier and SMF980A was ~1.5 dB. Similar to the previous case, we measured a series of SCG spectra at different pump powers (Figure 21(a)). The simulated SCG spectrum at an input power of 1600 mW (Figure 21(b)) again shows good agreement with experimental measurement. These two simulations and experimental results (for F1060c and SMF980A fibers) indicate smooth spectra of SC generated this way.

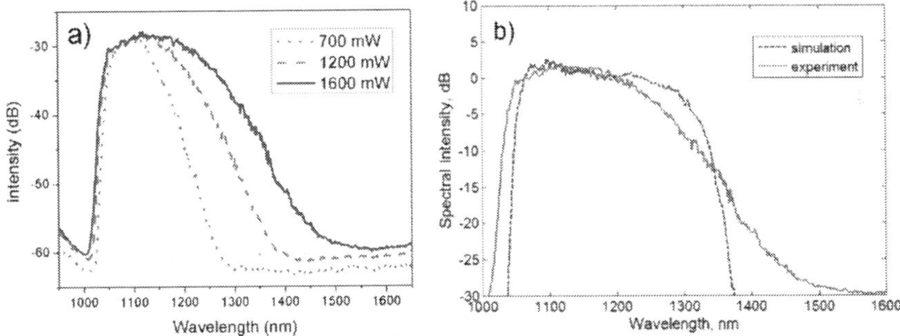

Figure 21. (a) The experimental spectra of SCG in SMF980A fiber at different input powers. (b) The comparison between experimental and simulated spectra of SCG in SMF980A fiber at 1600 mW input power.

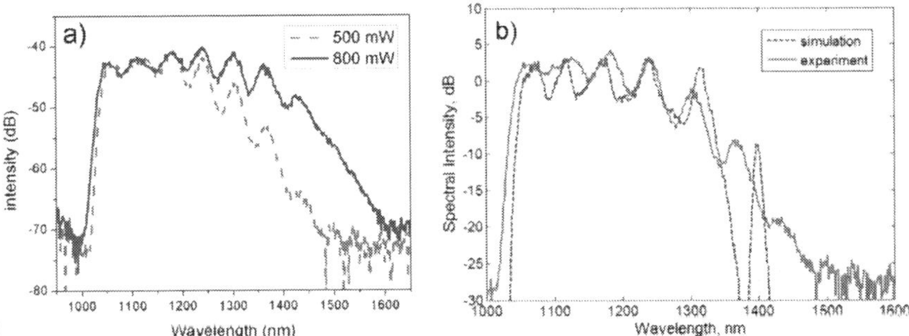

Figure 22. (a) The experimental spectra of SCG in UHNA3 fiber at different input powers. (b) The comparison between experimental and simulated spectra of SCG in UHNA3 fiber at 500 mW input power.

We also studied SCG by the UHNA3 fiber using our NLPs. The total splice loss was ~3 dB. The spectra of SCG generated in this fiber at different input powers are plotted in Figure 22(a). Spectral oscillators and broadened bandwidth in the SCG spectrum are evident and in good agreement with the simulated results (Figure 22(b)).

Examining Figures 20-22, we find that SCG spectra are different for the three types of fiber used and pumping conditions. Obviously, in order to get larger spectral broadening using SMFs with relatively small nonlinear coefficients, relatively longer fiber length is required (50 m for F1060c and SMF980A). Herein, the fiber with a larger nonlinear coefficient provides more broadband response (~260 nm vs. ~160 nm width at −10 dB level and ~1.5 W input power for SMF980A and F1060c, respectively) (Figures 20(a) and 21(a)). Applying UHNA3 fiber with high nonlinearity (γ = 52.3 W^{-1} km^{-1}) allows achieving significant spectral broadening (~440 nm at −10 dB level) with a shortened fiber length (10 m) and reduced input average power (~800 mW) of NLP. But we should mention that the shape of SCG spectra (Figure 22(a)) becomes modulated. The simulation results also indicate such behavior related with cascaded Raman scattering. We believe that such behavior can be explained by significantly increased doping concentration of germanium in UHNA3 fiber (~20 mol.%). This results not only in nonlinearity enhancement [45] but also in modification of Raman response function [43]. It is known that Raman gain in Germatate glass is much stronger than that in silica glass. As a result, the damping time of delayed Raman response in highly doped germanium fiber is longer (Table 3), which causes spectral oscillations.

In summary, we have reported a study of SCG with a scheme that uses a SMF pumped in the normal dispersion regime (~1 μm) by NLPs from a dispersion-mapped fiber laser. We showed SCG results using different SMFs and demonstrated that using a higher doping concentration of Germanium (~20 mol.%) in SMF results in the enhancement of optical fiber nonlinearity, reducing SCG threshold, and change in the Raman response function. The last one is found to be responsible for the appearance of oscillations in the SC spectrum. The achieved SC exhibits low threshold (21 nJ) and a broadband spectrum over 1030–1470 nm with a relatively short fiber length (10 m).

We believe that the achieved low energy threshold and flat SC spectrum are caused by unique properties of NLP (broadband spectral range and ability to propagate over a long distance). Higher level of spectral oscillation is caused by higher Raman gain in germanium doped fiber, which results in increasing the damping time of Raman response function.

Naturally, it is possible to use NLPs from the ANDi laser for SCG. Examples of such studies will be described in Sec. 4 of this chapter.

4. Applications for OCT

OCT is a technique of optical imaging that is extensively used in biomedical and clinical studies. The first OCT images were demonstrated by Huang *et al.* [46]. The authors confirmed that OCT is superior for obtaining images in highly scattering media. In 1993, the first *in vivo* OCT images were obtained by Fercher [47] and Swanson [48]. Thereafter, OCT technology advanced rapidly and was readily accepted for use in ophthalmology. In 1996, the first commercial ophthalmic OCT instrument by Carl Zeiss Meditec made its appearance [49]. Another significant milestone is the spectral-domain interferometric technique proposed theoretically in 1995 by Fercher *et al.* [50]. Soon afterwards, Wojtkowski showed that the technique was workable by imaging the retina [51]. Comparative studies of TD-OCT and SD-OCT were first reported by De Boer [52] and Leitgeb [53].

OCTs typically employ near-infrared light in order to achieve deep penetration in biological samples or scattering media. The axial resolution of an OCT is determined by the width of the field autocorrelation function, or the inverse of the spectral bandwidth of the light source. Thus, a broadband light source is required for ultra-high-resolution OCT.

Superluminescent diode (SLD), amplified spontaneous emission (ASE), and SC sources have been widely used as the light source for OCT systems. OCTs with SLDs typically have an axial resolution of 10–15 μm, due to the limited gain bandwidth (~100 nm or less). The bandwidth of ASE sources is also around several tens of nanometers. Thus, ultra-high-resolution OCT typically employs SC sources. In 2001, Hartl *et al.* reported the first OCT system based on SC generated with a PCF pumped by a femtosecond Ti: sapphire laser [54]. They achieved an axial resolution of ~2.5 μm for OCT imaging using the 370-nm-wide broadband source.

In addition, the operation wavelengths for OCT are also critical. For applications of OCTs in ophthalmology, light sources with central wavelength around 800 nm are widely used. As a result of the relatively short wavelength, ultra-high-resolution OCT can be easily obtained. However, a longer wavelength is desirable for increasing the penetration depth of the OCT. Ultra-high-resolution OCT in bio-tissue has been demonstrated with the SC source at the central wavelength of approximately 1, 1.3, 1.5, and 1.7 μm [55–57]. Water exhibits a broad absorption peak at 1.5 μm. Since biological materials are water rich, 1.5-μm light sources, highly developed for the optical communication industry, are not so suitable for OCT. *In vivo* OCT imaging in bio-tissues is thus typically performed at 1300 nm to take advantage of reduced scattering and enhanced penetration depth.

Moreover, for imaging of the cornea and anterior segment of the eye, OCT operated at 1300 nm has important potential advantages for glaucoma evaluation and refractive surgery [58]. The window around 1300 nm is particularly well suited for evaluation in detail of the response of the angle structures to light accommodation, which is an important part of the glaucoma workup.

Using rare-earth-doped fiber lasers, especially Yb-doped fiber lasers operating at high pulse energy with ultrashort pulse width (from tens to hundreds of femtoseconds) as the pump laser, new SC sources with a broad spectral bandwidth and high output power were reported by several groups [23–25]. Typically, PCF or highly nonlinear fiber is employed to generate SC in these schemes. The former approach provides higher nonlinearity and allows one to blue-shift the zero-dispersion wavelength [19].

Later, our group and others demonstrated [38, 39] that amplified NLPs could be successfully used for SCG through standard SMFs. NLPs [5, 9, 10, 59–62] exhibit attractive features such as a very smooth and broadband spectrum, a double-scaled autocorrelation trace with a sub-picosecond peak riding, and a wide sub-nanosecond pedestal, suggesting their low temporal coherence. It is conceivable that a SC source based on NLPs is potentially attractive for applications based on low-coherence interferometry such as OCT. In this work, we report for the OCT imaging with a new and improved SC source pumped by NLPs from a compact ANDi Yb-doped fiber laser.

NLPs from the ANDi Yb-doped fiber laser with an average power of 200 mW (pulse energy of 62.5 nJ) are first amplified through an Yb-doped fiber amplifier. The amplified NLPs are then used to excite the SC in a 50-m length of the standard SMF. At a pump power of 4.5 W, we are able to generate a flat SC spectrum at the central wavelength of 1320 nm with a spectral bandwidth of up to 420 nm (in 3-dB bandwidth), as shown in Figure 23. The average output power reaches up to 560 mW (pulse energy of 175 nJ).

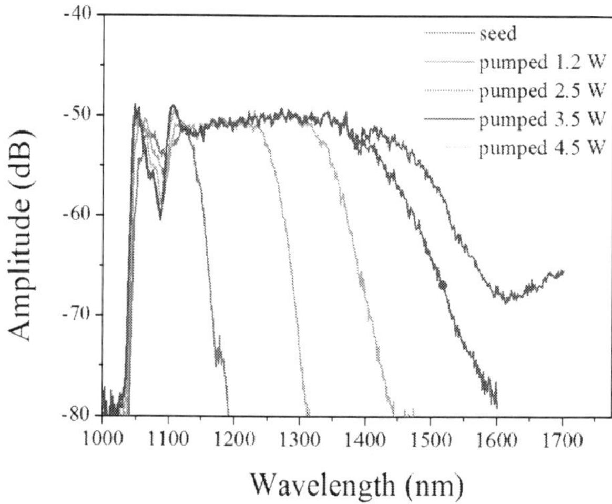

Figure 23. The SC spectra generated by an all-normal dispersion fiber laser system.

4.1. Time-domain OCT

To test whether our SC source is applicable to ultra-high-resolution OCT, we set up a simple OCT system. The system setup is similar to the Michelson interferometer as shown in Figure 24(a). There is a movable mirror in the reference arm, which directly reflects the light and changes the relative optical path length compared with the sample arm. Finally, the detector will receive the reflected light from two arms. At the sample arm, we assume the sample to have an inner structure with many layers, i.e., different refractive index variations, so the light will reflect back to the beam splitter and then be focused to the photodetector by focusing lens with different time delays. Using a pulse as the light source, we can observe the interferogram by oscilloscope. The interferogram provides the information about the position of the interface of the sample by different time delays corresponding to different positions of a movable mirror. The measured results can be understood from Figure 24(b).

4.2. Spectral-domain OCT

The spectral-domain approach for detection of OCT images is also referred to as the Fourier-domain OCT. The schematic of a spectral-domain OCT in shown in Figure 25(a) [49]. It can be immediately recognized that the movable mirror in TD-OCT is replaced by a stationary one. The photodetector is replaced by the spectrometer. The interferogram will be spectrally modulated and the wavelength dependence can be measured using a spectrometer (Figure 25(b)). We note that the optical path length mismatch ΔL must not be equal to zero during detection.

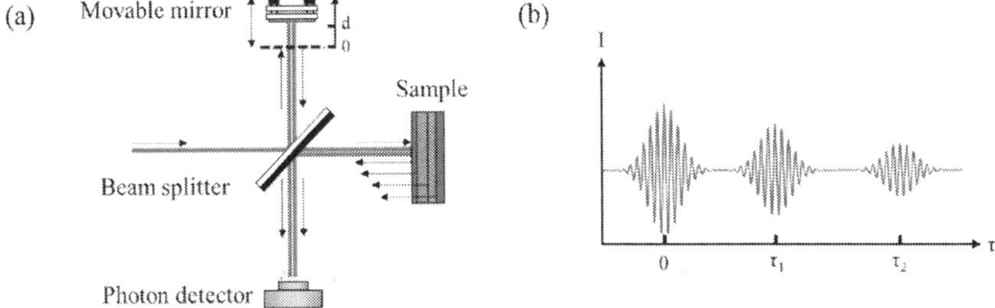

Figure 24. (a) The interferometer within an OCT system, and (b) the interferogram from an OCT system in time domain detection.

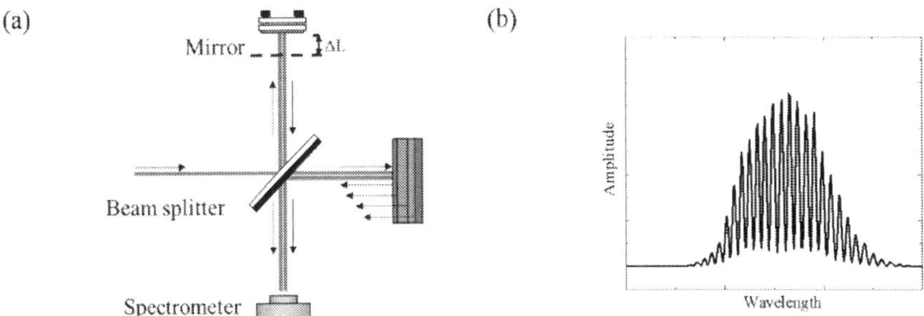

Figure 25. (a) The interferometer within an OCT system, and (b) the interferogram from an OCT system in spectral-/ Fourier domain detection.

4.3. OCT performance

Coherence length is a critical parameter for the OCT light source. The coherence length can be defined as the maximum path difference between reference and sample arms for which signals from both arm can interfere with each other. The coherence length of a source is related to the axial resolution of the OCT through

$$l_c = \frac{2\ln 2}{\pi} \frac{\lambda_0^2}{\Delta\lambda} \approx 0.44 \frac{\lambda_0^2}{\Delta\lambda} \tag{5}$$

where $\Delta\lambda$ is the 3 dB bandwidth of optical spectrum and λ_0 is the central wavelength.

In this experiment, the SC light source generated an output power of around 345 mW, which is much higher than that of typical SC sources. With the SC source operated at a bandwidth of 365 nm center around 1320 nm, the theoretical axial resolution of the instrument is estimated to be ~2.1 µm. Figure 26(a) shows A-scan result of the TD-OCT. This corresponds to a measured axial resolution of ~2.3 μm (Figure 26(b)). In spite of the ultra-high axial resolution provided by the TD-OCT, its signal-to-noise ratio (SNR) was only 12.5 dB. SD-OCT, on other hand, provides much better SNR and stability and quick scanning speed. Figure 27(a) shows the point-spread function (PSF) of the SD-OCT. An axial resolution of ~2.8 μm was achieved. In Figure 27(b), we show the PSFs of a free-space SD-OCT obtained at different depths. The data indicate that the system can provide OCT images at a depth of ~800 μm.

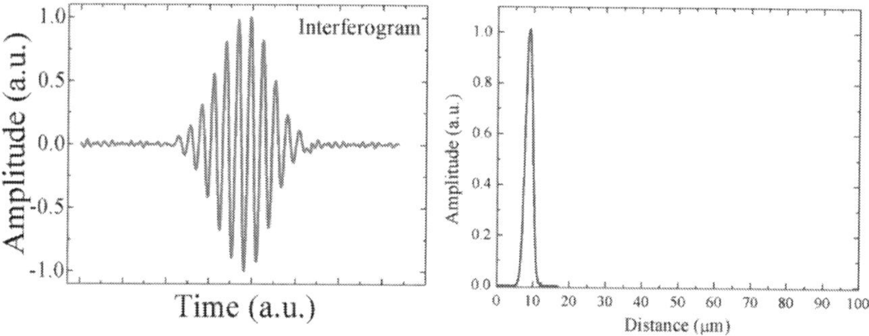

Figure 26. (a) A-scan result of the TD-OCT and (b) PSF.

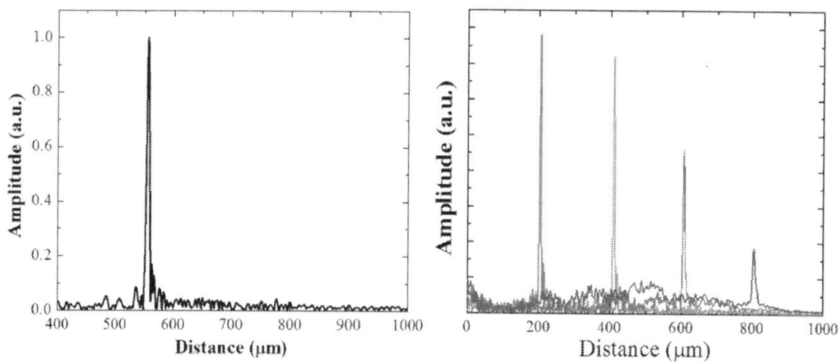

Figure 27. (a) PSF of SD-OCT and (b) PSFs at different depths.

Comparing with the free-space OCT, the fiber-based OCT is more compact and stable. Thus, results for the fiber-based spectral domain OCT system are also presented herein. Configuration of the fiber-based SD-OCT is shown schematically in Figure 28. The SC source was sent into a fiber coupler where it had a splitting ratio of 1:99. After the first fiber coupler, two optical fiber circulators are used in each arm. One arm serves as the reference, while the other arm delivers the signal to the sample. Afterwards, reflected light from the reference mirror and the sample combine together using a second fiber coupler with a splitting ratio of 1:1. Finally, the interference signal is detected using an optical spectrum analyzer.

Our broadband SC source can achieve an ultra-high-resolution of ~ 2 μm. That of the all-fiber SD-OCT is limited, however, by the optical bandwidth of fiber components of the OCT system, such as optical circulators and optical couplers designed for the central wavelength of 1310 nm. Owing to that limitation, the effective spectral bandwidth for the fiber OCT system was only 102 nm. Therefore, the theoretical axial resolution of our all-fiber SD-OCT is ~5.3 μm. Figure 29(a) shows the PSF with and without dispersion compensation. The resolutions are 18.6 and 6.4 μm, respectively. It is apparent that the resolution is improved by dispersion compensation. Figure 29(b) shows the A-scan results obtained at different depths. The data were processed by applying the inverse Fourier transform to the measured interferograms. Our results indicate that the scanning depth can reach 1 mm. The SNR also improved significantly, reaching as high as 126.7 dB.

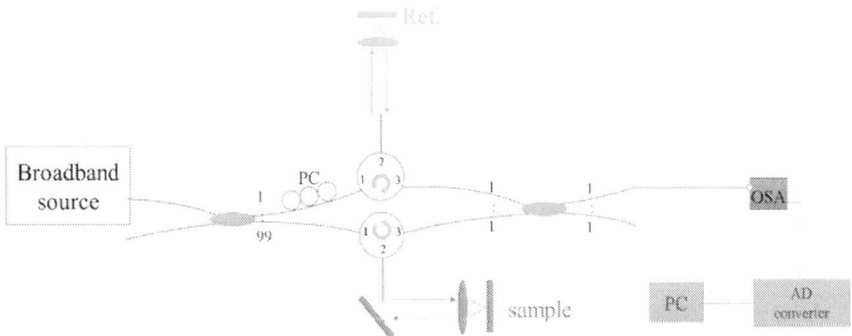

Figure 28. The schematic of SD-OCT system. PC, polarization controller; OSA, optical spectrum analyzer.

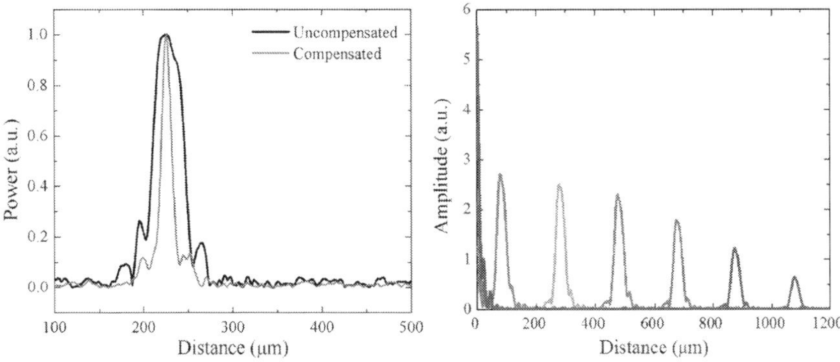

Figure 29. (a) The PSF with and without dispersion compensation. (b) A-scans at different depths.

4.4. OCT images

The surface tomography of a part of the one-dollar (new Taiwan dollar) coin is shown in Figure 30. In this case, the wavelength range of the source was ~45 nm, and the axial resolution was experimentally determined to be <17 μm. The OCT resolution is sufficient to get a clear image of the Chinese character, meaning "middle". The scanning step was 50 μm.

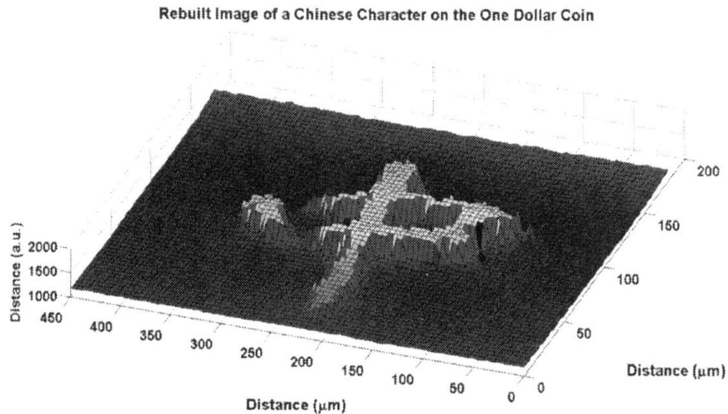

Figure 30. Image of a Chinese character "" on the coin of new Taiwan dollar.

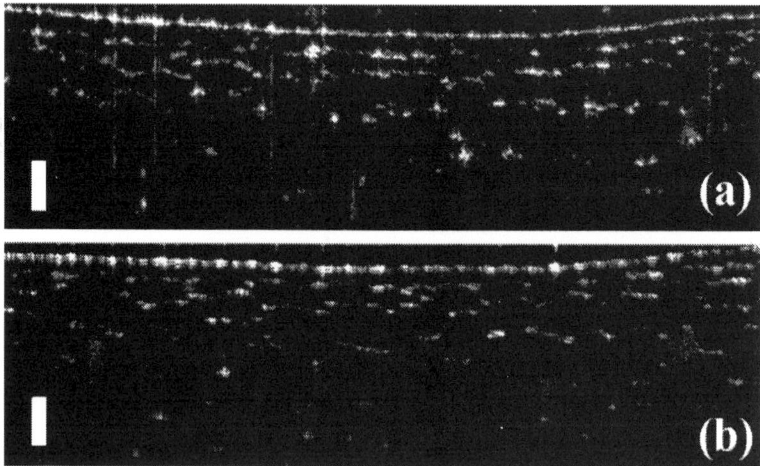

Figure 31. (a) SD-OCT image of an onion skin using a NLP laser. (b) SS-OCT image using a commercial swept source made by Santec. The scale bars represent 200 μm.

Figure 31 shows the OCT images of a layer of peeled onion skin using different light sources. Figure 31(a) is an image obtained with the fiber-based SD-OCT system using our noise-like SC light source, for which spectral bandwidth was limited by the fiber components as 108 nm. For comparison, Figure 31(b) is the OCT image obtained using the swept-source OCT (SS-OCT) system with a commercial swept source by Santec. The 2-D depth profile of the onion skin is obtained using both light sources. Several cell walls on onion skin can be easily identified. The best OCT images of the onion cell were obtained when we applied noise-like SC as a scan light source. The result of Figure 31(a) is the original image without dispersion compensation. This showed that higher image quality can be achieved through further data processing.

5. Summary

In this chapter, we describe our work on generation and amplification of medium- and high-energy NLPs with Yb-doped fiber laser systems. In the dispersion-mapped fiber laser system, the central wavelength and spectral bandwidth can be tuned. The tuning range was 33 nm and the bandwidth can be varied from 6 to 54 nm. Stable NLPs can be generated with output power from 0.1 to 1.6 W, limited by damage of the optical fiber components used. Pulse energy as high as 1. 25 µJ can be generated directly from the laser oscillator with a repetition rate of ~1 MHz. On the other hand, the ANDi noise-like fiber laser is more compact and can generate broadband NLPs with a bandwidth as high as ~80 nm. We also demonstrate SCG pumped by NLPs in easily available single-mode optical fibers. SC with a flat spectrum can be generated over 420 nm in 3-dB bandwidth, centered around 1.3 µm. Such broad spectrum arises from cascaded stimulated Raman scattering and self-phase modulation in the optical fibers. SMFs with high nonlinearity can be used to generate even broader spectra. Significant modulations due to the nonlinear effects rendered such sources unsuitable OCT. Experimental results are in good agreement with theoretical analysis and numerical simulation results. SC pumped by NLPs has been employed successfully in OCT systems. An ultra-high axial resolution of 2.3 µm was demonstrated. High-resolution fiber-based spectral domain OCT is also demonstrated. Scanned images show that a SC source pumped by NLPs is a viable alternative to SLD and SC pumped by femtosecond lasers as the broadband light source for applications such as OCT.

Acknowledgements

This work was supported in part by the Ministry of Science and Technology, Taiwan, under Grant 103-2622-E-007-006-CC2 and by the National Tsing Hua University Research Program Grant 104N2711E1. C. L. Pan is also supported by the Air Force Office of Scientific Research FA2386-13-1-4086. Some of the experimental results presented in this chapter were obtained through collaboration with Prof. Ping Xue and Mr. Chengming Wang, Tsinghua University, Beijing. We also gratefully acknowledge Yi-Lun Lin and Kuo-Sung Huang for their contributions to the work summarized in this chapter.

Author details

Ci-Ling Pan*, Alexey Zaytsev, Yi-Jing You and Chih-Hsuan Lin

*Address all correspondence to: clpan@phys.nthu.edu.tw

National Tsing Hua University, Hsinchu, Taiwan, R.O.C.

References

[1] Barnard C, Myslinski P, Chrostowski J, Kavehrad M. Analytical model for rare-earth-doped fiber amplifiers and lasers. IEEE J. Quantum Electron. 1994;30:1817–1830. DOI: 10.1109/3.301646

[2] Stone J, Burrus CA. Neodymium-doped fiber lasers: room temperature CW operation with an injection laser pump. Appl. Opt. 1974;13:1256–1258. DOI: 10.1364/AO.13.001256

[3] Barnes WL, Poole SB, Townsend JE, Reekie L, Taylor DJ, Payne DN. "Er^{3+}-Yb^{3+} and Er^{3+} doped fiber lasers. J. Lightwave Technol. 1989;7:1461–1465. DOI: 10.1109/50.39081

[4] Pottiez O, Grajales-Coutino R, Ibarra-Escamilla B, Kuzin EA, Hernandez-Garcia JC. Adjustable noiselike pulses from a figure-eight fiber laser. Appl. Opt. 2011; 50:E24–E31. DOI: 10.1364/AO.50.000E24

[5] Kobtsev S, Kukarin S, Smirnov S, Turitsyn S, Latkin A. Generation of double-scale femto/pico-second optical lumps in mode-locked fiber lasers. Opt. Exp. 2009;17:20707–20713. DOI: 10.1364/OE.17.020707

[6] An Y, Shen D, Zhao W, Long J. Characteristics of pulse evolution in mode-locked thulium-doped fiber laser. Opt. Commun. 2012;285:1949–1953. DOI: 10.1016/j.optcom.2011.12.001

[7] Kobtsev S, Smirnov S. Fiber lasers mode-locked due to nonlinear polarization evolution: golden mean of cavity length. Laser Phys. 2011;21:272–276. DOI: 10.1134/S1054660X11040050

[8] Zhao L, Tang D, Wu J, Fu X, Wen S. Noise-like pulse in a gain-guided soliton fiber laser. Opt. Exp. 2007;15:2145–2150. DOI: 10.1364/OE.15.002145

[9] Horowitz M, Barad Y, Silberberg Y. Noiselike pulses with a broadband spectrum generated from an erbium-doped fiber laser. Opt. Lett. 1997;22:799–801. DOI: 10.1364/OL.22.000799

[10] Smirnov S, Kobtsev S, Kukarin S, Ivanenko A. Three key regimes of single pulse generation per round trip of all-normal-dispersion fiber lasers mode-locked with nonlinear polarization rotation. Opt. Exp. 2012;20:27447–27453. DOI: 10.1364/OE.20.027447

[11] Tang D, Zhao L, Zhao B. Soliton collapse and bunched noise-like pulse generation in a passively mode-locked fiber ring laser. Opt. Exp. 2005;13:2289–2294. DOI: 10.1364/OPEX.13.002289

[12] Smirnov SV, Kobtsev SM, Kukarin SV. Efficiency of non-linear frequency conversion of double-scale pico-femtosecond pulses of passively mode-locked fiber laser. Opt. Exp. 2014;22:1058–1064. DOI: 10.1364/OE.22.001058

[13] Aguergaray C, Runge A, Erkintalo M, Broderick NG. Raman-driven destabilization of mode-locked long cavity fiber lasers: fundamental limitations to energy scalability. Opt. Lett. 2013;38:2644–2646. DOI: 10.1364/OL.38.002644

[14] Zaytsev A, Lin CH, You YJ, Tsai TF, Wang CL, Pan CL. A controllable noise-like operation regime in a Yb-doped dispersion-mapped fiber ring laser. Laser Phys. Lett. 2013;10:045104. DOI: 10.1088/1612-2011/10/4/045104

[15] Lim H, Ilday FÖ, Wise FW. Generation of 2-nJ pulses from a femtosecond ytterbium fiber laser. Opt. Lett. 2003;28:660–662. DOI: 10.1364/OL.28.000660

[16] Agrawal GP. Nonlinear fiber optics. 5th ed. Waltham: Elsevier/Academic; 2013. DOI: 10.1016/B978-0-12-397023-7.00011-5

[17] Pan CL, Zaytsev A, Lin CH, You YJ, Wang CH. Progress in short-pulse Yb-doped fiber oscillators and amplifiers. In: Lee CC, editor. The current trends of optics and photonics. Dordrecht: Springer; 2015. pp. 61–100. DOI: 10.1007/978-94-017-9392-6_3

[18] Chong A, Buckley J, Renninger W, Wise FW. All-normal-dispersion femtosecond fiber laser. Opt. Exp. 2006;14:10095–10100. DOI: 10.1364/OE.14.010095

[19] Dudley JM, Genty G, Coen S. Supercontinuum generation in photonic crystal fiber. Rev. Mod. Phys. 2006;78:1135–1184. DOI: 10.1103/RevModPhys.78.1135

[20] Bizheva K, Považay B, Hermann B, Sattmann H, Drexler W, Mei M, Holzwarth R, Hoelzenbein T, Wacheck V, Pehamberger H. Compact, broad-bandwidth fiber laser for sub-2-μm axial resolution optical coherence tomography in the 1300-nm wavelength region. Opt. Lett. 2003;28:707–709. DOI: 10.1364/OL.28.000707

[21] Diddams SA, Jones DJ, Ye J, Cundiff ST, Hall JL, Ranka JK, Windeler RS, Holzwarth R, Udem T, Hansch TW. Direct link between microwave, and optical frequencies with a 300 THz femtosecond laser comb. Phys. Rev. Lett. 2000;84:5102–5105. DOI: 10.1103/PhysRevLett.84.5102

[22] Boivin L, Collings BC. Spectrum slicing of coherent sources in optical communications. Opt. Fiber Technol. 2001;7:1–20. DOI: 10.1006/ofte.2000.0342

[23] Song R, Hou J, Chen S, Yang W, Lu Q. High power supercontinuum generation in a nonlinear ytterbium-doped fiber amplifier. Opt. Lett. 2012;37:1529–1531. DOI: 10.1364/OL.37.001529

[24] Hu X, Zhang W, Yang Z, Wang Y, Zhao W, Li X, Wang H, Li C, Shen D. High average power, strictly all-fiber supercontinuum source with good beam quality. 2011;36:2659–2661. DOI: 10.1364/OL.36.002659

[25] Chen H, Chen Z, Chen S, Hou J, Lu Q. Hundred-watt-level, all-fiber-integrated supercontinuum generation from photonic crystal fiber. Appl. Phys. Exp. 2013;6:032702-1-3. DOI: 10.7567/APEX.6.032702

[26] Alfano RR, Shapiro SL. Emission in the region 4000 to 7000 Å via four-photon coupling in glass. Phys. Rev. Lett. 1970;24:584–587. DOI: 10.1103/PhysRevLett.24.584

[27] Hernandez-Garcia JC, Pottiez O, Estudillo-Ayala JM, Rojas-Laguna R. Numerical analysis of a broadband spectrum generated in a standard fiber by noise-like pulses from a passively mode-locked fiber laser. Opt. Commun. 2012;285:1915–1919. DOI: 10.1016/j.optcom.2011.12.069

[28] Petropoulos P, Ebendorff-Heidepriem H, Finazzi V, Moore RC, Frampton K, Richardson DJ, Monro TM. Highly nonlinear and anomalously dispersive lead silicate glass holey fibers. Opt. Exp. 2003;11:3568–3573. DOI: 10.1364/OE.11.003568

[29] Ebendorff-Heidepriem H, Petropoulos P, Asimakis S, Finazzi V, Moore RC, Frampton K, Koizumi F, Richardson DJ, Monro TM. Bismuth glass holey fibers with high nonlinearity. Opt. Exp. 2004;12:5082–5087. DOI: 10.1364/OPEX.12.005082

[30] Liao M, Yan X, Qin G, Chaudhari C, Suzuki T, Ohishi Y. A highly non-linear tellurite microstructure fiber with multi-ring holes for supercontinuum generation. Opt. Exp. 2009;17:15481–15490. DOI: 10.1364/OE.17.015481

[31] Qin G, Yan X, Kito C, Liao M, Chaudhari C, Suzuki T, Ohishi Y. Ultrabroadband supercontinuum generation from ultraviolet to 6.28 µm in a fluoride fiber. Appl. Phys. Lett. 2009;95:161103. DOI: 10.1063/1.3254214

[32] El-Amraoui M, Gadret G, Jules JC, Fatome J, Fortier C, Désévédavy F, Skripatchev I, Messaddeq Y, Troles J, Brilland L, Gao W, Suzuki T, Ohishi Y, Smektala R. Microstructured chalcogenide optical fibers from As_2S_3 glass: towards new IR broadband sources. Opt. Exp. 2010;18:26655–26665. DOI: 10.1364/OE.18.026655

[33] Xiao L, Demokan MS, Jin W, Wang Y, Zhao CL. Fusion splicing photonic crystal fibers and conventional single-mode fibers: Microhole collapse effect. J. Lightwave Technol. 2007;25:3563–3574. DOI: 10.1109/JLT.2007.907787

[34] Pourbeyram H, Agrawal GP, Mafi A. Stimulated Raman scattering cascade spanning the wavelength range of 523 to 1750~nm using a graded-index multimode optical fiber. Appl. Phys. Lett. 2013;102:201107. DOI: 10.1063/1.4807620

[35] Yin K, Zhang B, Yang W, Chen H, Hou J. Over an octave cascaded Raman scattering in short highly germanium-doped silica fiber. Opt. Exp. 2013;21:15987–15997. DOI: 10.1364/OE.21.015987

[36] Sayinc H, Hausmann K, Morgner U, Neumann J, Kracht D. Picosecond all-fiber cascaded Raman shifter pumped by an amplified gain switched laser diode. Opt. Exp. 2011;19:25918–25924. DOI: 10.1364/OE.19.025918

[37] Ilev I, Kumagai H, Toyoda K, Koprinkov I. Highly efficient wideband continuum generation in a single-mode optical fiber by powerful broadband laser pumping. Appl. Opt. 1996;35:2548–2553. DOI: 10.1364/AO.35.002548

[38] Hernandez-Garcia JC, Pottieza O, Estudillo-Ayalab JM. Supercontinuum generation in a standard fiber pumped by noise-like pulses from a figure-eight fiber laser. Laser Phys. 2012;22:221–226. DOI: 10.1134/S1054660X1123006X

[39] Zaytsev A, Lin CH, You YJ, Chung CC, Wang CL, Pan CL. Supercontinuum generation by noise-like pulses transmitted through normally dispersive standard single-mode fibers. Opt. Exp. 2013;21:16056–16062. DOI: 10.1364/OE.21.016056

[40] Brückner V. Elements of optical networking. Basics and practice of optical data communication. Wiesbaden: Springer; 2011. DOI: 10.1007/978-3-8348-8142-7_8

[41] Rottwitt K, Povlsen JH. Analyzing the fundamental properties of Raman amplification in optical fibers. J. Lightwave Technol. 2005;23:3597–3605. DOI: 10.1109/JLT.2005.857776

[42] Blow KJ, Wood D. Theoretical description of transient stimulated Raman scattering in optical fibers. IEEE J. Quantum Electron. 1989;25:2665–2673. DOI: 10.1109/3.40655

[43] Kato T, Suetsugu Y, Nishimura M. Estimation of nonlinear refractive index in various silica-based glasses for optical fibers. Opt. Lett. 1995;20:2279–2281. DOI: 10.1364/OL.20.002279

[44] Lin C, Nguyen VT, French WG. Wideband near-I.R. continuum (0.7–2.1 μm) generated in low-loss optical fibers. Electron Lett. 1978;14:822–823. DOI: 10.1049/el:19780556

[45] Santhanama J, Agrawal GP. Raman-induced spectral shifts in optical fibers: general theory based on the moment method. Opt. Commun. 2003;222:413–420. DOI: 10.1016/S0030-4018(03)01561-X

[46] Huang D, Swanson EA, Lin CP, Schuman JS, Stinson WG, Chang W, Hee MR, Flotte T, Gregory K, Puliafito CA. Optical coherence tomography. Science. 1991;254:1178–1181. DOI: 10.1126/science.1957169

[47] Fercher AF, Hitzenberger CK, Drexler W, Kamp G, Sattmann H. In vivo optical coherence tomography. Am J. Ophthalmol. 1993;116:113–114. DOI: 10.5858/arpa.2012-0252-SA

[48] Swanson EA, Izatt J, Lin C, Fujimoto J, Schuman J, Hee M, Huang D, Puliafito C. In vivo retinal imaging by optical coherence tomography. Opt. Lett. 1993;18:1864–1866. DOI: 10.1364/OL.18.001864

[49] Izatt JA, Choma MA. Theory of optical coherence tomography. In: Drexler W, Fujimoto J, editors. Optical coherence tomography. Berlin Heidelberg: Springer; 2008. pp. 47–72. DOI: 10.1007/978-3-540-77550-8_2

[50] Fercher AF, Hitzenberger CK, Kamp G, El-Zaiat SY. Measurement of intraocular distances by backscattering spectral interferometry. Opt. Commun. 1995;117:43–48. DOI: 10.1016/0030-4018(95)00119-S

[51] Wojtkowski M, Leitgeb R, Kowalczyk A, Bajraszewski T, Fercher AF. In vivo human retinal imaging by Fourier domain optical coherence tomography. J. Biomed. Opt. 2002;7:457–463. DOI: 10.1117/1.1482379

[52] de Boer JF, Cense B, Park BH, Pierce MC, Tearney GJ, Bouma BE. Improved signal-to-noise ratio in spectral-domain compared with time-domain optical coherence tomography. Opt. Lett. 2003;28:2067–2069. DOI: 10.1364/OL.28.002067

[53] Leitgeb R, Hitzenberger C, Fercher A. Performance of fourier domain vs. time domain optical coherence tomography. Opt. Express. 2003;11:889–894. DOI: 10.1364/OE.11.000889

[54] Hartl I, Li XD, Chudoba C, Ghanta RK, Ko TH, Fujimoto JG, Ranka JK, Windeler RS. Ultrahigh-resolution optical coherence tomography using continuum generation in an air–silica microstructure optical fiber. Opt. Lett. 2001;26:608–610. DOI: 10.1364/OL.26.000608

[55] Kodach VM, Kalkman J, Faber DJ, van Leeuwen TG. Quantitative comparison of the OCT imaging depth at 1300 nm and 1600 nm. Biomed. Opt. Express. 2010;1:176–185. DOI: 10.1364/BOE.1.000176

[56] Lim H, Jiang Y, Wang Y, Huang YC, Chen Z, Wise FW. Ultrahigh-resolution optical coherence tomography with a fiber laser source at 1 µm. Opt. Lett. 2005;30:1171–1173. DOI: 10.1364/OL.30.001171

[57] Ishida S, Nishizawa N, Ohta T, Itoh K. Ultrahigh-resolution optical coherence tomography in 1.7 µm region with fiber laser supercontinuum in low-water-absorption samples. Appl. Phys. Express. 2011;4:052501. DOI: 10.1143/APEX.4.052501

[58] Choma MA, Hsu K, Izatt JA. Swept source optical coherence tomography using an all-fiber 1300-nm ring laser source. J. Biomed. Opt. 2005;10:044009. DOI: 10.1117/1.1961474

[59] Lecaplain C, Grelu PH. Rogue waves among noiselike-pulse laser emission: an experimental investigation. Phys. Rev. A. 2014;90:013805. DOI: 10.1103/PhysRevA.90.013805

[60] Hernandez-Garcia JC, Pottiez O, Grajales-Coutiño R, Ibarra-Escamilla B, Kuzin EA, Estudillo-Ayala JM, Gutierrez-Gutierrez J. Generation of long broadband pulses with a figure-eight fiber laser. Laser Phys. 2011;21:1518–1524. DOI: 10.1134/S1054660X11150114

[61] Jeong Y, Vazquez-Zuniga LA, Lee S, Kwon Y. On the formation of noise-like pulses in fiber ring cavity configurations. Opt. Fiber Technol. 2014;20:575–592. DOI: 10.1016/j.yofte.2014.07.004

[62] Suzuki M, Ganeev RA, Yoneya S, Kuroda H. Generation of broadband noise-like pulse from Yb-doped fiber laser ring cavity. Opt. Lett. 2015;40:804–807. DOI: 10.1364/OL.40.000804

Chapter 11

45°-Tilted Fiber Gratings and Their Application in Ultrafast Fiber Lasers

Zhijun Yan, Chengbo Mou, Yishan Wang, Jianfeng Li, Zuxing Zhang, Xianglian Liu, Kaiming Zhou and Lin Zhang

Additional information is available at the end of the chapter

http://dx.doi.org/10.5772/61739

Abstract

This chapter reviews the recentachievements of 45°-tilted fiber gratings (45°-TFGs) in all fiber laser systems, including the theory, fabrication, and characterization of 45° TFGs and 45° TFG-based ultrafast fiber laser systems working in different operating regimes at the wavelength of 1 μm, 1.5 μm, and 2 μm.

Keywords: Tilted fiber grating, nonlinear polarization rotation, mode-locked fiber laser

1. Introduction

Recently, ultrafast fiber lasers giving rise to ultrashort light pulses have attracted much more attention in modern scientific and industrial communities owing to their wide applications and unique advantages, such as compact configuration, high reliability, high output power, better beam quality, and low cost. It is a complex physical mechanism to generate ultrashort pulses, which is a result of the interplay of group velocity dispersion (GVD), self-phase modulation (SPM), gain saturation, cavity loss, and higher-order dispersion in the laser cavity. Especially, in a passively mode-locked ultrafast fiber laser system, the key to achieve ultrashort output is to implement a saturated absorption mechanism (SAM) in the laser cavity. There are two main SAMs: (1) saturated absorption material-based physical intensity SAM (semiconductor- or nano-material-based saturable absorber) and (2) fiber nonlinearity-based artificial SAM (nonlinear polarization rotation (NPR), nonlinear optical loop mirror (NOLM)). Compared with the other techniques, NPR overcomes the limitation on optical damage threshold and modulation capability of those physical absorbers. In the early development stage, most of the fiber laser systems used some bulk components which greatly affected the integration

and stability and also induced extra insertion loss. With the advent of multifarious in-fiber components, fiber laser systems with all-fiber configuration are becoming possible, which have boomed the development of ultrafast fiber lasers. Moreover, fiber laser system constructed by in-fiber devices benefits from advantages such as no collimation, low insertion loss, high stability, and zero maintenance. In-fiber components are hence of critical importance for ultrafast fiber lasers in order to exhibit the above-mentioned plethora of merits. The most used in-fiber devices in a fiber laser include rare-earth doped gain fiber [1], fused silica-based in-fiber beam combiner [2], and fiber grating-based components [3–5]. In NPR technique, it is important to employ a linear polarizer to induce polarization-dependent loss in the laser cavity. As an effective in-fiber polarizer, 45°-tilted fiber grating (45°-TFG) was first reported by Zhou et al. in 2005 [6]. Compared with other commercial in-fiber polarizers, 45°-TFGs own many unique advantages such as high polarization extinction ratio (PER), broadband responsivity, low insertion loss, flexible wavelength adjustability, and simple fabrication method, and it can be adapted to most types of fiber [7].

Because of their high PER and broadband response, fiber lasers constructed using 45°-TFGs exhibit desirable features, including high signal-to-noise ratio (SNR), good stability, and particularly high degree of polarization (DOP) output laser beam. The first trial of using a 45°-TFG as an in-fiber polarizer in fiber laser application was demonstrated by Mou et al. in 2009 [8], in which the grating was applied to achieve a single polarization continuous wave (CW) output with high DOP (>99%). Use of a 45°-TFG as the main functional device to achieve mode-locked fiber laser was reported in 2010 [9]. To date, several types of pulsed fiber laser systems utilizing 45°-TFGs with various operation regimes and wavelength ranges have been reported, which are listed in Table 1.

Year	Achievements	Authors	References
2010	Conventional soliton pulse with 600 fs duration, ~1 nj output pulse energies and 10.34 MHz repetition rate in passively mode-locked erbium-doped fiber laser	Mou et al.	[9]
2012	Dissipative soliton pulse with 4 ps duration, 29.5 MHz repetition in normal-dispersion passively mode-locked ytterbium-doped laser at 1 μm	Liu et al.	[10]
2013	Bound dissipative pulse with 5.7 ps duration and 29.6 MHz repetition rate in the all-normal dispersion mode-locked ytterbium-doped fiber laser	Liu et al.	[11]
2013	Stretched pulse with 90 fs duration, 1.68 nJ pulse energy, and 47.8 MHz repetition rate in passively mode-locked erbium-doped fiber laser	Zhang et al.	[12]
2014	Conventional soliton pulse with 2.2 ps duration, 74.6 pJ pulse energy, and 1.902 MHz repetition rate in mode-locked Thulium-doped fiber laser	Li et al.	[13]
2015	Single polarization, dual-wavelength mode-locked Yb-doped fiber laser	Liu et al.	[14]
2015	Dissipative soliton pulses with 96.7 fs duration and 251.3 MHz repetition rate in passively mode-locked erbium-doped fiber laser	Zhang et al.	[15]

Table 1. List of 45°-TFG-based all-fiber mode-locked laser systems

This chapter will review the theory, fabrication, and characterization of 45°-TFGs and their applications in all fiber mode-locked laser systems. The chapter will include three main parts. The first part will give a general introduction and fundamental background on the development of 45°-TFGs with a particular emphasis on fiber laser applications; second, the theory, fabrication, and characterization of 45°-TFGs will be discussed; finally, 45°-TFG-based mode-locking fiber laser systems will be reviewed including mode-locked fiber laser working at different operation regimes [9, 12, 15] and mode-locked fiber lasers operating at the wavelength ranges of 1 μm, 1.5 μm, and 2 μm [10, 13].

2. Theory, fabrication, and spectral characteristics of 45°-TFGs

Ultraviolet (UV)-inscribed fiber grating devices that have been mostly developed during the last two decades show many advantages, such as simple and mature fabrication process, wide availability in a range of optical fibers and broad operating wavelength range from visible to mid-IR [16]. The periodic structure of a fiber grating offers unique function to control light from the forward propagating core mode into the backward propagating core/cladding modes or to the forward propagating cladding and radiation modes, depending on the grating structure. As a mature technique, UV inscription has been used for producing many types of in-fiber grating devices, including standard fiber Bragg grating (FBG)-based reflectors [17], chirped fiber Bragg grating (CFBG)-based dispersion compensator [18], long period grating (LPG)-based mode convertor [19], and tilted fiber grating (TFG)-based polarizer and polarization-dependent loss equalizer [20].

2.1. Theory

The tilted fiber grating with asymmetric structure can induce high polarization-dependent mode coupling among core, cladding, and radiation modes. The 45°-TFG allows strong coupling of *s*-polarization (TE) from the forward propagating core mode into radiation modes, while the residual *p*-polarization (TM) light propagates along the fiber core with a minimal loss, which forms an ideal in-fiber polarizer. The physical principle behind this phenomenon may be well explained by the Brewster's law. As we know, the light incident at Brewster's angle on an optical interface will partially cease its TE component. In a typical UV-inscribed fiber grating structure, the UV-induced refractive index modulation is very small ($\Delta n \sim 10^{-1}$), far less than the index of fiber core. The Brewster angle for a UV-inscribed grating plane may be calculated as $\theta_{\text{Brewster}}=\arctan(n_{\text{core}}/(n_{\text{core}} + \Delta n)) \cong 45°$. Thus, when the grating structure inscribed into a fiber core at 45° with respect to the normal of fiber axis could be regarded as a series of optical interfaces at Brewster's angle, the grating will totally radiate TE light out from the fiber core, similar to the pile-of-plate polarizer, acting as an in-fiber polarizer (Figure 1).

In ref. [6], the transmission spectra of TFGs with various tilting angles for TE and TM light have been simulated (see Figure 2). The transmission loss of TM is almost zero, when the tilted angle of grating is at 45°.

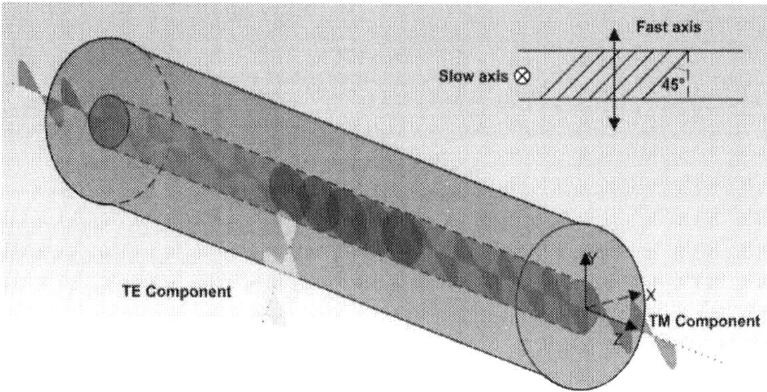

Figure 1. Schematic of 45°-TFG-based in-fiber polarizer.

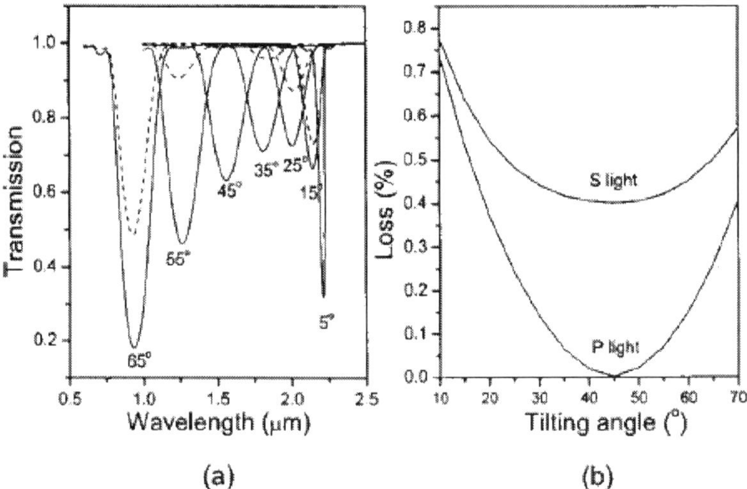

Figure 2. (a) Simulated transmission spectra of TFGs with various tilting angles. TM-light (dashed curves); TE-light (solid curves); (b) simulated transmission losses of TFGs versus tilting angles for s-light (TE) and p-light (TM). The peak wavelength is set as 1.55 μm and the grating period is varied accordingly[6].

2.2. Theoretical analysis of 45°-TFG

2.2.1. Phase matching condition

The phase matching condition (PMC) originates from the conservation of momentum, which provides a clear way to understand the mode coupling mechanism of the fiber grating. Under

PMC, the energy of one mode in the optical fiber can be transferred to another mode by a fiber grating structure. The general vector expression of PMC of a fiber grating can be written as follows:

$$\vec{K}_x = \vec{K}_{core} + \vec{K}_G$$
$$\vec{K}_x = n_x \frac{2\pi}{\lambda}; \quad \vec{K}_{core} = n_{core} \frac{2\pi}{\lambda}; \quad \vec{K}_G = n_{core} \frac{2\pi}{\Lambda} \cos\theta \tag{1}$$

where subscript x represents core/cladding/radiation mode, θ is the tilt angle of fiber grating, n_{core} is the refractive index of the fiber core.

In Equation 1, the refractive index difference between fiber core and cladding could be neglected due to the weakly guiding condition. Hence, the strongest coupling wavelength of a 45°-TFG occurs at

$$\lambda_{strongest} = \frac{n\Lambda_G}{\cos 45°} \tag{2}$$

where n is the modal index of fundamental core mode and Λ_G is the normal period of the grating.

2.2.2. Numerical simulation of 45°-TFG

So far, there are three main theories that can be used to simulate the TFG spectral response: (1) coupled mode theory (CMT) [21, 22], (2) volume current method (VCM) [23], and (3) beam tracing method (BTM)[24]. Each method has its own pros and cons: BTM is the simplest method and better understood by the readers, but it only gives a rough estimation, because the impact of the waveguide structure has been ignored in this method; CMT is more accurate, but the simulation procedure can be quite complex, and analytical solutions are usually required. Above all, VCM is a better method especially in calculation of the distribution of radiation modes in the near and far fields a 45°-TFG, although from which the cladding boundary of waveguide is neglected. These distributed radiation modes are of great interest due to the intrinsic mode-coupling feature from the tilted fiber grating structure. From VCM theory, the loss coefficient per unit length of a 45°-TFG can be expressed as follows [6, 23]:

$$\alpha = -\frac{k_o^3 \delta n^2}{4n(1+(\frac{u^2}{w^2}))} \frac{K_1^2(aw)}{K_0^2(aw)} \oint [1 - \sin^2\xi \cos^2(\chi-\phi)]$$
$$\times \left[\frac{R_s J_0(au) J_1(aR_s) - u J_0(aR_s) J_1(au)}{R_s^2 - u^2} \right]^2 d\phi \tag{3}$$

where $k_0 = 2\pi/\lambda_0$ is the wave vector of light in vacuum, δn and n are the modulated and original refractive indices of the fiber core, a is the core radius and u and w are the waveguide parameters, J and K are the first kind Bessel function and the second kind modified Bessel function, respectively.

In Equation 3, $R_s = (R_t^2 + k_0^2 n_{cl}^2 \sin\xi^2 + 2R_t k_0 n_{cl} \sin\xi \times \cos\varphi)^{1/2}$, where ξ is the angle between the radiation beam and the fiber axis, which satisfies $R_g - i_{eff} k_0 + k_0 n_{cl} \cos\xi = 0$; χ denotes the polarization of the core mode; R_t and R_g are wave vectors of grating along the fiber axis and across the fiber cross section and are defined as $R_t = 2\pi/\Lambda_g \sin45°$ and $R_g = 2\pi/\Lambda_g \cos45°$, where Λ_g is the period of grating.

For a specific type of fiber, the waveguide parameters of fiber are fixed; thus the loss coefficient per unit length will solely depend on the UV-induced index change and azimuthal angle of polarization state. So, the transmission loss of a 45°-TFG may be calculated as follows:

$$T = \mathrm{Exp}\left(-\alpha\left(\phi,\ \delta n, \lambda\right) * l\right) \tag{4}$$

where l is the length of grating, λ is the wavelength of incident light, and φ denotes the polarization state of incident light. According to the coordinator system in the analysis, when $\varphi = 0°$, T represents the transmission loss of TM polarization light; when $\varphi = 90°$, T represents the transmission loss of TE polarization light.

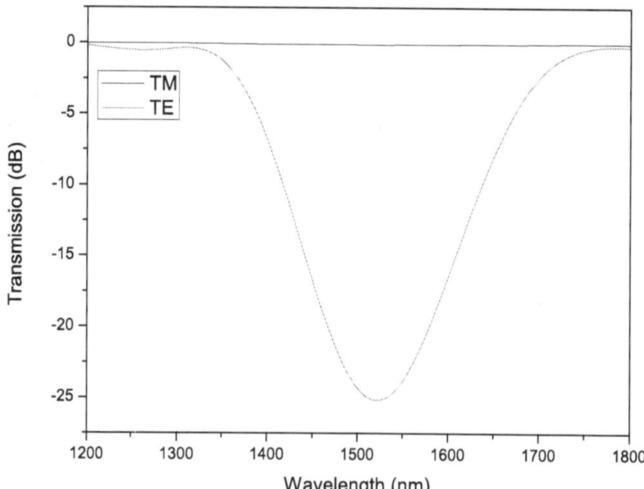

Figure 3. The simulation results of the transmission loss against the wavelength for TM polarization (black line) and TE polarization (red line) of a 45°-TFG with 24-mm length [7].

Figure 3 displays the simulated transmission spectra of a 45°-TFG with 24-mm length, which clearly shows that the TM polarization light has near-zero loss when passing through the 45°-TFG, while the TE polarization light has a very broad loss band with a maximum loss of 27 dB at 1520 nm. In the simulation, the fiber parameters were set the same as the SMF-28 single-mode fiber; the period and the length of grating, and the refractive index modulation were 748 nm, 24 mm, and 0.0013, respectively.

2.2.3. The design of tilt angles

In contrast to the inscription of FBG, the interference fringes of UV beam are tilted at an angle with respect to the fiber axis during the inscription of TFGs. Due to the cylindrical shape of optical fiber, the tilt angle of the interference fringes outside of the fiber is different from that inside of the fiber, as depicted in Figure 6.

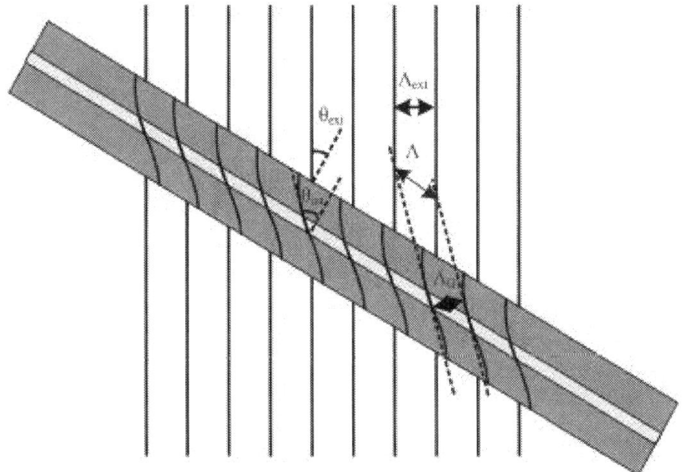

Figure 4. Illustration of fringes of a TFG outside and inside of the fiber with external angle θ_{ext} and internal angle θ_{int}.

The relationship between θ_{ext} and θ_{int} is given by ref. [29] as follows:

$$\theta_{int} = \frac{\pi}{2} - \tan^{-1}\left[\frac{1}{n_{UV}\tan(\theta_{ext})}\right] \tag{5}$$

where n_{UV} is the refractive index of the fiber at wavelength of UV laser (here, it is around 1.52). According to Equation 4, to inscribe 45°-TFGs, the tilted angle of interference pattern outside the fiber is 33.7° with respect to the fiber axis. As shown in the relation between the period of internal interference fringe and external fringe can be given as:

$$\Lambda = \frac{\Lambda_{ext}}{\cos\theta_{ext}} = \frac{\Lambda_G}{\cos\theta_{int}} \tag{6}$$

where Λ_G and Λ_{ext} are the periods of grating and the UV interference fringes, respectively, and Λ_{ext} is half of the period of phase mask (Λ_{PM}). The relationship between the strongest coupling wavelength of a 45°-TFG and the period of phase mask is given as follows:

$$\lambda_{strongest} = \frac{n\Lambda_{PM}}{2\cos 33.7°} \tag{7}$$

2.3. Fabrication of 45-TFGs

There are three main grating fabrication methods: two-beam interferometric technique (holographic method)[25], point-by-point inscription [26], and phase mask scanning technique [27, 28]. Among them, due to the limitation of UV beam size, the point-by-point technique is not suited for short-period grating inscription, the holographic method has the grating length limitation, and only the phase-mask scanning technique is more suitable for the fabrication of longer and stronger 45°-TFGs. A typical 3-D schematic of the grating inscription system is shown in Figure 4, which includes a 244-nm CW frequency doubled Ar+ laser with computer-controlled high-precision air-bearing stage. The phase mask used for 45°-TFG fabrication has 33.7° tilted pattern with respect to the fiber axis, so as to make sure the grating pattern inside of the fiber core will have a 45° tilt angle [7].

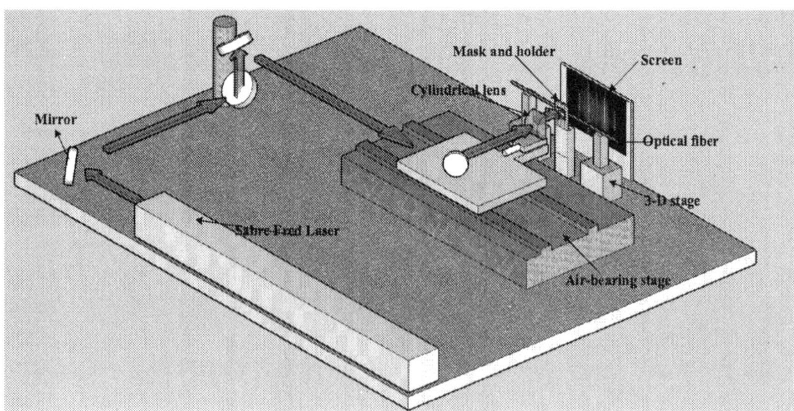

Figure 5. Typical fiber grating inscription system using the phase-mask scanning technique

To enhance the photosensitivity of standard single-mode fiber, the fiber samples are usually hydrogen loaded at 150 bar at 80°C for two days. The phase mask for inscribing 45°-TFG has a uniform pitch and 33.7° tilted angle with respect to the fiber axis (can be purchased from Ibsen Photonics). For standard FBGs with 30-mm length, a very small refractive index modulation (~10^{-4}–10^{-5}) can produce 40-dB reflectance. However, for 30-mm-long 45°-TFGs, to achieve 40 dB PER, the refractive index modulation induced by UV beam needs to be greater than 10^{-3}. The refractive index modulation level depends on the intrinsic photosensitivity of fiber and the UV exposure condition. For normal nonphotosensitive fiber, to induce higher index modulation, the fiber needs to be exposed under high dose of UV radiation for longer time. Figure 5 shows the micro-image of a 45°-TFG inscribed into SMF-28 single-mode fiber with 8.65-µm fiber core diameter. Ten pitch periods of the grating is 7.59 µm, which indicates the central response wavelength of this 45°-TFG is at ~1550 nm.

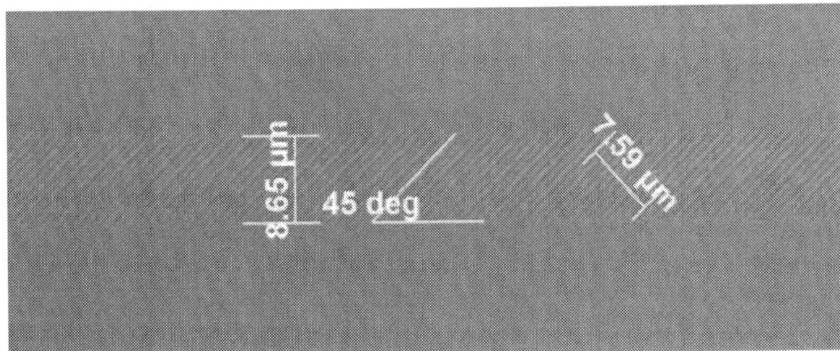

Figure 6. The micro-image of the grating structure of a UV-inscribed 45°-TFG in SMF-28 fiber

2.4. PER characterization

As a polarization-dependent device, the PER is a key parameter to be evaluated for a 45°-TFG, which is the peak-to-peak difference in transmission with respect to all possible states of polarization [30]. Quantitatively, the PER is the ratio of the transmission with respect to all polarization states. For a 45°-TFG, the PER is the ratio of the transmission of the TM and TE polarization states, which can be expressed as follows:

$$\text{PER} = 10 \times \log \frac{T_{\text{TM}}}{T_{\text{TE}}} = 10 \times l\big(\alpha(0°) - \alpha(90°)\big)\log(e) \tag{8}$$

From Equations 3 and 9, the PER of 45°-TFG is linearly proportional to the length of grating and the square of index modulation.

As an ideal in-fiber polarizer, the PER, operation bandwidth, and output polarization state are very important parameters. The typical experimental setup for measuring PER is shown in Figure 7, which usually consists of a light source, a power meter (or an optical spectrum analyzer), and a commercial fiber polarizer and polarization controller (PC). The light source is usually a single-wavelength source; the polarizer is used to generate polarized light with a high degree of polarization (it is not necessary to use a polarizer, if the light source gives out polarized light) and the PC is employed to change the polarization state for the output. The maximum and minimum transmission through the component can directly be measured using this system by adjusting PC. The PER can then be calculated using Equation 9.

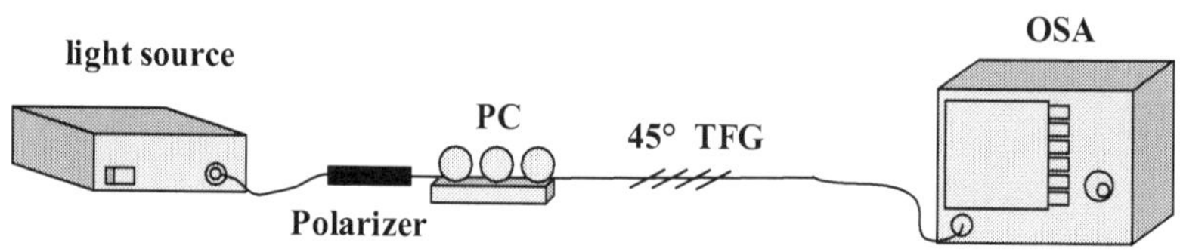

Figure 7. The experiment setup of PER measurement

Figures 8 (a) and (b) show the typical PER results of a 24-mm-long 45°-TFG measured by the difference of maximum and minimum transmission loss using a 1530-nm single-wavelength laser with respect to all polarization states. It can be seen from these figures that the entire PER profile is near-Gaussian-like and covers a broad wavelength range. During the fabrication, one may observe that under the same UV inscription condition, the PER increases monotonically with the grating length. Figure 8(c) plots the PERs of four 45°-TFGs with different lengths (5, 15, 24, and 48 mm), clearly showing that the PER is near-linearly proportional to the grating length, which is in good agreement with previous numerical analysis. The degree of linear polarization of a 45°-TFG can be evaluated by measuring its polarization distribution. Figure 8(d) shows the polarization distributions for three 45°-TFGs with PERs of 10 dB, 20 dB, and 40 dB and a bare fiber for comparison. As shown in the figure, the higher the PER, the lower the output power at the azimuth angles of 90° and 270°, and a strong 45°-TFG has a near-perfect figure of "8" shape with a very narrow waist indicating an ultra-high PER, while the bare fiber gives a circular shape without showing any polarization dependency.

3. 45°-TFG-based all-fiber ultrafast laser systems

Ultrafast fiber laser is defined as the pulsed laser with duration from picoseconds to femto-seconds that are usually derived from a mode-locked fiber laser. Mode-locked fiber laser could be sorted as active and passive mode locking. The former relies on a physical phase modulator and the latter utilizes a saturable absorption mechanism-based modulation, from which the high-intensity pulse center experiences low loss while the low-intensity pulse wing experien-ces high loss. Passive mode-locking techniques are efficient to generate ultrafast pulses, simply due to the fast perturbation of the cavity. The ultra-short pulses result from the interaction of various physical effects, including GVD, SPM, saturable gain, filtering effect, and cavity loss. By using an artificial saturable absorber, the NPR-based mode-locked fiber laser was first demonstrated by K.Tamura in 1992 [32].

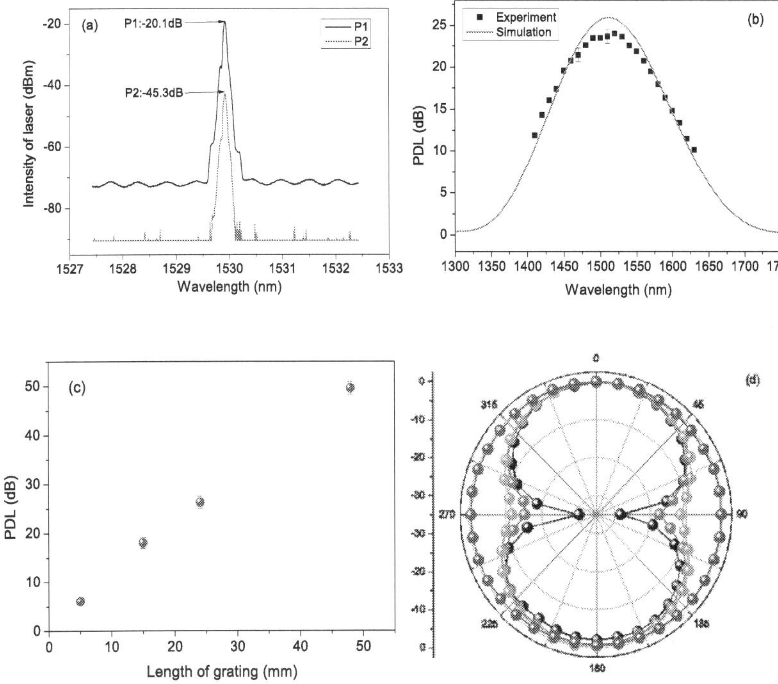

Figure 8. (a) The transmission spectra of a 45°-TFG based in-fiber polarizer at 1550nm at two orthogonal polarization states (P1 and P2); (b) the PER response of a 45°-TFG over 90nm from 1520nm to 1610nm; (c) the PER response of 45°-TFG with different grating length at 153onm; (d) the polarization distribution profiles for three 45°-TFGs with ● (green) 10dB PER, ● (red) 20 dB PER and ● (black) 40 dB PER and a bare fiber with ● (blue) 0dB PER [31].

3.1. NPR effect

In a more physical description, NPR can be regarded as the "interference" of right- and left-hand circular polarization light experiencing different nonlinear phase shifts induced by Kerr effect. The working principle of NPR is similar to that of nonlinear optical loop mirror(NOLM) both of them are based on interference between two non-equal intensity light beams that induce different nonlinear phase shifts. The phase shift in NOLM is induced from a non-balanced coupler, while in NPR it is generated from an elliptical polarization light, which resolves into two orthogonal circularly polarized components with different intensities. The NPR-based configuration usually consists of a linear polarizer and two PCs, in which the transmission intensity through the linear polarizer will be power dependent.

In the NPR-based laser system, a linear polarizer is a critical element. So far, there are two types of commercial in-fiber linear polarizer: (1) the evanescent field coupling-based in-fiber linear polarizer which is made by coating the exposed guiding region with a birefringent material (TM pass polarizer) [33], or a metal film (TE pass polarizer) [34], or an anisotropic absorption material (TM pass polarizer) [35]; (2) polarizing fiber-based linear polarizer which is actually a high birefringence polarization-maintaining fiber [36]. Both types of in-fiber polarizer have been applied to construct mode-locked fiber lasers [37]. Recently, a picosecond mode-locked fiber laser has been reported using a micro-fiber-based polarizer [38]. However, all these types of in-fiber polarizers have their intrinsic disadvantages, such as low power tolerance, complex fabrication process for evanescent field absorption-based in-fiber polarizer, and long fiber requirement and narrow operation bandwidth and limitation of operation fiber type. As aforementioned, a 45°-TFG is an ideal in-fiber polarizer, and it can overcome the most disadvantages associated with the two mentioned conventional types of polarizer. In this section, a detailed review of recent achievements in 45°-TFG-based all-fiber ultrafast fiber lasers will be given.

The most reported mode-locked fiber lasers using 45°-TFGs are based on a common ring laser cavity structure as shown in Figure 9, in which the pump laser is combined into laser cavity by a fiber combiner; the gain fiber could be chosen to ytterbium-/erbium-/thulium-doped fiber to achieve different operation wavelengths; a 90/10 optical coupler (OC) is employed to couple out the laser light; a 45°-TFG is sandwiched between two PCs (three different period 45°-TFGs should be used for ytterbium-/erbium-/thulium-based fiber systems, respectively); polarization-independent isolator is used to achieve unidirectional operation.

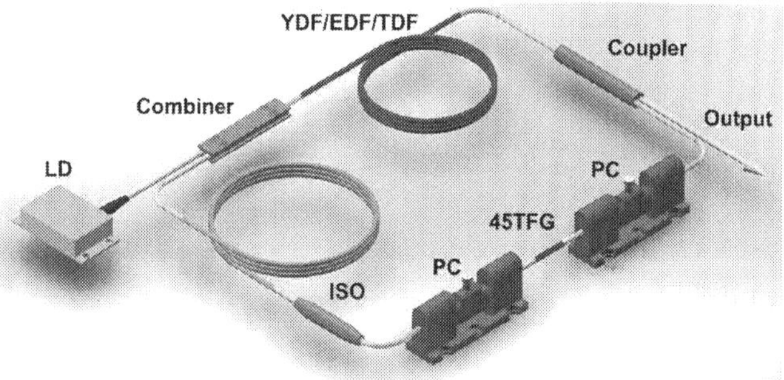

Figure 9. The configuration of a ring fiber laser cavity.

3.2. 45°-TFG-based mode-locked fiber laser at 1.5 μm

The mode-locked fiber laser in the 1.5-μm region is very important due to the potential applications in modern optical fiber communication and microwave photonics, which is also a convenient experimental platform for nonlinear science research because of wide availability of various components and types of fibers at 1.5 μm. Based on different cavity dispersion mappings, the mode-locked fiber laser could operate in the regime of conventional soliton, the stretched pulse, the similariton, and the dissipative soliton, and operating in this regime depends on the net cavity dispersion (anomalous or normal dispersion). So far, 45°-TFG-based mode-locked fiber lasers at 1.5 μm have been demonstrated as the conventional solitons, the stretched pulses, and the dissipative solitons, which will be described in detail in the following sessions.

3.2.1. Conventional soliton pulse mode-locked fiber laser

One of the features of conventional soliton pulses is that they preserve their shape during propagation over very long distance and the pulses are still reproducible even after various perturbations, which have been used in a range of applications. The generation of conventional soliton pulses results from the balance of nonlinearity and dispersion. The first NPR-based soliton mode-locked fiber ring laser was reported by Tamura in 1992 [32]. The first 45°-TFG-based soliton fiber mode-locking laser was demonstrated in 2010 [9], in which the ring laser cavity consists of an ~6-m erbium-doped fiber (EDF) with nominal absorption coefficient of ~12 dB/m at 1530 nm and nominal dispersion –8.6 ps/nm/km, a 12-m SMF-28 fiber with an anomalous dispersion of ~+18 ps/nm/km, and a 0.5-m B/Ge fiber with dispersion of ~+10 ps/nm/km. The net cavity GVD of the cavity is around –0.0143 ps^2, which guarantees the laser working at soliton regime. The mode-locked short pulses can be achieved by properly adjusting the two PCs in the cavity (shown in Figure 9).

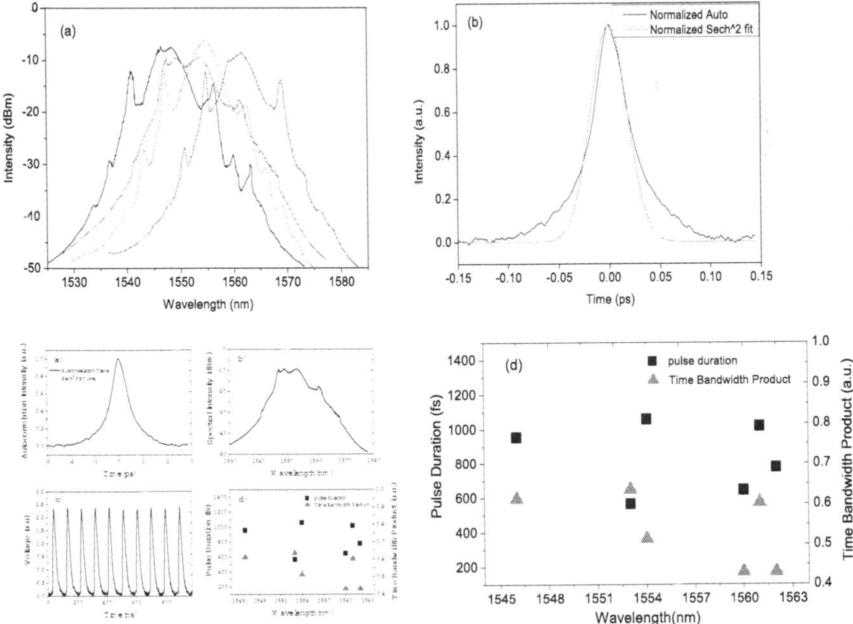

Figure 10. Measured characteristics of 45°-TFG-based all-fiber soliton mode-locking laser: (a) optical spectra of mode-locked fiber laser under different polarization status; (b) recorded autocorrelation trace; (c) output pulse train; and (d) pulse width and time-bandwidth products as a function of the wavelength [9]

Figure 10(a) shows the spectra of stable soliton pulse at different central wavelengths, from which we can observe the Kelly sidebands on the spectrum that are caused by periodic perturbation during the cavity trip. The stable soliton pulses can only be sustainable at relatively low pump power. With the increasing pump power, the pulses become unstable, and the Kelly sideband becomes much stronger. Figure 10(b) shows that the pulse duration of soliton mode-locking laser at the central wavelength of 1553 nm is ~600 fs. The pulse train was recorded by an oscilloscope as shown in Figure 10(c), which indicates the pulse separation is ~90 ns and the repetition rate is calculated around 11.1 MHz, which agrees very well with the value calculated from the total cavity length (18.5 m). Figure 10(d) shows the duration and time–bandwidth product (TBP) of soliton pulses at different central wavelengths. In the experiment, 12 mW average output power was obtained giving an ~1-nJ peak pulse energy [9].

3.2.2. Stretched pulse mode-locked fiber laser

Most important applications of ultrafast lasers are the supercontinuum generation and micromachining. The high-energy ultrashort pulses are preferred in these applications. However, the periodic perturbation for solitons may limit the duration of pulse and energy scalability. Using low anomalous dispersion fiber could overcome the periodic perturbation-induced pulse instability, but the low anomalous dispersion fiber limits the pulse energy [39]. To overcome these restrictions, one of the most efficient methods is to construct laser cavity with both normal and anomalous GVD segments, in which the pulses will be compressed at the normal GVD segment and broadened at the anomalous GVD segment. Such pulses experience something like the soliton "breathing" during the cavity round trip, so they are also called as stretched pulses [40]. In 2013, Zhang et al. reported a sub-100-fs stretched pulse by using a 45°-TFG-based ring laser system [12]. The configuration of this laser is similar to the one in Figure 9, which has a net cavity GVD of ~0.013 ps^2, contributed by an ~1.16-m EDF with nominal absorption coefficient of ~80 dB/m at 1560 nm and normal dispersion ~–53.4 ps/nm/km, 2.65-m SMF-28 fiber with anomalous dispersion ~18 ps/nm/km and 0.43 m HI1060 fiber with anomalous dispersion ~5.5 ps/nm/km.

Similarly, the mode locking in Zhang's laser was achieved by adjusting the PCs in the cavity. The pulse duration was measured by a commercial optical autocorrelator with <1 fs resolution, and the result shows an ~90-fs Gaussian fit trace (see in Figure 11(a)). From Figure 11(b), we see ~54 nm pulse spectra centered at 1575 nm that gives a TBP of ~0.58, indicating that the compressed Gaussian pulse is a little beyond the transform limit. From Figure 11(b), it is also noticed that the output spectra from the different ports have almost the same profile and width. Figures 11(c) and (d) show the pulse train is with an ~21-ns interval between two adjacent pulses and mode-locked pulse laser has a fundamental repetition rate of ~47.8 MHz, which are in good agreement with the 4.24-m cavity length. The 55-dB SNR shows that the mode-locked laser is working at a stable state [12]. It is also believed that with further optimization of the cavity parameters, shorter pulse duration could be obtained.

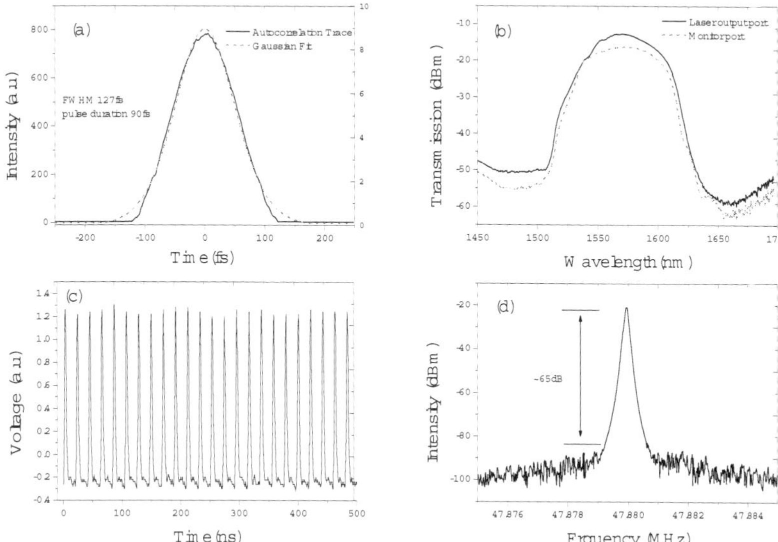

Figure 11. Characterization of 45°-TFG-based stretched pulse laser: (a) autocorrelation trace (black line) and its Gaussian fitting (red dotted line); (b) optical spectra of both output (black line) and monitor (red dotted line) ports; (c) pulse trains; (d) RF spectrum[12]

3.2.3. Dissipative soliton pulse mode-locked fiber laser

In theory, the generation of the dissipative soliton pulse is not only the balance of nonlinearity and dispersion but also the balance between gain and loss. The concept of dissipative soliton is thought as a fundamental extension of conventional soliton theory. However, there is a significant difference between conventional solitons and dissipative solitons, as the latter owns a nonuniform phase, whereas the phase of the former is constant. The nonuniform phase of dissipative pulses is caused by the fact that solitons continually exchange energy with the environment [41]. In contrast to the conventional solitons, dissipative solitons are largely positively chirped with high pulse energy. Adapting external cavity compression, <100 fs duration is easily achieved.

In 2015, Zhang et al. demonstrated a 250-MHz high fundamental repetition rate dissipative soliton laser using a 5-cm-long 45°-TFG as an intra-cavity mode locker [15]. The total cavity length of this laser is around 0.8 m, which is constituted by a 0.3-m-long EDF with normal dispersion ~−53.4 ps/nm/km at 1560 nm and peak absorption of around 150 dB/m at 1530 nm used as the gain medium, and 0.5-m SMF-28 fiber included 5-cm-long 45°-TFG with anomalous dispersion ~18 ps/nm/km. The net GVD of the laser cavity is ~0.009 ps². Two 976 nm laser diodes combined via a polarization beam combiner are injected into the pump port to generate enough nonlinearity. When pump power was greater than 700 mW, the stable, self-start dissipative soliton mode locking was achieved by adjusting the PCs to an appropriate

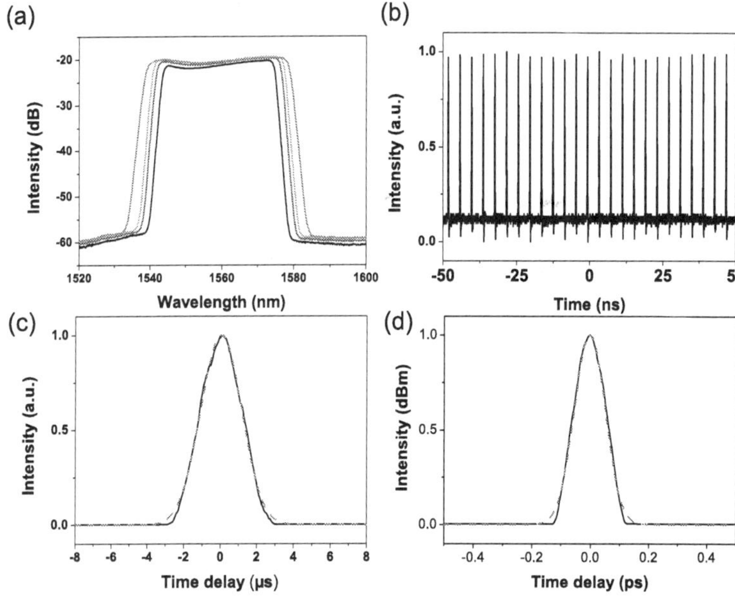

Figure 12. (a) Output spectra at different pump powers; (b) pulse train of the mode-locked pulses, (c) and (d) autocorrelation traces before and after out-cavity compression (red dashed lines are Gaussian fitting)[15].

polarization state. When the pump power was <700 mW, the laser was operating at a Q-switching regime. As shown in Figure 12(a), the full width at half maximum (FWHM) of output pulse spectra is around 37.5 nm and with increasing pump power, the FWHM of the spectra becomes broader. At 900 mW, maximal pump power, the 3-dB spectral bandwidth reaches to 41.2 nm. During the whole pump power range, no multiple pulse phenomena occurred. Once the mode-locked pulse has been generated, the laser was stable and the average output power was 43.4 mW. The output pulse train was measured by an oscilloscope, which has shown 4 ns pulse interval and 250 MHz repetition rate which agree very well with the values calculated from the cavity length (see in Figure 12 (b)). The pulse duration was measured by autocorrelation and assumed from Gaussian fitting, showing the durations were 1.8 ps and 96.7 fs before and after out-cavity compression with SMF, respectively. The obtained results were limited by the available pump power; it is expected that with higher pump power, high repetition rate fiber laser system up to GHz with increased output and shortened pulse width may be achievable.

3.3. 45°-TFG-based YDF mode-locked fiber laser at 1 μm

Ytterbium-doped fiber (YDF) has a number of interesting properties, such as simple electronic level structure, less quantum defect, upper-state lifetimes, and relative broad gain bandwidth

[42, 43]. Such properties potentially allow YDF-based laser systems to have very high slope efficiency, low thermal effect, high output power, and ultrashort pulses. YDF-based laser systems have been widely applied in micro-machining area due to their high photon energy, extra high power output, easy beam delivery, and a robust setup. In contrast to the EDF-based laser system, both active and passive fibers at the 1-μm region have normal dispersion. Recent research has reported that the hollow core fiber has an anomalous dispersion at ~1 μm and been employed in a laser cavity to achieve dispersion management [44]. By using the NPR technique, ultrashort, self-starting mode-locking laser operating at ~1-μm region could be achieved by using YDF and 45°-TFG. According to the phase matching condition, 45°-TFG could also work at 1-μm region by properly designing the period of tilted grating. Figure 13(a) shows the transmission spectra of a 50-mm-long 45°-TFG with 563 nm pitch period at two orthogonal polarization states at 1055.7 nm. The PER of 45°-TFG is quite high, at around 34.9 dB at 1055.7 nm. Figure 13(b) shows the simulated full PER profile of a 45°-TFG at the central wavelength of 1 μm, in which the FWHM is around 180 nm that is far broader than the gain bandwidth of YDF.

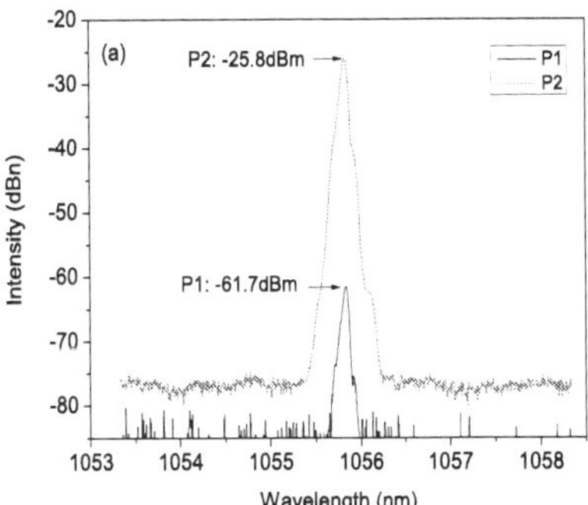

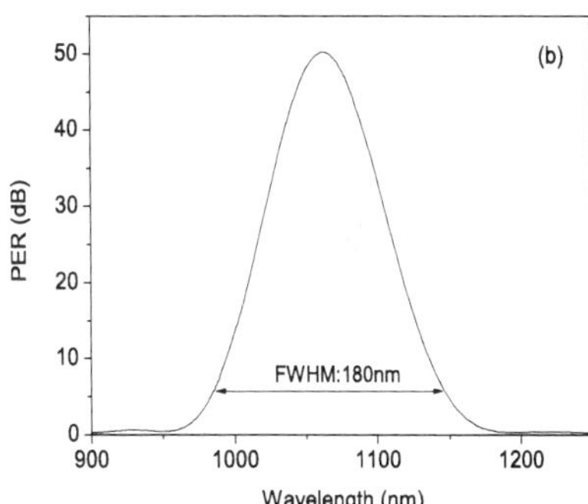

Figure 13. (a) The measured transmission spectra of a 45°-TFG-based in-fiber polarizer at 1055.7 nm at two orthogonal polarization states (P1 and P2); (b) the simulation result of a 50-mm-long 45°-TFG with 568 nm period and 0.0013 index change.

3.3.1. Dissipative soliton mode-locked fiber laser

In 2012, Liu et al. for the first time demonstrated a 45°-TFG-based all-fiber normal-dispersion mode-locked laser at the 1-μm region [10]. In this system, a 48-mm-long 45°-TFG with 33 dB PER at 1040 nm was used to build up NPR mechanism and generate dissipative soliton pulses. The all-fiber laser structure was based on a ring cavity oscillator (similar to the one in Figure 9), in which the total laser cavity length was 7 m including 0.7-m-long YDF with an absorption coefficient of 500 dB/m at 976 nm and 20 ps^2/km nominal dispersion and 6.3 m HI1060 fiber with 22.1 ps^2/km at 1050 nm. The net GVD was 0.153 ps^2. The 45°-TFG-based NPR technique was employed as a mode locker to generate the dissipative soliton pulses.

The stable mode-locking state was achieved when the pump power reached the threshold of 196 mW. Figure 14(a) shows the optical spectrum with 9 nm FWHM at the central wavelength of 1050 nm. The steep spectrum profile edges are caused by the effect of gain spectral filtering and the physical filter bandwidth. The uniform pulse train measured by an oscilloscope with pulse–pulse separation of 33.7 ns (see in Figure 14 (b) and the radio frequency (RF) show that the mode-locked laser was working under fundamental mode with 29.646 MHz repetition rate, which is in good agreement with the value calculated from the total cavity length. The RF spectra in the range of 300 MHz bandwidth have shown there is no Q-switching or harmonic solitons. Figure 14(d) represents the pulse profile and phase in temporal domain, in which the pulse duration evaluated from the Gaussian fit is around 4 ps. The TBP of around 10 and parabolic phase profile indicated that the pulses have positive linear chirp. The DOP of output pulse was also measured around 26 dB, which indicated the output pulse was nearly single polarization state [10].

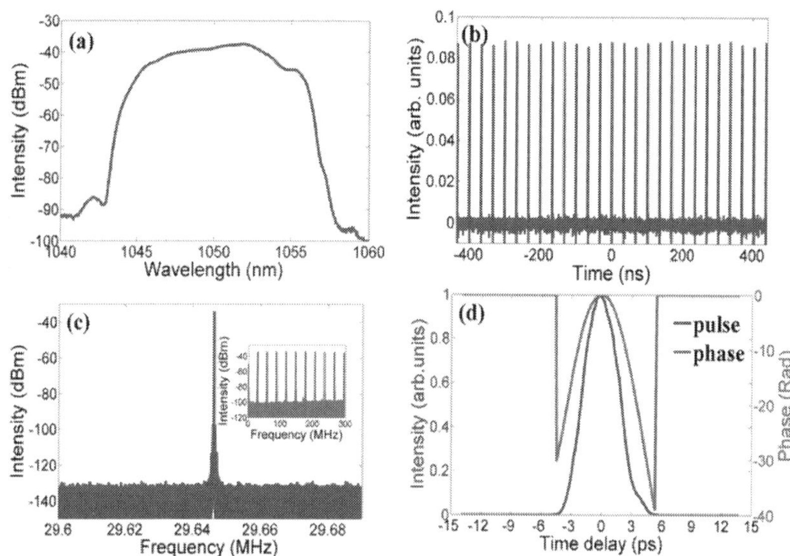

Figure 14. Characteristics of a 45°-TFG-based dissipative soliton YDF mode-locked fiber laser: (a) the optical spectra, (b) pulse train measured by an oscilloscope, (c) radio frequency spectra and (d) pulse duration and phase profile of output pulse at 196 mW pump power[10].

3.3.2. Bound-state mode-locked fiber laser

Soliton mode-locked lasers with multiple pulse output have attracted more interests in optical communications, plasma accelerators, and cosmetic surgery. There are two main operation regimes to generate multiple pulses: soliton interaction [45] and pulse splitting [46]. The former has been investigated for the limitation of soliton application in long-distance fiber optic communications. As for specific multiple-pulse solitons, the bound-state solitons have fixed and discrete soliton separations, which have been extensively investigated in anomalous dispersion EDF-based fiber laser systems. Recent research work has shown that the bound-state solitons also exist in the normal-dispersion lasers. Ortaç et al. have reported a two-soliton molecule in an all-polarization-maintaining YDF fiber laser operating in the normal dispersion regime [47]. The bounded dissipative pulses in 45°-TFG-based all-fiber YDF laser were also observed [11], in which the design of laser cavity is the same as the one described in the previous section (see in section 3.3.1). Using the coupled Ginzburg–Landau equation (GLE) model, the theoretical analysis has indicated that the dissipative pulses in the normal disper‐sion domain also follow the energy quantization effect. The mechanism of formation of multiple pulse is based on soliton interaction through the dispersive waves in the all-normal dispersion region, in which the dispersive waves with discrete spectra can be generated when the propagating soliton in the laser cavity experiences periodical loss and amplification. In this YDF laser, the physical filtering of isolator also plays a critical role in the formation of bound-state pulses. Due to the soliton energy quantization effect, the bound-state solitons have the identical parameters, that is, the peak-to-peak separation is fixed and will not change with the pump power.

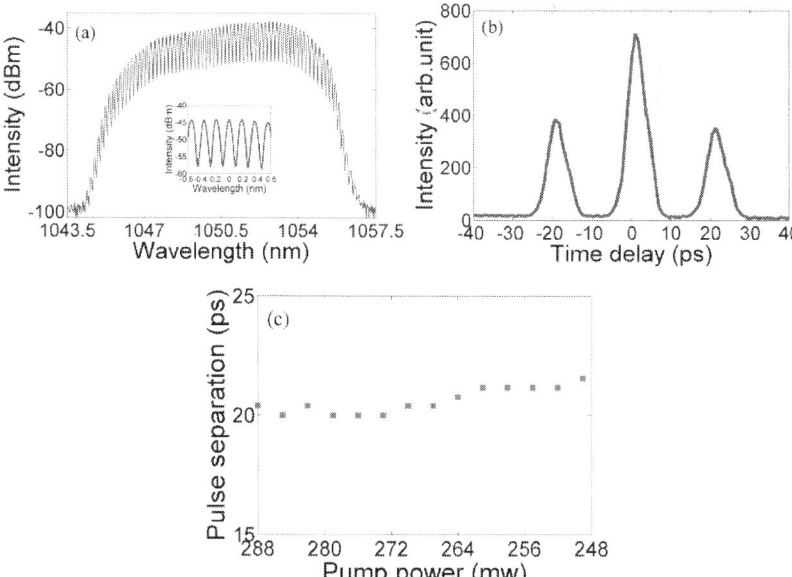

Figure 15. (a) Optical spectrum and (b) autocorrelation trace of bound-state dissipative mode-locked laser at pump of 249 mW; (c) the pulse separation influence as the pump power changes from 288 mW to 249 mW[11].

Since the dispersive waves in a laser system are pump strength dependent, by increasing the pump power, stable bound-state pulses could be formed. In the experiment, when the pump power reached to 214 mW, the stable single pulse was first generated by adjusting the PCs. After the pump power increased to 260 mW, the bound-state pulses (multiple pulses) were observed, which resulted from the peak power clapping effect under the stronger pump power. Once the formation of multiple pulses was established, by changing the pump power between 249 mW and 288 mW, the bound state of the pulses was still sustained. Reducing the pump power to 249 mW, the laser output reverted to the stable single pulse, and increasing the pump power above 288 mW, the laser operation evolved to CW regime. A pump power of 249 mW was verified as the threshold for bound-state pulse operation in this laser configuration. Figure 15(a) shows the output spectrum of bound-state pulse at the pump power of 249 mW. The pulse duration trace is shown in Figure 15(b), in which the Gaussian shape fitted pulse has a 4.2-ps duration and 21.6-ps separation (5 times larger than the pulse duration). As shown in Figure 15(b), three pulse peaks with a height ratio of 1:2:1 indicate that there are two identical bound-state pulses. By changing the pump power between 249 mW and 288 mW, the pulse separation remains almost unchanged as shown in Figure 15(c)).

3.3.3. Dual-wavelength mode-locked fiber laser at 1 μm

As shown in Figure 13(b), the 45°-TFG with central response around 1μ m also has a very broad PER profile. By using a 45°-TFG in the ring cavity, a dual-wavelength YDF mode-locked fiber has also been demonstrated [14]. The configuration of this dual-wavelength fiber laser consists of a segment of 45°-TFG in HI1060 fiber, a 980/1053-nm wavelength-division-multiplexed (WDM) coupler, a 0.73-m-long YDF with an absorption coefficient of 500 dB/m at 976 nm and 20 ps^2/km, an OC)with 30% output, two PCs, and 5.57-m HI1060 fiber with 22.1 ps^2/km at 1050 nm. The net GVD of the cavity is ~0.14 ps^2, which indicates the laser working in the dissipative regimes.

The stable self-started dual-wavelength mode-locked dissipative soliton was easily obtained by adjusting two PCs at a pump power higher than the mode-locking threshold. Figure 16(a) clearly shows the spectrum of the dual-wavelength mode-locked laser at the pump power of 339 mW, in which the two-peak central wavelengths are at 1033 nm and 1053 nm with approximately 10 nm FWHM. Due to the different round trip times, the dual-wavelength pulses have different repetition rates. The pulse train spectra of dual-wavelength pulse measured by a 6-GHz bandwidth oscilloscope had a very slightly different separation, making it difficult to measure the pulse train gap difference. As shown in Figure 16(b), there are two pulse trains, when one pulse train is triggered, the other is still moving. The uniform intensity pulse train shown in Figure 16(b) denotes that the dual-wavelength pulses have equal pulse energy. The RF spectrum in Figure 16(c) shows clearly that the dual-wavelength mode-locked laser has a 6-kHz difference of fundamental repetition rate, as one pulse at 1033 nm has 32.804 MHz repetition rate and the other at 1050 nm has 32.798 MHz repetition rate (see in Figure 16(c). The 65-dB SNR measured from RF spectrum indicates the laser was working at stable mode-locked state. The average power of pulse was 31 mW under 339 mW pump power. To further confirm pulse performance, the output port was connected with a filter to filter out the

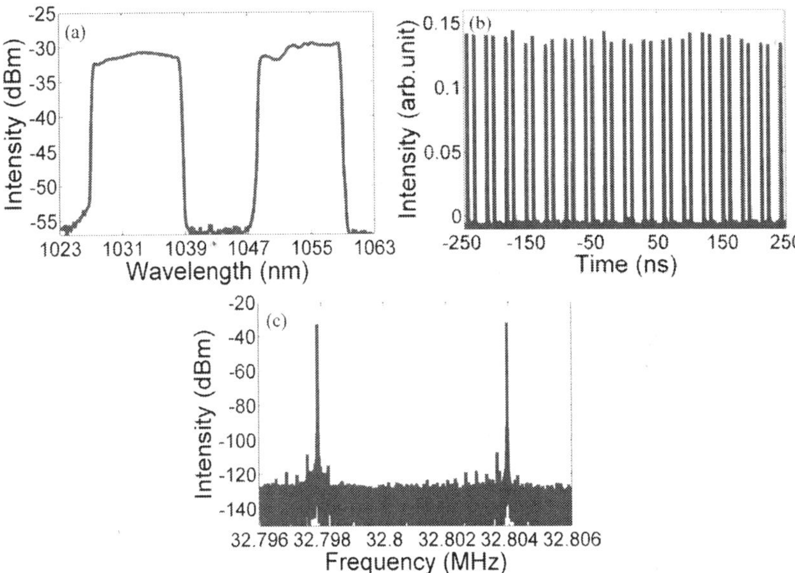

Figure 16. (a) Output spectrum, (b) oscilloscope trace, and (c) RF spectrum of dual-wavelength mode-locked laser at 1μ m [14].

pulse at 1033 nm. The optical spectrum, pulse train in oscilloscope, and RF spectra corresponding to the pulse at 1033 nm all disappeared, only the ones for the 1053-nm pulse remained, as can be seen in Figures 17(a), (b), (c). The average pulse power at 1053 nm measured was 15 mW, which is half of the dual-wavelength pulse power, further proving that the two pulses have the same pulse energy. The pulse duration shown in Figure 17(d) is ~4.3 ps fitted with Lorentz shape, and the corresponding TBP of pulse is calculated as 12.5, indicating that the pulse is highly chirped.

Through further investigation, it was observed that the dual-wavelength pulse laser was only sustainable at the pump power between 339 mW and 389 mW. Once the pump power increased to above 389 mW, the operation at short wavelength became CW. With further increase in pump power to 441 mW, the CW emission totally ceased and the fiber laser operated in stable single-wavelength mode locking [14].

3.4. 45°-TFG-based mode-locked fiber laser at 2 μm

Recently, as several types of thulium-doped fiber (TDF) are available, the all-fiber mode-locked fiber lasers have attracted increasing interest in the 2-μm region, as mid-IR fiber lasers are desirable not only for military applications in guidance and light detection and ranging (LIDAR) systems, but also for civil applications in biology, remote gas sensing, and free-space communication. Ultrafast fiber laser at 2 μm also serves as the seed source to generate the

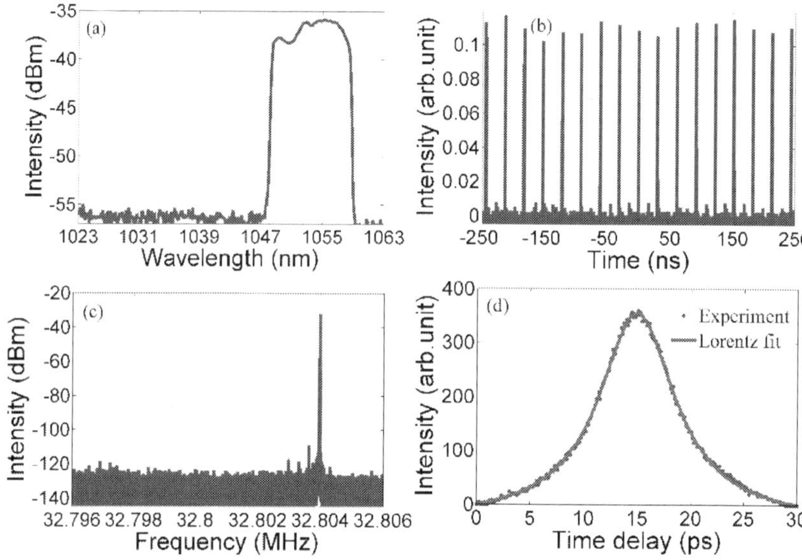

Figure 17. (a) the output spectra, (b) pulse train, (c) radio frequency and (d) autocorelation of output dual-wavelength pulse after a 1033-nm filter [14].

supercontinuum in the mid-IR region. There are a good number of reports on 2-μm mode-locked fiber lasers utilizing nanomaterial-based SA, NOLM, and NPR techniques [13, 48, 49]. Sun et al. have recently reported the fabrication of FBGs and 45°-TFGs in the 2-μm region. Figure 18 shows that a 25-mm-long 45°-TFG with 980-nm period has a strong PER response with FWHM over 400 nm and a peak value of PER around 25 dB at the central wavelength of 2 μm. By using the 45°-TFG-based NPR technique, Li et al. have demonstrated a TDF fiber mode-locked laser working at conventional soliton and noise-like soliton regimes [13].

The ring cavity of TDF laser emitting light around 2μ m reported by Li et al. has a total of 105-m-long cavity length, constructed by 7-m double-clad Tm^{3+}-doped fiber (Coractive, DCF-TM-10/128) with −84 ps²/km at 2 μm, 95.0-m SM2000 fiber with −84 ps²/km, and 3.0-m SMF-28e pigtail fiber from the pump combiner, isolator, and coupler estimated to be −80 ps²/km. The use of 95-m SM2000 is to increase the nonlinearity of the cavity as the 2-μm fiber has a low nonlinear coefficient. The fiber combinations give a total net GVD of ~−7.76 ps², indicating this laser is operating at a large anomalous dispersion. The TDF 2-μm laser was pumped by two 793-nm diode lasers (Lumics, Germany) combined by a (2 + 1) × 1 pump combiner (ITF, Canada).

3.4.1. Conventional soliton mode locking

The conventional soliton results from the balance between nonlinearity and dispersion of laser cavity. The optical fiber at 2 μm has relatively low nonlinear coefficient. To enhance the

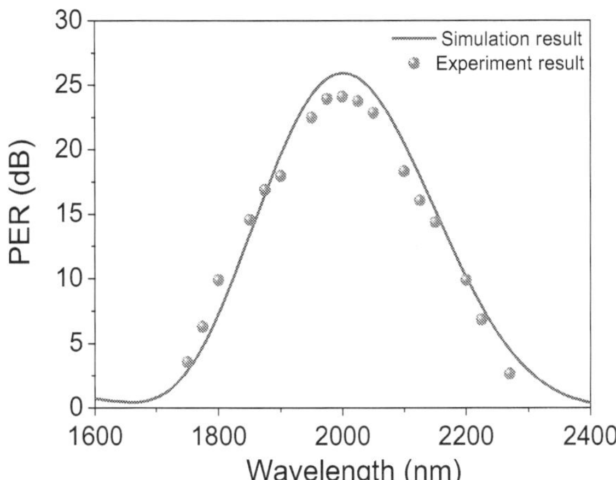

Figure 18. Measured (dot) and simulated (line) PER profile of 45°-TFG with a broad response at the 2-μm region [13].

nonlinear effect of laser cavity, the laser system, as mentioned above, has employed extra 95-m SM2000 in the ring cavity. However, due to the anomalous dispersion of active and passive fibers at 2 μm, the net GVD of the laser cavity is quite large.

During the experiment, the stable single pulse was not self-started, and the laser was first operating at multi-pulse mode-locking regime at the pump power of 1.320 W. Upon reducing the pump power to 1.232 W, a stable single-pulse mode-locked output was achieved. In the experiment, it was found that the single-pulse mode locking was working at a very narrow pump power range between 1.230 W and 1.232 W. Below 1.230 W, the laser operation evolved to CW regime, while above 1.232 W, the laser came back to multi-pulse operation. Figure 19(a) shows the output spectrum under single-pulse operation. As shown in the figure, there are very strong Kelly sidebands on the soliton spectrum, and the intensity of +/–1 sideband is even higher than the intensity of center value of the soliton. The Kelly sidebands are generated from the interference between the soliton and dispersive waves and usually exist in the fiber laser system associated with high dispersion and nonlinearity. The appearance of strong Kelly sidebands also indicates that the pulse duration may be close to the minimum possible value. The pulse train recorded by oscilloscope is shown in Figure 19(b). The fundamental repetition rate of mode-locked pulse was measured at 1.902 MHz by an RF spectrum analyzer, which matches well with the theoretically cavity length dependent value, and also indicates only one pulse was generated per round trip (see in Figure 19(c). Figure 19(d) displays the autocorrelation trace of the mode-locked pulses with a scanning range of 10 ps, in which the FWHM is about 3.4 ps corresponding to the pulse duration of 2.2 ps when using sech^2-pulse fitted. According to 2.2-nm FWHM of the optical spectrum of soliton mode-locked laser at central wavelength of 1992.7 nm, the TBP is calculated to be around 0.335, which is almost the transform-limited value (0.315 for sech^2 shape pulse).

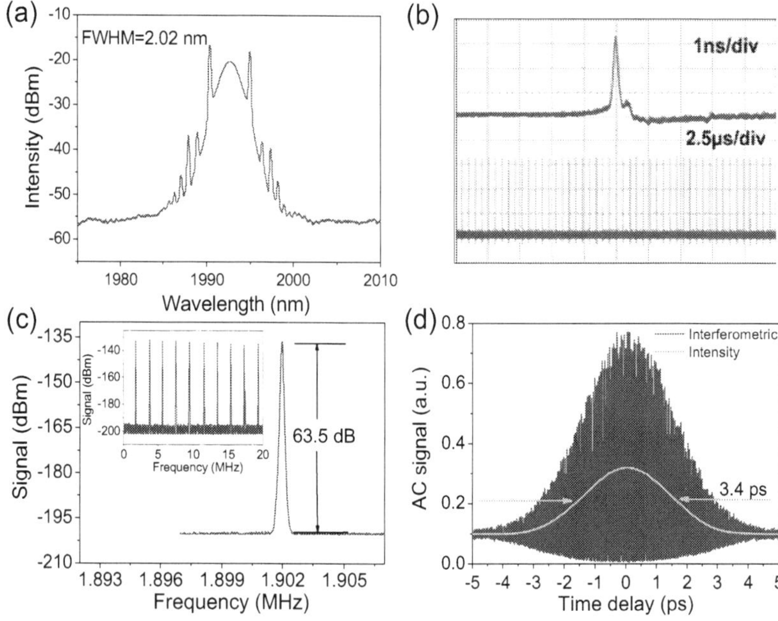

Figure 19. Thulium-doped fiber-based mode-locked fiber laser at conventional soliton regime: (a) output optical spectrum; (b) pulse train on oscilloscope; (c) RF spectrums with scanning range of 10 kHz to 20 MHz (inset); (d) intensity of autocorrelation with sech2-pulse fitting [13].

3.4.2. Noise-like mode locking

In contrast to the stable soliton mode-locked laser, the noise-like mode-locked laser emits a group of pulses with randomly varying duration and peak power, which was first reported by Horowitz et al. in 1997 [50]. The feature of low coherence length and broad spectral bandwidth of noise-like pulse lasers may make them useful for optical spectrum slicing and fiber optical sensing. Due to the weak distortion after propagating in the dispersion medium, the noise-like pulse may also be utilized for supercontinuum generation [51].

Li et al. observed noise-like pulse emission from the stable soliton mode-locked laser when the pump power increased to 1.82 W. The autocorrelation trace they observed showed a narrow peak riding on a broad pedestal extended over 150-ps measurement window, as shown in Figure 20(a). The relative broad and smooth output spectrum of noise-like pulse shown in Figure 20(b) reveals the central wavelength is at 1994.2 nm and FWHM is quite broad, around 18.1 nm. Both output and autocorrelation trace have indicated clearly the laser operates at the noise-like mode-locked regime. The RF spectrum has shown a repetition rate of 1.902 MHz, which proves the noise-like pulse operated at the fundamental mode-locking regime. Comparing with the stable soliton mode-locked laser, the noise-like laser showed a lower SNR around 47.3 dB. The noise-like mode-locking state could be still sustained at the maximum

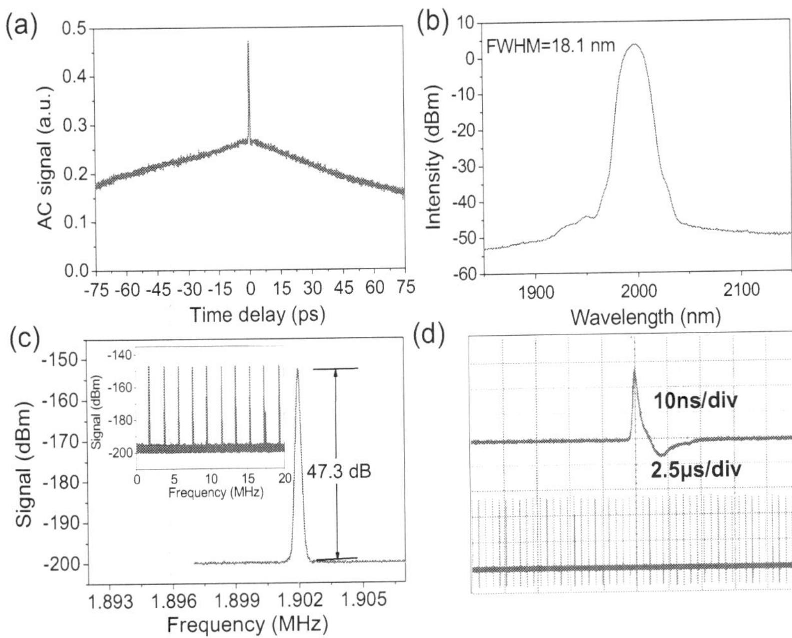

Figure 20. Thulium-doped fiber-based mode-locked fiber laser operating at the noise-like regime: (a) autocorrelation trace; (b) optical spectrum; (c) RF spectra; (d) output pulse train recorded by oscilloscope [13].

launched pump power of 7.56 W with a 7.5% slope efficiency and the maximum output power of 475.8 mW, corresponding to 250.1 nJ pulse energy. Such high pulse energy could be used to generate mid-IR supercontinuum.

Overall, the above-described mode-locked lasers at 2 μm have shown good stability under laboratory conditions within 24 hours. Further investigation of cavity length has revealed that adding or reducing the length of SM2000 fiber has not improved the condition to revert to soliton state mode locking. In fact, upon reducing the SM2000 fiber length from 95 m to 85 m, even higher pump power was required to generate noise-like pulses, while increasing the SM2000 fiber length to 120 m, the laser system was operating at the CW regime and no longer supported the mode-locking regime[13].

4. Summary and future work

In summary, we have presented the detailed theory, fabrication, and characterization of 45°-TFGs, which reveal clearly their high polarization function. 45°-TFGs as ideal in-fiber polarizers have many advantages, such as high PER performance, broad working bandwidth, low insertion loss, and high handling power. We have also reviewed the recent achievements of applications of 45°-TFGs in all-fiber ultrafast fiber lasers at three distinctive wavelength ranges (1 μm, 1.5 μm, and 2 μm) operating in different mode-locking regimes. For future applications, we foresee that 45°-TFGs may be utilized in fiber laser systems to extend the operation wavelength to visible and deeper mid-IR range. One significant property of 45°-TFGs is their fiber compatibility and low insertion loss; thus they could be suitable in-fiber devices for high power all-fiber ultrafast laser systems. Overall, we expect the 45°-TFG-based fiber laser systems to play an important role to generate frequency comb and multi-wavelength operation for potential applications in high-speed optic communications, gas and environment sensing, social security, and medical diagnosing and treatment systems.

Author details

Zhijun Yan[1,2*], Chengbo Mou[1], Yishan Wang[2], Jianfeng Li[3], Zuxing Zhang[4], Xianglian Liu[5], Kaiming Zhou[1,2] and Lin Zhang[1]

*Address all correspondence to: yanz3@aston.ac.uk

1 Aston Institute of Photonic Technologies (AIPT), School of Engineering and Applied Science, Aston University, Aston Triangle, Birmingham, United Kingdom

2 State Key Laboratory of Transient Optics and Photonics, Xi'an Institute of Optics and Precision Mechanics, Chinese Academy of Sciences, Xi'an, China

3 State Key Laboratory of Electronic Thin Films and Integrated Devices, School of Optoelectronic Information, University of Electronic Science and Technology of China (UESTC), Chengdu, China

4 School of Optoelectronic Engineering, Nanjing University of Posts and Telecommunications, Nanjing, China

5 Key laboratory of Advanced Transducers and Intelligent Control system, Ministry of Education, College of Physics and Optoelectronics, Taiyuan University of Technology, Taiyuan, China

References

[1] Miniscalco, W.J.. Erbium-doped glasses for fiber amplifiers at 1500 nm. Journal of Lightwave Technology. 1991;9 (2): 234–250.

[2] Kawasaki, B.S., Hill, K.O., Lamont, R.G.. Biconical-taper single-mode fiber coupler. Opt. Lett. 1981;6 (7):327–328.

[3] Lee, B.. Review of the present status of optical fiber sensors. Optical Fiber Technology. 2003;9 (2):57–79.

[4] Rao, Y.J.. In-Fibre Bragging grating sensors. Meas. Sci. Technol. 1997;8:355–375..

[5] Bennion, I., Williams, J.A.R, Zhang, L., Sudgen, K., Doran N. J.. UV-written in-fibre Bragg gratings. Optical and Quantum Electronics. 1996;28 (2):93–135.

[6] Kaiming Zhou, G.S., Xianfeng Chen, Lin Zhang, Ian Bennion. High extinction ratio in-fiber polarizers based on 45° tilted fiber Bragg gratings. Opt. Lett.. 2005;30 (11): 1285–1287.

[7] Yan, Z., Mou, C., Chen, X., Zhou, K., Zhang, L.. UV-Inscription, Polarization-Dependant Loss Characteristics and Applications of 45°Tilted Fiber Gratings. J. Lightwave Technol. 2011;29 (18):2715–2724.

[8] Chengbo Mou, Kaiming Zhou, Lin Zhang, Ian Bennion. Characterization of 45°-tilted fiber grating and its polarization function in fiber ring laser. Journal of Optical Society of America. 2009;26 (10):1905–1911.

[9] Mou, C., Wang, H., Bale, B.G., Zhou, K, Zhang, L, Bennion, I. All-fiber passively mode-locked femtosecond laser using a 45°-tilted fiber grating polarization element. Optics Express. 2010;18 (18):18906–18911.

[10] Liu, X., et al.. All-fiber normal-dispersion single-polarization passively mode-locked laser based on a 45°-tilted fiber grating. Optics Express. 2012;20 (17):19000–19005.

[11] Xianglian, L., et al.. Bound dissipative-pulse evolution in the all-normal dispersion fiber laser using a 45° tilted fiber grating. Laser Physics Letters. 2013;10 (9):095103.

[12] Zhang, Z., et al.. Sub-100 fs mode-locked erbium-doped fiber laser using a 45 degree-tilted fiber grating. Optics Express. 2013;21 (23):28297–28303.

[13] Li, J., et al.. Thulium-doped all-fiber mode-locked laser based on NPR and 45° tilted fiber grating. Optics Express. 2014;22 (25):31020–31028.

[14] Xianglian, L., et al.. Single-polarization, dual-wavelength mode-locked Yb-doped fiber laser by a 45°-tilted fiber grating. Laser Physics Letters. 2015;12(6):065102.

[15] Zuxing, Z., et al.. All-fiber 250 MHz fundamental repetition rate pulsed laser with tilted fiber grating polarizer. Laser Physics Letters. 2015;12 (4):045102.

[16] Hill, K.O, Fujii Y., Johnson, D. C., Kawasaki, B.S.. Photonsensitivity in optical fibre waveguides: Application to reflection filrer fabrication. Appl. Phys. Lett.. 1978;32:647–649.

[17] G. Meltz, W.W.M., W.H.G. 1990. TuG1. In-Fibre Bragg grating tap. Optical Fibre Communications, San Francisco, California, USA, 1990 OSA Technical Digest Series 1. 1990;:TuG1.

[18] Williams, J. A. R., Sugden, I.B., K., Doran, N. J.. Fibre dispersion compensation using a chirped in-fibre Bragg grating. Electronics Letters. 1994;30:985–987.

[19] Ramachandran, S., Wang, Z., Yan, M.. Bandwidth control of long-period grating-based mode converters in few-mode fibers. Opt. Lett.. 2002;27 (9):698–700.

[20] Mihailov, S. J., R.B.W., Lu, P., Ding, H., Dai, X., Smelser, C., Chen, L.. UV-Induced polarisation-dependant loss(PDL) in tilted fibre Bragging gratings: application of a PDL equaliser. IEE Proc. Oproelectronics. 2002;149:211–216.

[21] Sipe, T.E., J.E.. Tilted fibre phase gratings. J. Opt. Soc. Amer. A. 1996;13:296–313.

[22] Brown, Y.L., T.G.. Radiation modes and tilted fiber gratings. Journal of Optical Society of America. 2006;23 (8):1544–1555.

[23] Yufeng Li, M.F., T.E.. Volume Current Method for Analysis of Tilted Fibre Gratings. Journa of Ligthwave Technology. 2001;19(10):1580–1591.

[24] Yoshino, T.. Theoretical analysis of a tilted fiber grating polarizer by the beam tracing approach. J. Opt. Soc. Am. B. 2012;29(9):2478–2483.

[25] G. Meltz, W.W.M., W.H.G.. Formation of Bragg Grating in optical fibres by a transverse holographic method. Optics letters. 1989;14:823–825.

[26] Hill, K.O., et al.. Photosensitivity in Eu2+ :Al2O3-Doped-Core Fiber: Preliminary Results and Application to Mode Converters. Optical Society of America. 1991;

[27] Hill, K. O., B.M., Bilodeau, F., Johnson, D. C., Albert, J.. Bragging gratings fabricated in monomode photosensitive optical fibre by UV exposure through a phase mask. Appl. Phys. Lett. 1993;62:1035–1037.

[28] Anderson, D. Z., V.M., Erdogan, T., White, A.. Production of in-fibre gratings using a diffractive optical element. Electron. Lett.. 1993;29:566–567.

[29] Mihailov, S. L., R.B.W., Stocki, T. J., Johnson, D. C.. Fabrication of tilted fibre-grating polarisation-dependent loss equliser. Electron. Lett.,. 2001;37:284–286.

[30] Technologies, A.. Polarization Dependent Loss Measurement of Passive Optical Components. Application Note. 2001;5988–1232EN.

[31] Yan, Z., Zhou, K., Zhang, L.. In-fiber linear polarizer based on UV-inscribed 45o tilted grating in polarization maintaining fiber. Opt. Lett.. 2012;37 (18):3819–3821.

[32] Tamura, K., Haus, H.A., Ippen, E.P. Self-starting additive pulse mode-locked erbium fibre ring laser. Electronics Letters. 1992;28:2226–2228.

[33] Bergh, R.A., Lefevre, H.C., Shaw, H.J.. Single-mode fiber-optic polarizer. Opt. Lett.. 1980;5 (11):479–481.

[34] Feth, J.R., Chang, C.L.. Metal-clad fiber-optic cutoff polarizer. Opt. Lett.. 1986;11 (6): 386–388.

[35] Bao, Q., et al.. Broadband graphene polarizer. Nat Photon. 2011;5(7):411–415.

[36] Okamoto, K., Hosaka, T., Noda, J.. High-birefringence polarizing fiber with flat cladding. Journal of Lightwave Technology. 1985;3 (4):758–762.

[37] Villanueva, G.E. Pérez-Millán, P.. Dynamic control of the operation regimes of a mode-locked fiber laser based on intracavity polarizing fibers: Experimental and theoretical validation. Opt. Lett.. 2012;37 (11):1971–1973.

[38] Zhang, Z., et al.. All-fiber mode-locked laser based on microfiber polarizer. Optics Letters. 2015;40(5):784–787.

[39] Tamura, K., et al.. 77-fs pulse generation from a stretched-pulse mode-locked all-fiberring laser. Optics Letters. 1993;18(13):1080–1082.

[40] Turitsyn, S.K., Bale, B.G. Fedoruk, M.P.. Dispersion-managed solitons in fibre systems and lasers. Physics Reports. 2012;521(4):135–203.

[41] Grelu, P. Akhmediev, N.. Dissipative solitons for mode-locked lasers. Nat Photon. 2012;6 (2):84–92.

[42] DeLoach, L.D., et al.. Evaluation of absorption and emission properties of Yb^{3+} doped crystals for laser applications. Quantum Electronics, IEEE Journal of. 1993;29(4):1179–1192.

[43] Payne, S.A., et al.. Ytterbium-doped apatite-structure crystals: A new class of laser materials. Journal of Applied Physics. 1994;76(1):497–503.

[44] Ruehl, A., et al.. Similariton fiber laser with a hollow-core photonic bandgap fiber for dispersion control. Optics Letters. 2007;32 (9):1084–1086.

[45] Smith, K., Mollenauer, L.F.. Experimental observation of soliton interaction over long fiber paths: discovery of a long-range interaction. Optics Letters. 1999;14(22):1284–1286.

[46] Ranka, J.K., Schirmer, R.W., Gaeta, A.L.. Observation of Pulse Splitting in Nonlinear Dispersive Media. Physical Review Letters. 1996;77(18):3783–3786.

[47] Ortaç, B., et al.. Observation of soliton molecules with independently evolving phase in a mode-locked fiber laser. Optics Letters. 2010;35 (10):1578–1580.

[48] Li, J., et al.. All-fiber passively mode-locked Tm-doped NOLM-based oscillator operating at 2-μm in both soliton and noisy-pulse regimes. Optics Express. 2014;22 (7): 7875–7882.

[49] Solodyankin, M.A., et al.. Mode-locked 1.93 μm thulium fiber laser with a carbon nanotube absorber. Optics Letters. 2008;33 (12):1336–1338.

[50] Horowitz, M., Barad, Y., Silberberg, Y.. Noiselike pulses with a broadband spectrum generated from an erbium-doped fiber laser. Optics Letters. 1997;22 (11):799–801.

[51] Zaytsev, A., et al.. Supercontinuum generation by noise-like pulses transmitted through normally dispersive standard single-mode fibers. Optics Express. 2013;21(13):16056–16062.

Chapter 12

Numerical Simulation of Fiber Laser Operated in Passively Q-Switched and Mode-Locked Regimes

Sorin Miclos, Dan Savastru, Roxana Savastru and Ion Lancranjan

Additional information is available at the end of the chapter

http://dx.doi.org/10.5772/61882

Abstract

The aim of this chapter is to highlight the role of simulation methods as tools for analysis of low and medium average power fiber laser operated in passively Q-switched and/or mode-locking regimes into the design of various applications such as materials micro-processing of sensor applications. The chapter's purpose consists in making available to specialists in the field of lasers, electro-optics and even nano-photonics improved procedures for designing high-accuracy remote sensors dedicated to large range of laboratory, industrial and military applications. The reason that this chapter deals with passive optical Q-switching and mode-locking techniques tailored for fiber lasers is the high percentage of sensing devices operating in this regime. Numerical simulation results obtained for this class of laser emitters can be used for other types of lasers, such as optical fiber lasers. There are briefly presented the two main mathematical methods used to analyze solid laser oscillators in passive optical Q-switching regime: the coupled rate equations approach and the iterative approach. The validation of the presented numerical simulation methods is done by comparison with experimental results.

Keywords: Fiber laser, all-fiber passive optical Q-switching, all-fiber mode-locking numerical simulation

1. Introduction

Erbium, ytterbium and ytterbium/erbium co-doped fiber lasers operated in mode-locking regime (ML) are an extremely useful tool for an increasing number of researches, medical and industrial applications. Their main characteristic is their output consisting of a phase coherent train of very short pulses, less than a picosecond. The application range of ML-operated fiber lasers spans from micro-machining metals [1] all the way to the most precise frequency measurements ever made [2]. Based on the many proposals for new technologies that utilize

mode-locked lasers [3, 4], it is clear that these lasers will be an invaluable tool for future technologies.

Here, we present simulation results obtained in analyzing a particular type of mode-locked fiber laser, namely those using erbium (Er), ytterbium (Yb) or ytterbium/erbium-doped or co-doped single-mode (SM) optic fiber as active medium and/or saturable absorber [5-13]. These types of lasers are intensively investigated because of their advantages, i.e. low cost, low power consumption, long term of use, robustness, and ease of long-distance transmission (through single-mode fiber). The performed analysis is pointing to a special topic of such lasers, namely the stabilization of the repetition frequency of these lasers. Initial demonstrations of these lasers showed large amounts of high frequency noise in these systems [14, 15]. We then turned our attention towards investigation of reduction of frequency noise in these systems for design improvement, synchronizing remotely located fiber lasers using this fast actuator in conjunction with a stabilized fiber link.

The performed simulation of Er, Yb or Yb/Er fiber lasers operated in ML regime has the scope of developing a software toolbox dedicated to an optimal laser design for micromachining glass and bare SM optic fiber.

2. Theory

The ML technique is based on phase locking many different frequency modes of a laser cavity. For practical reasons, as their bulk solid state counterparts fiber lasers are used in many applications which require high peak power and pulse energy [16-23]. There are two most well-established techniques for pulsing fiber lasers: Q-switching and mode locking in either active or passive form [17, 19-21]. Because of its "in principia" simplicity, one of the most used device for pulsing fiber lasers is the passive Q-switch cell manufactured, in essence, of a saturable absorber material. Q-switching is an effective method to obtain giant short pulses from a laser by spoiling the cavity loss periodically with a modulator inside the resonator cavity [17, 19]. In this technique, the pump delivers constant power all time, where the energy is stored as accumulated population inversion during the OFF times (high loss). During the ON time, the losses are reduced and the accumulated population difference is released as intense pulse of light. Q-Switching allows the generation of pulses of mJ energy, ns duration and few Hz to hundreds of kHz repetition rate [17-21]. For shorter pulses in picoseconds or sub-picoseconds ranges, mode- locking is the main mechanism [21-23]. The phase locking of different frequency modes of a fiber laser cavity imposes the laser to produce a continuous train of extremely short pulses rather than a continuous wave (CW) of light [16-18]. In principle, a continuous train of extremely short pulses can be generated from a passively Q-switched laser. It is also possible that the laser generates a continuous train of extremely short pulses which have amplitude modulated by the saturable absorber and having an envelope of the laser pulses peaks similar to Q-switched pulses [21-23]. The difference between these two scenarios of fiber laser emission obtained under continuous pumping of doped optic fiber active media lies in the optical phase of the pulses. The mode-locked pulses are phase-coherent with each other, while the Q-switched pulses are not [24-26]. This simple fact has massive consequences regarding the application of these two types of lasers. This technique induces a fixed phase relationship between the longitudinal modes of the laser cavity. Interference between these modes causes the laser light to produce a train of pulses. Locking of mode phases enables a periodic variation in the laser output which is stable over time, and with periodicity given by the round trip time of the cavity [27-33].

To understand the mode-locking process, even if it appears too basic, it is useful to begin by investigating a CW pumped fiber laser, which is supposed to have a CW output in the frequency domain. For a single longitudinal mode CW laser, considered as having a Fabry-Perot cavity and a frequency defined as

$$v = \frac{c}{2nL} \tag{1}$$

it can be considered that one resonant mode of the laser cavity overlaps in frequency with the gain medium. Thus, this hypothetical fiber laser emits a CW beam with a narrow range of frequencies defined as

$$E(t) = E_1 \exp\left[i\left(\omega_1 t + \varphi_1\right)\right] \tag{2}$$

In general, however, the gain medium could overlap with several modes. We can describe the output of such a laser in the time domain as

$$E(t) = \sum_1^N E_n \exp\left[i\left(\omega_n t + \varphi_n\right)\right] \tag{3}$$

where the sum is over all of the lasing cavity modes, E_n is the amplitude of the nth mode, ω_n is the angular frequency of the nth mode, and ϕ_n is the phase of the nth mode. For the single-mode laser, this sum just has one term as given above. As we will see, the phase term plays the key role in the difference between incoherent multimode lasing and mode locking. It can be considered an increasing range of overlap between the gain bandwidth and with fiber laser cavity modes. In this case, there are N terms in Equation (3). The output of such a laser depends critically if there is a well-defined phase relationship between the N modes. If each mode has a randomly varying phase with respect to the other modes, then a time domain detector on the output would show us that the laser is emitting a CW beam with a large amount of intensity noise, while a frequency domain detector would show us that the energy was contained in narrow spikes (with lots of intensity noise) spaced evenly by the free spectral range (FSR) of the cavity. However, if we can fix the relative phases to a clearly defined value, then the situation changes dramatically. With fixed phase relationships, the N modes can combine to interfere in such a way as to constructively interfere at multiples of the roundtrip time of the cavity, while they destructively interfere elsewhere. This process creates shorter pulses as the number of phase-locked modes increases [27-33].

An obvious question to be asked concerns how this well-defined relation between fiber laser cavity longitudinal modes is exactly obtained, implicitly how the phase locking of the longitudinal modes is obtained. The answer to this question could be obtained from the time domain picture of mode-locking. It is clearly experimentally established that a mode-locked laser produces ultra-short pulses at a rate equal to the round trip time of the optical cavity. The explanation of this experimental observation is that there has to be some part of the fiber laser that allows the pulse emission over CW radiation. This statement equates to say that there is needed of some element providing high loss at low intensity (CW radiation) and lower loss at high intensity (pulsed operation). Such a device is a saturable absorber. The operational principles of atoms or molecules as saturable absorbers are straightforward: low-intensity light is absorbed by the atoms and re-emitted into 4π steradians (i.e. out of the laser cavity), while high-intensity light fully excites the atoms and passes most of its photons through the medium. The main feature of the saturable absorber is its decreasing loss with increasing intensity. As is seen immediately, this behavior can be mimicked with optical processes that have nothing to do with actual atomic or molecular resonance absorption.

It is interesting to note that the history of mode-locked lasers began not long after the first demonstration of a continuous wave lasing in 1960, and it is intimately connected with search for obtaining substances having a saturable absorption. Maiman's [34] ruby laser was created at Hughes Research Laboratory in California; the creation of the first mode-locked laser would occur at Bell Laboratories in New Jersey. In 1964, Hargrove et al. [35] used an extremely clever acousto-optic technique to provide a loss modulation in a helium-neon laser cavity, which led to the laser being actively mode locked. In 1965, Mocker and Collins showed that they could achieve transient locking of the modes of a multimode Q-switched laser using a saturable Q-switching dye (cryptocyanine in methanol) [17]. Since only a few modes were involved in this process, the pulse widths were on the order of tens of nanoseconds. Their technique, however, required no active modulator, and thus was the first demonstration of passive mode locking. The drawback of this dye was that it required the laser to be Q-switched in order to saturate and thus the laser emitted mode-locked pulses only at the Q-switched intervals. The transient nature of the mode locked pulses proved to be problematic in practical applications (ultrafast spectroscopy, nonlinear optics, etc.). This problem was solved in 1972 when Ippen et al. introduced a laser based on the saturable dye (Rhodamine 6G) that could mode lock continuously [18]. The pulses from this laser were found to have pulse widths of only 1.5 picoseconds. After this demonstration, many researcher efforts were done in order to push the gain bandwidth further with other types of saturable absorbers, including, in the first stages, different types of dyes, and after that period, solid-state saturable absorbers such as the ones made of semiconductor materials (SESAM) and of crystals or glasses doped with different types of ions such as Co^{2+}, Cr^{4+}, Zn^{2+} and others. In the case of fiber laser, one newly appearing mode-locking technique consists of using optic fiber doped with Sm or Tm ions or even with the same type of ions as the optic fiber active medium.

Regarding the ML operation of fiber laser, it is worth to analyze with a type of saturable absorber known as an effective saturable absorber. It is a special class of saturable absorbers, using processes other than atomic/molecular absorption. These saturable absorbers do not have to rely only on actual atomic transitions. This means that the recovery time for the saturable absorber can be much faster than for atomic transitions. Slow saturable absorbers can produce pulses with duration less than a picosecond by shortening the leading edge of the pulse via saturable absorption and the trailing edge via gain saturation. However, if the atomic transition of saturable absorber recovers fast enough, it can shorten both sides of the pulse using the saturable absorber effect which is achieved by exploiting the intensity- dependent index of refraction defined as

$$n(I) = n_0 + n_2 I \tag{4}$$

where n_0 is the index of refraction, n_2 is the nonlinear index coefficient and I is the optical intensity. The recovery time for a saturable absorber based on this effect is essentially instantaneous because non-resonant optical processes are extremely fast inducing a nonlinear index to respond on the order of a few optical cycles. Two types of mode locking based on effective saturable absorbers are mostly used and reported in literature: Kerr lens mode locking (KLM) and additive pulse mode locking (APM).

KLM is based on lens creation into the gain medium as an effect of the nonlinear index of refraction, lens which causes self-focusing of the beam [36, 37]. KLM is combined with an intra-cavity aperture, this effect creates a situation where the cavity prefers pulsed operation because if the laser is in CW operation, there is a high loss due to the aperture, while in pulsed operation the beam focuses through the aperture with minimal loss.

APM is realized by interference of circulating pulses. In the first realization of APM [38, 39], this interference was between pulses propagating in two coupled cavities. The main cavity has the gain medium and an output coupler, while the secondary cavity has a nonlinear section, an optical fiber. Pulses that are coupled to the nonlinear cavity experience an intensity-dependent phase shift. When these pulses are coupled back to the main cavity they can be made to overlap with the normal pulses in such a way as to constructively interfere at their peaks, while destructively interfering at their wings. Thus, the addition of multiple pulses results in pulse shortening on every round trip, just like a real saturable absorber. One special type of APM is based on nonlinear polarization rotation (P-APM) [39] and is particularly useful in a fiber laser cavity. The basic idea is that the added pulses are not from separate cavities, but are co-propagating with different polarization. Elliptically polarized pulses propagate in a Kerr medium to produce nonlinear polarization rotation. Experimentally, this situation can be produced by inserting a quarter-wave plate into the fiber cavity, so that linear polarization can be turned into elliptical. The highest intensity part of the pulse (i.e. the peak) undergoes a nonlinear phase shift and thus rotates its polarization some amount. The wings of this pulse, which have low intensity, do not undergo this phase shift and thus experience no rotation. A quarter-wave plate and linear polarizer at the output of the Kerr medium (fiber) turn the intensity-dependent polarization into an intensity-dependent transmission. This type of mode locking can produce pulse widths that are close to the gain bandwidth limit of Er ($\approx$100 fs) [39].

Mode-locked Er or Yb/Er fiber lasers have many advantages to be considered regarding various applications. One main advantage consists in the fact that an all fiber cavity needs no realignment. Also, the components needed to build a mode-locked all fiber laser are relatively cheap due to their mass production in the telecommunications industry. As an example, a nonlinear polarization mode-locked all fiber laser could be built for an expense of less than 4, 000 USD. For comparison, a typical solid-state titanium doped sapphire mode-locked laser could be purchased from a vendor for around 100, 000 USD. While the typical output power of a Ti:Sapphire system is roughly an order of magnitude larger than that of a mode-locked Er fiber laser, it is straightforward and inexpensive to build an Er amplifier that allows the Er-based system to reach average power levels close to those of the mode locked Ti:Sapphire oscillator. Using a frequency doubling crystal, one can even transform the 1, 550 nm centered Er laser to Ti:Sapphire wavelengths around 750 nm. Finally, the relatively small gain bandwidth of the Er gain medium can easily be converted into an octave of spectrum using highly nonlinear fiber. All of these factors have played a part in the rapid emergence of fiber lasers in the world of ultrafast physics in the past 10 years. Erbium-doped fiber is particularly useful over other rare-earth doped fibers (i.e. ytterbium, neodymium, thulium, etc.) due to silica glass's low loss window in the telecommunications C band (conventional band: 1, 530–1, 565 nm).

In order to obtain a better understanding of the operation of an Er or Yb/Er-doped mode-locked fiber laser, the rate equation approach is useful. This approach starts with the Yb and Er ions energy level diagrams, schematically presented in Figure 1. The Er ion is a quasi-3 level system, meaning that although the lowest state in the lasing scheme is not the true ground state, it is still low energy enough that it has some population due to thermal excitation.

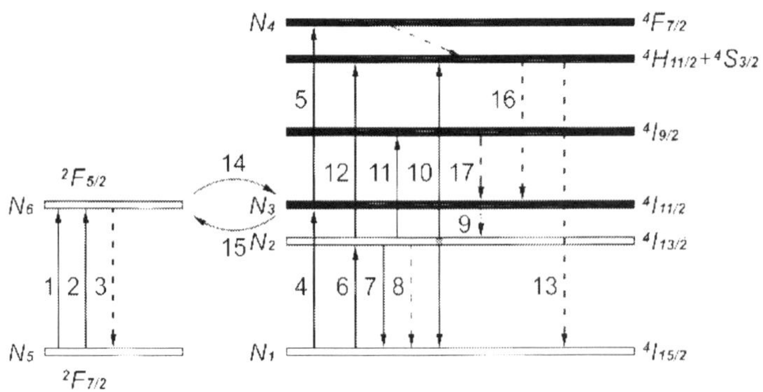

Figure 1. The Yb and Er active ions energy levels diagrams.

The pumping of Er ions is accomplished by either 980 nm or 1, 450 nm light generated by a semiconductor diode laser. In Figure 2, an example of mode-locked Er or Yb/Er-doped all-fiber laser is presented. Here, pumping diodes are laser pumping diodes with CW or pulsed-chopped emission at 976 nm at powers up to 25 W, combiner and WDM are coupling the pump power into a double clad Yb-doped fiber used as active medium, Yb-DCF is Yb-doped double clad fiber active medium with 15 μm diameter core and 15 m length, HR-FBG is a high reflectivity (99 %) FBG used as rear laser mirror and OC-FBG is a low reflectivity (15 %) FBG used as laser output coupler.

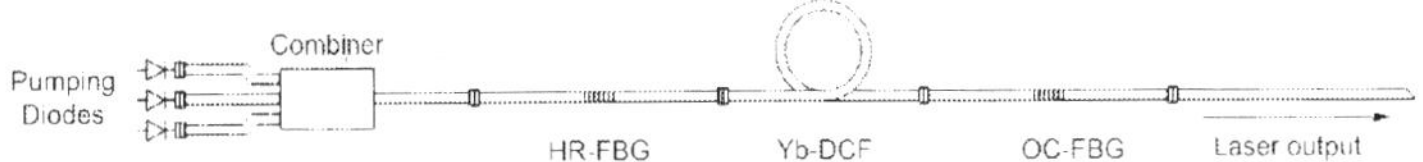

Figure 2. Configuration diagram of passively Q-switched all-fiber laser using an "excess" of fiber.

Regarding the pumping efficiency using 980 nm diode laser radiation, Yb/Er co-doped has better performances in comparison with simple Er-doped fiber laser. The role of Yb ions is the one of absorbing the pump radiation at 975-980 nm and due to its intense fluorescence at wavelength 980 nm, a wavelength closed to Er ions absorption band, to transfer optical excitation to Er ions. For single Er-doped optic fiber active medium, the 980 nm and the 1, 450 nm schemes result in similar efficiencies.

The trivalent erbium ion, when pumped with 980 nm light, is excited to the $^4I_{11/2}$ state, which decays to $^4I_{13/2}$ (see Figure 1). The decay between $^4I_{11/2}$ and $^4I_{13/2}$ is non-radiative (multiple phonon decay) and occurs within a few μs, while the metastable state ($^4I_{13/2}$) has a lifetime of about 10 ms. Since the $^4I_{11/2}$ state has such a short lifetime, we can make the approximation that this highest excited state has zero steady-state population (i.e. no population accumulates). This approximation reduces the number of participating energy levels to two. We can now write down the relevant rate equations that describe the number of erbium ions in the upper (N_2) and lower (N_1) energy levels

$$\frac{dN_1(t)}{dt} = A_{21}N_2(t) + \left(N_2(t)\sigma_e^s - N_1(t)\sigma_e^s\right)\frac{I_s}{h\nu_s} + \left(N_2(t)\sigma_e^p - N_1(t)\sigma_a^p\right)\frac{I_p}{h\nu_p} \tag{5}$$

$$\frac{dN_2(t)}{dt} = -A_{21}N_2(t) + \left(N_1(t)\sigma_a^s - N_2(t)\sigma_e^s\right)\frac{I_s}{h\nu_s} + \left(N_1(t)\sigma_a^p - N_2(t)\sigma_e^p\right)\frac{I_p}{h\nu_p} \tag{6}$$

where A_{21} is the Einstein A coefficient (inverse lifetime) for spontaneous emission, σ_a^s is the cross section for stimulated absorption at the signal wavelength, σ_a^s is the cross section for stimulated emission at the signal wavelength, σ_a^p is the cross section for stimulated absorption at the pump wavelength, σ_a^p is the cross section for stimulated emission at the pump wavelength, I_s is the signal intensity, I_p is the pump intensity, $h\nu_s$ is the energy of each individual signal photon and hv_p is the energy of each individual pump photon. By dividing the beam intensity by the photon energy of that beam, we get the total number of photons passing through a given area (i.e. photon flux).

To achieve lasing, we must have population inversion such that $N_2 > N_1$. The threshold for this condition occurs when the ion density in N_2 just equals N_1. By setting the Equations (5) and (6) equal and solving for the pump intensity, we find the threshold intensity for population inversion

$$I_{pth} = \frac{h \cdot v_p}{\tau\left(\sigma_a^p - \sigma_e^p\right)} \tag{7}$$

For a pump wavelength of 980 nm, this intensity is roughly 6 kW/cm². Since the mode field area of a single-mode Er fiber is around 20 µm², the pump power needed to achieve inversion is on the order of a few mW. This calculation is for a lossless cavity, however. Due to losses in fiber splices, the output coupler and losses in the coupling of the pump diode to the fiber, the actual pump power required for lasing threshold is of the order of tens of mW (typical 980 nm pump diodes reach average powers beyond 600 mW). It is also instructive to look at the evolution of the signal beam as it propagates through the gain (amplifying) section of the laser cavity. A simple differential equation governs the signal in the presence of a 2-level gain medium

$$\frac{dI_s(z)}{dz} = \left(N_2(t) \cdot \sigma_e^s - N_1(t) \cdot \sigma_a^s\right) \cdot I_s(z) \tag{8}$$

with the solution

$$I_s(z) = I_0 \exp(gl) \tag{9}$$

$$I_s(z) = I_0 \exp\left[\left(N_2(t) \cdot \sigma_e^s - N_1(t) \cdot \sigma_a^s\right) \cdot l\right] \tag{10}$$

where I_0 is the intensity entering the gain section, l is the total length of the gain section and g is the gain defined as

$$g(t) = \left(N_2(t) \cdot \sigma_e^s - N_1(t) \cdot \sigma_a^s\right) \tag{11}$$

For our analysis, we will consider the absorption of the signal beam to be zero, thus

$$g(t) = N_2(t) \cdot \sigma_e^s \tag{12}$$

The gain is then dependent only on the density of excited atoms N_2 and the emission cross section of the excited Er atoms at the signal wavelength σ_a^s. The emission cross section is a constant, thus to determine the gain we only have to find N_2. Using Equation (8), we have

$$\frac{dN_2(t)}{dt} = -A_{21}N_2(t) + \left(-N_2(t)\sigma_e^s\right)\frac{I_s}{hv_s} + \left(N_1(t)\sigma_a^p\right)\frac{I_p}{hv_p} \tag{13}$$

In the small signal limit, the pump intensity is much larger than the signal intensity ($I_p \gg I_s$). Using this approximation along with the fact that we are analyzing a steady-state scenario (d/dt → 0), we can ignore the I_s term and set the left-hand side of Equation (13) equal to zero. Solving for N_2 yields

$$N_2(t)\left(I_s \ll I_p\right) = \left(N_1(t)\sigma_a^p\right)\frac{I_p}{A_{21}hv_p} \tag{14}$$

$$N_2(t)\left(I_s \ll I_p\right) = \tau\left(N_1(t)\sigma_a^p\right)\frac{I_p}{hv_p} \tag{15}$$

$$N_2(t)\left(I_s \ll I_p\right) = \tau R \tag{16}$$

where τ is the lifetime of the excited state, defined as

$$\tau = \frac{1}{A_{21}} \tag{17}$$

and R is excitation rate which is defined as

$$R = \left(N_1(t)\sigma_a^p\right)\frac{I_p}{hv_p} \tag{18}$$

This equation shows that the density of excited atoms in the small-signal limit is simply given by the lifetime of the exited state (τ) multiplied by excitation rate R. Using the fact that g is defined as

$$g(t) = N_2(t) \cdot \sigma_e^s \tag{19}$$

the small signal gain is g_0 is defined by the relation

$$g_0(t) = \tau \cdot N_2(t) \cdot \sigma_e^s \cdot R \tag{20}$$

As the signal beam is increased to higher intensity, however, we must take into account the term in Equation (13) that involves I_s. Solving for N_2 yields

$$N_2(t) = \frac{N_2\left(I_s \ll I_p\right)}{1 + \dfrac{I_s}{I_{sat}}} \tag{21}$$

And the large signal gain is thus

$$g(t) = \frac{g_0}{1 + \dfrac{I_s}{I_{sat}}} \tag{22}$$

where I_{sat} is the saturation intensity defined as

$$I_{sat} = \frac{1}{\sigma_e^s \tau} \tag{23}$$

And finally, the differential change in signal intensity per length of gain in the strong pump regime is

$$\frac{dI_s(z)}{dz} = \frac{I_s \cdot g_0}{1 + \dfrac{I_s}{I_{sat}}} \tag{24}$$

The picture of the signal evolution is now complete. At low signal levels, there is an exponential increase in the number of signal photons in the gain medium. However, as the signal level is increased further, the gain begins to saturate and asymptotically approaches a value defined by

$$g(t) \approx I_{sat} \cdot g_0 = R \tag{25}$$

Thus, as expected, at high signal levels, the signal intensity increases linearly with the pump intensity. The fundamental characteristics of lasing, namely, the small-signal gain and the gain saturation have now been covered.

It is worth to mention an important aspect related to the mode-locking theory, namely, the frequency comb. With the advent of the frequency comb [40] in the late 1990s, mode-locked lasers, including the fiber ones, began to receive much attention concerning the frequency metrology applications. The frequency comb appears as a simultaneous solution for two important purposes separated by a vast gap to be achieved using mode-locking laser, especially fiber lasers. These two purposes are: on one side, the field of precision measurements imposes creation of actuated lasers that would have the narrowest possible spectral linewidth and, on the other side, the field of ultrafast spectroscopy was mainly interested in creating extremely short time domain bursts of electric field, which necessarily require that the pulses have a large spectral bandwidth. These two goals, which seem to be in direct opposition of each other, can be achieved simultaneously with a frequency comb.

As was mentioned, the frequency comb is based on mode-locked lasers [40-42]. In fact, mode-locking and frequency comb are definitions used interchangeably. This is not quite right, however, since technically a frequency comb really refers to a mode-locked laser that has been carrier-envelope phase stabilized. The frequency comb can be understood by using and by combining its description in time and frequency domains [40-42]. The time domain output of the laser can be viewed as the multiplication of the fast electric field oscillations and an envelope function, representing a modulation of electric field amplitude by the envelope function. The ultimate limit on the width of this envelope would be an envelope that encompasses only 1 cycle of the electric field, in other words, corresponding to one round trip along laser cavity. It can be shown that the envelope travels at a speed known as the group velocity v_g defined as

$$v_g = \frac{c}{n - \lambda \dfrac{dn}{d\lambda}} \tag{26}$$

while the electric field fast oscillations travel at the phase velocity v_p defined as

$$v_p = \frac{c}{n} \tag{27}$$

These two velocities are, in general, not equal and thus lead to a walk-off or slippage between the two entities, known as carrier-envelope offset phase. Using the shift theorem of Fourier transforms [42], we see that the Fourier transform turns this time domain phase slip into a frequency offset, f_0. Thus, the optical frequencies of the comb can be defined in terms of two frequencies as

$$v_n = n \cdot f_{rep} + f_0 \tag{28}$$

where v_n is the optical frequency of the nth comb mode and f_{rep} is the repetition frequency of the laser, which is related to the optical length of its cavity. Clearly, a random variation of the offset frequency would smear out the comb in frequency space and make it useless for any sort of precision measurement. An analogy to this sort of measurement would be like trying to measure the length of something with a ruler that is always moving back and forth slightly. Thus, it is clear that to do any sort of precision measurement with a mode-locked laser, we need to stabilize this offset frequency (and thereby produce a frequency comb).

The first technique that achieved the ability to measure (and thus stabilize) f_0 relied on the so-called f-$2f$ interferometer, a complicated experimental setup which is quite a technical achievement by itself. This technique is based on a simple manipulation of Equation (28). In this scheme, light from a Ti:sapphire laser was sent through a highly nonlinear fiber with low net dispersion to broaden the bandwidth of the pulses to an octave [40]. The octave spanning pulses were then coupled into an interferometer where in one arm the light was passed through a second harmonic crystal and underwent sum-frequency-generation (SFG). The two beams were then recombined on a beam-splitter, sent through an optical filter, and detected onto a photodetector to produce a heterodyne beat at f_0. The octave spanning pulse bandwidth ensures that we have optical frequencies present in a range from v_n to v_{2n}, while the second harmonic arm converts the v_n light to v_{2n} light via SFG. Filtering out the highest frequencies with the optical filter, and using the frequency comb equation, we can thus write the frequencies present in the two arms as

$$v_{2n} = 2n \cdot f_{rep} + 2v_n \cdot f_0 = 2n \cdot f_{rep} + 2f_0 \tag{29}$$

Once these two beams form a heterodyne beat on the photodetector, we can take the difference frequency which is

$$2v_n - v_{2n} = \left(2n \cdot f_{rep} + 2f_0\right) - \left(2n \cdot f_{rep} + f_0\right) = f \tag{30}$$

The first demonstration of this method [40] opened the door for experiments involving the frequency comb. Precision metrology benefited dramatically from the compact all-in-one nature of the frequency comb, while new techniques such as broadband cavity-ring down spectroscopy [43-44] have been developed based on the comb.

Another theoretical issue of mode-locked fiber laser concerns the consequence of nonlinear effects produced in the optic fiber during ultrashort light pulses propagation. It can be noticed that Equations (5–25) describe what should be defined as the energetic part of the mode-locked ultrashort laser propagation phenomena. In Equations (5–25), the cycle pump absorption –

population inversion – light energy at the laser wavelength is analyzed. Equations (26–29) are helpful for defining the aspects related to the repetition frequency of ultrashort light pulses. It is to be observed that because of the fiber optic guiding effect, the ultrashort light pulses propagation, even with energies on the nJ scale and full width half measure time duration of ps or fs is happening through core transverse area, meaning extremely large light intensity. Due to the square variation law of light intensity versus its electric field amplitude, it can be concluded that nonlinear effects can be produced. Quantitatively, all these qualitative comments can be developed starting from the system of Maxwell equations describing electromagnetic pulses propagation through optic fibers. In the system of Maxwell equations, the polarization vector of the dielectric propagation medium can be split into two terms, one linear and the other corresponding to its nonlinear variation.

$$\frac{\partial A}{\partial z} + \beta_1 \frac{\partial A}{\partial t} + \frac{i}{2}\beta_2 \frac{\partial^2 A}{\partial t^2} = i\gamma |A|^2 A \tag{31}$$

where A represents the propagating electromagnetic field potential, the β_1-term is responsible for the group velocity (with the transformation $T=t-\beta_1 z$ which transforms to the moving frame of the pulse), β_2 is often called the group-velocity dispersion (GVD) parameter. This equation is an example of the nonlinear Schrödinger equation (NLSE).

3. Simulation Results

As previously mentioned, the simulation of the Er- or Yb/Er-doped all-fiber laser was performed aiming to the development of a set of toolbox useful for evaluation of passively Q-switched and/or mode-locked laser dedicated to glass and optic fiber micro-processing. For accomplishing this task, sense previous experience in this field constitutes an advantage [45-47]. The simulation scripts can be grouped into two sets: the first one developed on the bases of Equations (1-3) and describing the parameters of a ML fiber laser at the first level and the second, the more complex one, using Equations (5-22) which treats the propagation of ultrashort laser pulses through laser resonator. The second set is an attempt to solve nonlinear Schroedinger equation.

In Figures 3, 4, 5 and 6 are presented results obtained in investigation of the noise factor role in definition of mode-locked laser pulses emitted by an Yb/Er fiber laser as schematically presented in Figure 2, using the simulation codes of the first set. A CW pump power of 55 W was considered. The simulated mode-locked Yb/Er laser is composed of a double clad fiber having a length in excess (9 m). The value of noise factor, d, is increased from zero up to 1.5. Figures 3 to 6 represented the overlapping between gain bandwidth and fiber laser resonator modes. In Figure 3 d is considered as zero and the mode-locked laser pulses are clearly defined.

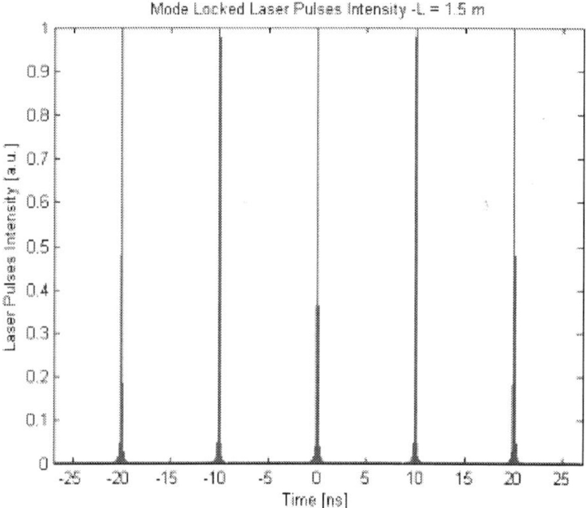

Figure 3. Laser pulse intensity versus time simulated for an Yb/Er fiber laser with 9 m length, "excess" of fiber, considering a noise factor $d = 0$.

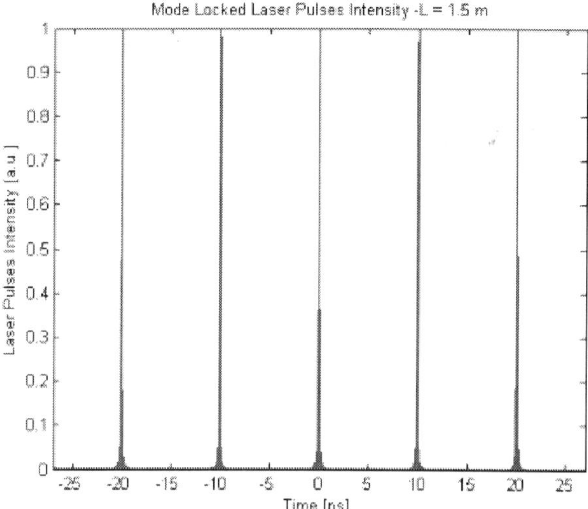

Figure 4. Laser pulse intensity versus time simulated for an Yb/Er fiber laser with 9 m length, "excess" of fiber, considering a noise factor $d = 0.05$.

In Figure 4, a small increase of d value, up to 0.05 was considered and some noise appearing on the wings of the mode-locked pulses can be observed. In Figure 5, the d factor has a value of 0.5 and the noise on the wings of mode-locked pulses is more evident. In Figure 6, the noise factor is increased up to 1.5 and its effects on the wings of mode-locked pulses are visible. This process creates shorter pulses as the number of phase-locked modes increases.

In Figures 7, 8 and 9, the simulation results are obtained using the second set of codes in the case of the same previously defined mode-locked Yb/Er all-fiber laser. In Figure 7, as can be observed, the mode-locked laser pulses are presented as analyzed in time domain. In Figure 8 results of the same analysis obtained in defining the fiber laser pulse spectrum are presented considering frequency as input are presented. In Figure 9, the simulated pulse spectrum is presented but considering wavelength as the argument.

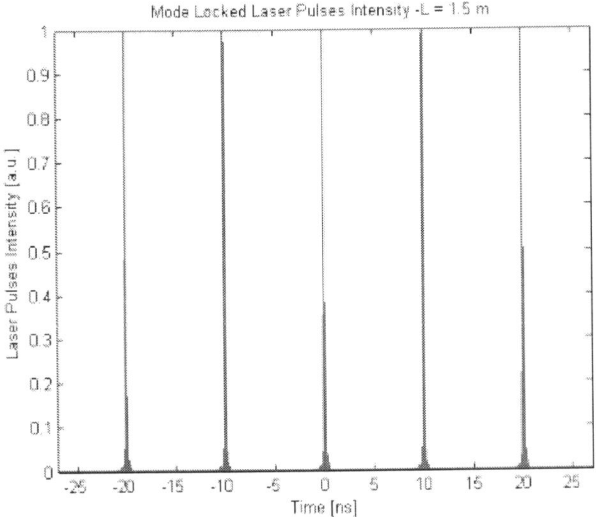

Figure 5. Laser pulse intensity versus time simulated for an Yb/Er fiber laser with 9 m length, "excess" of fiber, considering a noise factor $d = 0.5$.

In each Figure (7 to 9), the simulation was performed for three moments: t_0 – the initial moment, $t_0 + 40\ t_r$ and $t_0 + 80\ t_r$, where t_r is the laser cavity round trip time.

Simulations of the propagation equation (nonlinear Schrödinger equation) presented in Figures 7, 8 and 9 are helpful for the understanding of pulse propagation in optical fibers. Especially chromatic dispersion, nonlinear effects and their interplay, leading to physical phenomena like solitons, can be investigated with these simulations. The possibility to take a characterized pulse and to let it propagate through various fibers (as long as the fiber parameters are known) is helpful for the understanding of the pulse compression after the amplifier.

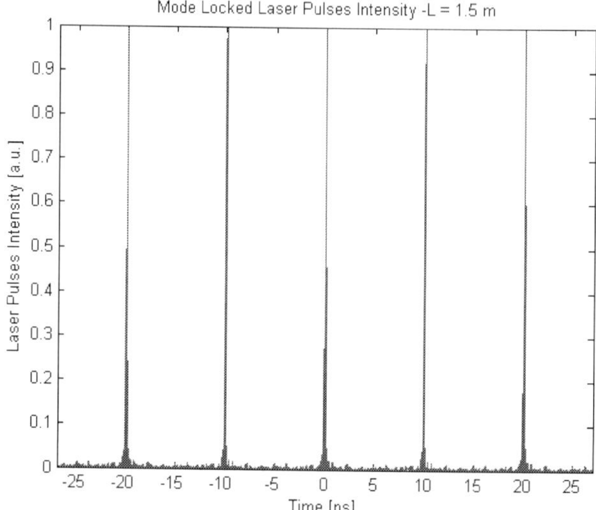

Figure 6. Laser pulse intensity versus time simulated for an Yb/Er fiber laser with 9 m length, "excess" of fiber, considering a noise factor $d = 1.5$.

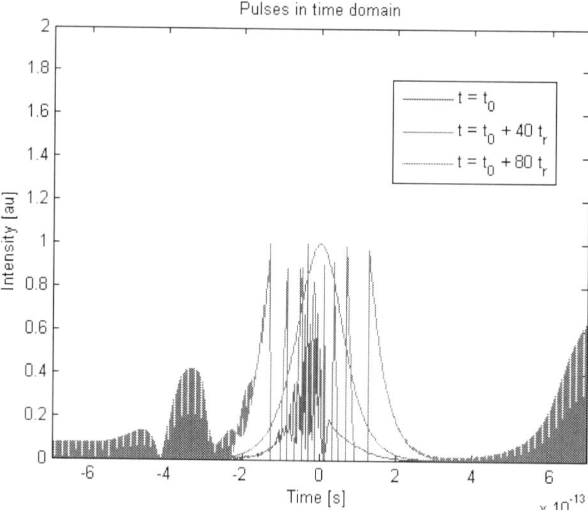

Figure 7. Laser pulse intensity analyzed in time domain in the case of an Yb/Er fiber laser with 9 m length, "excess" of fiber.

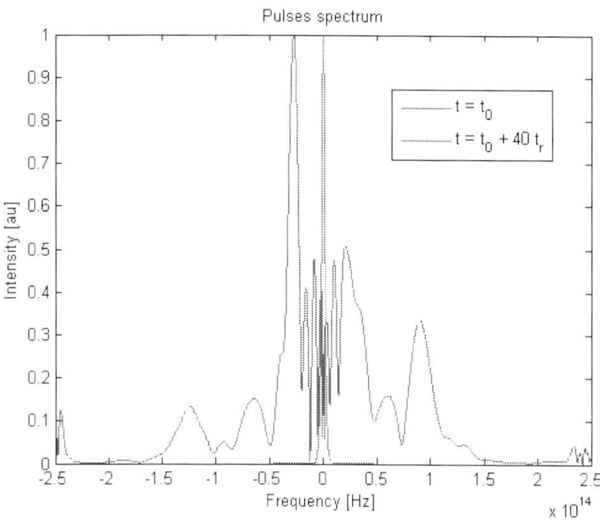

Figure 8. Laser pulse spectrum simulated for an Yb/Er fiber laser with 9 m length, "excess" of fiber, considering frequency as input.

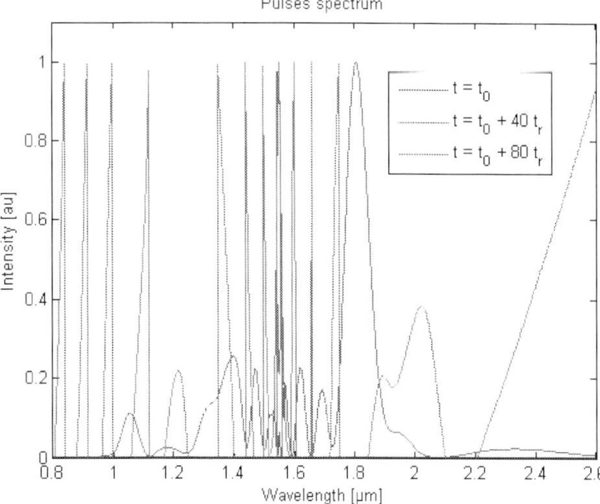

Figure 9. Laser pulse spectrum simulated for an Yb/Er fiber laser with 9 m length, "excess" of fiber, considering wavelength as input.

The simulation also made it possible to investigate the spectral broadening of an ultrashort laser pulse in a highly nonlinear fiber. The results show that it is theoretically possible to create an octave spanning supercontinuum with the pulse from our laser system. The simulation makes possible the design of a measuring system for the carrier-envelope frequency based on a f-$2f$ interferometer. It is possible to investigate the spectrum that barely spans a factor of two into an attempt to detect a carrier-envelope set frequency beat above the noise.

The performed simulation has as a future objective, the construction of mode-locked Er all-fiber laser using a special technique, the polarization additive-pulse mode locking (P-APM) procedure. It is based on the fact that due to the nonlinear refraction of the optic fiber, different intensities see a different index of refraction. For elliptically polarized radiation, the net result is polarization rotation. The investigated P-APM mechanism works like this: a pulse with a strongly elliptic polarization is sent into a Kerr-medium; dependent on the intensity there will be more or less polarization rotation; combined with a polarizer which only transmits the rotated part (higher intensity), this acts as a "pulse-shortener". The mode-locking elements are located in the free-space path of the fiber laser. The polarizer is represented by a polarizing beam-cube and a Faraday-isolator or by an optic fiber polarization controller. The quarter-wave plate (QWP) right, or the optic fiber which have this function, makes the polarization elliptical; the other two wave plates change the polarization to maximize the transmission at pulsed operation. It can be noticed that the mode-locking is self-starting, meaning that the pulse builds up from initial CW fluctuations, just as it is the case of an excess Er fiber laser (see Figure 2) and is analyzed in Equations (1-3).

4. Conclusions

The main purpose of this chapter is to present a possible development of simulation codes applicable for Er or Yb/Er all-fiber lasers operated in mode-locking regime with glass and fiber optic micro-machining application. In this sense, the basic theoretical notions are defined in Section 2: Theory and examples of simulation codes use could be observed in Section 3: Simulation Results. A fairly good agreement between the simulated data and the similar ones reported in literature can be noticed.

Author details

Sorin Miclos*, Dan Savastru, Roxana Savastru and Ion Lancranjan

*Address all correspondence to: miclos@inoe.ro

National Institute R&D of Optoelectronics - INOE , Magurele, Ilfov, Romania

References

[1] Ostendorf A, Kamlage G, Chichkov B. Precise deep drilling of metals by femtosecond laser pulses. RIKEN Review: Focused on Laser Precision Microfabrication. 2003;50:87-9.

[2] Osky W, et al. Single-atom optical clock with high accuracy. Phys Rev Lett 2006;97:020801.

[3] Ye J. Absolute measurement of a long, arbitrary distance to less than an optical fringe. Opt Lett 2004;29:1153-5.

[4] Schibli T. Combs for dark energy. Nat Photonics 2008;2:712-3.

[5] Tsai T-Y. et al. All-fiber passively Q-switched erbium laser using mismatch of mode field areas and a saturable-amplifier pump switch. Opt Lett 2009;34:2891-3.

[6] Wu B, Chu PL. Fast optical switching in Sm^{3+}-doped fiber. IEEE Phot Technol Lett 1996;8:230-2.

[7] Fotiadi AA, et al. All-fiber passively Q-switched Ytterbium laser. Proc CLEO/Europe 29B, CJ2-3, 2005.

[8] Tsai T-Y, et al. Saturable absorber Q-and gain-switched all-Yb^{3+} all-fiber laser at 976 and 1064 nm. Opt. Expr 2010;18: 23523-8.

[9] Moore S W et al. 400 µJ 79 ns amplified pulses from a Q-switched fiber laser using an Yb^{3+}-doped fiber saturable absorber. Opt Expr 2012;20:23778-9.

[10] Soh DBS, Bisson SE, et al. () High-power all-fiber passively Q-switched laser using a doped fiber as a saturable absorber: numerical simulations. Opt. Lett 2011;36:2536-8.

[11] Fotiadi AA, et al. () Dynamics of All-Fiber Self-Q-switched Ytterbium/Samarium Laser. Proc CLEO/QELS/PhAST 15, CMC4, 2007.

[12] Fotiadi AA, et al. All-fiber coherent combining of Er-doped amplifiers via refractive index control in Yb-doped fibers by two-wavelengths optical signal. Proc CLEO/Europe, CJ3-5, 2009.

[13] Huang JY, et al. High-power 10-GHz self-mode-locked Nd:LuVO4 laser. Appl Opt 2008;47:2297-302.

[14] Hong F, Minoshima K, Onae A, et al. Broad-spectrum frequency comb generation and carrier-envelope offset frequency measurement by second-harmonic generation of a mode-locked fiber laser. Opt Lett 2003;28:1516-8.

[15] Hudson DD, Holman KW, et al. Mode-locked fiber laser frequency-controlled with an intracavity electro-optic modulator. Opt Lett 2005;30:2948-50.

[16] Hargrove LE, Fork RL, Pollack MA. Locking of HeNe laser modes induced by synchronous intracavity modulation. Appl Phys Lett 1964;5:4-5.

[17] Mocker HW, Collins RJ. Mode competition and self-locking effects in a Q-switched ruby laser. Appl Phys Lett 1965;7:270-3.

[18] Ippen EP, Shank CV. Dienes A, Passive mode locking of the CW dye laser. Appl Phys Lett 1972;21:348-50.

[19] Rüdiger P. Encyclopedia of Laser Physics and Technology, 1ed. John Wiley & Sons, 2008.

[20] Digonnet MJF. Rare-earth-doped fiber lasers and amplifiers, 2d ed. Marcel Dekker, 2001, vol. 25.

[21] Svelto O, Hanna DC. Principles of Lasers. Springer, 1998.

[22] Saleh BEA, Teich MC, Masters BR. Fundamentals of Photonics, Second Edition. John Wiley & Sons, 1991.

[23] Wang Y, Xu C-Q. () Actively Q-switched fiber lasers: Switching dynamics and nonlinear processes. Prog Quantum Electronics. 2007;31:131-216.

[24] Schreiber T, Nielsen CK, et al. Microjoule-level all-polarization-maintaining femtosecond fiber source. Opt Lett 2006;31:574-6.

[25] Seung BC, Song H, Gee S, Dug YK. Self-starting passive mode-locked ytterbium fiber laser with variable pulse width. Proc. SPIE - Fiber Lasers VII: Technology, Systems, and Applications. 2010;7580:75802C.

[26] Ortac B, Plotner M, et al. Experimental and numerical study of pulse dynamics in positive net-cavity dispersion mode locked Yb-doped fiber lasers. Optics Exp 2007;15:15595-602.

[27] Jang GH, Yoon TH. Environmentally-Stable All-normal-dispersion Picosecond Yb-doped Fiber Laser with an Achromatic Quarter-wave-plate. Laser Phys 2010;20:1463-8.

[28] Lian FQ, Fan ZW, et al. Ytterbium doped all-fiber-path all-normal dispersion mode-locked laser based on semiconductor saturable mirror. Laser Phys 2011;21:1103-7.

[29] Turchinovich D, Liu X, Laegsgaard J. () Monolithic all-PM femtosecond Yb-fiber laser stabilized with a narrow-band fiber Bragg grating and pulse-compressed in a hollow-core photonic crystal fiber. Optics Express. 2008;16:14004-14.

[30] Tian X, Tang M, et al. High-energy wave-breaking-free pulse from all-fiber mode-locked laser system. Optics Exp 2009;17:7222-7.

[31] Song R, Chen H, et al. A SESAM passively mode-locked fiber laser with a long cavity including a band pass filter. J Optics. 2011;13:035201.

[32] Liu J, Xu J, Wang P. High Repetition-Rate Narrow Bandwidth SESAM Mode-Locked Yb-Doped Fiber Lasers. IEEE Photonics Technol Lett 2012;24:539-41.

[33] Prochnow O, Ruehl A, et al. All-fiber similariton laser at 1 μm without dispersion compensation. Optics Exp 2007;15:6889-93.

[34] Maiman TH. Stimulated optical radiation in ruby. Nature. 1960;187:493-4.

[35] Hargrove LE, Fork RL, Pollack MA. Locking of HeNe laser modes induced by synchronous intracavity modulation. Appl Phys Lett 1964;5:4-5.

[36] Spence DE, Kean PN, Sibbett W. 60-fsec pulse generation from a self-mode-locked Ti:sapphire laser. Opt Lett 1991;16:42-4.

[37] Ell R, et al. Generation of 5-fs pulses and octave-spanning spectra directly from a Ti:sapphire laser. Opt Lett 2001;26:373-5.

[38] Mollenauer LF, Stolen RH. The soliton laser. Opt Lett 1984;9:13-5.

[39] Stolen RH. Nonlinearity in fiber transmission. Proc IEEE 1980;68:1232-6.

[40] Jones DJ, Diddams SA, et al. Carrier-envelope phase control of femtosecond mode-locked lasers and direct optical frequency synthesis. Science. 2000;288:635-40.

[41] Cundiff S, Ye J, Hall J. Optical frequency synthesis based on mode-locked lasers. Rev Sci Instrum 2001;72:3749-59.

[42] Walker R, Udem T, et al. Frequency dependence of the fixed point in a fluctuating frequency comb. Appl Phys B 2007;89:535-43.

[43] Cundiff S, Ye J. Colloquium: Femtosecond optical frequency combs. Rev Mod Phys 2003;75:325-32.

[44] Everson KM, Wells JS, et al. Accurate frequencies of molecular transitions used in laser stabilization: the 3.39-μm transition in CH4 and the 9.33- and 10.18-μm transitions in CO_2. Appl Phys Lett 1973;22:192-9.

[45] Lancranjan I, Miclos S, Savastru D. Numerical simulation of a DFB-fiber laser sensor (I). J Opt Adv Mat 2010;12:1636-45.

[46] Lancranjan II, Savastru D, et al. Analysis of a passively q-switched Nd:YAG slab laser oscillator/amplifier system. Proc SPIE 2012;8547:854712.

[47] Savastru D, Vlase A, et al. Numerical simulation of laser techniques for art conservation – Part I – Fiber laser analysis. UPB Sci Bull-Ser. A 2011;73:167-78.

Chapter 13

Gain Saturation in Optical Fiber Laser Amplifiers

Maryam Eilchi and Parviz Parvin

Additional information is available at the end of the chapter

http://dx.doi.org/10.5772/62136

Abstract

This chapter describes the determination of amplifying parameters in rare–earth–doped optical fiber laser amplifiers. In the context of this review, the system will be analyzed under both continuous–wave (CW) and pulse conditions. A comprehensive analysis has been implemented using the set of coupled propagation rate equations based on the atomic energy structure of dopant as well as the absorption and emission cross sections.

Regarding the spectral line broadening associated with a suitable amplification relation, gain and saturation can be acquired for various pumping schemes including co– or coun-ter–propagation and the bidirectional modes.

It was shown that the gain and saturation properties are strongly dependent on the pump power, dopant concentration, and fiber length mainly due to the dominant effect of the overlapping factor of dual–clad fibers.

Keywords: Gain, Saturation, ASE parasitic noises, Broadening, Amplification relation, Filling (overlapping) factor

1. Introduction

Recently, intense activities have been devoted to characterize rare–earth–doped double–clad fiber laser amplifiers. Owing to their compactness, superb beam quality, great thermal control, and high efficiency, they demonstrate to be important light sources in medicine, modern telecommunication [1], and industries [2, 3]. Progress has been made in developing a fiber–based source in which the mean power scales up to several kilowatts.

The purpose of the fiber amplifier is to intensify a less powerful optical beam, which propagates either inside another fiber or in free space. In this way, the amplifier is "seeded" with a low–power laser beam. As it is known from the main principles of laser physics, the laser oscillator

is a device that contains an optical amplifier and puts inside a resonator to provide positive feedback.

The fiber amplifiers are fundamentally divided into core–pumped and cladding–pumped [4] according to the environment of the pump propagation. In the former classification, the amplifying light propagates in the same volume where laser gain media is located, while in the latter, the pump is coupled and propagates outside the core of the doped fiber. The same classification is valid for fiber laser oscillators. Cladding–pumped amplification technique is a hot topic in laser science and technology for high–power regimes, which is the focus of this chapter.

After the invention of the dual–clad fibers, the output powers of the doped fiber lasers have been lifted by several orders of magnitude, and immense activities have been devoted to relevant topics. These systems demonstrate several inherent features including [5] non–uniformly distributed population inversion due to end pumping as well as very high gain even in single–pass amplification by a large ratio of gain length to cross–sectional area.

In addition, both neodymium (Nd) and ytterbium (Yb) are suitable doping elements for high–power regime. Despite the fact that Nd has a relatively low laser threshold, Yb–doped double–clad fiber amplifiers present the prospects for a number of interesting applications due to the broad gain bandwidth with excellent beam quality, renowned as potential sources accessible to high–pump absorption, leading to ultimate efficiency [5, 6]. Furthermore, the benefits from efficient performance, small quantum defects, and superb spectroscopic characteristics as well as free competition processes are competitive with other rare–earth dopants. Particularly, the capability of generating high powers makes them very attractive wherein the output power up to 10/50 kW at CW single–mode/multi–mode schemes has been recently reported.

Besides, a fiber amplifier is characterized by a couple of significant coefficients, i.e., the small signal gain and the saturation quantity. In this chapter, a theoretical consideration of the CW and pulsed fiber arrays is reviewed, followed by the derived main formulas that are important for practical design. In Section 2, a brief description has been devoted to various types of laser amplifiers, operating based on the stimulated emission and optical nonlinearities. An extensive analysis is given in Section 3, regarding the set of coupled propagation rate equations accompanied with the effect of amplified spontaneous emission (ASE) parasitic noises for various pumping modes based on the atomic energy structure of dopants. The serious spectral line broadening mechanisms of gain media is described in Section 4. Eventually, the main issue of relevant amplification relation has been investigated in order to find the corresponding gain saturation parameters.

2. Various types of fiber laser amplifiers

Different kinds of laser amplifiers operate based on the stimulated emission amplification principle, or established upon optical nonlinearities, e.g., Raman or Brillouin amplifiers as well as optical parametric amplifiers (OPAs).

Raman–based fiber laser amplifier is one approach that has attracted several groups [7–13]. Here, the governed amplification process is the stimulated Raman scattering (SRS) in the fiber which causes an energy transfer from the pump to the signal. In fact, the vibrational spectrum of core material defines the Raman shift. If the wavelength of laser beam signal is known, then the optimum wavelength for pump signal can be calculated.

Commonly, Raman fiber lasers and amplifiers are not considered as ways for generating high–power narrow–linewidth lasers. For high–efficient operations, typically hundreds of meters of fiber are necessary to provide enough gain.

On the contrary, the optical gain is obtained by stimulated Brillouin scattering (SBS) in the Brillouin fiber amplifiers. It is also pumped optically, and a part of the pump power is transmitted to the signal through SBS. Physically, each pump photon creates a signal photon, and the remaining energy is used to excite an acoustic phonon [14]. Classically, the pump beam gets scattered from an acoustic wave moving through the medium at the speed of sound.

Moreover, the fiber–based OPA is a well–known technique offering a wide gain bandwidth using only a few hundred meters of fiber [15]. This phenomenon is strictly operated based on four–wave mixing (FWM) and phase–matching nonlinearities [16] in which a couple of interacting photons over two wavelength bands, also called pump and idler, usually travel collinearly through a nonlinear optical crystal, creating signal and idler output photons. During this phenomenon, pump light beam becomes weaker and amplifies the idler wave.

For birefringent nonlinear crystals, the collinear incident beams may non–collinearly scatter outside the crystal. Non–collinear OPAs were developed to have additional degree of freedom for central wavelength selection, allowing constant gain up to second order in wavelength.

3. Theory

3.1. Absorption and emission cross sections

All rare–earth–doped fiber amplifiers intensify the seed signal light through stimulated emission. The local rate equations describe the dynamics of the emission and absorption processes of doping ions within its host material by making use of its atomic energy structure as well as spectroscopic properties [17]. The relation between the emission and absorption cross sections for two energy–level gain media is determined using McCumber theory, which satisfies the following formula [18]:

$$\sigma_e(v) = \sigma_a(v)\exp\left(\frac{\varepsilon(T) - hv}{K_B T}\right) \tag{1}$$

where σ_a and σ_e are the absorption and emission cross sections, h, v, K_B, and T depict the Planck constant, frequency of light, Boltzmann constant, and the absolute temperature in

Kelvin, respectively. Moreover, $\varepsilon(T)$ shows the mean transition energy between two manifolds at temperature T. This quantity, which depends on the temperature but not on the optical frequency, can be calculated from the energies of the single Stark levels. Otherwise, one can obtain an estimation regarding the assumption that the energy levels within each Stark manifold are equidistant [19]. Alternatively, $\varepsilon(T)$ can be calibrated, e.g., using the reciprocity method or the Füchtbauer–Ladenburg equation. On the contrary,

$$\exp\left(\frac{\varepsilon(T)}{K_B T}\right) = \frac{Z_1}{Z_2}\exp\left(\frac{E_2 - E_1}{K_B T}\right) \tag{2}$$

where Z_1 and Z_2 are the partition functions, and E_1 (E_2) is the first (second) energy level. For Yb–doped gain mediums, $\varepsilon(T)$ is often close to the photon energy of the zero–phonon transition, i.e., the transition between the lowest sublevels of both manifolds. Furthermore, at $\lambda = 975$ nm, $\varepsilon(T) = h\nu$ becomes equal to 1.27 eV, which leads to $\sigma_e = \sigma_a$. An additional formula is provided by the McCumber analysis, linking the radiative lifetime and the emission cross section as follows:

$$\frac{1}{\tau} = \frac{8\pi n^2}{C^2}\int v^2 \sigma_e(v)dv \tag{3}$$

Here, τ indicates the lifetime of the excited state level of atoms, n is the refractive index of the gain media, and C denotes the speed of light in vacuum. As an example, Figure 1 demonstrates a model for the Yb^{3+} ions energy–level structure, consisting of two manifolds: a ground state $^2F_{7/2}$ (with four Stark levels labeled L_0, L_1, L_2, and L_3) and a well–separated excited state $^2F_{5/2}$ (with three Stark levels labeled U_0, U_1, and U_2), which is seated ~10000 cm^{-1} above the ground level. This is the reason why there is no excited state absorption (ESA) at either the pump or laser wavelengths [6]. This large energy gap also precludes the concentration quenching and the nonradiative decay via multiphonon emission from $^2F_{5/2}$ even in a host of high phonon energy such as silica. The energy band diagram for the lasing material is a quasi–three–level system [20].

The first transition between two Yb manifolds is the absorption and emission of the pump. Afterward, the spontaneous decay from the $^2F_{5/2}$ manifold, as well as the absorption, and stimulated emission of the signal simultaneously appear.

Absorption and stimulated emission cross sections of Yb–doped silica glass optical fibers are shown in the spectroscopic diagram in Figure 2. The absorption or fluorescence peak at 975 nm (A) represents the zero–line transition between the lowest energy levels of the ground state (L_0) and the excited state (U_0) in the manifold. The laser operation at 975 nm is a three–level process because the emission is due to a transition to the lowest Stark level. The absorption peak at shorter wavelength (B) corresponds to a transition from the ground level L_0 to either of the excited level U_1 and U_2. The absorption peak at longer wavelength (C) is attributed to

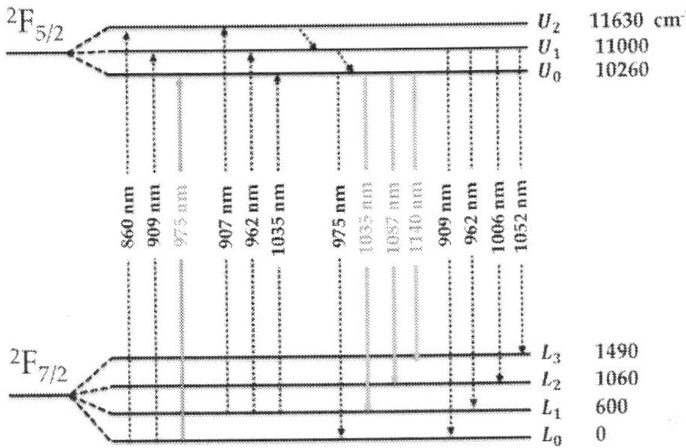

Figure 1. Energy–level diagram of Yb:silica glass

the transition from level L_1 which can produce reabsorption and leads to higher thresholds in the Yb^{3+} laser systems working around 1 µm.

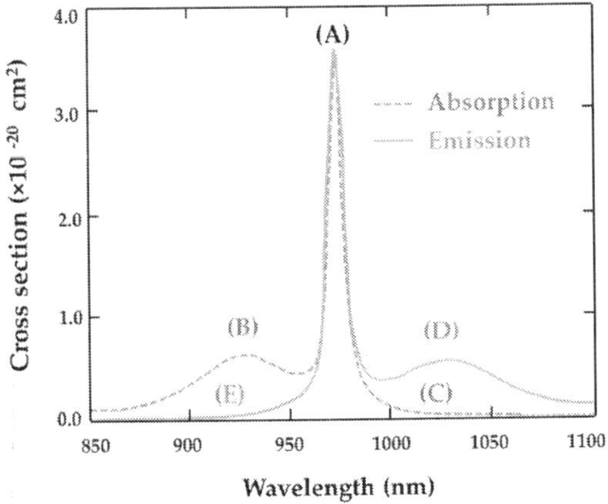

Figure 2. Typical absorption and emission spectra of Yb^{3+} ions in germanosilicate host [21]

In addition, the emission spectrum peak (D) corresponds to the energy transfer from level U_0 to the levels L_1, L_2, and L_3, while that at (E) belongs to the transition from level U_1, generating weak emissions around 900 nm. The broad absorption spectrum of Yb^{3+} ions enables the easy configuration of the pump wavelength. Depending on the requirement of the laser system, the laser signal wavelength can be configured in the range from 970 nm to 1200 nm due to the wide emission spectrum of Yb.

3.2. Rate equations

Amplifier characteristics, comprising the operating wavelength and the gain bandwidth, are determined by the dopants rather than by the host medium. However, because of the tight confinement of light provided by guided modes, fiber amplifiers can provide high optical gains at moderate pump intensity levels over relatively large spectral bandwidths, making them suitable for many telecommunications and signal–processing applications [22].

Currently, the rate equations are known as the most powerful tools to foresee the laser amplifier features associated with the non–uniform pumping along the laser length [23–31]. Here, the rate equations for the quasi–three–energy–level structures are typically introduced with the definition of the relevant parameters. Using the fourth–order Runge–Kutta method, the effect of ASE parasitic noises and gain can be considered to assess the efficiency. The performance of amplifiers also depends on the rates of radiative and nonradiative decays caused by several mechanisms related to lattice vibrations, ion–ion interactions, and cooperative up–conversion to higher levels [32].

In the case of fiber amplifier as a single–pass array without reflectors, the backward signal intensity, $I^-(z, t, \lambda)$, is equal to zero. Therefore, the set of coupled propagation rate equations accompanied with ASE in the pulsed regime can be organized as follows [33–35]:

$$\frac{\partial N_2(z,t,\lambda)}{\partial t} = \frac{\Gamma_p(\lambda)\lambda_p}{hC}\left[\sigma_{ap}(\lambda)N - \sigma_p^{tot}(\lambda)N_2(z,t,\lambda)\right]I_p^{tot}(z,t,\lambda)$$

$$+\frac{1}{hC}\cdot\sum_j\Gamma_A(\lambda_j)\lambda_j\left[\sigma_{aA}(\lambda_j)N - \sigma_A^{tot}(\lambda_j)N_2(z,t,\lambda_j)\right]I_A^{tot}(z,t,\lambda_j) \qquad (4)$$

$$+\frac{\Gamma_s(\lambda)}{hC}\int\left[\sigma_a(\lambda)N - \sigma^{tot}(\lambda)N_2(z,t,\lambda)\right]I^+(z,t,\lambda)\lambda d\lambda - \frac{N_2(z,t,\lambda)}{\tau} - C_{up}N_2^2(z,t,\lambda)$$

$$\pm\frac{dI_p^\pm(z,t,\lambda)}{dz} = \pm\Gamma_p(\lambda)\left[\left\{\sigma_p^{tot}(\lambda) - \sigma_{24}(\lambda)\right\}N_2(z,t,\lambda) - \sigma_{ap}(\lambda)N\right]I_p^\pm(z,t,\lambda) \mp \alpha_p I_p^\pm(z,t,\lambda) \qquad (5)$$

$$\frac{dI^+(z,t,\lambda)}{dz} = \Gamma_s(\lambda)\left[\sigma^{tot}(\lambda)N_2(z,t,\lambda) - \sigma_a(\lambda)N\right]I^+(z,t,\lambda) - \alpha_s I^+(z,t,\lambda) \qquad (6)$$

$$\pm\frac{dI_A^\pm(z,t,\lambda_j)}{dz} = \pm\Gamma_A(\lambda_j)\left[\sigma_A^{tot}(\lambda_j)N_2(z,t,\lambda_j) - \sigma_{aA}(\lambda_j)N\right]I_A^\pm(z,t,\lambda_j)$$

$$\mp\alpha_A I_A^\pm(z,t,\lambda_j) \pm \Gamma_A(\lambda_j)\sigma_{eA}(\lambda_j)N_2(z,t,\lambda_j)I_0(\lambda) \pm S.\alpha_{RS}(\lambda)I_A^\mp(z,t,\lambda_j) \qquad (7)$$

where the superscripts "±" describe the traveling directions of the light waves along the fiber propagation axis (z). Furthermore, p, s, and A subscript indexes define the parameters at the pump, laser, and ASE wavelengths. The notation I denotes the intensity of the pump, lasing, or ASE radiations, and $I^{tot}(z, t, \lambda)$, which stands for the total intensity, is the summation of forward, $I^{+}(z, t, \lambda)$, and backward, $I^{-}(z, t, \lambda)$, beams. Hence, some pumping modes are considered, including co–propagation, since both the signal and pump intensities are injected at $z = 0$ and propagate in the same direction or counter–propagation since the pump intensity is entered at $z = L$ (the signal and pump propagate in the opposite directions). When the pump injects in both fiber ends, it results in the bidirectional configuration.

Furthermore, $N_1(z, t, \lambda)$ and $N_2(z, t, \lambda)$ represent the population of the ground state as well as the upper lasing level of atoms, accordingly. $N(z, t, \lambda) = N_1(z, t, \lambda) + N_2(z, t, \lambda)$ describes the dopant concentration per unit volume, σ^{tot} is equal to $(\sigma_a + \sigma_e)$, and λ is the wavelength of the light. The absorption of excited state level at the pump wavelength is indicated by σ_{24} which is independent on the population of the upper level of laser.

In order to solve these equations, the cooperative up–conversion coefficient (C_{up}) must be known. According to the model mentioned in [36], it can be expressed as

$$C_{up} = \frac{4\pi}{3} \frac{R_0^6}{R^3 \tau} \tag{8}$$

where R shows the average separation between various uniformly distributed ions, R_0 demonstrates a critical interaction distance, and $\tau = 1 / A_{21}$. Neglecting the formation of ion pairs, we can approximately consider

$$R_0^6 = \frac{9}{16\pi^2} \frac{C_{up}\tau}{N_{doped\ ions}} \tag{9}$$

On the contrary, the local intensity per unit wavelength, $I_0(\lambda) \cong 2mh\, C^2(1 / A_{core})\Delta\lambda_s / \lambda_s^3$ [37], is a constant parameter accounting for the contribution of the spontaneous emission into the propagating single transverse laser mode with the spectral bandwidth $\Delta\lambda$, where the factor two arises from a couple of orthogonal polarization directions. The parameter m exhibits the number of transverse modes inside the fiber which is unity for monomode fibers.

In low–power communication fiber amplifiers, the core and cladding structures are centro-symmetric having circular cross sections. In order to remove this power limitation, an intelligent solution was proposed in 1988 by Snitzer and coworkers, namely double–clad fibers. At this model, the fiber core is off–centered or the inner cladding form is triangular, rectangular, hexagonal, elliptical, D–shaped, and so on, where the rays scatter in many accessible directions [25, 38, 39]. Hereupon, the pump light helically propagates through the clad and interacts with the active core. The pump overlapping factor (Γ_p) states that how much of the pump mode intensity actually coupled to the gain media. Assuming a spatially uniform pump distribution over the inner clad cross section, Γ_p can be estimated approximately by

$$\Gamma_p(\lambda) = \int_0^{2\pi}\int_0^a \Gamma_p(r,\varphi,\lambda)rdrd\varphi = \int_0^{2\pi}\int_0^a \frac{rdrd\varphi}{A_{clad}} \cong \frac{A_{core}}{A_{clad}} \tag{10}$$

where a stands for the core radius of active fiber and A_{core} and A_{clad} for the mode field area of the core and the multimode first cladding, respectively. On the contrary, the signal filling factor provides a figure of interaction between the optical modes at the signal radiation and the dopant ions profile. The spatial overlap integral due to the signal indicates that [25, 40, 41]

$$\Gamma_s(\lambda) = \int_0^{2\pi}\int_0^a \Gamma_s(r,\varphi,\lambda)rdrd\varphi = \int_0^{2\pi}\int_0^a \frac{\psi(r,\varphi,\lambda)rdrd\varphi}{\int_0^{2\pi}\int_0^\infty \psi(r,\varphi,\lambda)rdrd\varphi} = \frac{I^{core}}{I^{core} + I^{clad}} \tag{11}$$

For single–clad fibers, the dimensionless coefficient $\Gamma_s(\lambda)$ is equal to unity, while it is smaller than unity (0.6–0.86) in double–clad fibers. Apparently, Γ_s is a function of the pump intensity which approaches a saturated value at high pumping rates [42]. Furthermore, I^{core} and I^{clad} are defined as the intensity confined in the core and inner cladding, respectively [25], which can be theoretically expressed as follows [41]:

$$I^{core} = \frac{1}{A_{core}}\int_0^a J^2\left[U(r/a)\right]J^{-2}(U)rdr \tag{12}$$

$$I^{clad} = \frac{1}{A_{core}}\int_a^\infty K^2\left[W(r/a)\right]K^{-2}(W)rdr \tag{13}$$

Here, $\psi(r, \varphi, \lambda)$ shows the normalized transverse envelope of the modal field intensity, and $J(K)$ demonstrates the Bessel function (modified Bessel function) inside (outside) the fiber core. Parameters U and W are determined by the characteristic equations that satisfy the boundary conditions at $r = a$, namely [43]:

$$U = a\sqrt{n_{core}^2 k_0^2 - \beta^2} \tag{14}$$

$$W = a\sqrt{\beta^2 - n_{clad}^2 k_0^2} \tag{15}$$

where n_{core} and n_{clad} denote the refractive indices of the fiber core and inner cladding, respectively. The propagation constant β of any mode of fiber is limited within the interval $n_{core}k_0 \geq \beta \geq n_{clad}k_0$ which can be defined as $\beta = n_{eff}k_0$. In this equation, n_{eff} is the effective refractive index and $k_0 = 2\pi / \lambda_0$ is the wave number in free space. The intensity profile $|\psi(r, \varphi, \lambda)|^2$ resembles the linearly polarized $L P_{lm}$ modes. In terms of normalized electromagnetic field distribution of the fiber [44], we find that

$$\int_0^{2\pi}\int_0^{\infty}\psi\left(r,\varphi,\lambda\right)rdrd\varphi = Re\left[\int_0^{2\pi}\int_0^{\infty}E\left(r,\varphi,\lambda\right)\times H^*\left(r,\varphi,\lambda\right).\hat{z}\ rdrd\varphi\right] = 1 \tag{16}$$

where E and H depict the electric and magnet fields of the propagation mode, respectively, and the versor of the z propagation direction is indicated by $\hat{z}$. In the case of a Gaussian envelope approximation [45], the distribution of signal overlapping factor is as follows:

$$\Gamma_s\left(\lambda\right) \cong 1 - \exp\left[-2\left(\frac{b}{\omega_s\left(\lambda\right)}\right)^2\right] \tag{17}$$

where b shows the radius of doped area, and $\omega_s(\lambda)$ is $1/e^2$ of the intensity profile radius at the given wavelength calculated from the polynomial form [46]:

$$\omega_s\left(\lambda\right) = a\left[A_1 + A_2 / V^{1.5}\left(\lambda\right) + A_3 / V^6\left(\lambda\right)\right] \tag{18}$$

On the one hand, the normalized frequency is defined as $V = 2\pi a.NA / \lambda_s$, where NA represents the corresponding numerical aperture. For single–mode fibers, V could be smaller than 2.405 which hints that the core supports only the fundamental transverse $L P_{01}$ mode.

On the other hand, along with the signal, the spontaneous emission is also intensified. In fact, the parasitic ASE noises cause the degradation of the signal–to–noise ratio (SNR) [47–49]. It is due to the lack of reflectors such as fiber Bragg gratins (FBGs) or dichroic mirrors in the amplifier stage, while the selective wavelength oscillates in the oscillator. The evolution formula for the ASE intensities propagating in a given direction is written according to Eq. (7). The total ASE noise at point z along the fiber is the consequence of the intensity of ASE from the previous sections of the fiber accompanying the local intensity $I_0(\lambda)$ due to sponta-neous emission. The modified population inversion deals with the ASE terms too.

The positive scattering parameter $\alpha_{RS}(\lambda) = B / \lambda^4$ includes the losses due to Rayleigh scattering. The Rayleigh scattering coefficient B of pure silica is found to be ~0.63 dBμm^4/km, and S denotes the capture fraction by the fiber, such that a fraction $S.\alpha_{RS}(\lambda) = 1.3 \times 10^{-7} \text{m}^{-1}$ [50] is recaptured by the fiber. The capture fraction is defined as the proportion of the total energy scattered at z which is recaptured by the fiber in the return direction that increases with $(NA)^2$ [48, 51, 52].

In addition, Γ_A is defined as ASE filling factor which arises from the interaction between ASE modes and core dopant concentration, and α_A displays the ASE scattering loss. For simplicity, Γ_A and α_A for various ASE peaks presume to be identical to the main signal in the vicinity of λ_s. The ASE spectrum is divided into several channels with spectral width $\Delta\lambda$, whereas the signal channel width $\Delta\lambda_s$ is much smaller than the former, i.e., $\Delta\lambda_s \ll \Delta\lambda$. The ASE intensities $I_A^{\pm}(z, t, \lambda_j)$ actually propagate in the positive and the negative z –directions, both in co– and counter-propagation with the signal. Hence, the total ASE parasitic noise is stated by introducing forward and backward ASE components at any point of the fiber, namely

$$I_A^{tot}\left(z,t,\lambda_j\right) = I_A^{+}\left(z,t,\lambda_j\right) + I_A^{-}\left(z,t,\lambda_j\right) \tag{19}$$

It is worth noting that the boundary conditions for the ASE channels are given by $I_A^{+}(0) = I_A^{-}(L) = 0$.

Furthermore, Rayleigh backscattering (RBS) is another important issue that affects the performance of the signal source, unless the optical isolators are well employed [48, 49, 53]. When ASE and RBS are intense enough, these may restrict the amplifier gain leading to a drop in the efficiency in many applications. In the case of strongly pumped condition, RBS at the ASE wavelengths can be ignored mainly due to the reduction of ASE to get significantly weaker than the signal [54]. Here, we assume that

$$I^{\pm}\left(z,t,\lambda\right) \cong I_s^{\pm}\left(z,t\right)\delta\left(\lambda - \lambda_s\right) \tag{20}$$

$$\frac{dI^{\pm}\left(z,t\right)}{dz} \equiv \frac{\partial I^{\pm}\left(z\right)}{\partial z} \pm \frac{n}{C}\frac{\partial I^{\pm}\left(t\right)}{\partial t} \tag{21}$$

Thus, the rate equations at signal wavelength can be simplified as follows:

$$\begin{aligned}
\frac{\partial N_2\left(z,t\right)}{\partial t} &= \frac{\Gamma_p \lambda_p}{hC} \cdot \left[\sigma_{ap}N - \sigma_p^{tot}N_2\left(z,t\right)\right]I_p^{tot}\left(z,t\right) + \\
&+ \frac{1}{hC} \cdot \sum_j \Gamma_A \lambda_j \left[\sigma_{aA}(\lambda_j)N - \sigma_A^{tot}N_2\left(z,t\right)\right]I_A^{tot}\left(z,t\right) + \\
&+ \frac{\Gamma_s \lambda_s}{hC}\left[\sigma_{as}N - \sigma_s^{tot}N_2\left(z,t\right)\right]I_s^{+}\left(z,t\right) - \frac{N_2\left(z,t\right)}{\tau} - C_{up}N_2^2\left(z,t\right)
\end{aligned} \tag{22}$$

$$\frac{\partial I_p^{\pm}\left(z\right)}{\partial z} \pm \frac{n}{c}\frac{\partial I_p^{\pm}\left(t\right)}{\partial t} = \pm\Gamma_p\left[\left(\sigma_p^{tot} - \sigma_{24}\right)N_2\left(z,t\right) - \sigma_{ap}N\right]I_p^{\pm}\left(z,t\right) \mp \alpha_p I_p^{\pm}\left(z,t\right) \tag{23}$$

$$\frac{\partial I_s^+(z)}{\partial z} \pm \frac{n}{c}\frac{\partial I_s^+(t)}{\partial t} = \Gamma_s\left[\sigma_s^{tot}N_2(z,t)-\sigma_{as}N\right]I_s^+(z,t)-\alpha_s I_s^+(z,t) \tag{24}$$

$$\frac{\partial I_A^\pm(z)}{\partial z} \pm \frac{n}{c}\frac{\partial I_A^\pm(t)}{\partial t} = \pm\Gamma_A\left[\sigma_A^{tot}N_2(z,t)-\sigma_{aA}N\right]I_A^\pm(z,t)$$
$$\mp\alpha_A I_A^\pm(z,t)\pm\Gamma_A\sigma_{eA}N_2(z,t)I_0(\lambda)\pm S.\alpha_{RS}(\lambda)I_A^\mp(z,t) \tag{25}$$

However, in the case of CW fiber amplifiers, disregarding ESA and cooperative up–conversion, the set of first–order coupled nonlinear time–independent steady–state ($\partial/\partial t \to 0$) differential equations of an end–pumped configuration is realized by [28, 34, 37]

$$\frac{N_2(z)}{N}=\frac{\Gamma_p\lambda_p\sigma_{ap}P_p^{tot}(z)+\sum_j\Gamma_A\lambda_j\sigma_{aA}(\lambda_j)P_A^{tot}(z)+\Gamma_s\lambda_s\sigma_{as}P_s^+(z)}{\Gamma_p\lambda_p\sigma_p^{tot}P_p^{tot}(z)+\dfrac{hCA_{core}}{\tau}+\sum_j\Gamma_A\lambda_j\sigma_A^{tot}P_A^{tot}(z)+\Gamma_s\lambda_s\sigma_s^{tot}P_s^+(z)} \tag{26}$$

$$\pm\frac{dP_p^\pm(z)}{dz}=\pm\Gamma_p\left[\left(\sigma_p^{tot}-\sigma_{24}\right)N_2(z)-\sigma_{ap}N\right]P_p^\pm(z)\mp\alpha_p P_p^\pm(z) \tag{27}$$

$$\frac{dP_s^+(z)}{dz}=\Gamma_s(\lambda)\left[\sigma_s^{tot}N_2(z)-\sigma_{as}N\right]P_s^+(z)\;\alpha_s P_s^+(z) \tag{28}$$

$$\pm\frac{dP_A^\pm(z)}{dz}=\pm\Gamma_A\left[\sigma_A^{tot}N_2(z)-\sigma_{aA}N\right]P_A^\pm(z)$$
$$\mp\alpha_A P_A^\pm(z)\pm\Gamma_A\sigma_{eA}N_2(z)P_0(\lambda)\pm S.\alpha_{RS}P_A^\mp(z) \tag{29}$$

It is worth mentioning that there are no significant nonlinear optical effects such as SBS and SRS as well as thermal damages up to 50 W–CW single–mode pump powers [5]. In addition, the dipole–dipole interaction, clustering, and quenching are serious only in highly doping elements.

3.3. Gain saturation

Tremendous efforts were made to develop the fiber amplifiers during the recent decade. Indeed, the amplifier is characterized by very important and unique gain and saturation parameters [55–60] that are often used to analyze and compare different gain mediums in certain applications.

The advantages of operating in the gain saturation regime are mainly summarized as follows [4]:

1. Small fluctuations in the input signal do not reflect the same extent in the output amplified signal.

2. The fiber amplifier, which has multiple spectrally close input signals with varied intensity, may work as a gain equalizer because smaller input signal powers have higher gain (through less saturation), and higher input powers have lower gain (due to a higher degree of saturation).

3. A saturated optical amplifier demonstrates a high–energy extraction efficiency; therefore, the overall efficiency of the system is high.

According to theoretical analysis, high–input optical fluence does not increase indefinitely in an amplifier but rather saturates, leading to amplifier gain reduction. Obviously, even from the general consideration of energy conservation, the gain of the amplifier has to saturate, because one cannot extract more power from the amplifier than it was excited (pumped) with.

The emission and absorption cross sections are used to introduce the gain coefficient. According to Eq. (28), the gain for signal radiation after simplification can be defined as follows:

$$\gamma_s(z) = \Gamma_s \left[\sigma_{es} N_2(z) - \sigma_{as} N_1(z) \right] \tag{30}$$

In a fiber amplifier, it depends on the distance z from its input end. By substituting Eq. (26) into Eq. (30) and ignoring ASE noises, the gain is rearranged in the following formalism:

$$\gamma_s(z) = \frac{\left(hCA_{core} \right)^{-1} \Gamma_s \Gamma_p \lambda_p N\tau \left(\sigma_{ap}\sigma_{es} - \sigma_{ep}\sigma_{as} \right) P_p^{tot}(z) - N\Gamma_s \sigma_{as}}{\left(hCA_{core} \right)^{-1} \Gamma_p \lambda_p \tau \sigma_p^{tot} P_p^{tot}(z) + 1 + \left(hCA_{core} \right)^{-1} \Gamma_s \lambda_s \tau \sigma_s^{tot} P_s^+(z)} \tag{31}$$

Due to high signal power, which saturates the gain, the first term of the dominator is small enough with respect to others. Assuming $\sigma_{as} \ll \sigma_{es}$ in the numerator, $\gamma_s(z)$ is given by [5, 26]

$$\gamma_s(z) = \frac{\left(N\Gamma_p\sigma_{ap} \right)\left(\lambda_p / \lambda_s \right)\left[P_p^{tot}(z) / P_s^{sat}(z) \right] - N\Gamma_s \sigma_{as}}{1 + P_s^+(z) / P_s^{sat}(z)} \tag{32}$$

where the signal saturation power is introduced as follows:

$$P_s^{sat}(z) = \frac{hCA_{core}}{\Gamma_s \lambda_s \left(\sigma_{as} + \sigma_{es} \right)\tau} \tag{33}$$

It decreases the small–signal gain by a factor of one–half equivalent to a 3 dB reduction in gain [49]. It is worth mentioning that not only $P_s^{sat}(z)$ relies on the photon energy of the seed signal, the mode field area, the upper state lifetime, the absorption, and stimulated emission cross sections of the gain media, but it also strongly correlates to the signal overlapping factor. Combining Eqs (11) and (33), we imply that $P_s^{sat}(z)$ is a linear function of pump and signal powers in CW regime, namely

$$P_s^{sat}\left(z\right) = \frac{hCA_{core}}{\lambda_s\left(\sigma_{as} + \sigma_{es}\right)\tau}\frac{P^{core} + P^{clad}}{P^{core}} \tag{34}$$

From another point of view, the gain coefficient in the steady–state condition can be explained by

$$\gamma_s\left(z\right) = \frac{\gamma_0\left(z\right)}{1 + P_s^{+}\left(z\right)/P_s^{sat}\left(z\right)} \tag{35}$$

where $\gamma_0(z)$ is the small–signal gain. The role of the term $P_s^{+}(z)/P_s^{sat}(z)$ is to reduce the gain as signal power increases in the fiber. This phenomenon is common to all amplifiers and is referred to as gain saturation. If the amplifier operates at power levels such that $P_s^{+}(z)/P_s^{sat}(z) \ll 1$ for all z, the amplifier is said to operate in the unsaturated regime. Moreover, by comparing Eqs (32) and (35), the theoretical term for the maximum value of the gain at distance z is given by [5, 26]

$$\gamma_0\left(z\right) = N\Gamma_p\sigma_{ap}\left(\lambda_p / \lambda_s\right)\left[P_p^{tot}\left(z\right)/P_s^{sat}\left(z\right)\right] - N\Gamma_s\sigma_{as} \tag{36}$$

Let us assume a doped dual–clad fiber amplifier with length L , such that pumping light is injected into the inner clad, and the laser seed is transmitted into the active medium. In strongly pumped single–pass fiber amplifier, $N_2(z)$ is presumed to be much smaller than N over a significant part of the cavity length, such that Eq. (27) becomes analytically integrable to yield the axial backward and forward propagation pump powers along the fiber core according to the Lambert–Beer (L–B) law:

$$P_p^{tot}\left(z\right) = P_p^{+}\left(z\right) + P_p^{-}\left(z\right) = P_p^{+}\left(0\right)e^{-\alpha z} + P_p^{-}\left(L\right)e^{\alpha(z-L)} \tag{37}$$

where $P_p^{+}(0)$ and $P_p^{-}(L)$ are the pump powers injected in both fiber ends at $z=0$ and $z=L$, respectively. In fact, the total attenuation α includes the scattering and pump absorption losses, namely

$$\alpha = \alpha_p + N\Gamma_p\sigma_{ap} \tag{38}$$

Here, α_p depicts the absorption coefficient of the core at the pump wavelength due to scattering loss. Taking average over the fiber length, the effective small–signal gain coefficient per unit length is written in terms of α consequently [56]:

$$\gamma_0 = \frac{1}{L}\int_0^L \gamma_0(z)\,dz \tag{39}$$

Hence

$$\gamma_0 = (1/L)(\lambda_p / \lambda_s)\left(1 - \frac{\alpha_p}{\alpha}\right)\left(1 - e^{-\alpha L}\right)\left(\frac{P_p^+(0) + P_p^-(L)}{P_s^{sat}}\right) - N\Gamma_s\sigma_{as} \tag{40}$$

In the general case of the varied inversion along the gain fiber, the absolute gain factor should be integrated along the entire fiber length [4]:

$$g = \exp\left(\int_0^L \gamma_0(z)\,dz\right) \tag{41}$$

Accordingly, the gain and saturation properties are strongly dependent on the pump power, dopant concentration, fiber length, as well as the pumping modes (co–/counter–propagation and bidirectional pumping) mainly due to the dominant effect of the filling factors [26, 61, 62].

In return, if optical pulses are amplified, the gain coefficient associates with the elapsed time t due to $N_2(z, t)$. Therefore, the gain can be expressed as follows:

$$\gamma_s(z) = \frac{\gamma_0(z)}{1 + I_s^+(z)/I_s^{sat}(z)} \tag{42}$$

where $I_s^+(z)$ is the amplified light intensity, and $I_s^{sat}(z)$ is the so–called gain saturation parameter, defined as an intensity of the amplifying beam in active medium at which the small–signal (unsaturated) gain coefficient is reduced by 50% (i.e., half). In turn, the gain saturation parameter with good approximation can be clarified as the following general formula:

$$I_s^{sat}(z) = \frac{hC}{\Gamma_s\lambda_s(\sigma_{as} + \sigma_{es})\tau} \tag{43}$$

4. The spectral line broadening

The nature of spectral line broadening plays a very important role in the performance of the fiber laser amplifiers. The stimulated emission cross section is conventionally rewritten by

$$\sigma_e(v) = \frac{\lambda^2}{8\pi n^2 \tau} g(v)$$
(44)

where $g(v)$ indicates the spectral line shape. Using Eqs (33), (40), and (43), one can realize that the amplifying parameters, i.e., the small–signal gain and the saturation intensity/power are believed to be correlated to the line shape and the consequent broadened linewidth.

The spectral line broadening at room temperature (300 K) smooths the overall line shape, which becomes resolved only at low temperatures [4]. With a temperature decrease, Stark level structure becomes more and more evident and determines the line shape characteristic profile.

In the case of rare–earth–doped fiber oscillator and amplifier, different types of broadenings exist. These include strong homogeneous as well as the inhomogeneous broadenings [56, 63]. The designation of whether the transition line shape is homogeneously or inhomogeneously broadened is based on whether the lines are from the same type of centers or from different sorts.

Usually, a homogeneous broadening is explained by the random perturbation of a similar type of optical centers, while that from different kinds is responsible for inhomogeneously broadened lines [4]. Such kinds of interaction result in the shortening of the excited state lifetime of the optical center.

Therefore, the homogeneous broadening is inherent to each atom in a medium as depicted in Figure 3(a). It can be spontaneous broadening, Stark broadening, collision broadening, thermal and dipolar, and so on. The spontaneous broadening is a primary broadening yielded from the lifetime of a pumping level and the frequency uncertainty. The Stark broadening is due to the degeneration of energy level caused by external electric field, making the lifetime very short and the spectral width very large. Moreover, the collision broadening is dominant for gas lasers as excimer lasers operate with high gas pressure discharge.

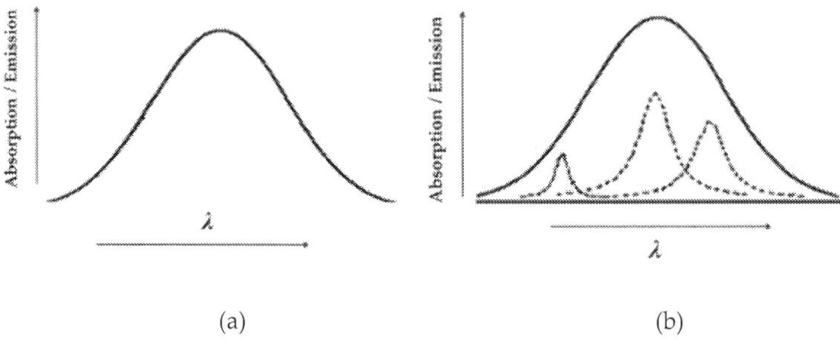

Figure 3. (a) Homogeneous and (b) inhomogeneous broadenings

Conversely, the inhomogeneous broadening of the optical center's line shape during a transition between energy levels originates from a local site–to–site variation in the optical center's surrounding field in the lattice environment. The strength and symmetry of the field around the rare–earth ion determine the spectral properties of the optical transitions, as well as the transition strength. According to Figure 3(b), in the case of inhomogeneous broadening, the overall shape of the spectral line is a superposition of all individual, homogeneously broadened lines corresponding to different types of the optical centers. Hence, this is a phenomenon that various frequencies are generated by a variety of influences, and the spectrum is broadened.

The inhomogeneous broadening can be a strain caused by lattice defect or inhomogeneity of magnetic or electric field (crystalline field) for solid–state lasers and Doppler broadening for gas lasers.

In a crystal of solid–state laser, the crystalline field is not uniform since the crystal is imperfect. As a result, a large inhomogeneous broadening is yielded. In a glass doped with rare–earth ions, the inhomogeneous width is relatively largely broadened compared with the crystalline solid–state laser. The amorphous nature of the glass, random distribution of dopants, and host material imperfection lead to site–to–site variations of the local electric field, producing a degree of inhomogeneous broadening in the system which results in spectral hole burning (SHB).

Indeed, the optical transition line shape in the case of homogeneous broadening is described by a Lorentzian function given by

$$g_{\mathrm{L}}(v) = \frac{1}{2\pi^2} \frac{(\Delta v_{\mathrm{h}}/2)}{(v_{\mathrm{h}} - v_0)^2 + (\Delta v_{\mathrm{h}}/2)^2} \tag{45}$$

where v_0 is the central frequency of the optical transition, and Δv_{h} represents the transition spectral line full width at half maximum (FWHM) as well as $\int_{-\infty}^{+\infty} g_{\mathrm{L}}(v, v_0)dv = 1$. Figure 4 shows Lorentzian/Gaussian function in the case of homogeneous/inhomogeneous broadening.

Subsequently, the line shape of the inhomogeneously broadened optical transition corresponds to the Gaussian line profile, which can be expressed as the following function of frequency:

$$g_{\mathrm{G}}(v) = \frac{2}{\Delta v_{\mathrm{inh}}} \sqrt{\frac{\ln 2}{\pi}} \exp\left[-\ln 2 \left(\frac{v_{\mathrm{inh}} - v_0}{\Delta v_{\mathrm{inh}}/2}\right)^2\right] \frac{(\Delta v_{\mathrm{h}}/2)}{(v - v_0)^2 + (\Delta v_{\mathrm{h}}/2)^2} \tag{46}$$

where Δv_{inh} is the FWHM of the inhomogeneously broadened line. Concluded by the optical center surrounding fields, the inhomogeneous linewidth is insensitive to the temperature of the host material.

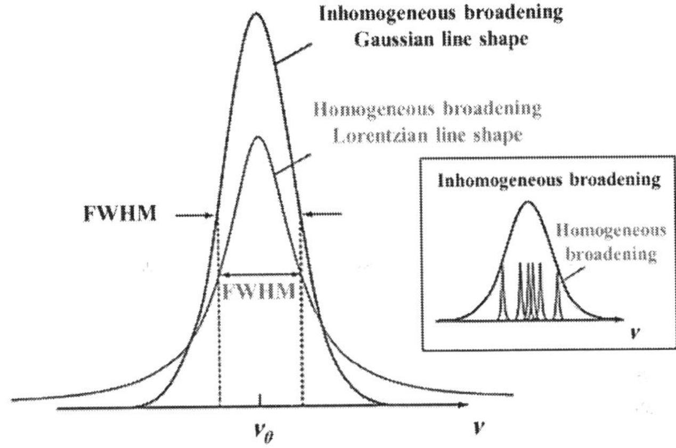

Figure 4. Lorentzian–shaped and Gaussian–shaped spectra [64]

5. Amplification relations

The state–of–the–art fiber amplifiers are significantly related to the determination of amplifying parameters (gain and saturation) which are not explicitly introduced into the rate equations; however, these can be obtained by using a suitable amplification relation. It is applicable for the homogeneous line broadening in Yb:silica fiber amplifiers [63] as well as the Er–doped silica fiber lasers with inhomogeneous line broadening.

In general, the laser amplification consists of a couple of distinct transient (short pulse) and steady–state (long pulse) regimes. The gain and saturation values can be estimated by numerically solving the amplification relations performing the best fitting based on least square method (LSM).

5.1. Pulsed regime

In the transient form of the amplification relation, the input pulse is considerably shorter than the fluorescence lifetime of the medium, such as Q–switched or mode–locked pulses with inhomogeneous gain media. The propagation rate equations are the best procedure to simulate the amplified signal energy released from the stimulated emission. This process depends on the energy stored in the upper laser level prior to the synchronization situation of the pumped atoms and incident seed signal. Furthermore, the time–dependent photon transport equation can be utilized to evaluate how an inverted population affects the distribution of a pulse traversing through the amplifier. Hopf [65] and Frontz–Nodvik [66] solved the nonlinear equations for various types of input pulse shapes to obtain the inverted electron population and the photon flux, while the input signal to the amplifier was taken as a square pulse of duration τ. Finally, the transient amplification relation is given by

$$E_{\text{out}} = E_s^{\text{sat}}\ln\left\{1+\left[\exp\left(\frac{E_{\text{in}}}{E_s^{\text{sat}}}\right)-1\right]\exp(\gamma_0 L)\right\}$$

(47)

where E_{in} and E_{out} denote the input and output signal energy and E_s^{sat} the saturation energy derived from the instantaneous rate equations of the optical pulse, respectively.

For the pulses whose shapes are sufficiently near the eigen–function for the propagation in the amplifying medium, the change in the average temporal width of the pulse can be reasonably expected to be negligible. In such cases, the approximation $E_n = \tau \times I_n$ is justified where n can be designated as input, output, or saturation. With this substitution, Eq. (47) can be rewritten as [59]

$$I_{\text{out}} = I_s^{\text{sat}}\ln\left\{1+\left[\exp\left(\frac{I_{\text{in}}}{I_s^{\text{sat}}}\right)-1\right]\exp(\gamma_0 L)\right\}$$

(48)

where I_s^{sat}, is the effective saturation intensity:

$$I_s^{\text{sat}} = \frac{t_2}{2\tau}I_s^{\text{sat}}$$

(49)

Here, t_2 is the upper level lifetime of the transition.

5.2. CW regime

For the seed signal, which is long compared to the fluorescent time including the free running oscillator or the CW seed laser [5, 67, 68] under homogeneous line broadening, a Hargrove [69] steady–state gain characterizes the amplification mechanism as follows [5, 6]:

$$-\ln\left(P_{\text{out}}/P_{\text{in}}\right) = \frac{P_{\text{out}}-P_{\text{in}}}{P_s^{\text{sat}}} - \gamma_0 L$$

(50)

Here, P_{in} (P_{out}) represents the input (output) signal power to the amplifier. This equation can be simplified using mathematical derivations such that [70]

$$P_{\text{out}} = P_s^{\text{sat}} \times W\left[\frac{P_{\text{in}}\exp\left(\gamma_0 L+\dfrac{P_{\text{in}}}{P_s^{\text{sat}}}\right)}{P_s^{\text{sat}}}\right]$$

(51)

where W shows the Lambert W –Function to be defined by the series expansion [71]:

$$W(x) = \sum_{n=1}^{\infty} \frac{(-1)^{n-1} n^{n-2}}{(n-1)!} x^n \tag{52}$$

In addition, in the limiting cases

$$P_{out} = P_{in} \exp(\gamma_0 L) \qquad P_{in} \ll P_s^{sat} \tag{53}$$

$$P_{out} = P_{in} + P_s^{sat} \gamma_0 L \qquad P_{in} \gg P_s^{sat} \tag{54}$$

These relations indicate that for low–input powers, the overall gain ($G = P_{out}/P_{in}$) is nonlinearly proportional to gain–length product, and the gain coefficient becomes equal to the small–signal gain. In contrast, the gain reduces to the saturated value for the high–input signals.

A straightforward integration shows that the CW gain of the amplifier is considered by

$$G(v) = \exp\left[\int_0^L g(z,v)dz\right] \tag{55}$$

The origin of gain saturation lies in the power dependence of the gain coefficient. Since $g(z, v)$ is diminished when P_{in} becomes comparable to P_s^{sat}, the amplification factor G is also expected to decrease.

Author details

Maryam Eilchi[1*] and Parviz Parvin[2]

1 Laser and Optics Research Institute, Nuclear Science and Technology Research Institute, Tehran, Iran

2 Photonic Engineering Group, Department of Physics, Amirkabir University of Technology, Tehran, Iran

References

[1] F. J. Duarte. Tunable laser applications. 2nd ed. CRC Press. Taylor & Francis group, LLC; 2008. 480p.

[2] C. Labaune, D. Hulin, A. Galvanauskas, and G. A. Mourou. On the feasibility of a fiber-based inertial fusion laser driver. Optics Communications. 2008; 281(15–16): 4075–4080.

[3] G. Mourou, B. Brocklesby, T. Tajima, and J. Limpert. The future is fibre accelerators. Nature Photonics. 2013; 7: 258–261.

[4] Valerii (Vartan) Ter-Mikirtychev. Fundamentals of Fiber Lasers and Fiber Amplifiers. Switzerland: Springer International Publishing; 2014.

[5] N. S. Kim, T. Hamada, M. Prabhu, C. Li, J. Song, K. Ueda, A. Liu, and H. J. Kong. Numerical analysis and experimental results of output performance for Nd-doped double-clad fiber lasers. Optics communications. 2000; 180: 329–337.

[6] H. M. Pask, R. J. Carman, D. C. Hanna, A. C. Tropper, C. J. Mackechnie, P. R. Barber, and J. M. Dawes. Ytterbium-doped silica fiber lasers: versatile sources for the 1-1.2 µm region. IEEE Journal of Selected Topics in Quantum Electronics. 1995; 1: 2–13.

[7] Y. Feng, S. Huang, A. Shirakawa, and K. Ueda. 589nm light source based on Raman fiber laser. Japanese Journal of Applied Physics.2004; 43(6A): L722–L724.

[8] D. Georgiev, V. P. Gapontsev, A. G. Dronov, M. Y. Vyatkin, A. B. Rulkov, S. V. Popov, and J. R. Taylor. Watts-level frequency doubling of a narrow line linearly polarized Raman fiber laser to 589nm. Optics Express. 2005; 13(18): 6772–6776.

[9] D. Bonaccini Calia, W. Hackenberg, S. Chernikov, Y. Feng, and L. Taylor. AFIRE: fiber Raman laser for laser guide star adaptive optics. In: Proceedings of SPIE 6272, 62721M; 2006.

[10] P. Dupriez, C. Farrell, M. Ibsen, J. K. Sahu, J. Kim, C. Codemard, Y. Jeong, D. J. Richardson, and J. Nilsson. 1W average power at 589 nm from a frequency doubled pulsed Raman fiber MOPA system. In: Proceedings of SPIE 6102, 61021G; 2006.

[11] Y. Feng, L. Taylor, and D. Bonaccini Calia. Multiwatts narrow linewidth fiber Raman amplifiers. Optics Express. 2008; 16(15): 10927–10932.

[12] Y. Feng, L. Taylor, and D. Bonaccini Calia. 20W CW, 4MHz linewidth Raman fiber amplifier with SHG to 589nm. In: Photonics West; Proceedings of SPIE 7195, xvii–xviii; 2009.

[13] L. Taylor, Y. Feng, and D. B. Calia. High power narrowband 589 nm frequency doubled fibre laser source. Optics Express. 2009; 17(17):14687–14693.

[14] J. H. Franz, V. K. Jain. Optical Communications: Components and Systems. CRC Press. Narosa Publishing House; 2000. 717p.

[15] J. Hansryd and P. A. Andrekson. Broad-band continuous-wave-pumped fiber optical parametric amplifier with 49-dB gain and wavelength-conversion efficiency. IEEE Photonics Technology Letters. 2001; 13: 194–196.

[16] J. Hansryd, P. A. Andrekson, M. Westlund, J. Li, and P. O. Hedekvist. Fiber-based optical parametric amplifiers and their applications. IEEE Journal of Selected Topics in Quantum Electronics. 2002; 8(3).

[17] O. G. Okhotnikov. Fiber Lasers. Wiley-VCH Verlag GmbH & Co. KGaA; 2012. 294p.

[18] D. E. McCumber. Theory of phonon-terminated optical masers. Physical Review. 1964; 134: A299–A306.

[19] W. J. Miniscalco and R. S. Quimby. General procedure for the analysis of Er^{3+} cross-section. Optics Letters. 1991; 16: 258–260.

[20] H. M. Pask, R. J. Carman, D. C. Hanna, A. C. Tropper, C. J. Mackechnie, P. R. Barber and J. M. Dawes. Ytterbium–doped silica fiber lasers: Versatile sources for the 1~1.2μm region. IEEE Journal of Selected Topics in Quantum Electronics.1995; 1: 2–13.

[21] M. R. A. Moghaddam, S. W. Harun and H. Ahmad. Comparison between Analytical Solution and Experimental Setup of a Short Long Ytterbium Doped Fiber Laser. Optics and Photonics Journal.2012; 2(2): 65–72.

[22] M. Premaratne and G. P. Agrawal. Light Propagation in Gain Media, Optical Amplifiers. Cambridge: Cambridge University Press; 2011.

[23] L. Pan, I. Utkin, and R. Fedosejevs. Experiment and numerical modeling of high-power passively Q-switched ytterbium-doped double-clad fiber lasers. IEEE Journal of Quantum Electronics. 2010; 46: 68–75.

[24] Z. Duan, L. Zhang, and J. Chen. Analytical solutions to rate equations including losses describing threshold pumped fiber lasers. Optik-International Journal for Light and Electron Optics. 2008; 119: 395–399.

[25] M. Gong, Y. Yuan, C. Li, P. Yan, H. Zhang, and S. Liao. Numerical modeling of transverse mode competition in strongly pumped multimode fiber lasers and amplifiers. Optics Express. 2007; 15: 3236–3246.

[26] M. Eichhorn. Numerical modeling of Tm-doped double-clad fluoride fiber amplifiers. IEEE Journal of Quantum Electronics. 2005; 41: 1574–1581.

[27] G. Hu, C. Shan, X. Deng, J. Zhang, Y. Pan, and L. Wang. Threshold characteristics of linear cavity Yb^{3+}-doped double-clad fiber laser. Optics & Laser Technology. 2005; 37: 3–7.

[28] L. Xiao, P. Yan, M. Gong, W. Wei, and P. Ou. An approximate analytic solution of strongly pumped Yb-doped double-clad fiber lasers without neglecting the scattering loss. Optics Communications. 2004; 230: 401–410.

[29] C. Jiang, W. Hu, and Q. Zeng. Numerical analysis of concentration quenching model of Er^{3+}-doped phosphate fiber amplifier. IEEE Journal of Quantum Electronics. 2003; 39: 1266–1271.

[30] I. Kelson and A. Hardy. Optimization of strongly pumped fiber lasers. Journal of Lightwave Technology.1999; 17: 891–897.

[31] I. Kelson and A. A. Hardy. Strongly pumped fiber lasers. IEEE Journal of Quantum Electronics. 1998; 34: 1570–1577.

[32] M. J. F. Digonnet. Rare-Earth-Doped Fiber Lasers and Amplifiers. 2nd ed. CRC Press. Marcel Dekker, Inc; 2001. 798p.

[33] R. Oron and A. Hardy. Approximate analytical expressions for signal amplification in strongly pumped fiber amplifiers. IEEE Proceedings of Optoelectronics. 1998; 145(2): 138–140.

[34] M. Ilchi-Ghazaani and P. Parvin. Impact of cavity loss on the performance of a single-mode Yb:silica MOFPA array. Optics & Laser Technology. 2015; 65: 94–105.

[35] N. A. Brilliant. Ytterbium-Doped, Dual-Clad Fiber Amplifiers. PhD Thesis, University of Mexico. 2001.

[36] M. Federighi and F. Di Pasquale. The effect of pair-induced energy transfer on the performance of silica fiber amplifiers with high Er^{3+}/Yb^{3+} concentrations. IEEE Photonics Technology Letters. 1995; 7(3): 303–305.

[37] A. Hardy and R. Oron. Signal Amplification in strongly pumped fiber amplifiers. IEEE Journal of Quantum Electronics. 1997; 33: 307–313.

[38] Y. Yuan, M. Gong, C. Li, and P. Yan. Theoretical and experimental study on transverse mode competition in a partial-coiled multimode fiber laser. Laser Physics. 2008; 18: 52–57.

[39] M. Iilchi-Ghazaani, P. Parvin, and S. Mohammadian. Determination of amplifying parameters of LMA Yb:silica fiber amplifier. In: 2013; Czech Republic. Proceedings of SPIE 8781, Integrated Optics: Physics and Simulations. 878116; 2013.

[40] P. C. Becker, A. Lidgard, J. R. Simpson, and N. A. Olsson. Erbium-doped fiber amplifier pumped in the 950-1000 nm region. IEEE Photonics Technology Letters. 1990; 2(1): 35–37.

[41] Z. Duan, J. Chen, Z. Gao, and Y. Huang. Theoretical analysis of multimode fiber amplifiers and lasers containing spatial filters. Optik-International Journal for Light and Electron Optics. 2011; 122(24): 2211–2215.

[42] M. A. Rebolledo and S. Jarabo. Erbium-doped silica fiber modeling with overlapping factors. Applied Optics. 1994; 33(24): 5585–5593.

[43] D. Gloge. Weakly Guiding Fibers. Applied Optics. 1971; 10(10): 2252–2258.

[44] A. D'Orazio, M. De Sario, L. Mescia, V. Petruzzelli, and F. Prudenzano. Refinement of Er^{3+}-doped hole-assisted optical fiber amplifier. Optics Express. 2005; 13: 9970–9981.

[45] T. J. Whitley and R. Wyatt. Alternative Gaussian spot size polynomial for use with doped fiber amplifiers. IEEE Journal of Photonics Technology Letters. 1993; 5(11): 1325–1327.

[46] P. Myslinski and J. Chrostowski. Gaussian-mode radius polynomials for modeling doped fiber amplifiers and lasers. Microwave and Optical Technology Letters. 1996; 11: 61–64.

[47] E. Desurvire and J. R. Simpson. Amplification of spontaneous emission in erbium-doped single-mode fibers. IEEE Journal of Lightwave Technology. 1989; 7: 835–845.

[48] M. N. Zervas and R. I. Laming. Rayleigh scattering effect on the gain efficiency and noise of erbium-doped fiber amplifiers. IEEE Journal of Quantum Electronics. 1995; 31: 468–471.

[49] A. A. Hardy and R. Oron. Amplified spontaneous emission and Rayleigh backscattering in strongly pumped fiber amplifiers. Journal of Lightwave Technology. 1998; 16: 1865–1873.

[50] Q. Han, J. Ning, H. Zhang, and Z. Chen. Novel shooting algorithm for highly efficient analysis of fiber Raman amplifiers. Journal of Lightwave Technology. 2006; 24: 1946–1952.

[51] A. H. Hartog and M. P. Gold. On the theory of backscattering in single-mode optical fibers. Journal of Lightwave Technology. 1984; LT-2: 76–82.

[52] E. G. Neumann. Single Mode Fibers Fundamentals. New York: Springer-Verlag; 1988.

[53] R. Oron and A. A. Hardy. Rayleigh backscattering and amplified spontaneous emission in high-power ytterbium-doped fiber amplifiers. JOSA B. 1999; 16: 695–701.

[54] P. C. Becker, N. A. Olsson, and J. R. Simpson. Erbium-Doped Fiber Amplifiers: Fundamentals and Technology. New York: Lucent Technologies; 1999.

[55] S. Mohammadian, P. Parvin, M. Ilchi-Ghazaani, R. Poozesh, and K. Hejaz. Measurement of gain and saturation parameters of a single-mode Yb:silica fiber amplifier. Optical Fiber Technology. 2013; 19: 446–455.

[56] P. Parvin, M. Ilchi-Ghazaani, A. Bananej, and Z. Lali-Dastjerdi. Small signal gain and saturation intensity of a Yb: Silica fiber MOPA system. Optics & Laser Technology. 2009; 41(7): 885–891.

[57] S. Behrouzinia, R. Sadighi-Bonabi, P. Parvin, and M. Zand. Temperature dependence of the amplifying parameters of a copper vapor laser. Laser Physics - Lawrence. 2004; 14(8): 1050–1053.

[58] S. Behrouzinia, R. Sadighi, and P. Parvin. Pressure dependence of the small-signal gain and saturation intensity of a copper vapor laser. Applied Optics. 2003; 42(6): 1013–1018.

[59] P. Parvin, M. S. Zaeferani, K. Mirabbaszadeh, and R. Sadighi. Measurement of the small-signal gain and saturation intensity of a XeF discharge laser. Applied Optics. 1997; 36(6): 1139–1142.

[60] A. Saliminia, P. Parvin, A. Zare, and R. Sadighi. The small signal gain and the saturation intensity measurement of the nitrogen-ion laser. Optics & Laser Technology. 1996; 28(3): 207–211.

[61] A. Escuer, S. Jarabo, and J.M. Alvarez. Analysis of theoretical models for erbium-doped silica fibre lasers. Optics Communications. 2001; 187: 107–123.

[62] M. A. Rebolledo and S. Jarabo. Erbium-doped silica fiber modelling with overlapping factors. Applied Optics. 1994; 33: 5585–5593.

[63] R. Paschotta, J. Nilsson, A. C. Tropper, and D. C. Hanna. Ytterbium-doped fiber amplifiers. IEEE Journal of Quantum Electronics. 1997; 33(7): 1049–1056.

[64] Laser Tutorial. Kokyo Inc; 2014–2015. Available from: http://laser.photoniction.com/tutorial/

[65] F. Hopf. High Energy Lasers and Their Applications. Addison: Wesley; 1974.

[66] L. M. Frantz, J. S. Nodvik. Theory of pulse propagation in a laser amplifier. Journal of Applied Physics.1963; 34: 2346–2349.

[67] A. A. Hardy, R. Oron. Signal amplification in strongly pumped fiber amplifiers. IEEE Journal of Quantum Electronics.1997; 33: 307–313.

[68] I. Kelson, A. Hardy. Strongly pumped fiber lasers. IEEE Journal of Quantum Electronics.1998; 34: 1570–1577.

[69] W. Koechner. Solid-State Laser Engineering. Springer-Verlag New York; 2006. 748p.

[70] O. Mahran. Gain and noise figure of ytterbium doped lead fluoroborate optical fiber Amplifiers. Australian Journal of Basic and Applied Sciences. 2010; 4(8): 4020–4028.

[71] R. M. Corless, G. H. Gonnet, D. E. G. Hare, D. J. Jeffrey, and D. E. Knuth. On the Lambert W function. Advances in Computational Mathematics. 1996; 5(1): 329–359.

Chapter 14

Heat Generation and Removal in Fiber Lasers

Maryam Eilchi and Parviz Parvin

Additional information is available at the end of the chapter

http://dx.doi.org/10.5772/62102

Abstract

The present chapter looks at heat generation and heat removal in fiber lasers, particularly if high–power or high–energy operation is required. In the context of the review, for the purpose of calculation of heat dissipation for different parts of the active gain media and providing effective cooling procedures, thermal loading as well as longitudinal and transverse temperature profiles of dual–clad fibers are comprehensively investigated both inside and outside of the doped fiber core. Considering numerical analysis, the heat deposited in the fiber due to pump and laser power is determined via the steady–state equations and also transient conductive, convective, as well as radiative heat transfer equations. Besides this, important features regarding how to mitigate thermal effects are stated. On the other hand, we will show that chilling mechanisms are very efficient methods for dissipating heat which is extensively adopted in high–power regimes. Finally, the concept of a cryogenic laser is discussed after propounding a novel cooling system, namely the dry–ice chiller.

Keywords: Thermal effects, heat dissipation, temperature treatment, conductive, convective and radiative heat transfer equations, cryogenic lasers, cooling

1. Introduction

Due to eminent efficiency, good compactness and reliability, outstanding spatial beam quality, efficient heat dissipation, and freedom from thermal lensing, fiber lasers are now competing with their bulk solid–state counterparts for interesting scientific and industrial applications [1–3] such as material processing, defense, remote sensing, free–space communication, etc. With the availability of high–power and high–brightness laser diodes accompanied with cladding pumped architecture, a rise in output power from ytterbium (Yb)–doped, double–clad fiber laser sources has been dramatic recently, maturing to the point of hundreds of Watts [4–6], even in the case of continuous–wave (CW) regimes, 10/50 kW for single–transverse–mode/ multi–mode operations [7], and beyond.

Thanks to the long and thin fiber geometries, stress fracture and beam distortion, which are major problems for bulk solid–state lasers, can both be alleviated in fiber lasers. However, heat management and nonlinear effects, i.e., stimulated Brillouin and Raman scatterings (SBS and SRS), are still the most critical issues for scaling higher output powers [8]. By utilizing large mode area (LMA) fibers and broadening the spectral bandwidth of the seeding signal, the latter can be effectively suppressed.

Fiber lasers inherently exhibit exceptional capacity for heat dissipation, facilitated by very long and thin fiber cylinders with a large surface area to active volume ratio. In practice, the physical design to eliminate nonlinearities deteriorates thermal management in high–power regimes [9–11] and hence increases the threshold to thermal lensing features. One should keep in mind that double–clad optical fibers are surrounded by a low index polymer coating with limited tolerance to heat (~150–200 °C) [12], while the core temperature is always located below the melting point of quartz (1982 K) [13].

The majority of the heat converted from the optical power takes place in the fiber core, where most of the pump power is absorbed. The maximum temperature is hence expected to appear at the fiber axis. The fraction of power turned into heat due to quantum defect is defined as the ratio of pump–to–signal wavelengths, however, the actual heat fraction depends on the detailed kinetics of the system being used [14].

It is worth nothing that, whilst the fiber laser is immune to thermal effects to a large extent at low powers, there are significant characteristics and major restrictions in a kilowatt power domain [15–16] which influence laser performance and cannot be ignored. In one embodiment, pump–induced heating can cause a number of serious problems, comprising;

- formation of thermal cracks due to internal thermal stress and expansion;

- shortening of fiber lifetime owing to damage of fiber coatings, even melting of the glass;

- degradation of laser beam quality due to thermal lensing;

- deterioration of optical coupling efficiency affected by the undesirable temperature–induced motion of mechanical parts;

- enhancement of mode instability [17–18] and mode distortion [19];

- reduction of laser quantum efficiency and gain coefficient; and finally

- intensification of threshold power.

Another approach implies that fiber fractures may be distinguished at high average power under pulsed operation.

From another point of view, the question of how to optimize fiber and pump conditions in order to facilitate heat dissipation is critical at any kilowatt level. Once an appropriate distribution of operating temperatures is considered for the whole fiber, other issues like thermo–optic effects, fiber lifetime, and mechanical stability are easily diminished [20]. Thus, it is important to investigate an accurate model for estimating temperature along the fiber, and then to supply appropriate cooling techniques.

There are some solutions to drastically lower the operating temperature for double–clad fiber lasers including active or passive efficient cooling [21]. Increasing of gain, beam quality, thermal conductivity, as well as efficiency, together with reducing the thermal expansion and thermo–optic coefficients, temperature gradients, thermal lensing, self–pulsing [22], thermally induced broadening, saturation, and threshold powers are the main approaches used in such chilling systems but at the cost of raising spectral linewidth [23]. Synthesizing those factors in cryogenically cooled systems allows for strong improvements in master–oscillator and power–amplifier performance.

This chapter emphasizes the understanding of heat generation and removal concerned with fiber laser amplifiers. A comprehensive review is provided to predict thermal treatment along the active media in Section 2. Particular focus is placed on theoretically analyzing the pump–induced temperature change, applicable for a couple of CW and pulsed modes for optical fiber lasers and amplifiers. Additionally, three–dimensional (3D) simulation is implemented both for axial and transverse thermal distributions by means of conductive, convective, as well as radiative heat transfer relations. In the final section, significant chilling procedures are introduced, including air and liquid cooling, thermoelectric heat sinks, dry–ice chillers, and the concept of cryogenic lasers, etc.

2. Theory

In order to improve laser performance and decrease thermal destructive effects, tempera-ture evolution must be determined within the fiber laser which relies on the pump beam intensity profile, thermal properties (glass fiber and cladding materials), geometry, and cooling medium [24].

In all optical doped fibers, thermal effects are associated with absorbing a finite amount of optical power by the active gain media. If the electronic relaxation of the dopant involves non–radiative processes, heat is generated. For high–power regimes, thermal effects can limit the maximum pump power that can be delivered to the fiber and therefore the maximum output power [25] which can be extracted. In turn, this can mitigate the maximum seed signal injected in booster amplifiers and fiber attenuators.

Herein, explicit expressions for the thermal behavior made by the pump or lasing power, as well as heat deposited both inside and outside of the fiber core, are derived by analytically solving the heat diffusion equation [26–28]. In general terms, we consider only the case where the core and cladding regions are concentric. This assumption can also be readily modified in a more advanced treatment of the scaling effects and is not expected to influence the substance of our conclusions.

By presuming circular cross–sectional areas seen in Figure 1, there are three distinct regions for double–clad fibers that need to be addressed: (I) the core, (II) the inner cladding, and (III) the outer cladding zones. The quantities a, b, and c indicate the core, inner, and outer cladding radii, respectively.

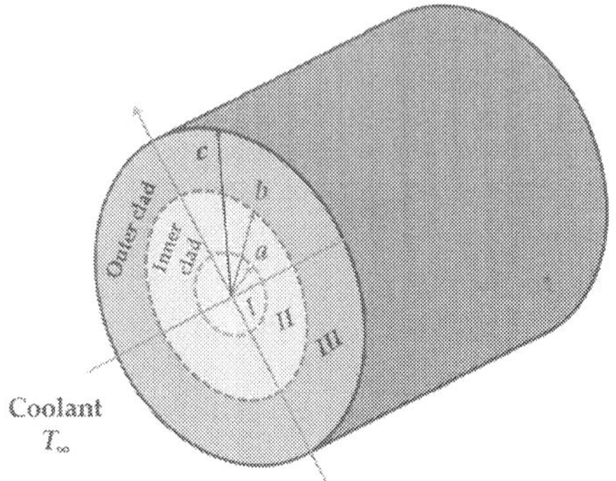

Figure 1. The geometry of various layers of a typical dual–clad fiber.

Commonly, the general form of the heat conduction equation in an isotropic medium in order to determine the time–dependent 3D temdperature distribution can be written as:

$$\rho c_v \frac{\partial T(r,\varphi,z,t)}{\partial t} - \kappa \nabla^2 T(r,\varphi,z,t) = Q(r,\varphi,z,t) \tag{1}$$

Here, T illustrates the temperature, the radial/axial coordinate is r/z, φ the azimuthal angle, and t denotes the time. Moreover, κ shows the thermal conductivity of the fiber and the dissipated heat density per unit volume that is created in the fiber core is displayed as $Q(r,\varphi,z,t)$. In addition, ρ and c_v demonstrate the mass density of the material and the specific heat capacity at a constant volume, respectively.

2.1. CW Mode

2.1.1. Transverse thermal analysis

In what follows, we focus on the case where the deposition of heat density into the fiber is uniform, $Q(r,\varphi,z,t)=Q_0$ [29–30] which is a good approximation in long, weakly doped fibers. Note that, any azimuthal variations can be ignored for the symmetrical structure of the fiber in cylindrical coordinates. On the other hand, no z variation is taken into account ($\partial^2/\partial z^2=0$), where z is the propagation direction along the fiber.

To model thermal effects, we refer to the cladding with identifying separated layers. In the first approach, let us suppose that the core and two cladding regions are composed of similar glass, with analogous thermal and mechanical properties [26] as well as comparable temperature–independent parameters comprising Poisson's ratio, Young's modulus, thermal expansion, and refractive index variation. Moreover, the dominant heat profile is deposited only in the uniform core.

Regarding Figure 1, under the steady–state operation ($\partial/\partial t = 0$), the heat equation for an isotropic medium Eq. (1) can be rewritten as follows:

$$\nabla^2 T(r) = \frac{-Q_0}{\kappa} \tag{2}$$

Therefore, applying a simplifier hypothesis, the above equation can be expounded for three distinct areas:

$$\frac{1}{r}\frac{\partial}{\partial r}\left[r\frac{\partial T_1(r)}{\partial r}\right] = \frac{-Q_0}{\kappa_1} \qquad 0 \leq r \leq a \tag{3}$$

$$\frac{1}{r}\frac{\partial}{\partial r}\left[r\frac{\partial T_2(r)}{\partial r}\right] = 0 \qquad a \leq r \leq b \tag{4}$$

$$\frac{1}{r}\frac{\partial}{\partial r}\left[r\frac{\partial T_3(r)}{\partial r}\right] = 0 \qquad b \leq r \leq c \tag{5}$$

The temperatures and their derivatives must be continuous across the borders [31]. Therefore, the multipoint boundary conditions of the thermal conductive equations are given by:

$$T_1(r = a) = T_2(r = a) \text{ and } T_2(r = b) = T_3(r = b) \tag{6}$$

$$\kappa_1 \left.\frac{\partial T_1(r)}{\partial r}\right|_{r=a} = \kappa_2 \left.\frac{\partial T_2(r)}{\partial r}\right|_{r=a} \tag{7}$$

$$\kappa_2 \left.\frac{\partial T_2(r)}{\partial r}\right|_{r=b} = \kappa_3 \left.\frac{\partial T_3(r)}{\partial r}\right|_{r=b} \tag{8}$$

In addition, the temperature in the center of the core ($r = 0$) must satisfy:

$$\left.\frac{\partial T_1(r)}{\partial r}\right|_{r=0} = 0 \tag{9}$$

Yet, another boundary condition is that for ($r = c$), where Newton's law of cooling is governed, T_3 actualizes the following formula [26]:

$$\kappa_3 \left.\frac{\partial T_3}{\partial r}\right|_{r=c} = h\left[T_c - T_3(r=c)\right] \tag{10}$$

where h is the total of the convective and radiative heat transfer coefficients and T_c ascertains the coolant temperature. Straightforward solution of Eqs (3)–(5), subject to the aforementioned boundary conditions, results in the following expressions for the temperatures in regions I, II, and III [13];

$$T_1(r) = T_0 - \frac{Q_0 r^2}{4\kappa_1} \qquad 0 \le r \le a \tag{11}$$

$$T_2(r) = T_0 - \frac{Q_0 a^2}{4\kappa_1} - \frac{Q_0 a^2}{2\kappa_2}\ln\left(\frac{r}{a}\right) \qquad a \le r \le b \tag{12}$$

$$T_3(r) = T_0 - \frac{Q_0 a^2}{4\kappa_1} - \frac{Q_0 a^2}{2\kappa_2}\ln\left(\frac{b}{a}\right) - \frac{Q_0 a^2}{2\kappa_3}\ln\left(\frac{r}{b}\right) \qquad b \le r \le c \tag{13}$$

Equation 11 shows that in the pumped core region, the temperature varies quadratically with r, while Eqs (12) and (13) reveal that the temperature logarithmically falls off. Furthermore, T_0 (the temperature of core's center), which is related to the coolant temperature, can be described by use of [13]:

$$T_0 = T_c + \frac{Q_0 a^2}{2hc} + \frac{Q_0 a^2}{4\kappa_1} + \frac{Q_0 a^2}{2\kappa_2}\ln\left(\frac{b}{a}\right) + \frac{Q_0 a^2}{2\kappa_3}\ln\left(\frac{c}{b}\right) \tag{14}$$

Besides, the average temperature T_{av} [26] can be calculated thus:

$$T_{av} = \frac{\int_0^a T_1(r)\,dr + \int_a^b T_2(r)\,dr + \int_b^c T_3(r)\,dr}{\int_0^c dr} \tag{15}$$

2.1.2. 3D thermal analysis

2.1.2.1. Analytical approach

The obtained results in the previous section can be easily extended to the non–uniform heat deposition or pump light absorption [32–33]. Therefore, Eq. (3) modifies to the following form [34–35];

$$\frac{1}{r}\frac{\partial}{\partial r}\left[r\frac{\partial T_1(r,z)}{\partial r}\right] = \frac{-Q(r,z)}{\kappa_1} \qquad 0 \leq r \leq a \tag{16}$$

The absorbed pump power within the fiber length is expressed as follows:

$$\Delta P_{abs} = P_P(z) - P_P(z+dz) = \frac{dP_P(z)}{dz}.\Delta z \tag{17}$$

$$\Delta Q(r,z) = (1-\eta)\Delta P_{abs} \tag{18}$$

Therefore, utilizing the definition of forward and backward pump powers [36–38], we can conclude that:

$$Q(r,z) \simeq Q(z) = \frac{\alpha(z)P_P(z)}{\pi a^2}(1-\eta) \tag{19}$$

where

$$\alpha(z) = \alpha_a(z) + \alpha_s \tag{20}$$

Here, $Q(z)$ represents the heat power density along the axial direction of the fiber, η is the quantum efficiency, $\alpha_a(z)$ denotes the absorption coefficient and α_s indicates the signal scattering loss. Hence, the temperatures in the core and claddings can be derived by employing steady–state thermal conductive equations subject to boundary conditions, as below [39]:

$$T_1(r,z) = T_0 - \frac{Q(z)r^2}{4\kappa_1} \qquad 0 \leq r \leq a \tag{21}$$

$$T_2(r,z) = T_0 - \frac{Q(z)a^2}{4\kappa_1} - \frac{Q(z)a^2}{2\kappa_2}\ln\left(\frac{r}{a}\right) \qquad a \leq r \leq b \tag{22}$$

$$T_3(r,z) = T_0 - \frac{Q(z)a^2}{4\kappa_1} - \frac{Q(z)a^2}{2\kappa_2}\ln\left(\frac{b}{a}\right) - \frac{Q(z)a^2}{2\kappa_3}\ln\left(\frac{r}{b}\right) \qquad b \leq r \leq c \tag{23}$$

In general terms, the temperature profiles are significantly affected by pump evolution along the fiber length. For example, the temperature conducts along the radial and axial directions of the fiber in forward and bidirectional pumping modes shown in Figure 2(a) and (b), respectively. It can be seen that the temperature distribution for the forward pump mode is uneven along the fiber. At the fiber axis ($r=0$), the maximum (minimum) temperature of 357.8 °C (33.0 °C) is attained at the fiber input (output) side. Therefore, some active cooling procedures have to be taken at the input side of the fiber. As would be expected, the radial distribution of the temperature profile can be ignored while significant heat is traveling through the axial direction of the fiber.

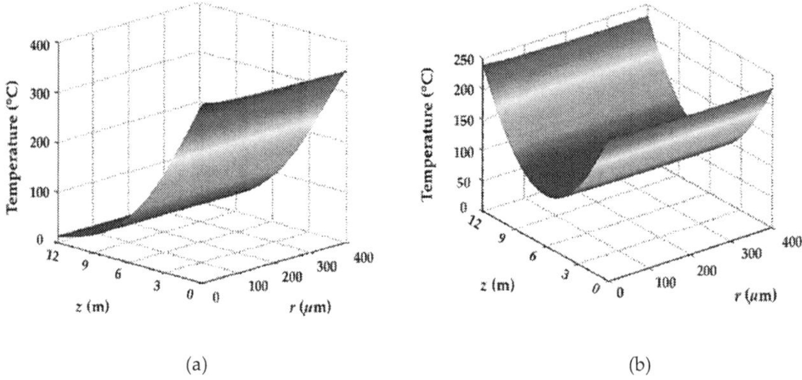

(a)   (b)

Figure 2. Evolution of temperature along both radial and axial directions of the fiber for (a) forward pump mode and, (b) bidirectional pumping configuration [35].

Analogously, according to Figure 2(b), at the fiber axis ($r=0$), a maximum (minimum) temperature of 237.4 °C (84.2 °C) is reached at the fiber input and output sides (middle of the fiber). The maximum temperature difference in the radial direction is 7.6 °C, which is obtained at the output side.

Comparing the results of the forward pump mode with that of the bidirectional pumping configuration, one can be said that the temperature evolutions in axial and radial directions of the fiber for the two–end pump mode are more even, and the maximum temperature in the fiber is decreased by 120.4 °C. Therefore, the bidirectional pumped array is preferred here.

2.1.2.2. Numerical approach

The scope of this section relates to the case where $\partial^2/\partial z^2 \neq 0$. Here, the injected pump light is absorbed by doped ions in the fiber core. Hence, there is a heat source in the core that does not exist in the claddings. In this case, according to the non–uniformity (exponential evolution) of the pump power in the axial direction, the Q value is not constant. On the other side, since the fiber length is much larger than the cross section, the capability of heat dissipation from the fiber end facets is much lower than from the fiber sides.

The heat dissipation as well as both transverse and longitudinal temperature distributions in a typical rare–earth doped dual–clad fiber is expressed by the following time–independent thermal conduction equation in symmetric cylindrical coordinate (r, z) [9, 15, 20, 27]:

$$\frac{1}{r}\frac{\partial}{\partial r}\left[r\frac{\partial T_1(r,z)}{\partial r}\right]+\frac{\partial^2 T_1(r,z)}{\partial z^2}=\frac{-q(r,z)}{\kappa_1} \qquad 0 \leq r \leq a \tag{24}$$

$$\frac{1}{r}\frac{\partial}{\partial r}\left[r\frac{\partial T_2(r,z)}{\partial r}\right]+\frac{\partial^2 T_2(r,z)}{\partial z^2}=0 \qquad a \leq r \leq b \tag{25}$$

$$\frac{1}{r}\frac{\partial}{\partial r}\left[r\frac{\partial T_3(r,z)}{\partial r}\right]+\frac{\partial^2 T_3(r,z)}{\partial z^2}=0 \qquad b \leq r \leq c \tag{26}$$

The boundary condition at the surface between the fiber cladding and the ambient environment is given by Newton's and Stefan–Boltzmann's laws, as below [35]:

$$\kappa_3 \left.\frac{\partial T_3}{\partial r}\right|_{r=c} = h(z)\left[T_c - T_3(r=c)\right] \tag{27}$$

Here, h is a function of z too [10] that can be determined using the following relations:

$$h(z)=0.5 \times N_u(z).\kappa_c.c^{-1} \tag{28}$$

$$N_u(z).exp\left[\frac{-2}{N_u(z)}\right]=0.16 \times \left[G_r(z).P_r\right]^{1/3} \tag{29}$$

$$G_r(z)=8g.d_c^2.c^3.\mu_c^{-2}.T_c^{-1}\left[T(r=c,z)-T_c\right] \tag{30}$$

where T indicates the temperature in the fiber, and the heat sink temperature is represented by T_c. Also, $q(r, z)$ exhibits the heat dissipated in a unit volume, chiefly generated in the fiber core, where most of the pump power is absorbed. This can be obtained by calculating all the input and output optical powers flowing into and out of a unit volume at (r, z). Furthermore, $h(z)$ shows the convective coefficient, g ascertains the acceleration of gravity as well as N_u, G_r, and P_r which denote the Nusselt, Grashof, and Prandtl numbers, respectively. Moreover, d_c, μ_c, and κ_c accordingly depict the density, viscosity, and thermal conductivity of the coolant. Thus, the analytical expression for the Nusselt number is expressed below [40]:

$$N_u(z) = \left\{ 0.60 + \frac{0.387 \left[G_r(z) P_r \right]^{1/6}}{\left[1 + \left(0.559 / P_r \right)^{9/16} \right]^{8/27}} \right\}^2 \tag{31}$$

Since the highest temperature occurs at the fiber axis, it needs to pay more attention to the temperature behavior along the fiber axis. We then introduce T_{ave} as the longitudinally averaged temperature along the fiber axis, namely [27]:

$$T_{ave} = \frac{1}{L} \int_0^L T(r = 0, z) dz \tag{32}$$

At last, the dissipated heat across the whole fiber cross section is realized by [20]:

$$Q(z) = 2\pi \int_0^c q(r, z) r dr \tag{33}$$

Therefore, the heat converted from the optical power per unit length of fiber can be compared under different pumps and fiber conditions. Herein, there is no any analytical solution and Eqs (24)–(26) can be solved numerically using the finite element method (FEM).

In 2004, Wang et al. [10] showed that lower operating temperature and more uniform heat dissipation in fibers can be obtained by optimizing the arrangement of pump powers, pump absorption coefficients, and fiber lengths through the distributed side–pumping mode. As a beneficial solution for a traditional end–pumped scheme, the arrangement of uneven pump absorption coefficients along the cavity can improve laser efficiency and reduce fiber temperature.

Figures 3 and 4 show the calculated temperature evolution in a typical Yb–doped double–clad fiber at the fiber axis ($r = 0$) and at the inner/outer cladding boundary ($r = 125$ μm) with uniform as well as non–uniform pump absorption coefficients. From these figures, one can reach the conclusion that the temperature difference in the radial direction is much smaller than in the axial direction. However, the average temperature is very high at the fiber end faces.

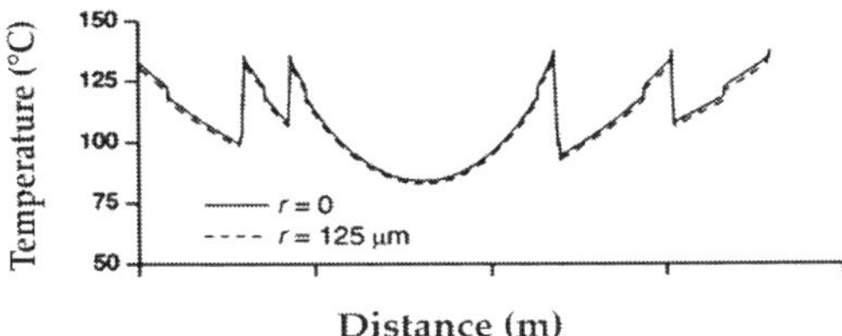

Figure 3. Temperature distribution in a 32–m fiber under an end–pump scheme with non–uniform pump absorption coefficients in five segments [10].

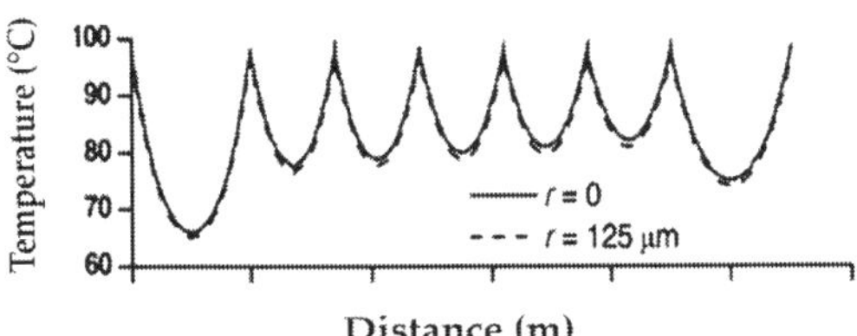

Figure 4. Temperature distribution in a 55–m fiber under a distributed side–pumped scheme with a uniform pump absorption coefficient of 1.2 dB/m, in seven segments [10].

In addition, the uneven temperature distribution results from non–uniform pump absorption in the fiber.

2.1.3. Radiative heat transfer

When the temperature of a body rises, three dominant thermal effects can be seen in it: convection, conduction, and the radiation. The latter process is in the form of electromagnetic radiation emitted by a heated surface in all directions. It travels directly to its point of absorption at the speed of light [35]. The total radiant heat energy radiated by a surface at a temperature greater than absolute zero is proportional to the Stefan–Boltzmann law. Usually, at high pump powers, the process of heat transfer is not only dominated by convection, but also being employed with radiation. The radiative heat transfer has to be considered in the thermal, stress, and thermo–optic analyses of any type of high–powered fiber lasers.

The procedure of heat dissipation from the fiber core to the fiber periphery is illustrated in Figure 5. At first, the heat generated in the fiber core is transferred to the surface by thermal conduction, and then dissipated to the ambient air by convective and radiative heat transfer. There are no heat sources in both the fiber inner and outer claddings.

For isotropic medium, the heat transfer between the surface of the fiber and the surrounging medium is given by [41]:

$$Q_c = 2\pi c L h_c \left(T\big|_{r=c} - T_\infty \right) \tag{34}$$

where the convective heat transfer coefficient (h_c), can be obtained from the following equation:

$$h_c = 0.5 N_u \kappa c \tag{35}$$

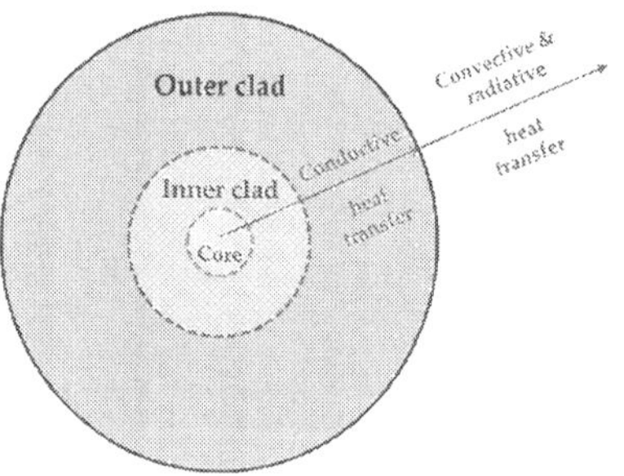

Figure 5. Model of heat dissipation in a double–clad fiber.

Also, the radiative heat transfer can be calculated using:

$$Q_r = 2\pi c L \varepsilon \sigma_b \left(T\big|_{r=c}^4 - T_\infty^4 \right) \tag{36}$$

where L is the fiber length, $T\big|_{r=c}$ ascertains the temperature at the surface of the fiber, T_∞ the environmental temperature, ε the surface emissivity, and σ_b denotes the Stefan–Boltzmann constant of the fiber material.

For a more convenient course of calculation, the radiative heat transfer coefficient (h_r) can also be described as:

$$Q_r = 2\pi c L h_r \left(T\big|_{r=c} - T_\infty \right) \tag{37}$$

In other words, the value of (h_r) can be clarified from:

$$h_r = \frac{\varepsilon \sigma_b \left(T\big|_{r=c}^4 - T_\infty^4 \right)}{T\big|_{r=c} - T_\infty} \tag{38}$$

Eventually, the total heat transfer is a summation of the convection and radiation heat transfers as follows:

$$Q = Q_c + Q_r = 2\pi c L h \left(T\big|_{r=c} - T_\infty \right) \tag{39}$$

where

$$h = h_c + h_r \tag{40}$$

It can see from Figure 6 that radiative heat transfer strictly increases with temperature, while convection remains almost constant during pumping. For this reason, the total heat transfer coefficient is affected more by radiation [40]. From the data in the curve, the h_c coefficient possesses non–linear variation when the temperature difference between the surface of the fiber and the environment is below 100 °C. This means that the effect of convection heat transfer is dominant at low temperature domains.

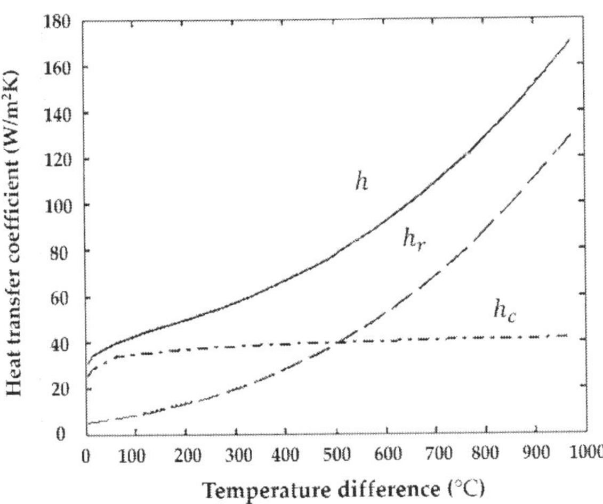

Figure 6. Heat transfer coefficients h_c, h_r, and h as the temperature difference between the surface of the fiber and the environment varies from 5 to 975 °C [40].

2.2. Pulsed regime

Contrary to CW mode, the analytical expressions for the temperature portion in individual fiber regions for the pulsed operating system are given by:

$$\rho c_v \frac{\partial T_1(r,\varphi,z,t)}{\partial t} - \kappa_1 \nabla^2 T_1(r,\varphi,z,t) = \eta P_v(r) \qquad 0 \leq r \leq a \tag{41}$$

$$\rho c_v \frac{\partial T_2(r,\varphi,z,t)}{\partial t} - \kappa_2 \nabla^2 T_2(r,\varphi,z,t) = 0 \qquad a \leq r \leq b \tag{42}$$

$$\rho c_v \frac{\partial T_3(r,\varphi,z,t)}{\partial t} - \kappa_3 \nabla^2 T_3(r,\varphi,z,t) = 0 \qquad b \leq r \leq c \tag{43}$$

where η represents the quantum efficiency and the average pump power absorbed per unit volume is demonstrated by $P_v(r)$. Here, the heat produced by multiple pulses or pump, simulates utilizing the time–dependent convection heat transfer equation in the dynamic regime.

At the first approximation, we assume that the absorbing region of the fiber is long enough and the rate of change of the absorption product is small enough, so that the longitudinal temperature gradient within the doped region is small. The ratio of the rate of heat flows out of the ends of the doped zone to that of the sides of the doped region after an instantaneous heating process effectively scales as the ratio of the dopant radius to the length of gain fiber. Consequently, at a given location z along the fiber, heat flows mostly radially, and at all times, $t>0$, the z –dependence of $T(r, \varphi, z, t)$ is the same as that of the initial temperature profile $T(r, \varphi, z)$. In the following, we therefore omit the z –dependence in our notation for the sake of simplicity. As the fiber cools down, by supposing the azimuthal symmetry, the spatial and temporal evolution of the core temperature distribution is described by the homogeneous heat conduction equation [42] with a source term:

$$\rho c_v \frac{\partial T(r,t)}{\partial t} - \kappa \nabla^2 T(r,t) = \eta P_v(r) \tag{44}$$

The general solution is proposed in the following form [16]:

$$\Delta T(r,t) = \sum_{m=1}^{\infty} a_m(t) J_0\left(\frac{p_m r}{b}\right) exp\left(\frac{-t}{\tau_m}\right) \tag{45}$$

where $\Delta T(r, t) = T(r, t) - T_0$ explains the temperature rise profile, T_0 denotes the equilibrium temperature of the surrounding medium, and the time constants τ_m namely:

$$\tau_m = \frac{\rho c_v}{\kappa} \frac{b^2}{p_m^2} \tag{46}$$

The coefficients $a_{\mathrm{m}}(t)$ and p_{m} are determined by two boundary conditions. The first one is energy conservation; presuming that cooling is due to natural air convection and, the heat flowing out of the fiber (at $r=b$) is proportional to the temperature difference between the fiber and the periphery air. The proportionality factor is the heat transfer coefficient h. This condition can be expressed as:

$$-\kappa\nabla T\left(r,t\right)\big|_{\mathrm{r=b}} = h\left[T\left(r=b,t\right)-T_{0}\right] \tag{47}$$

Substituting the general solution Eq. (45) into Eq. (47) yields:

$$p_{\mathrm{m}}J_{1}\left(p_{\mathrm{m}}\right) = \frac{hb}{\kappa}J_{0}\left(p_{\mathrm{m}}\right) \tag{48}$$

which determines the values of p_{m}.

The second boundary condition is at $t=0$; supposing the temperature distribution at any time equals that of the initial time. Multiplying both sides of Eq. (45) by $rJ_{0}(p_{\mathrm{n}}r\,/b)$ and integrating in r across the fiber while applying Eq. (48) yields the time–dependent amplitudes $a_{\mathrm{m}}(t)$ which are the solutions of:

$$\frac{\partial a_{\mathrm{m}}}{\partial t} = \frac{\displaystyle\int_{0}^{b}\frac{P_{\upsilon}\left(r\right)}{\rho c^{\upsilon}}J_{0}\left(\frac{p_{\mathrm{m}}r}{h}\right)rdr}{\dfrac{b^{2}}{2}J_{0}^{2}\left(p_{\mathrm{m}}\right)\left[1+\left(\dfrac{bh}{\kappa p_{\mathrm{m}}}\right)^{2}\right]}\exp\left(\frac{t}{\tau_{\mathrm{m}}}\right) \tag{49}$$

with $a_{\mathrm{m}}(t=0)=0$.

In the particular case of a step pump profile of radius s $[P_{\upsilon}(r)=P_{\upsilon}$ for $r<s$ and 0 for $r>s]$, this explanation becomes:

$$\Delta T\left(r,t\right) = \frac{2sb\eta P_{\upsilon}}{\kappa}\sum_{m=1}^{\infty}\frac{J_{1}\left(\dfrac{p_{\mathrm{m}}s}{b}\right)}{p_{\mathrm{m}}^{3}J_{0}^{2}\left(p_{\mathrm{m}}\right)}\frac{1-\exp\left(\dfrac{-t}{\tau_{\mathrm{m}}}\right)}{1+\left(\dfrac{bh}{\kappa p_{\mathrm{m}}}\right)^{2}}J_{0}\left(\frac{p_{\mathrm{m}}r}{b}\right) \tag{50}$$

Assuming $\eta=1$, the radial temporal evolution of the temperature profile at different times is shown in Figure 7. Shortly after the pump is turned on, the heat and the temperature increment are mostly confined to the vicinity of the core. As time goes on, the temperature at the center of the fiber grows, and as heat flows outward, the temperature rise spreads toward the fiber edge. Eventually, the temperature profile becomes nearly uniform.

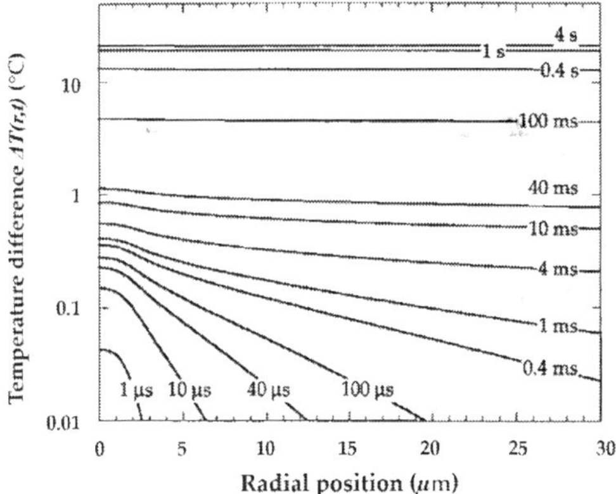

Figure 7. Radial temperature profile in a fiber at various times, $t>0$, assuming that at time $t=0$, the fiber begins to absorb 10 mW/cm of power in a doped region of 2 μm radius [16].

In the cladding regions, a_m is a time–independent coefficient which is determined as follows:

$$a_\mathrm{m} = \frac{2}{b^2 J_0^2\left(p_\mathrm{m}\right)\left(1+\left(bh/\kappa p_\mathrm{m}\right)^2\right)} \int_0^b \Delta T\left(r,0\right) J_0\left(\frac{p_\mathrm{m} r}{b}\right) r\,dr \tag{51}$$

3. Cooling mechanisms

Fiber lasers and amplifiers have proven themselves as reliable systems with excellent beam quality and high output power. From the numerical and analytical analyses, one can conclude that the thermal effects must be considered at high–power regimes. Therefore, effective heat dissipation is a significant factor, with the aim of preventing damage to the fiber ends, interfaces, and coating. In order to suppress thermal issues, a suitable cooling method can be considered. This section remarks on the fast chilling of optical components and splices in order to modify their practical design, therefore, obtaining an optimum situation.

Heat generated in high–power fiber laser amplifiers is the source of increased temperature and stress inside the gain medium, which causes poor beam quality and restricts average output power. To solve this problem, fiber structures with enhanced mode areas have been suggested. These novel high–power schemes rely on multimode fibers with large diameter cores [43–44] or the amplification process takes place in a fiber cladding [45]. However, even if a high–quality beam is required, they are ultimately restricted by thermal effects.

Furthermore, a significant limiting factor in fiber laser amplifiers under strong pumping conditions is the temperature increase, growing the unsaturable loss mechanism. Effective heat extraction can reduce the temperature–dependent unsaturable losses especially in bismuth–doped fibers, resulting in increased laser performance [46]. In the case of rare–earth ions, unsaturable absorption losses can produce quenching processes such as cross–relaxation and up–conversion [47]. This can lead to wasting of the pump energy, raising the laser threshold, as well as reducing the conversion efficiency.

It is worth mentioning that cryogenically cooled systems promise a revolution in power scalability while maintaining good beam quality because of significant improvements in efficiency and thermo–optic properties. This is particularly true for Yb lasers due to their relatively low quantum defect as well as their broadband absorption spectrum even at cryogenic temperatures [48]. Amplification in Yb–doped fibers is generally possible from 976 nm to 1200 nm, but below 1030 nm it becomes more challenging, since the absorption cross section grows towards shorter wavelengths [49]. This thermal population and thus absorption in the wavelength range above 1000 nm can be significantly alleviated by cooling the fiber to low temperatures [50].

In addition to this, any surface cooling creates a thermal gradient that strains the laser medium as well as distorts optical waves. Lowering of the doping concentration and increasing of the fiber length make the cooling easier, but enhance non–linear effects such as SBS and SRS, which can deplete the amplified signal. An appropriate way to overcome this challenge is by using the anti–Stokes cooling method for spontaneous and stimulated emission radiation–balanced lasers, mentioned by Steven R. Bowman in 1999 [51–52]. Hereupon, the thermal load generated from stimulated emission can be dissipated thoroughly, which permits lasing without detrimental heating of the laser medium.

From another point of view, lower operating temperatures and more uniform heat dissipation at ambient temperature can be achieved by optimizing the arrangement of pump powers, pump absorption coefficients, and fiber lengths [10], presuming a distributed side–pumping mode in passive cooling systems. Additionally, forced chilling methods comprising passive air cooling [53] and active liquid cooling [54–55] through convectional processes, as well as conductive thermoelectric (Peltier) effects [56] using cold plates or heat sinks are noticeable techniques, and thereby a number of detrimental thermal effects can be effectively suppressed.

The liquid chilled thermal management is commonly performed by means of cold water, fluorocarbon refrigerants, ethylene glycol, commercial silicone fluids, and any electronic coolants or the like [54]. Although water cooling has proved to be very efficient at dissipating heat and is extensively adopted for most high–power, solid–state lasers, it cannot be directly applied to most Yb–doped double–clad arrays. For fibers with limited chemical stability such as, fluoride fibers, water cooling should be avoided. However, air cooling is a good candidate for many applications due to its compact structure and moderate dissipating efficiency [20].

In general, for efficiently cooling a system, the exothermic components which have any significant heat load, including high–power laser diodes, integrated combiners, splice points, optical reflectors, as well as doped fibers can be immersed or placed in contact with a thermal sink of appropriate temperature [57]. In 2011, Fan et al. [13] introduced a copper heat sink with various geometries in order to chill a doped fiber, something which was critical for the reliability of their high–power operation. The thermal contact resistance per unit surface $R_{tc}^{''}(m^2K/W)$ between the fiber and the heat sink is expressed as [58]:

$$R_{tc}^{''} = \frac{T_s - T_\infty}{q''}$$

(52)

where $T_s(K)$ is the fiber surface temperature, $T_\infty(K)$ the heat sink temperature, and $q''(W/m^2)$ the heat flux. In fact, Eq. (52) suggests the concept for the treatment of the heat flow through a fiber layer by analogy to the diffusion of electrical charge [12], where the temperature difference is analogous to the electrical voltage which drives the heat flow through a thermal resistant. For an active fiber, the heat generation, $q_0(W/m^3)$, the heat load $q'(W/m)$, and the pump absorption $\alpha(dB/m)$ are related to each other [59]:

$$q_0 = \left(1 - 10^{-\alpha \cdot \frac{dL}{10}}\right) = \frac{P\left(1 - \lambda_p/\lambda_s\right)}{\pi a^2} = \frac{q'dL}{\pi a^2}$$

(53)

where P represents the pump power through a section of length dL, and λ_s and λ_p display the signal and pump wavelengths, respectively. Moreover, q'' is defined by $q'/perimeter$.

However, at moderate temperatures higher than ~-30 °C, a novel dry–ice chiller [60–61] is preferred. The practical scheme of a dry–ice heat exchanger being used as fast–cooling equipment for hot optical components is based on the cryogenic powder mixture using ethylene glycol to supply a high–capacity chilling bath, attaining thermal equilibrium for a long while. At the beginning of this process, the dry–ice/glycol bath creates a boiling gel to reduce the temperature to a few degrees as long as the dry–ice debris is available in the coolant. When thermal exchange between dry–ice and glycol takes place, the boiling goes on and the gel temperature drops from -30 °C to ~300 K [62], depending on the dry–ice loading.

Otherwise, while scaling up the power in high–power or high–energy domains, cryogenic lasers [63] may be required where in general terms, the gain medium is operated at cryogenic temperatures through liquefied gases including ammonia (~240 K), liquid CO_2 (~195 K), methane (~111 K), liquid nitrogen (~77 K), liquid neon (~27 K), or even liquid helium (~4 K).

The spectral properties of the amorphous glass host material are strongly affected by temperature [50] and one can illustrate how cryogenic cooling methods are employed to dramatically raise the efficiency and the stability of fiber laser amplifiers [64] as well as to efficiently diminish self–pulsing at the price of a spectrally broad emission spectrum. The justification is that the gain cross section of dopant ions becomes greater for various inversions under cryogenic chilling by means of optical refrigeration. In addition, this is believed to be related to the reduction of the thermal population in the upper stark levels of manifold, so that reabsorption at the signal wavelength completely eradicates. This enables the use of a considerably longer fiber length, which could entirely absorb the pump light.

Besides this, spectral density decreases as linewidth increases. The rising in linewidth is believed to be a result of the alleviating the homogeneous broadening due to the lowered operating temperature. To prevent this detrimental problem and achieve a temporally stable, narrow linewidth, highly efficient laser, a volume Bragg grating [23] can be used according to Figure 8. By efficiently cooling an active gain media, the thermal population and therefore the reabsorption losses drop, shifting the preferred lasing beam toward shorter wavelengths [11]. For cryogenically cooling fiber sources, the gain material can easily be submerged in the coolant.

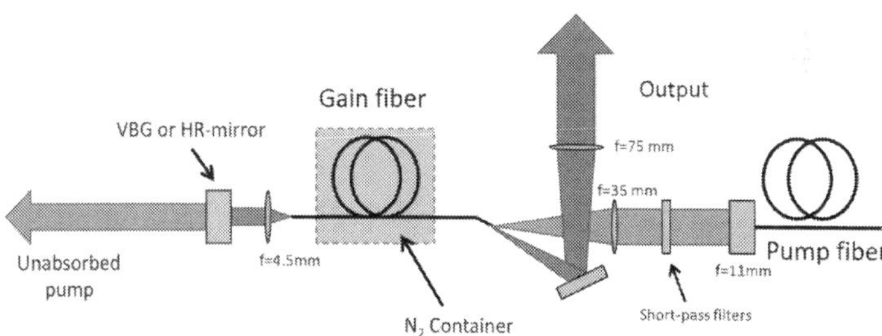

Figure 8. Cryogenically cooled fiber laser [64].

In 2006, Seifert et al. intensely reduced the thermal population of the lower laser levels of a Yb–doped fiber amplifier by using cryogenic cooling [65]. One year later, the temperature effect on the emission properties of Yb–doped optical fibers was investigated by Newell et al. They revealed that chilling efficiently eradicates the absorption tail above 1 μm [50]. Moreover, cooling of gain media using liquefied gases drastically limits the destructive third–order non-linear self–pulsing [22] effect. These results show that cryogenic cooling of Yb–doped fiber lasers is a simple way to stabilize the temporal output as well as substantially enhance the efficiency.

Author details

Maryam Eilchi[1*] and Parviz Parvin[2]

*Address all correspondence to: m_ilchi@aut.ac.ir

1 Laser and Optics Research Institute, Nuclear Science and Technology Research Institute, Tehran, Iran

2 Photonic Engineering Group, Department of Physics, Amirkabir University of Technology, Tehran, Iran

References

[1] S. Yin, P. Yan, and M. Gong. End-pumped 300 W continuous-wave ytterbium-doped all-fiber laser with master oscillator multi-stage power amplifiers configuration. *Opt. Express.*2008; 16(2): 17864–17869.

[2] J. Richardson, J. Nilsson, and W. A. Clarkson. High-power fiber lasers: Current status and future perspectives. *J. Opt. Soc. Am. B.* 2010; 27(11): B63–B92.

[3] Y. Jeong, J. Sahu, D. Payne, and J. Nilsson. Ytterbium-doped large-core fiber laser with 1.36 kW continuous wave Output power. *Opt. Express.* 2004; 12(25): 6088–6092.

[4] E. Stiles. New developments in IPG fiber laser technology. In: 5th International Workshop on Fiber Lasers: 2009.

[5] B. He, J. Zhou, Q. Lou, Y. Xue, Z. Li, W. Wang, J. Dong, Y. Wei, and W. Chen. 1.75 killowatt continuous wave output fiber laser using homemade ytterbium-doped large-core fiber. *Microw. Opt. Technol. Lett.*2010; 52(7): 1668–1671.

[6] C. Wirth, O. Schmidt, I. Tsybin, et al. 2 kW incoherent beam combining of four narrow-linewidth photonic crystal fiber amplifiers. *Opt. Express.* 2009; 17(3): 1178–1183.

[7] Available from: *http://www.IPGPhotonics.com*

[8] M. Ilchi-Ghazaani and P. Parvin. Distributed scheme versus MOFPA array on the power scaling of a monolithic fiber laser. *IEEE J. Quant. Electron.* 2014; 50(8): 698–708.

[9] D. C. Brown and H. J. Hoffman. Thermal, stress, and thermal-optic effects in high average power double-clad silica fiber lasers. *IEEE J. Sel. Top. Quantum Electron.* 2001; 37(2): 207–217.

[10] Y. Wang, C. Q. Xu, and H. Po. Thermal effects in kilowatt fiber lasers. *IEEE Photon. Technol. Lett.*2004; 16(1): 63–65.

[11] N. A. Brilliant and K. Lagonik. Thermal effects in a dual-clad ytterbium fiber laser. *Opt. Lett.*2001; 26(21): 1669–1671.

[12] M. A. Lapointe, S. Chatigny, M. Piché, M. Cain-Skaff, and J. N. Maran. Thermal effects in high power CW fiber lasers. In: *Proc. SPIE. 719511U:* 2009. DOI: 10.1117/12.809021.

[13] Y. Fan, B. He, J. Zhou, J. Zheng, H. Liu, Y. Wei, J. Dong, and Q. Lou. Thermal effects in kilowatt all-fiber MOPA. *Opt. Express.* 2011; 19(16): 15162–15172.

[14] Y Huo and P. Cheo. Thermomechanical properties of high-power and high-energy Yb-doped silica fiber lasers. *IEEE Photon. Technol. Lett.* 2004; 16(3): 759–761.

[15] L. Zenteno. High-power double-clad fiber lasers. *J. Lightwave Technol.* 1993; 11: 1435–1446.

[16] M. K. Davis, M. J. F. Digonnet, and R. H. Pantell. Thermal effects in doped fibers. *J. Lightwave Technol.*1998; 16: 1013–1023.

[17] Rumao Tao, Pengfei Ma, Xiaolin Wang, Pu Zhou and Zejin Liu. 1.3 kW monolithic linearly polarized single-mode master oscillator power amplifier and strategies for mitigating mode instabilities. *Photon. Res.* 2015; 3(3): 86–93.

[18] B. Ward, C. Robin, and I. Dajani. Origin of thermal modal instabilities in large mode area fiber amplifiers. *Opt. Express.* 2012; 20: 11407–11422.

[19] W. W. Ke, X. J. Wang, X. F. Bao, and X. J. Shu. Thermally induced mode distortion and its limit to power scaling of fiber lasers. *Opt. Express.* 2013; 21: 14272–14281.

[20] Y. Wang. Heat dissipation in Kilowatt fiber power amplifiers. *IEEE J. Quantum Electron.*2004; 40(6): 731–740.

[21] V. Ashoori and A. Malakzadeh. Explicit exact three-dimensional analytical temperature distribution in passively and actively cooled high-power fibre lasers. *J. Phys. D. Appl. Phys.* 2011; 44(35): 1–6.

[22] K. Sumimura, H. Yoshida, H. Okada, H. Fujita, and M. Nakatsuka. 2007. Suppression of self-pulsing in Yb-doped fiber lasers with cooling by liquid nitrogen. In: Lasers and Electro-Optics – Pacific Rim; 2007; CLEO/Pacific Rim. pp. 1–2.

[23] P. Jelger, K. Seger, V. Pasiskevicius, and F. Laurell. Highly efficient temporally stable narrow linewidth cryogenically cooled Yb-fiber laser. *Optics express.* 2009; 17(10): 8433–8438.

[24] M. El H. Assad and D. C. Brown. Thermodynamic analysis of end-pumped fiber lasers subjected to surface cooling. *IEEE Journal of Quantum Electronics.* 2013; 49(1): 100–107.

[25] D. C. Hanna, M. J. McCarthy, and P. J. Suni. Thermal considerations in longitudinally pumped fibre and miniature bulk lasers. In: *Boston. Proc. Fiber Laser Sources and Amplifiers: SPIE 1171*; 1990. pp. 160–166. DOI: 10.1117/12.963149

[26] D. C. Brown and H. J. Hoffman. Thermal, stress, and thermo-optic effects in high average power double-clad silica fiber lasers. *IEEE J. Of Quantum Electron.* 2001; 37(2): 207–217.

[27] Y. Wang, C. Q. Xu, and H. Po. Analysis of Raman and thermal effects in kilowatt fiber lasers. *Opt. Commun.* 2004; 242(4): 487–502.

[28] P. Elahi and N. Zare. The analytical solution of rate equations in end-pumped fiber lasers with minimum approximation and temperature distribution during the laser operation. *Acta Phys. Pol. A.* 2009; 116(4): 522–524.

[29] Y. Huang, H. Tsai, and F. Chang. Thermo-optic effects affecting the high pump power end pumped solid state lasers: Modeling and analysis. *Opt. Commun.* 2007; 273: 515–525.

[30] M. Sabaeian, H. Nadgaran, M. De Sario, L. Mescia, and F. Prudenzano. Thermal effects on double clad octagonal Yb:glass fiber laser. *Opt. Mater.* 2009; 31: 1300–1305.

[31] P. Li, C. Zhu, S. Zou, H. Zhao, D. Jiang, G. Li, and M. Chen. Theoretical and experimental investigation of thermal effects in a high power Yb^{3+}-doped double-clad fiber laser. *Opt. Laser Technol.* 2008; 40: 360–364.

[32] Z. Mohammed, H. Saghafifar, and M. Soltanolkotabi. An approximate analytical model for temperature and power distribution in high-power Yb-doped double-clad fiber lasers. *Laser Phys.* 2014; 24: 115107–115119.

[33] Z. Mohammed and H. Saghafifar. Optimization of strongly pumped Yb-doped double-clad fiber lasers using a wide-scope approximate analytical model. *Opt. Laser Technol.* 2014; 55: 50–57.

[34] D. Xue. Three-dimensional simulation of the temperature field in high-power double-clad fiber laser. *Optik.* 2011; 122: 932–935.

[35] J. Li, K. Duan, Y. Wang, X. Cao, W. Zhao, Y. Guo, and X. Lin. Theoretical analysis of the heat dissipation mechanism in Yb^{3+}-doped double-clad fiber lasers. *Journal of Modern Optics.*2008; 55(3): 459–471.

[36] P. Parvin, M. Ilchi-Ghazaani, A. R. Bananej, and Z. Lali-Dastjerdi. Small signal gain and saturation intensity of a Yb:Silica fiber MOPA system. *Journal of Optics & Laser Technology (JOLT).*2009; 41: 885–891.

[37] M. Ilchi-Ghazaani and P. Parvin. Impact of cavity loss on the performance of a single-mode Yb:silica MOFPA array. *Optics & Laser Technology.* 2015; 65: 94–105.

[38] S. Mohammadian, P. Parvin, M. Ilchi-Ghazaani, R. Poozesh, and K. Hejaz. Measurement of gain and saturation parameters of a single-mode Yb:silica fiber amplifier. *Optical Fiber Technology.* 2013; 19: 446–455.

[39] M. Abouricha, A. Boulezhar, and N. Habiballah. The comparative study of the temperature distribution of fiber laser with different pump schemes. *Open Journal of Metal.*2013; 3: 64–71.

[40] P. Yan, A. Xu, and M. Gong. Numerical analysis of temperature distributions in Yb-doped double-clad fiber lasers with consideration of radiative heat transfer. *Opt. Eng.* 2006; 45(12): 124201 1–4.

[41] J. P. Holman. Heat Transfer (8th edition).McGraw-Hill, New York, 1997.

[42] C. Kittel and H. Kroemer. Thermal Physics (2nd edition). W. H. Freeman, New York, 1980, pp. 424–430.

[43] R. I. Epstein, M. I. Buchwald, B. C. Edwards, T. R. Gosnell, and C. E. Mungan. Observation of laser-induced fluorescent cooling of a solid. *Nature* (London). 1995; 377: 500–502.

[44] D. Taverner, D. J. Richardson, L. Dong, J. E. Caplen, K. Williams, and R. V. Penty. 158-µJ pulses from a single-transverse-mode, large-mode-area erbium-doped fiber amplifier. *Opt. Lett.* 1997; 22(6): 378–380.

[45] G. Nemova and R. Kashyap. High-power long period grating assisted EDFA. *J. Opt. Soc. Am. B.*2008; 25: 1322–1327.

[46] M. P. Kalita, S. Yoo, and J. K. Sahu. Influence of cooling on a bismuth-doped fiber laser and amplifier performance. *Applied Optics.*2009; 48(30): G83–G87.

[47] M. E. Davis, M. J. F. Digonnet, and R. H. Pantell. Characterization of clusters in rare earth-doped fibers by transmission measurements. *J. Lightwave Technol.* 1995; 13: 120–126.

[48] Y. Fan, D. J. Roshan, L. Aggarwal, J. R. Ochoa, B. Chann, M. Tilleman, and J. Spitzberg. Cryogenic Yb^{3+}-doped solid-state lasers. *IEEE Journal of Selected Topics in Quantum Electronics.*2007; 13(3): 448–459.

[49] R. Paschotta, J. Nilsson, A. C. Topper, and D. C. Hanna. Ytterbium-doped fiber amplifiers. *IEEE J. Quantum Electron.* 1997; 33: 1049–1056.

[50] T. C. Newell, P. Peterson, A. Gavrielides, and M. P. Sharma. Temperature effects on the emission properties of Yb-doped optical fibers. *Opt. Commun.* 2007; 273: 256–259.

[51] S. R. Bowman. Laser without internal heat generation. *J. Lightwave Technol.* 1999; 35: 115–122.

[52] G. Nemova and R. Kashyap. Optimization of the dimensions of an Yb^{3+}:ZBLANP optical fiber sample for laser cooling of solids. *Optics Letters.*2008; 33(19): 2218–2220.

[53] S. Sakuragi, M. Saitoh, H. Kotani, and K. Imagawa. Cooling mechanism for optical fiber. US Patent. Patent Date: February 25, 1986; Patent No. US 4,572,609.

[54] J. A. Davis. Systems and methods of cooling a fiber amplifier with an emulsion of phase change material. US Patent. Patent Date: November 25, 2008; Patent No. US 7,457,502 B2.

[55] S. Tokita, M. Murakami, S. Shimizu, M. Hashida, and S. Sakabe. Liquid-cooled 24 W mid-infrared Er:ZBLAN fiber laser. *Opt. Lett.* 2009; 34(20): 3062–3064.

[56] L. Li, H. Li, T. Qiu, V. L. Temyanko, M. M. Morrell, A. Schülzgen, A. Mafi, J. V. Moloney, and N. Peyghambarian. 3-dimensional thermal analysis and active cooling of short-length high-power fiber lasers. *Opt. Express.* 2005; 13(9): 3420–3428.

[57] J. E. Rothenberg, S. J. Brosnan, and T. Epp. High efficiency, high power cryogenic laser system. Patent Application Publication. Publication Date: August 30, 2007;Pub. No. US 2007/0201518 A1.

[58] Y. Fan, S. Dai, B. He, C. Zhao, J. Zhou, Y. Wei, J. Zheng, and Q. Lou. Efficient heat transfer in high-power fiber lasers. *Chinese Optics Letters (COL)*.2012; 10(11): 111401 1–4.

[59] Lapointe, S. Chatigny, M. Piché, M. Cain-Skaff, J. Maran. Thermal effect in high power CW fiber lasers. In: *Proc. of SPIE*. 7195. 71951U-1; 2009. DOI: 10.1117/12.809021.

[60] P. Parvin and M. Ilchi-Ghazaani. Modified Dry Ice Heat Exchanger for Heat Removal of Portable Nuclear Reactors. Non Provisional US Patent. Filling Date: 05/14/2012 (Application No. US 13/470,547).

[61] M. Ilchi-Ghazaani and P. Parvin. Characterization of a dry ice heat exchanger. *International Journal of Refrigeration*. 2011; 34(3): 1085–1097.

[62] Do. W. Lee and C. M. Jensen. Dry-ice bath based on ethylene glycol mixtures. *Journal of Chemical Education*. 2000; 77(5): 629–629.

[63] R. Steinborn, A. Koglbauer, P. Bachor, T. Diehl, D. Kolbe, M. Stappel, and J. Walz. A Continuous wave 10 W cryogenic fiber amplifier at 1015 nm and frequency quadrupling to 254 nm. *Optics Express*.2013; 21(19): 22693–22698.

[64] P. Jelger. High Performance Fiber Lasers with Spectral, Thermal and Lifetime Control [thesis]. Stockholm, Sweden: Department of applied physics, KTH, Royal Institute of Technology,2009.

[65] A. Seifert, M. Sinther, T. Walther, and E. S. Fry. Narrow-linewidth, multi-Watt Yb-doped fiber amplifier at 1014.8 nm. *Appl. Opt*. 2006; 45: 7908–7911.

Chapter 15

Computational Dynamics of Anti-Corrosion Performance of Laser Alloyed Metallic Materials

Olawale S. Fatoba, Patricia A.I. Popoola, Sisa L. Pityana and Olanrewaju S. Adesina

Additional information is available at the end of the chapter

http://dx.doi.org/10.5772/62334

Abstract

Laser surface alloying (LSA) is a material processing technique that utilizes the high power density available from defocused laser beam to melt both reinforcement powders and a part of the underlying substrate. Because melting occurs solitary at the surface, large temperature gradients exist across the boundary between the underlying solid substrate and the melted surface region, which results in rapid self-quenching and resolidifications. Reinforcement powders are deposited in the molten pool of the substrate to produce corrosion-resistant coatings. These processes influence the structure and properties of the alloyed region. A 3D mathematical model is developed to obtain insights on the behavior of laser melted pools subjected to various process parameters. It is expected that the melt pool flow, thermal and solidification characteristics will have a profound effect on the microstructure of the solidified region.

Keywords: Laser Process Parameters, Computational Dynamics, Anti-Corrosion Performance, Martensitic stainless steels, Mild steel

1. Introduction

1.1. Laser surface treatment

Laser surface treatment has a strong impact on classical manufacturing and repair tasks, addressing markets such as turbo machinery, aeronautics, automotive, off-shore and mining as well as tool, die, and mould making and life science [1]. According to Steen and Mazumdar [2], laser has some distinctive properties for surface heating. For opaque materials, such as metals, the laser beam electromagnetic radiation is absorbed within the first few atomic layers and there are no associated eddy currents or hot gas jets. Moreover, there is no radiation spillage outside the optically defined beam area. Compared with other methods of surface modification, laser surface engineering is characterized by possibility of forming alloys of non-equilibrium compositions, formation of a fine microstructure, development of a metallurgical bond between the surface layer and the substrate, a small heat-affected zone and the combination of a controlled minimal dilution of the substrate by the coating material, and nevertheless, a very strong fusion bond between them. Characteristics and advantages such as high productivity, automation worthiness, non-contact processing, elimination of finishing operation, reduced processing cost, improved product quality, greater material utilization and minimum heat affected zone have led to increasing demand of laser in material processing [3–5].

1.2. Categories of lasers

Lasers can be classified according to their active medium, wavelength, and excitation mechanism. There are various types of lasers used in industries, but the common type of lasers used are gas, solid-state, dye, and diode lasers also known as semiconductor lasers classified according to their active medium.

1.2.1. Gas lasers

Gas lasers utilize gas or gas mixture as their active medium. Excitation usually is achieved by current flow through the gas. During operation, the gas is often in the state of plasma, containing a significant concentration of electrically charged particles. Frequently used gases include CO_2, argon, krypton, excimer, and gas mixtures such as helium–neon. The most commonly used gas laser in materials processing is the CO_2 laser. CO_2 lasers use a gas mixture of CO_2, helium (He), nitrogen (N_2), and possibly some hydrogen (H_2), water vapor, and/or xenon (Xe) for generating laser radiation. CO_2 lasers emit light with a wavelength of 10.6 µm with an overall efficiency of 10–13%. Regardless of the low efficiency, the CO_2 lasers have a good beam quality and focusability. They are widely used in engineering and material processing because of the high power that can be obtained (>5 kW) and the high speed accuracy for cutting, welding, and marking both ferrous and non-ferrous materials.

1.2.2. Solid-state lasers

Solid-state lasers, also called solid crystalline or glass lasers, consist of a host and an active ion doped in the solid host material. The active media used are rare earth ions such as neodymium, erbium, and holmium and transition metals such as chromium, titanium, nickel, and others. The most common utilized ions are Cr^{3+} and Nd^{3+} with the host as yttrium aluminum garnet (YAG), glass, and yttrium lithium fluoride (YLF). The beam has a wavelength of 1.06 µm. These lasers generate high output powers, or lower powers with very high beam quality, spectral purity, and stability. These lasers have found major applications in the automotive industry for its high speed welding of body components [6].

1.2.3. Diode lasers

Diode lasers, also known as semiconductor lasers, are based on semiconductor grain media, which are diodes that are electrically pumped. They operate based on electrical pumping with moderate voltages. High efficiency can be achieved particularly for high-power diode lasers and allows their use as pump sources for highly efficient solid-state lasers and diode-pumped lasers. Diode lasers are much smaller than gas or solid state lasers in the same power range. They have found major success in conduction welding, cladding, and laser hardening. Their applications are extremely widespread, including areas as diverse as optical data transmission, optical data storage, metrology, spectroscopy, and materials processing.

1.2.4. Dye lasers

Dye lasers use an organic dye as the gain medium with gain spectrum as available dye or a mixture of dyes. Dye lasers are normally pumped at short wavelengths with a green laser such as argon ion laser, a frequency doubled solid state laser, or an excimer laser emitting ultraviolet light. The most important feature in dye lasers is the output wavelength that can be adjusted. Today, they are still used in areas such as spectroscopy for chemical analysis of gaseous samples because of their distinct wavelengths that are hard to generate [7].

1.2.5. Fiber lasers

Fiber lasers belong to the solid-state laser group. Laser beams are generated by means of seed laser and magnified in specially designed glass fibers, which are supplied with energy through pump diodes. Fiber lasers with a wavelength of 1.064 μm produce an exceedingly small focal diameter; consequently, their intensity is up to 100 times higher than that of CO_2 lasers with the same emitted average power. Fiber lasers are optimally suited for metal marking via annealing, for high-contrast plastic markings, and for metal engraving. Fibers feature a long service life of at least 25,000 laser hours and are generally maintenance free.

1.3. Industrial applications of different lasers

Lasers are preferable tools compared with the other traditional ones. They are widely used in industry for cutting, welding, surface treatments, and drilling, especially in the automobile industry in developed countries [8–10]. Most car frames are produced by laser cutting on a programmed robot assembly line. Also many car components are laser treated or processed. Moreover, lasers are also functionally used for medical purposes such as short-sight correction and cancer operations. Other applications such as those used for communications, data transmission, internet backbones, and audio vision home appliances are increasingly used in daily life. Table 1 shows industrial application of different lasers.

1.4. Laser surface modification techniques

Laser surface techniques have attracted industries owing to the possibility of accurate control of the area where laser radiation is delivered, as well as the amount and rate of energy deposition. The flexibility of control of the beam's interaction with regard to wavelength, energy density, and interaction time and the wide choice of interacting environments has led to the significant developments in laser technology such as laser welding, drilling, alloying, and cladding [11]. The laser's ease of automation and robotic manipulation capability also

Type of laser	Wavelength	Areas of application
Carbon dioxide (CO_2)	10.6 µm	Material processing, surgery, etc.
Dye laser	390–640 nm	Medicine, birth mark removal
Nd: YAG	1.064 µm	Material processing
Nd: Glass	1.062 µm	Velocity and length measurement
Excimer	193 nm	Laser surgery
Ruby	694.3 nm	Tattoo removal, holography
Hydrogen fluoride	2.7–2.9 µm	Laser weapon
Helium-neon	632 nm	Holography, spectroscopy
Argon	454.6 nm	Lithography, spectroscopy

Table 1. Industrial applications of different lasers

make laser surface technique very suitable for repair activities in extreme or remote environments, such as under water or in areas with radioactive contamination [12].

The distinctive advantages of the laser surface alloying (LSA) technique for surface modification include the refinement of the grain size because of rapid quench rates and the generation of meta-stable structures with novel properties that are not feasible by competing methods and laser surface alloying (LSA) modifies the surface morphology and near surface structure of components and its alloys with perfect adhesion to the interface of the bulk steel [13,14]. With optimal laser processing parameters, a dependable coating that is free of pores and cracks can be produced on the matrix. LSA can rapidly provide a crack-free and thick layer in all instances with metallurgical bonds at the boundary between the substrate and alloyed layer [15]. In LSA, external alloying elements in the form of powder, paste, suspension, electrolytic coatings, and plasma or flame sprayed coatings are introduced into the surface of a substrate, as preplaced addition material or injected directly into the melt pool, treated by a high power laser beam [16]. The particles introduced in the interaction zone completely dissolve in the liquid phase, thereby modifying the surface layer chemical composition [17,18]. The result of this is a rapid self quenching and resolidification of new alloy because of the large temperature gradients between the substrate and melted surface region [6]. Evolution of a wide variety of microstructures is one of the consequences as a result of the rapid cooling from the liquid phase [19,20]. Hence, the synthesis of new alloy is possible by depositing a premixed ratio of elemental powders during laser alloying. Powders alloyed on worn or new working surfaces of components by LSA provide specific properties such as erosion resistance, corrosion resistance, high abrasive wear resistance, heat resistance, and combinations of these properties. Consequently, safety in automotive and aerospace applications and improvements in machinery performance can be realized by the method [21]. According to Poulon-Quintina et al. [22], laser beams can generate specific microstructures including nanocrystalline grains and metastable phases because of specific thermal characteristics induced by laser irradiation. Laser processing offers cost advantages and exceptional and important quality over traditional techniques. These include process compactibility, low porosity, high throughput speed, high process efficiency, and good surface homogeneity; in addition, the formation of a amorphous or non-equilibrium phase as well as refinement and homogenization of the microstructure, all without affecting the bulk properties of the substrate [23,24] as shown in Figure 1.

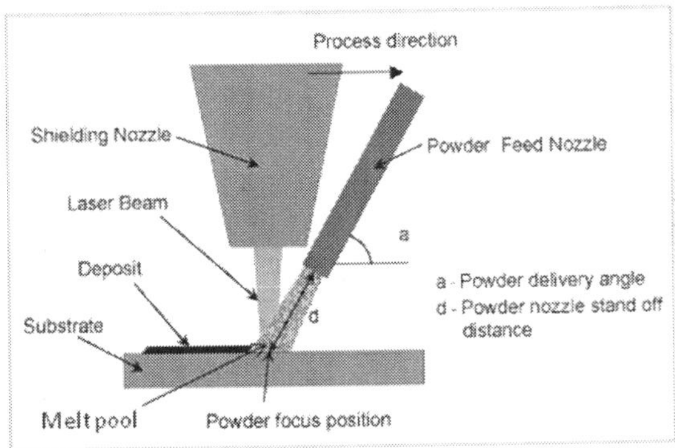

Figure 1. Schematic diagram of laser alloying experimental process (www.twi-global.com).

1.5. Laser beam characteristics

Laser beam characteristics play a very important role in laser material processing. Laser beam is characterized by several parameters such as laser beam mode, focusability, and polarization. The beam with low divergence angle produces a smaller focused spot and greater depth of focus [25]. The laser energy can be distributed in a uniform or Gaussian distribution over the laser beam spot area. To achieve a good quality beam, it is necessary to resonate the beam in a chamber where certain distributions of amplitude and phases of electromagnetic field can be produced because of repeated reflections between the mirrors [26]. These specific shapes produced in the resonator are called transverse electromagnetic modes (TEMs). Each TEM is a different energy distribution across the beam. TEM_{00} (Gaussian) and TEM_{01*} (created by oscillation between orthogonal TEM_{01} modes) are common in industrial lasers.

Another important issue is the reflectivity from the surface of the metal. The reflectivity is a strong function of laser wavelength and temperature, and it varies from metal to metal. As the temperature increases in the process zone, reflectivity decreases and absorptivity increases because of an increase in the photon population [27], and this indicates the potential for more energy absorption by hot material. However, this is only true when the surface conditions remain constant. In practice, there is often oxidation or phase change, which can alter this behavior of absorptivity. Laser absorption differs from one material to the other based on the wavelength of the laser. For example, CO_2 laser is very well absorbed in plastics and plywood, whereas Nd:YAG is poorly absorbed in the same materials. Nd:YAG has good absorption in steel and non-ferrous metals, whereas CO_2 laser is poorly absorbed in some non-ferrous metals [28]. Some metals and their absorptivity in different lasers are shown in Figure 2.

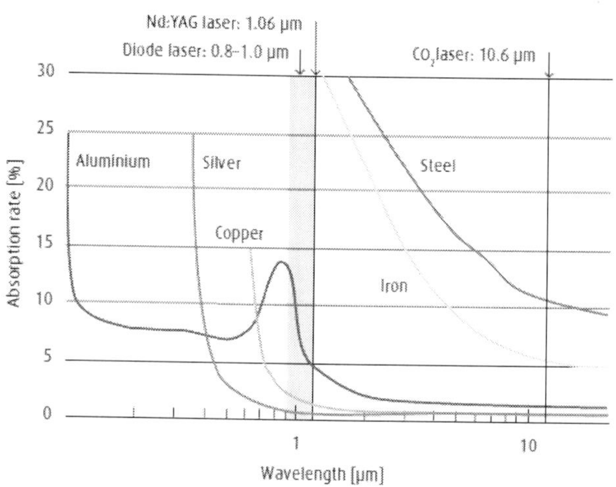

Figure 2. Absorption rate of laser radiations in cold metal [28].

2. Mechanism of heat transfer in laser processing

Heat is defined as energy transferred by virtue of a temperature difference. It flows from a high temperature region to a low temperature region. Heat transfer is used to predict the energy transfer taking place in the material bodies, which result from the temperature difference. There are three modes of heat transfer: conduction, convection, and radiation [29–32].

2.1. Conduction heat transfer

Conduction is transfer of the energy from high temperature region to the low temperature region in a body. In this situation, a temperature gradient will be formed, and heat is transferred by conduction. The rate of heat transfer per unit area is proportional to the normal temperature gradient:

$$q = -KA\frac{\partial T}{\partial x}$$

(1)

This is called Fourier's law of heat conduction. The positive constant K is the thermal conductivity of the material. The negative sign is included to ensure that heat flows in the direction of decreased temperature. q is the rate of heat transfer and $\frac{\partial T}{\partial x}$ is the temperature gradient in the direction of the heat flow. The unit of thermal conductivity K is $W/m/K$. Similarly, heat conduction rate equation can be written in y and z directions. In general, the heat flux is a vector quantity and is expressed as follows:

$$\vec{q} = -k\nabla T \tag{2}$$

$\bar{q}$ = Local heat flux density, W/m^2

k = Thermal conductivity of material, $W/m/k$

∇T = Temperature gradient, K/m

2.2. Convection heat transfer

Convection heat transfer is related to the transfer of heat from a bounding surface to a fluid in motion, or to the heat transfer across a flow plane within the interior of the flowing fluid. If the fluid motion is induced by the fan, blower, pump, or some other similar device, the process is called forced convection. If the fluid motion occurs as a result of the density difference produced by the temperature difference, the process is called free or natural convection [33]. The velocity of the fluid motion obviously influences the heat-transfer rate. Thus, the defining equation of convection heat transfer is:

$$q = hA(T_w - T_\infty) \tag{3}$$

The symbol h is called the convection heat-transfer coefficient. An analytical calculation of h may be made for some systems, but for complex situations it must be determined experimentally. The units of convection heat-transfer coefficient h are in watts per square meter per Celsius degree when the heat flow is in watts. Convection heat transfer will have a dependence on the viscosity of the fluid in addition to its dependence on the thermal properties of the fluid (e.g., thermal conductivity, specific heat, density).

2.3. Radiation heat transfer

In the conduction and convection heat transfer system, the energy is transferred through a material medium. However, in the radiation heat transfer system, heat energy can be transferred through the perfect vacuum regions. The mechanism involved is electromagnetic radiation that is propagated as a result of a temperature difference, which is called thermal radiation. Thermal radiation is electromagnetic radiation emitted by a body by virtue of its temperature and at the expense of its internal energy. Thermal radiation has same nature to the visible light, x-rays, and audio waves. The differences between these are their wavelengths and the source of generation. From thermodynamic consideration, an ideal thermal radiator or blackbody that emits energy, its rate is proportional to the fourth power of the absolute temperature of the body and directly proportional to its surface area. Thus,

$$q_{emitted} = \sigma A T^4 \tag{4}$$

q = Heat transfer per unit time (W)

σ = 5.669 × 10^{-8} W/m^{2}K^4

T = Absolute temperature, Kelvin (K)

This equation is the Stefan-Boltzmann law of thermal radiation. It governs only radiation emitted by a blackbody [33]. The equation is valid only for thermal radiation and may not be treated for other types of electromagnetic radiation so simply. Letter σ is the proportionality constant and is called the Stefan-Boltzmann constant with the value of 5.669 × 10^{-8} W/m^{2}K^4.

3. Mechanism of laser surface alloying

A focused laser beam as a heat source is used by laser alloying to create a melt pool on an underlying substrate. Powder material is then deposited through the nozzles into the melt pool. The deposited powder is then bonded with the substrate upon solidification. Laser alloying is a process similar to cladding except that another component of the alloy is injected into the molten pool of substrate. Alloying requires a greater laser power density than cladding [8]. The alloying process enables metallic and ceramic alloys [8] to be obtained. The process starts with melting of a substrate by laser irradiation. On the surface of a melt, there is temperature distribution, T, which results in the surface tension distribution, γ, as shown in Figure 3. The convection movement of the melt pool is caused by the surface tension, which pulls the materials from the center. When solid particles are injected into the melt pool, there is a good mixing of the particles with the substrate material. Melting of the particles and reaction with the substrate can then take place. As the laser beam moves to the next position, the reaction slows down and stops. Metastable phases are formed as a result of the subsequent rapid cooling of the melt. However, lowering the speed of the laser beam can also slow down the rate of cooling.

Solidification of remelted materials and convection motions in the laser melt pool decide the final distribution of the alloying element in the remelted zone and a big temperature gradient. The powder is either introduced directly by the nozzle during alloying or applied as paste that dries up on the specimen surface, and then subjected to alloying. This makes it possible to develop an alloy with bi- or multi-component structure. Moreover, a high degree of adhesion is obtained between the substrate and the coating, and the rapid self cooling made possible by heat removal to the cold substrate is responsible for the development of advantageous, fine-grained, and novel microstructures [34–36].

Figure 3 shows a schematic diagram of a typical laser surface alloying process. As shown in the figure, a part of the energy is absorbed when a laser beam with defined power moves with a steady scanning speed in a parallel direction and strikes the surface of a solid material. A melt pool is formed on the surface because of laser heating. Concurrently, particles fed into the melt pool mixes with the molten substrate by convention and diffusion. Resolidification and several complex phenomena of the region occur, as the laser beam moves away from one position to another, leading to a final microstructure of the alloyed surface.

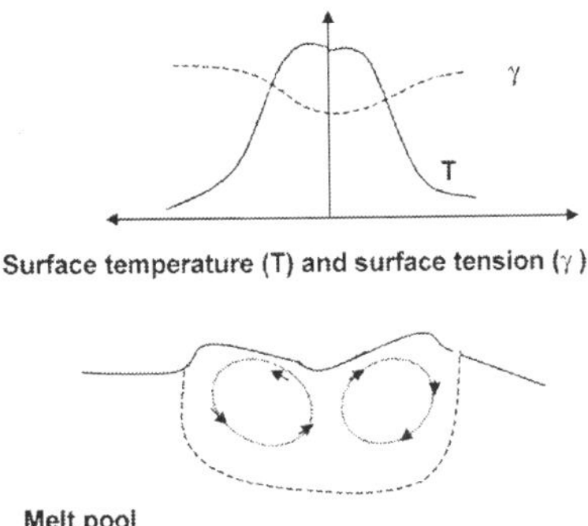

Figure 3. Surface temperature (*T*) and surface tension (γ) distribution across a laser melted pool [37].

4. Transport phenomena in laser alloying process

The characteristics of the molten pool were studied through experimental investigation and numerical modeling by Yang et al. [38]. The authors reported that the Marangoni flow takes hot fluid from the free surface toward the bottom of the melt pool. Likewise, Sarkar et al. [39] also studied heat, momentum, and mass transfer in a laser alloying process. The authors reported that concentration of particles in the molten pool could be predicted. Raj et al [40] reported that the melting of particles injected into the molten pool is not instant and the particles' mass flux boundary condition is difficult to model. Chung and Das [41] studied laser-induced vaporization, melting, and resolidification in metals. The authors derived the relationship for the times needed to commence melting, attain vaporization, and reach the maximum melting depth during the laser heating pulse. Li et al. [42] showed that the model integrating the volumetric heating source is more exact in the prediction of the melting process than the model of the surface heating source. Safdar et al. [43] investigated the geometry of the laser beam influence on the laser transformation hardening of steel. The authors found out that the best thermal history was produced by the triangular beam geometry in other to achieve improved highest hardness and transformation hardening without sacrificing the hardening depths and processing rate. Authors [44–47] studied surface tension gradient's imposition on the characteristics of the convention. However, several essential aspects still remain inexplicable. Among them are the instability of microflows and mechanisms of capillary thermal-concentration convection manifestation [44,45]. The thermal-capillary convection also known as Marangoni convection is one of the overriding factors dictating the quality of the laser alloying.

There are three important physical processes such as mass transfer, heat transfer, and convection transfer in laser molten pool. The rate of heating and cooling is determined by energy transfer, whereas the extent of mixing and final composition is determined by the convection and mass transfer. Specifically, convection transfer in laser molten pool can strongly affect the quality of laser alloying, cladding, and welding. Some computer simulations of heat and mass transfer have been reported [36,48]. To study the mechanism of convection in theory, 3D computer simulation of convection and heat transfer in laser molten pool are needed. The main physical process in laser molten pool requires some of the laser beam to be absorbed while the rest is reflected. If the threshold is exceeded by the heat absorbed, the molten pool will be developed. In stationary melting, molten pool shape and absorbability are constant. The buoyancy force and the surface tension gradient are the two driving forces for fluid flow in laser melt pool.

Safdar et al. [43] studied the effects of non-conventional laser beam geometries on the melting of metallic materials. The authors reported that the laser beam geometries did not have a significant effect on the resulting melt characteristics because of the high thermal conductivity of metals. Laser power intensity was the most significant processing parameter on the phase change in the irradiated region studied by Konrad et al. [49] when melting of a subcooled metal particle was subjected to a nanosecond laser heating. Bin-Mansoor and Yilbas [50] studied the laser heating and the phase change process in the irradiated region. However, the studies were limited to either two-dimensional axi-symmetric heating situations or moving heat source model without including the Marangoni effect. Moreover, the generated convective current in the molten pool because of the Marangoni flow modifies temperature found in the melt pool.

5. Numerical modeling in laser surface treatment

Didenko et al. [51–53] studied in detail the laser alloying process of high purity iron with 40 µm Cr electrolytically predeposited on the sample surface. The authors have used the CW CO_2 laser generating TEM10 Gaussian mode with an output power of 2 kW and 2.5 kW focused to the diameter of 3 mm and constant speed of the work table set to 18.4 mm/sec. The process was carried out in an argon atmosphere. For the process modeling, a multi-phase mathematical model of the laser remelting of high purity iron with a pre-deposited chromium layer was used, resulting from solution of the partial differential equations for conservation of energy, mass, and momentum. The FLUENT program was used for numerical modeling of the fluid flow and mass transfer in the molten pool during laser alloying. Finite element mesh used to simulate alloying process was prepared with the GAMBIT program. The numerical results predicted the final composition in the solidified alloy. Didenko et al. [52,53] compared with corresponding experimental results and the agreement they found was good. The non-uniform chromium distribution (the presence of high chromium concentration fields near the solid/ liquid interface) is caused by a multi-directional liquid material movement, which is due to the presence of few vortexes in the melted pool. The presence of vortexes in the liquid is caused by the non-uniform energy distribution in the laser beam (TEM10 mode), which directly influences the mass transport kinetics and gives rise to the final dimension and shape of the melted pool, its microstructure and, consequently, properties of the resolidified material.

Mondal et al. [54] studied Ni-Cr-Mo cladding on mild steel surface using CO_2 laser and process modeling with response surface methodology (RSM). An anti-corrosive powder mixture of Ni, Cr, and Mo with a selected ratio is deposited as a thin layer on the mild steel plate with the help of 3.5 kW CO_2 laser. Experiments were performed according to L_9 Taguchi orthogonal array. The study of the influence of process parameters on responses and process optimization to find the optimal input parameters combination by expecting the improved clad quality was also studied. Based on experimental data, a mathematical model was developed to find the relationship between process input parameter and responses. It was discovered that there is a high degree of approximation between the experimental results and the predicted one. The results of the experiment were extended to develop the regression model using response surface methodology (RSM). Multi-objective optimization was done to find out the optimal parametric setting to achieve desired clad bead dimension with aspect ratio ≤ 15, during laser cladding process. The optimization result showed that at laser power of 1.014 kW, scan speed of work table at 0.475 mm/min, and powder feed rate of 8.807 g/min, both the responses clad height and clad width are optimized at 0.25 mm and 3.85 mm, respectively. From the regression model, scan speed of work table and powder feed rate were the most significant parameters in laser cladding process. It was concluded that the range of these parameters should be selected carefully because the clad quality was very sensitive to these responses. The response surface methodology was found to be effective for the identification of key process parameters and development of significant relationship between the process variables and response.

Kochure and Nandurkar [55] applied the use of the Taguchi method of experimental design with L_9 orthogonal for selection of optimum process parameters of induction hardening of EN8 D steel. Orthogonal arrays L_9, signal to noise ratio, and analysis of variance (ANOVA) were applied to study performance characteristics of induction hardening process. Hardness and case depth were considered as performance characteristics. An analysis of variance (ANOVA) of response variables showed a significant influence on process variable power and heating time. The experimental investigation showed the effects of process parameters such as power, heating time on hardness, and case depth pattern achieved on work piece. The optimum parameters found were 14 kW power and heating time 4 sec, and power is the most influential parameter. Further multiple regression equations were formulated for estimating predicted values of hardness and case depths at various locations such as case depths at outer and inner vertical, top, and center portion of slots for a specified range. The results obtained by regression equations closely co-relate each other, which validate the regression equation developed.

A range of researchers carried out their extensive research work using CO_2 laser to investigate the laser coating performance on corrosion and wear behavior. Kathuria [56] presented a study of laser cladding process in both stationary and scanning beam modes with the laser cladding of satellite six on mild steel and Cr-Ni materials. The effects of the various parameters such as input power, beam interaction time, scanning frequency, and traverse speed were considered. Shepeleva et al. [57] presented a comparison between the laser cladding process in which the method of direct injection of cladding powder into the melt pool is used and plasma cladding process. They captured optical and SEM photographs of cross-section of clad-substrate interface. It was found that the laser-cladded zone has a smooth interface with the substrate, which prevents stress concentration at the clad-substrate interface during application. They also concluded that the laser-cladded zone, unlike the plasma treated surfaces, is free of micro-cracks and pores. Chryssolouris et al. [58] performed an experimental investigation on laser cladding with aluminum alloy as substrate and copper-based powder as cladding material. The process parameters of their experiments were powder feed rate (g/min), process speed (mm/min), and gas supply (l/h). They observed that the process speed did not affect dilution depth, while increasing powder feed rate might have a negative effect on performance. They concluded that to achieve an optimum clad result, in terms of increased clad depth and minimum alloying zone, powder feed rates should be kept low and process speed should be high. Meng et al. [59] conducted powder laser cladding experiments to improve wear resistance of titanium alloy (Ti-6Al-4V substrate) using NiCoCrAlY powder. The process parameters of this process were laser power (750 W), scanning speed (3–7 mm/s), and laser beam diameter (Φ3 mm). They observed that with high laser scanning speed, thick preplaced powder layer could not be melted completely and the quality of the coating was poor. They concluded that with the preplaced NiCoCrAlY powder, a laser cladding on Ti-6Al-4V surface without cracks and pores could be obtained, and micro-hardness of the surface is two times higher than that before cladding. Davim et al. [60] performed experimental study on geometric form of clad layer. They examined the effect of processing parameters such as laser power, scanning velocity, and powder mass flow rate on clad height, clad width, and depth penetration into the substrate. An analysis of variance (ANOVA) was performed to investigate the influence of processing parameters in the form of single cladding layer and hardness of coating. They also presented a prediction of laser clad geometry for coaxial laser cladding process (6 kW continuous CO_2 laser) through linear multiple regression analysis. They concluded that clad height increased with powder mass flow rate and laser power and decreased with scanning velocity. The depth of penetration increased with laser power and powder mass flow rate. The clad width increased with powder mass flow rate. The present work investigates the parametric effects of laser cladding parameters such as laser power, scan speed, and powder feed rate on performance evaluation parameters, namely clad height and clad width and a process optimization for the selection of optimal parameters combination using response surface methodology (RSM). The result of optimization can be used to set the process parameters at optimum level for the better clad quality during laser cladding operation. The result obtained through RSM technique can also be compared with other optimization method such as genetic algorithm and scatter search approach.

Mondal et al. [61] studied process optimization for laser cladding operation of alloy steel using genetic algorithm and artificial neural network: an investigation on single objective optimization for CO_2 laser cladding process considering clad height (H) and clad width (W) as performance characteristics. The equipment used for laser cladding was a 3.5 kW continuous wave CO_2 Laser Rapid Manufacturing (LRM) system. The LRM set-up consists of a high power laser system integrated with the beam delivery system, powder feeding system, and job/beam manipulation system. The key process parameters were laser power, scan speed, and powder feed rate. First, numerous single tracks were deposited at various machining conditions to obtain continuous and uniform tracks, which facilitated to determine the range of process parameters for control levels in Taguchi method. Then the actual experiments were performed as per the L_9 Taguchi orthogonal array. This optimization of multiple quality characteristics has been done using genetic algorithm (GA) approach. The aim of this work was to predict the performance characteristics (H and W) at optimized condition by applying back propagation method of artificial neural network (ANN). The essential input process parameters are identified as laser power, scan speed of work table, and powder feed rate. To validate the predicted result, an experiment as confirmatory test was carried out at the optimized cladding condition. It was observed that the confirmatory experimental result was showing a good agreement with the predicted one. It has also been found that the optimum condition of the cladding parameters for multi-performance characteristics varies with the different combinations of weighting factors.

Zuljan and Uran [62] studied the optimization of the laser wire cladding of tool steels using factor analysis. The aim was to establish reliable correlations between the input parameters and the parameters of laser wire cladding quality used in optimization. Using a Nd:YAG laser as the energy source, laser wire cladding was carried out on the seven most frequently used tool steels (1.2311, 1.2312, 1.2343, 1.2344, 1.2767, 1.2379, and 1.2550). The quality of a laser wire clad-weld was defined by their geometrical characteristics, mechanical properties, and minimum internal stresses in the area of the laser wire clad-weld. The importance of the general correlations between the laser wire cladding parameters was determined by means of a statistical factor analysis, which provides factor loadings with the variables. The laser wire cladding analysis was carried out through the purposeful selection of control parameters and monitoring of the laser wire clad-weld quality parameters. The results obtained and the laser wire cladding research procedure used provide a suitable tool for attaining the desired laser wire clad-weld quality and can be used by laser wire cladding designers and technologists.

Ermurat et al. [63] studied process parameters investigation of a laser-generated single clad for minimum size using design of experiments. The aim of the study was to investigate the effect of four important process parameters (i.e., laser focal distance, travel speed, feeding gas flow rate, and standoff distance) on the size of single clad geometry created by coaxial nozzle-based powder deposition by high power laser. Design of experiments (DOE) and statistical analysis methods were both used to find optimum parameter combinations to get minimum-sized clad, that is, clad width and clad height. Factorial experiment arrays were used to design parameter combinations for creating experimental runs. This procedure was somehow complicated in understanding the effects of the selected problem parameters on the outcome. Therefore, DOE methodologies were utilized so that the operation can be better modeled/ understood and automated for real life applications. The study also gives future direction for research based on the presented results. Taguchi optimization methodology was used to find out optimum parameter levels to get minimum sized clad geometry. Response surface method was used to investigate the non-linearity among parameters, and variance analysis was used to assess the effectiveness level of each problem parameters. The overall results showed that wisely selected four problem parameters had the most prominent effects on the final clad geometry. Minimum clad size was achieved at higher levels of gas flow rate, travel speed, and standoff distance and at minimum spot size level of the laser focal distance.

Influence of the process parameters was experimented to be able to produce minimum-sized clads created by laser-assisted direct metal part fabrication system using DOE and statistical analysis methods. Several process parameters affect the size of the clad geometry. Laser focal distance, standoff distance, gas flow rate, and travel speed were investigated, and the conclusions can be written as follow: Laser energy intensity is varying at different levels of laser focal distance because the size of the laser spot is changing at each level; results changing of the intensity of the penetrated energy to the substrate. Travel speed relates to the interaction time between laser spot and substrate material, which affects the clad size since dominating the size of the molten pool. Higher travel speeds shorten the interaction time and make the clad size small. Clad size reduces with the increase of the standoff distance. The effect of the standoff distance should be lowered as much as possible to build complex part geometry in good condition. The high level of feeding gas flow rate, minimum sized geometry was achieved because of reducing the powder-laser beam interaction time by increasing the powder particle speed. In addition, there is a powerful relation between standoff distance and gas flow rate. Standoff distance and feeding gas flow rate are the parameters that dominate the shape of the particle flow including particle speed. On top of that, shape of the laser beam waist has a connection about rate of the intensity of each particle moving through the beam of laser.

Mondal et al. [64] studied the application of artificial neural network for the prediction of laser cladding process characteristics at Taguchi-based optimized condition. An investigation on the optimization of multiple performance characteristics during CO_2 laser cladding process considering clad width and clad depth as performance characteristics has been presented. This optimization for multiple quality characteristics has been done using Taguchi's quality loss function. In the present work, a number of experiments have been performed to establish the interrelationship between process variables and response variables using the back propagation method of ANN. The essential input process parameters were identified as laser power, scan speed of work table, and powder feed rate. Moreover, the analysis of variance was also employed to determine the contribution of each control parameter on clad bead quality. To validate the predicted result, an experiment as confirmatory test was carried out at the optimized cladding condition. It was observed that the confirmatory experimental result was showing a good agreement with the predicted one. However, it has been found that the optimum condition of the cladding parameters for multi-performance characteristics varies with the different combinations of weighting factors.

Mondal et al. [65] studied the application of Taguchi-based gray relational analysis for evaluating the optimal laser cladding parameters for AISI1040 steel plane surface. The effect of various laser cladding process parameters such as laser power, scan speed, and powder feed rate on clad bead quality characteristics (or clad bead geometry) for AISI 1040 steel substrate have been studied by performing a number of experiments with L_9 orthogonal array. To find the process parametric setting for best quality clad bead based on experimental results, a multi-response optimization technique using gray relational analysis (GRA) was used. The GRA was applied on laser cladding process to find out the grey relational grade for each experiment. On optimization, power of 1.25 kW, scan speed of 0.8 m/min, and a powder feed rate of 11 gm/min had been found to be the best parametric setting for laser cladding operation of AISI 1040 steel substrate. Moreover, the analysis of variance was also performed to determine the contribution of each control factor on the clad quality characteristics. Finally, to ensure the robustness of GRA, a confirmatory test was performed at selected optimal parametric setting. An expression of gray relational analysis that directly integrates the multiple performance characteristics (i.e., laser power, scan speed, and powder feed rate) into a single performance characteristic is called gray relational grade. Therefore optimization of the complicated multiple performance characteristics can be greatly simplified to a single objective optimization problem through this approach. It was found that the performance characteristics of the laser cladding process such as clad height, clad width, and clad depth were improved together using this methodology. Furthermore, from the results of ANOVA, the contribution of each cladding factor on the cladding quality characteristics in decreasing order were laser power, scan speed of work table, and powder feed rate. Finally, the confirmation tests had ensured the robustness of the optimal combination of laser cladding process for AISI 1040 steel surface.

Babu et al. [66] carried out a systematic investigation on laser transformation hardening (LTH) process on high-strength low-alloy medium carbon steel using design of experiments (DOE). The effect of input process parameters such as laser power, travel speed over the response hardened width (HW), hardened depth (HD), and hardened area (HA) were analyzed. The experimental trials were conducted based on the design matrix obtained from the 3k full factorial design (FFD) using a 2 kW continuous wave Nd:YAG laser power system. A quadratic regression model was developed to predict the responses using response surface methodology (RSM). Based on the developed mathematical models, the direct and interaction effects of the process parameters on LTH were investigated. The optimal hardening conditions were identified to maximize the HW and minimize the HD and HA. The results of the validation test showed that the experimental values quite satisfactorily agreed with the predicted values of the mathematical models and hence, the models can predict the response adequately.

6. 3D-simulation of laser molten pool

Yang et al. [67] reported that to study the mechanism of convection in theory, 3D computer simulation of convection and heat transfer in laser molten pool is needed. The main physical process in laser molten pool requires some of the laser beams to be absorbed while the rest is reflected. If the threshold is exceeded by the heat absorbed, the molten pool will be developed. In stationary melting, molten pool shape and absorbability are constant. The buoyancy force and the surface tension gradient are the two driving forces for fluid flow in laser melt pool.

The surface tension gradient and the buoyancy force are defined by the following equations:

$$\frac{\partial \gamma}{\partial x} = \frac{\partial T}{\partial x} \cdot \frac{\partial \gamma}{\partial T} \tag{5}$$

$$F_b = -\rho \beta \Delta T_g \tag{6}$$

While the governing equations can be written as [68]:

Continuity:

$$\frac{\partial \rho}{\partial t} + \rho \nabla . \vec{V} = 0 \tag{7}$$

Momentum equation:

$$\rho[\frac{\partial \vec{V}}{\partial t} + (\vec{V}.\nabla)\vec{V}] = \mu \nabla^2 \vec{V} - \nabla P + F_b \tag{8}$$

Energy equation:

$$\frac{\partial \vec{T}}{\partial t} + (\vec{V}.\nabla)T = \alpha \nabla^2 T \tag{9}$$

Where $\vec{V} = u\vec{i} + v\vec{j} + w\vec{k}$

V = Total velocity of fluid, u, v, and w are components of V in x, y, and z direction, respectively.

μ = Viscosity

γ = Surface tension

β = Volumetric thermal expansion coefficient

g = gravitational acceleration

T = Temperature

P = Pressure

ρ = Mass density

α = Thermal diffusivity

Yang et al. [67] compared the effects of surface tension gradient and buoyancy force and their fluid fields in laser molten pool with computer simulation. It was discovered that in the center of molten pool, the fluid flow direction was from the bottom to the top. Likewise, on the surface of the molten pool, the fluid flow direction was from the center to the edge and in the interface of solid-liquid, the fluid flow direction was from top to bottom thereby producing a circular flow. As $\frac{\partial \gamma}{\partial T}$ liquid iron is negative, $\frac{\partial \gamma}{\partial x}$ in the pool center is lower than that in the pool edge;

therefore, liquid metal is drawn from center to edge. The Reynolds number (Re) was about 1200 (<critical Re=2000), which indicates a planar flow. The convection field pattern because of buoyancy force and surface tension are similar. This means they have the same flow direction from bottom to top in the pool center. The authors concluded that convection in laser molten pool is mainly induced by surface tension gradient. Finally, the authors concluded that there was a strong convection and heat transfer in laser molten pool. There are left and right flow cycles symmetrical to the plane center, which was perpendicular to the moving direction of laser beam. Convention and heat transfer caused the laser molten pool widen. The simulated results agreed with experimental results.

7. Conclusion

The following conclusions can be deduced:

- The application and effectiveness of LSA are highly dependent and sensitive to small changes in process parameters, and these process parameters play a significant role in the quality of the alloyed layer.

- When metal foam is in the phase changing environment, the heat transfer process is conduction dominated, irrespective of the heat source pulse width.

- Convection in laser molten pool is mainly induced by surface tension gradient. Also, the buoyancy force and the surface tension gradient are the two driving forces for fluid flow in laser melt pool.

- A little imbalance in the process parameters can result in large variations in the geometry, microstructure, and properties of the alloyed zone.

- Careful selection and control of process parameters through an optimization process are required to establish an appropriate laser power-scan speed combination for achieving defect-free alloyed layers.

Author details

Olawale S. Fatoba[1*], Patricia A.I. Popoola[1*], Sisa L. Pityana[1,2] and Olanrewaju S. Adesina[1]

*Address all correspondence to: fatobaolawale@yahoo.com; popoolaapi@tut.ac.za

1 Department of Chemical, Metallurgical and Materials Engineering, Tshwane University of Technology, Pretoria, South Africa

2 Centre for Scientific and Industrial Research-National Laser Centre, Pretoria, South Africa

References

[1] Kelbassa, I. (2011). Laser surface treatment and additive manufacturing: Basics and application examples.

[2] Steen, W.M., Mazumdar, J. (2010). Laser material processing. 4th edition, Springer-Verlag London Ltd., London.

[3] Lo, K.H., Cheng, F.T., Kwok, C.T., Man, H.C. (2003). Improvement of cavitation erosion resistance of AISI 316 stainless steel by laser surface alloying using fine WC powder. Surface and Coatings Technology. 165: 258–267.

[4] Oberlander, B.C., Lugscheider, E. (1992). Comparison of properties of coatings produced by laser cladding and conventional methods. Materials Science and Technology. 8: 657–665.

[5] Li, C., Wang, W., Han, B. (2011). Microstructure, hardness and stress in melted zone of 42CrMo steel by wide-band laser surface melting. Optics and Lasers in Engineering. 49: 530–535.

[6] Wirth, P. (2004). Introduction to industrial laser materials processing. Rofin, Germany.

[7] Labuschagne, K. (2006). Investigative study of martensite formation in laser transformation hardened steels. M-Tech. dissertation, Pretoria, Tshwane University of Technology.

[8] Steen, W.M., Powell, J. (1981). Laser surface treatment. Materials in Engineering (Surrey, England). 2(3): 157–162.

[9] Steen, W.M., Mazumdar, J. (2010). Laser material processing. 4th edition, Springer-Verlag London Ltd., London.

[10] Steen, W.M. (1998). Laser material processing. 2nd edition, Springer-Verlag London Ltd., London.

[11] Kusinki, J., Kac, S., Kopia, A., Radziswewska, A., Rozmus-Gornikowska, M., Major, B., Major, L., Marckar, J., Lisiecki, A. (2012). Laser modification of materials surface layer-review paper. Bulleting of the Polish Academy of Sciences. 60(4): 711–728.

[12] Mondal, A.K., Kumar, S., Blawert, C., Dahotre, N.B. (2008). Effect of laser surface treatment on corrosion and wear resistance of ACM720 Mg alloy. Surface and Coatings Technology. 202: 3187–3198.

[13] Kwok, C.T., Cheng, F.T., Man, H.C. (2006). Cavitation erosion and corrosion behaviour of laser –aluminized mild steel. Surface and Coatings Technology. 200: 3544–3552.

[14] Dobrzanski, L.A., Piec, M., Bonek, M., Jonda, E., Klimpel, A. (2007). Mechanical and tribological properties of the laser alloyed surface coatings. Journal of Achievements in Materials and Manufacturing Engineering. 20(1–2): 235–238.

[15] Fagagnolo, J.B., Rodrigues, A.V., Lima, M.S.F., Amigo, V., Caram, R. (2013). A novel proposal to manipulate the properties of titanium parts by laser alloying. Scripta Materialia. 68: 471–474.

[16] Brytan, Z., Bonek, M., Dobrzanski, L.A. (2010). Microstructure and properties of laser surface alloyed PM austenitic stainless steel. Journal of Achievements in Materials and Manufacturing Engineering. 40(1): 70–79.

[17] Li, J., Chen, C., Zhang, C. (2011). Phase constituents and microstructure of Ti_3Al/Fe_3Al + TiN/TiB_2 composite coating on titanium alloy. Surface Review and Letters. 18: 103–108.

[18] Kwok, C.T., Lo, K.H., Cheng, F.T., Man, H.C. (2003). Effect of processing conditions on the corrosion performance of laser surfaced melted AISI 440C martensitic stainless steel. Surface and Coatings Technology. 166: 221–230.

[19] Adebiyi, D.I., Popoola, A.P.I., Pityana, S.L. (2014). Microstructural evolution at the overlap zones of 12Cr martensitic stainless steel laser alloyed with TiC. Optics and Laser Technology. 61: 15–23.

[20] Wei, L., Huijun, Y., Chuanzhong, C., Diangang, W., Fei, W. (2013). Microstructures of hard coatings deposited on titanium alloys by laser alloying technique. Surface Review and Letters. 20: 1–6.

[21] Yakovlev, A., Bertrand, P., Smurov, I. (2004). Laser cladding of wear resistant metal matrix composite coatings. Thin Solid Films. 453: 133–138.

[22] Poulon-Quintina, A., Watanabe, I., Bertranda, C., Watanabc, E. (2012). Microstructure and mechanical properties of surface treated cast titanium with Nd:YAG laser. Dental Materials. 28: 945–951.

[23] Zhou, R., Sun, G.F., Chen, K.K., Tong, Y.Q. (2014). Effect of tempering on microstructure mechanical properties of cast iron rolls laser alloyed with C-B-W-Cr. Proceedings of the Global Conference on Polymer and Composite Materials.

[24] Sugioka, K., Cheng, Y. (2014). Ultrafast lasers-reliable tools for advanced materials processing. Light: Science & Applications. 3–30

[25] Toyserkani, E., Khajepour, A., Corbin, S. (2005). Laser cladding. CRC press, London.

[26] Svelto, O. (1998). Principles of lasers. 4th edition, Plenum Press, New York.

[27] Steen, W.M. (2003). Laser material processing. 3rd edition, Springer-Verlag London Ltd., London.

[28] Berkmanns, J., Faerber, M. (2010). Laser basics, BOC. Available from: https://boc.com.au/boc_sp/downloads/gas_brochures/BOC_216121_Laser%20Basics_v7.pdf.

[29] Callen, H.B. (1960). Thermodynamics, 31–58.

[30] Holman, J.P. (1989a). Heat transfer, SI Metric edition, 2–6.

[31] Müller, I. (1985a). Thermodynamics (Interaction of Mechanics and Mathematics Series), 1–3.

[32] Müller, I. (1985b). Thermodynamics (Interaction of Mechanics and Mathematics Series), 47–68.

[33] Holman, J.P. (1989b). Heat transfer, SI Metric edition, 10–15.

[34] Kwok, C.T., Cheng, F.T., Man, H.C. (2006). Cavitation erosion and corrosion behaviour of laser – aluminized mild steel. Surface and Coatings Technology. 200: 3544–3552.

[35] Dobrzanski, L.A., Piec, M., Bonek, M., Jonda, E., Klimpel, A. (2007). Mechanical and tribological properties of the laser alloyed surface coatings. Journal of Achievements in Materials and Manufacturing Engineering. 20(1–2): 235–238.

[36] Chande, T., Mazumder, J. (1983). Composition control in laser surface alloying. Metallurgical Transactions B. 14B: 181.

[37] Pawlowski, L. (1999). Thick laser coatings: A review. Journal of Thermal Spray Technology. 8(2): 279–295.

[38] Yang, L.X., Peng, X.P., Wang, B.X. (2001). Numerical modelling and experimental investigation on the characteristics of molten pool during laser processing. International Journal of Heat and Mass Transfer. 44: 4465–4473.

[39] Sarkar, S., Raj, P.M., Chakraborty, S., Dutta, P. (2002). Three dimensional computational modeling of momentum, heat and mass transfer in a laser surface alloying process. Numerical Heat Transfer Part A. 42: 307–326.

[40] Raj, P.M., Sarkar, S., Chakraborty, S., Dutta, P. (2001). Three dimensional computational modeling of momentum, heat and mass transfer in laser surface alloying with distributed melting of alloying element. International Journal of Numerical Methods for Heat & Fluid Flow. 11(6): 576–599.

[41] Chung, H., Das, S. (2004). Numerical modeling of scanning laser- induced melting, vaporization and resolidification in metals subjected to step heat flux input. International Journal of Heat and Mass Transfer. 47: 4153–4164.

[42] Li, J.F., Li, L., Stott, F.H. (2004). Comparison of volumetric and surface heating surfaces in the modeling of laser melting of ceramic materials. International Journal of Heat and Mass Transfer. 47: 1159–1174.

[43] Safdar, S., Li, L., Sheikh, M.A. (2007). Numerical analysis of the effects of non-conventional laser beam geometries during laser melting of metallic materials. Journal of Physics D: Applied Physics. 40: 593–603.

[44] Uglov, A.A., Smurov, I.Y., Taguirov, K.I., Guskov, A.G. (1992). Simulation of unsteady-state thermocapillary mass transfer for laser doping of metals. International Journal of Heat and Mass Transfer. 35(4): 783–793.

[45] Smurov, I., Covelli, L., Tagirov, K., Aksenov, L. (1992). Peculiarities of pulse laser alloying: Influence of spatial distribution of the beam. Journal of Applied Physics. 71(7): 3147–3158.

[46] Yuan, Z., Mukai, K., Huang, W. (2002). Surface tension and its temperature coefficient of molten silicon at different oxygen potentials. Langmuir. 18: 2054–2062.

[47] He, X., Fuerschbach, P.W., DebRoy, T. (2003). Heat transfer and fluid flow during laser spot welding of 304 stainless steel. Journal of Physics D: Applied Physics. 36: 1388–1398.

[48] Xichen, Y. (1990). Study on wide-band pattern of laser heat treatment [J]. Chinese Journal of Lasers. 17(4): 229–235.

[49] Konrad, C., Zhang, Y., Shi, Y. (2007). Melting and resolidification of a subcooled metal powder particle subjected to nanosecond laser heating. International Journal of Heat and Mass Transfer. 50: 2236–2245.

[50] Bin-Mansoor, S., Yilbas, B.S. (2006). Laser pulse heating of steel surface: Consideration of phase change process. Numerical Heat Transfer, Part A. 50(8): 787–807.

[51] Didenko, T. (2006). Laser surface melting – modelling and experimental verification of the melted zone shape and size, and chemical homogeneity. PhD Thesis, AGH University of Science and Technology, Kraków, 2006 (in Polish).

[52] Didenko, T., Kusinski, J., Kusinski, G. (2006). Multiphase model of heat and mass transport during laser alloying of iron with electrodeposited chromium layer. Proceedings of Multiscale and Functionally Graded Materials Conference. 1: 640–646.

[53] Didenko, T., Siwek, A., Kusinski, J. (2004). Numerical modelling of the laser alloying process. Proceedings of XI Conference: Informatics in Metals Technology. 1: 179–186.

[54] Mondal, S., Asish, B., Pradip, K. (2011). Ni-Cr-Mo cladding on mild steel surface using CO_2 laser and process modeling with response surface methodology. International Journal of Engineering Science and Technology. 3: 6805–6816.

[55] Kochure, P.G., Nandurkar, K.N. (2012). Taguchi method and ANOVA: An approach for selection of process parameters of induction hardening of EN8 D steel. International Journal of Advance Research in Science, Engineering and Technology. 1(2): 22–27.

[56] Kathuria, Y.P. (1997). Laser-cladding process: A study using stationary and scanning CO_2 laser beams. Surface & Coatings Technology. 97: 442–447.

[57] Shepeleva, L., Medres, B., Kaplan, W.D., Bamberger, M., Weisheit, A. (2000). Laser cladding of turbine blades. Surface & Coatings Technology. 125: 45–48.

[58] Chryssolouris, G., Zannis, S., Tsirbas, K., Lalas, C. (2002). An experimental investigation of laser cladding. CIRP Annals: Manufacturing Technology. 51(1): 145–148.

[59] Meng, Q., Geng, L., Ni, D. (2005). Laser cladding NiCoCrAlY coating on Ti-6Al-4V. Materials Letters. 59(22): 2774–2777.

[60] Davim, J.P., Oliveira, C., Cardoso, A. (2006). Laser cladding: An experimental study of geometric form and hardness of coating using statistical analysis. Proceedings of the Institute of Mechanical Engineers, Part B: Journal of Engineering Manufacture. 220(9): 1549–1554.

[61] Mondal, S., Tudu, B., Asish, B., Pradip, K. (2012). Process optimization for laser cladding operation of alloy steel using genetic algorithm and artificial neural network. International Journal of Computational Engineering Research. 2(1): 18–24.

[62] Zuljan, D., Uran, M. (2010). Optimization of the laser wire cladding of tool steels using factor analysis. Lasers in Engineering. 20: 21–38.

[63] Ermurat, M., Arslan, M.A., Erzincanli, F., Uzman, I. (2013). Process parameters investigation of a laser-generated single clad for minimum size using design of experiments. Rapid Prototyping Journal. 19(6): 452–462.

[64] Mondal, S., Asish, B., Pradip, K. (2014). Application of artificial neural network for the prediction of laser cladding process characteristics at Taguchi-based optimized condition. International Journal of Advanced Manufacturing Technology. 70: 2151–2158.

[65] Mondal, S., Paul, C.P., Kukreja, L.M. (2013). Application of Taguchi-based gray relational analysis for evaluating the optimal laser cladding parameters for AISI 1040 steel plane surface. International Journal of Advanced Manufacturing Technology. 66: 91–96.

[66] Babu, P.D., Buvanashekaran, G., Balasubramanian, K.R. (2013). Experimental investigation of laser transformation hardening of low alloy steel using response surface methodology. International Journal of Advanced Manufacturing Technology. 67: 1883–1897.

[67] Yang, B., Yang, X.C., Lei, J.B., Wang, Y.S. (2013). Computer simulation of physical transfer process in laser molten pool. Applied Mechanics and Materials. 341–342: 324–328.

[68] Brent, A.D., Voller, V.R., Reid, K.J. (1988). The enthalpy-porosity technique for modeling convection diffusion phase change: Application to the melting of a pure metal. Numerical Heat Transfer. 13: 297–318.

Chapter 16

Laser Surface Modification — A Focus on the Wear Degradation of Titanium Alloy

Olanrewaju Adesina, Patricia Popoola and Olawale Fatoba

Additional information is available at the end of the chapter

http://dx.doi.org/10.5772/61737

Abstract

Over the years, engineering materials are being developed due to the need for better service performance. Wear, a common phenomenon in applications requiring surface interaction, leads to catastrophic failure of materials in the industry. Hence, preventing this form of degradation requires the selection of an appropriate surface modification technique. Laser surface modification techniques have been established by researchers to improve mechanical and tribological properties of materials. In this chapter, adequate knowledge about laser surface cladding and its processing parameters coupled with the oxidation, wear and corrosion performances of laser-modified titanium has been reviewed.

Keywords: Surface modification technique, Laser surface cladding, Wear, Titanium

1. Introduction

Considering the swift increase in fuel consumption around the globe, it is essential to recognize the need for lightweight and high specific strength materials as a suitable approach to resolve the growing energy demand [1]. The energy efficiency of aero-engines and automobiles can be enhanced by reducing the engine weight [2]. In addressing the issue of energy efficiency, titanium and its alloys have been the prime materials for aerospace (with a weight share of about 36% being applied mainly in the fan and compressor sections for disks and blades; other areas include landing gear, window frames, galleys, and lavatories) and nonaerospace sectors such as automobiles (in the case of Rolls Royce and Jaguar cars where titanium is being used for suspension springs, connecting rods, valve strings, and underbody panels). In addition, Ti-6Al-4V alloy has the ability to substitute steel in friction and wear-critical diesel engine components like connecting rods, intake valves, pistons, suspension strings, and movable

turbocharger vanes. The distinctive properties of titanium alloy (Ti-6Al-4V) such as low density (from an economical point of view, a lower mass implies lower fuel consumption), its excellent combination of high specific strength ratio that is maintained at an elevated temperature, low modulus of elasticity, and corrosion resistance justify its applications in aerospace, automotive, and marine industries [3, 4]. For example, higher operating temperatures of cars and turbine engines allow more efficient energy (fuel) conversion with lower toxic emissions [5]. In addition, titanium also has a high melting point, 1678 °C, indicating that it shows good creep resistance over different range of temperatures [6, 7]. However, with the distinctive advantages, voiding catastrophic breakdown in application of titanium in higher service temperature and friction in aerospace and automobile industries justifies the need for surface engineering techniques for improving performance of engineering components, longer component life, and failure prevention. In this chapter, adequate knowledge about laser surface cladding (LSC), a type of laser surface modification, and its processing parameters coupled with the oxidation, wear and corrosion performances of laser-modified titanium will be discussed.

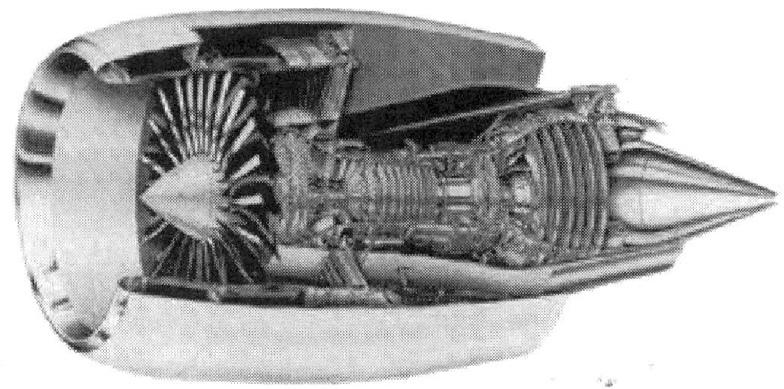

Figure 1. Titanium usage in the GE-90 aero-engine.

2. Titanium and its alloy

Titanium is richly available in the earth's crust at a level close to 0.6%, making it the fourth most abundant metal after the likes of aluminum, iron, and magnesium [8]. Rutile (TiO_2) and ilmenite ($FeTiO_3$) are important mineral sources of titanium alloy. Since their discovery in the early 1950s, titanium and its alloys have become choice materials for many (e.g., chemical, power generation, automobile, aerospace, and airframe) industries. Ti-6Al-4V is the most popular titanium-based alloy contributing to over 50% of global consumption [9], and it is 70% high in the United States. It has biocompatibility for biomedical implant applications, coupled with a good combination of mechanical and corrosive properties [10].

3. Properties and limitations

The unique properties of titanium and its alloys such as low density, excellent combination of high specific strength ratio, which is maintained at an elevated temperature, low modulus of elasticity, good compatibility, and good corrosion resistance, make them choice materials in a wide range of engineering applications such as aerospace, power generation, offshore, and chemical industries [5, 11]. While titanium possesses vast applications and excellent properties, its high relative cost, low hardness, poor oxidation resistance, its propensity to fail by galling, and poor tribological behavior in terms of high and unstable friction coefficient have retarded its engineering applications [2, 6, 12–15]. For example, numerous engineering components made of Ti-6Al-4V alloy are easily damaged at two surfaces in contact under load and in relative motion [4]. The atomic structure, crystal structure, and relatively low tensile and shear strength of titanium oxide film are fundamental causes for the high coefficient of friction and poor tribological properties of titanium [15].

4. Physical metallurgy of titanium and its alloys

Titanium is known to exist in two crystal states. Titanium and its alloys have a high melting point (1668 °C) and exist as a hexagonal closely packed (HCP) crystal structure alpha (α) phase below 882 °C, and above 882 °C they transform into a body-centered cubic (BCC) structure beta (β) at higher temperatures.

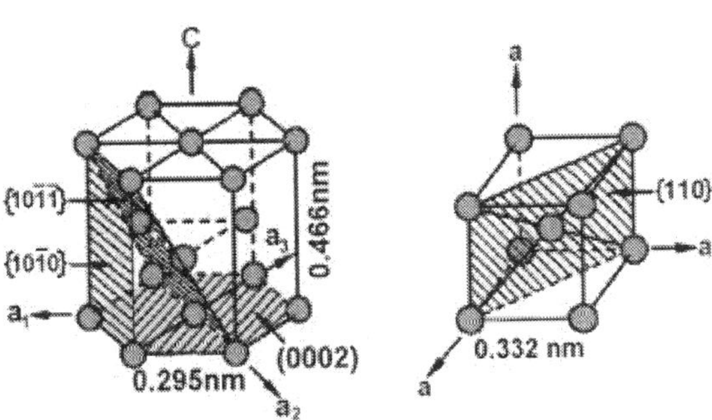

Figure 2. Crystal structure of titanium as (a) HCP α phase and (b) BCC β phase.

Titanium's properties are enhanced by its allotropic behavior characterized by alpha, alpha plus beta phase, and beta phases. They are microstructures and temperature dependent, which result from chemical composition and thermo-mechanical processing. The transformation temperature is strongly influenced by interstitial and substitutional elements and therefore depends on the purity of the metal [8]. The α-phase alloys of titanium are known to be characterized by hard, tough properties such as good corrosion resistance, good weld ability, creep resistance, and receptive to heat treatment coupled with ease of processing and fabrication. This justifies their application in fields such as aerospace for friction-related application. The α-phase alloy is stabilized using elements such as Al, O, N, and C, while the β-phase alloy is characterized by soft and malleable properties with low Young's modulus and superior corrosion resistance.

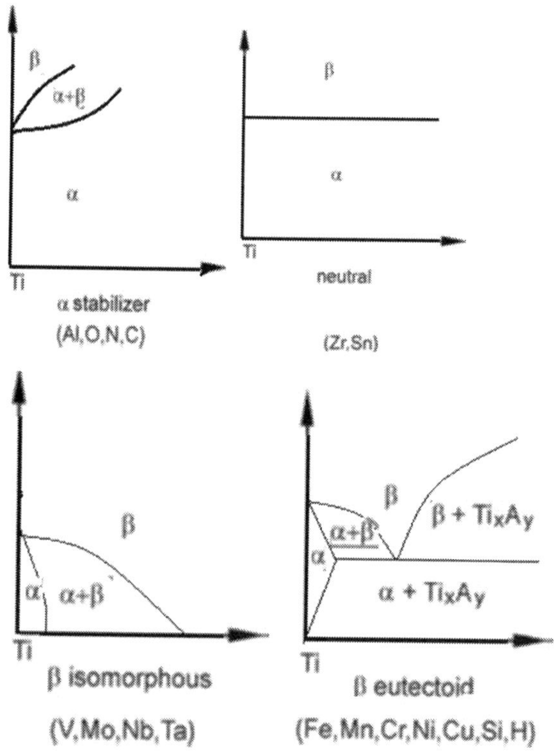

Figure 3. Equilibrium phase diagrams of Ti alloys with effects of alloying elements.

β-Phase alloys are stabilized in two stages, namely the isomorphus stage, using the elements Mo, V, Nb, and Ta; and the eutectoid stage, using the elements Fe, W, Cr, Si, Co, Mn, and H. In contrast, alpha + beta-phase alloys offer a combination of excellent ductility and strength when properly heat-treated, which makes them stronger than the alpha phase and even the beta-phase counterparts [10] due to the presence of both the α and β phases. $\alpha + \beta$ alloy is by far the most commonly used titanium alloy. The common type of Ti alloy containing α and β stabilizer is the Ti-6Al-4V alloy. In order to achieve the best combinations for a given application, an optimum control of the microstructure is essential. The properties of Ti-6Al-4V depend on various factors like composition, relative proportions of phases, heat treatment, and thermo-mechanical process conditions [11]. A comparison of Ti-6Al-4V alloy with the commercial pure Ti (Cp Ti) microstructure shows that the Cp Ti microstructure contains grain boundaries, fine acicular α, and Widmanstätten α structures, while Ti-6Al-4V alloy shows the presence of grains and acicular structures. The alloy is characterized by high strength and resistance to wear and corrosion. These advantageous properties make Ti-6Al-4V to be employed in various industries. In the aerospace industry, the alloy provides high strength at elevated temperatures. Titanium alloys are regarded weak at high temperatures, because they do not function well in these conditions.

5. Surface engineering

Titanium's limited use in engineering applications is due to its poor tribological properties, which are susceptible to failure by galling and high and unstable friction coefficients when rubbing against bearing materials [2]. Wear, a common phenomenon in applications requiring surface interaction, leads to catastrophic failure of materials in the industry. Hence, preventing this form of degradation requires the selection of an appropriate surface modification technique. A technique of achieving the specified requirement is by the development of high-temperature resistance, improved hardness, and high wear-resistant coatings suitable to protect the base material against corrosion, wear, and erosion – corrosion at high temperatures. Surface modification techniques can be applied to address these limitations [16, 17], such as improvement in the functionality of a solid surface by altering its chemical composition or microstructure leading to increase in the surface hardness, decrease in coefficient of friction, and enhanced wear resistance of titanium alloys without altering the desirable bulk properties of the substrate [2, 14].

5.1. Benefits of laser surface modification over conventional techniques

Various modification techniques are used to deposit the alloy layer onto the substrate such as pre-placing the alloy layer by electroplating, ion implantation, physical vapor deposition (PVD), chemical vapor deposition (CVD), carburizing, nitriding, thermal oxidation heat treatment, laser surface alloying (LSA), and laser surface cladding (LSC) [18]. The chemical heat treatment processes such as nitriding, carburizing, and boriding have some demerits such as long processing time and easy deformation of the substrate being treated [19]. In addition, thermal spray coatings possess low coating density and limited bond strength between the coating and the substrate. It is pertinent to note that these techniques give rise to many difficulties, such as poor adherence, lower bonding strength, and some defect at the interface [20]. However, laser surface modification techniques have been established by researchers to improve mechanical and tribological properties of materials. The main advantages of using laser as surface treatment are that the thermally affected regions are easily controlled in terms of depth, extent, and time above temperature [21]. Further, automation is possible due to lack of environmental disturbance while the radiant energy is delivered to the process.

6. Laser surface engineering

Laser was invented in 1960 as a new form of energy in industrial applications. Laser is one of the most flexible forms of energy that can be used to generate any required thermal experience on a substrate. Laser, an acronym for light amplification by stimulated emission of radiation, is known as a coherent, convergent, and monochromatic beam of electromagnetic radiation with wavelength ranging from ultraviolet to infrared [22]. Laser is an ideal tool for surface modification of metals in improving their corrosion and tribological properties [21, 23]. Laser generates radiant energy that is absorbed by top atomic layers of an opaque material, where it can either heat the surface or excite the surface atoms, leading to pyrolytic (thermo-chemical property of material at high temperatures) or photolytic processes (direct interaction of the photons by light or other radiant energy). If the photon energy is sufficiently high, the absorption of laser energy can result in phase transformations of the substrate [24]. The absorption process depends on the nature of the substrate and laser parameters used. A thin layer of the material could be heated, melted, or vaporized, and thereafter it solidifies to generate refinement or homogenization of the microstructure [25]. This takes place under various heat transfer processes such as conduction into the materials, convection, and radiation from the surface. This explains the advantage laser has over the available types of light sources, that is, the highly directional and high-intensity beam with an ability to focus on a small spot. The important properties that justify the use of laser in a wide variety of applications in manufacturing industries, electronics, medical, surveying, communication, and other industrial areas are: spatial and temporal coherence, low divergence, high continuous or pulsed power density, and monochromaticity [21].

7. Types of lasers

Lasers can be classified according to either the active medium, wavelength, or excitation mechanism into the following types: the CO_2 laser (with a wavelength of 10.6 μm), the neodymium yttrium aluminum garnet (Nd:YAG) laser (with a wavelength of 1.06 μm), the high-power diode laser (HPDL; 800–950 nm), and the excimer laser (248 nm for KrF).

7.1. The CO_2 laser

The CO_2 laser has a wavelength of 10.6 μm and output power can range from 1 W to more than 10 KW. They are widely used in engineering and material processing due to their ability to produce very high power with relative efficiency that can be obtained and high-speed accuracy for cutting, welding, and marking both ferrous and nonferrous materials.

7.2. Nd:YAG lasers (Solid-state type)

Nd:YAG laser consists of crystalline YAG with the chemical formula Y3Al5O12 and a wavelength of 1.06 μm. Shorter wavelength, temperatures, and nature of the surface usually lead to a higher absorptivity for metallic materials [25]. The advantage Nd:YAG laser has over CO_2 laser is that it couples better. This type of laser has found major application in the automotive industry for high-speed welding of body parts [26]. The main advantage of Nd:YAG laser over CO_2 laser relies on its shorter wavelength and its ability to deliver laser radiation through optical fibers. As a result of these advantages, the pulse Nd:YAG laser probably has a wider variety of applications (in various forms of material processing: alloying, cladding, drilling, spot welding, and laser marking) than any other type of laser.

7.3. Excimer lasers

These types of lasers are available only as pulsed lasers, which produce intense output in the ultraviolet and deep ultraviolet regions. They are used extensively in micromachining and medical applications. The main advantage of an excimer laser over other types is its very short wavelength. Excimer lasers have good beam quality and focused ability to spot the diameter that is approximately smaller than CO_2 laser beam with the same beam quality.

8. Laser surface modification

There are several laser surface modifications techniques, such as surface alloying, cladding, melting, hardening, direct deposition, physical deposition, and laser melt injection of laser surface. Laser surface modification techniques have been used extensively to improve the wear and corrosion resistance of mechanical components [27].

8.1. Laser surface hardening

Laser surface hardening (LSH) is a heat treatment technique used on a material surface domain to increase hardness of a material by the use of laser beam energy. The purpose of LSH is to increase the hardness of the boundary layer of the substrate by rapid heating and then by quenching. The heated zone is quenched by self-cooling, and this leads to the desired hardening effect due to a change in the microstructure [28]. In ferrous material, the primary basic mechanism of surface hardening is by phase transformation to form the relatively hard martensitic phase in the surface layer. The advantages of LSH over conventional methods of hardening include its flexibility, ability to automate the hardening process, and contactless local heat treatment with no need for additional cooling media such as oil or water [29].

8.2. Laser surface melting

Laser surface melting is a process where a thin layer of the substrate is melted by a high-power laser beam, which is then rapidly solidified without any attempt to modify the surface layer chemical composition [6]. The main advantage of this process over other laser modification processes is its ability to alter the microstructure without changing the composition [11].

8.3. Laser surface alloying

Laser surface alloying is one of the modification methods for improving surface-dependent properties such as wear and corrosion resistance 46. Laser surface alloying is a process of incorporating additional alloying elements into the surface of a material by a high-power laser beam to melt metal coatings and a small portion of underlying substrate). It involves high rate of melting, intermixing, and rapid solidification of the pre-placed or co-deposited alloying elements with part of the underlying substrate to form an alloyed layer. An LSA technique combines modification of both the microstructure and the chemical composition.

8.4. Laser surface cladding

Laser surface cladding is a rapid solidification technique that could overcome the aforementioned difficulties with many advanced features, such as thick coatings, low dilution ratio, high cooling rate, crack-free layer, reduction or elimination of porosity, limited heat affected zone with low thermal distortion, high refined microstructure, and strong metallurgical bond between coating and substrate [30]. In order to overcome the restriction of titanium alloys in machinery performance and safety at low cost with high value elements, laser cladding is employed to fabricate coatings with advanced tribological properties and high-temperature oxidation resistance [31]. Laser cladding process is a surface modification process in which a defocused laser beam is used to fuse an alloy or powder on a substrate [32]. In LSC, alloy may be introduced onto the surface of a substrate as powder or wire either during (direct injection) or prior to processing (pre-placed). The beam energy melts and solidifies rapidly both the pre-placed or injected powders and a thin layer of the opaque substrate [18]. Here, the powder and thin layer of the substrate rapidly reach their melting point causing homogenization to be achieved before solidification due to the photon energy absorption [33]. With this, vaporization is avoided due to rapid heating and solidification of the molten clad, which helps to inhibit long-range diffusion, avoid crystallization, achieve strong metallurgical bond with the substrate, and increase hardness [6, 34]. Laser beam–specific thermal characteristics induced by laser irradiation help generate specific microstructures, including metastable phases and nano-crystalline grains, which is an advantage over conventional techniques [35]. The most important factor to consider during cladding process is the melting of the alloying material to the substrate. This can be achieved by appropriate selection and control of laser processing parameter such as laser power (P), laser beam size (beam diameter D), laser scanning velocity (V), and thermal properties of the substrate, which also helps to achieve desirable properties such as degree of heating and phase transformation [18].

8.4.1. Single-track clad

During laser cladding process, laser energy emitted by the laser beam melts the injected powder causing fusion between the clad material and the substrate. Two vital factors, clad height and dilution in single track, are subject to laser power variation, laser scan rate, mass flow rate, and type of powder being deposited. Laser single-track cladding is conducted with incident laser beam on the working substrate for a single pass. Here, the width of the track is smaller or equal to the laser spot size with the clad height depending variably on laser working parameters [23].

8.4.2. Multi-track clad

In comparison to single-track cladding, multi-track laser cladding involves consecutive overlap of one track by the subsequent track. Morphologically, multi-tracks tend to exhibit dendrites compared to a single track, which is subsequently attributed to longer exposure periods at elevated temperatures [36]. To give yield to a clad with required thickness and larger surface area, single-track clad has to be repeated at several increments. This increment significantly affects the clad height and dilution ratio.

8.4.3. Forms of powder deposition: Cladding

Laser cladding process can be primarily differentiated by its means of introduction of clad material on the surface of the desired substrate. Powder and wire are two common forms in which clad material is introduced on the surface of the substrate. Powder deposition can be either coaxial or pre-placed where a mass of clad material is deposited on the substrate prior to introduction of a laser beam. Powder deposition approach uses a nozzle held separate from the laser beam that lays down powder mass ahead of the laser beam.

8.4.3.1. Pre-placed powder method

During high laser deposition rate (typically 15 pounds/h), pre-placed powder introduction method is preferable since it is possible to maximize the amount of powder being melted. This ensures that the powder width equals the width of the area scanned by the laser, as powder melting is also advantageously maximized, creating an optimized stable clad. The powder must be mixed with a chemical binder so that it can adhere to the substrate during laser scanning. The chemical binder must evaporate as a result of the high energy emitted by the laser beam. However, this can result in porosity of the clad [37]. When cladding with pre-placed powder, the melt pool is formed on top of the cladding material and proceeds downward to the substrate. Only when the substrate has been melted can a clad layer be formed. Therefore, it is difficult to control the depth of the melt pool, which results in a relatively high dilution.

8.4.3.2. Wire feeding

The clad material can be introduced by wire feeding on the substrate where it simultaneously melts with the substrate under the laser beam. This process can be hard to control, especially in objects of complex shapes and eventually results in high dilution rates.

8.4.3.3. Powder injection method

Powder delivery method requires a dedicated powder delivery system and a powder nozzle, which must direct the powder to the desired position. Powder injection method is a more flexible and easier method to control; however, changing the powder proved to be environmentally hazardous. The powder injection method resulted in good clad areas in recent research findings.

8.4.3.4. Coaxial cladding

The clad material is supplied in the powder form mainly coupled with shielding gas to prevent reaction with the surrounding gases. This method is widely used due to the ease of automation control and the minimal surface preparation it requires. It is divided into coaxial and side injection methods, differentiated by the method of powder injection into the process.

8.4.3.5. Side laser cladding

In the side cladding method, the powder is introduced from a nozzle together with the shielding gas at an angle to the laser beam such that the powder gets in contact with the laser energy before reaching the substrate. The powder is heated to ignition state before reaching the substrate that is already being heated together forming an alloy of the set depth.

9. Influence of laser processing parameters

There are a number of varying parameters controlling the laser processes. Individual parameter has its own distinctive functions, which can affect the processing results and operation outcomes. Laser power, laser beam size (beam diameter), powder flow rate, and laser scanning speed are some of the important factors in achieving a thin clad layer with low dilution but sufficient bonding strength [27]. However, these factors are responsible for the temperature distribution, high-quality microstructure, the shape of the melt pool, and the final geometry of the laser clad layer. Vaziri et al. studied the effect of laser parameters on properties of surface – alloyed Al substrate with Ni [38]. The influence of spot size and peak – power density of pulsed Nd:YAG laser on the depth of alloyed layer, the hardness and microstructure in LSA of Al with Ni. Results revealed that the hardness obtained was found to be 10–15 times the value of base Al. Reduction in the power density resulted in a decrease in the alloyed layer thickness, while increasing the peak power density increased the alloyed pool depth. The effect of laser beam diameter on the depth of the alloyed pool was examined; it showed that by increasing the laser beam diameters, the depth of the alloyed pool decreased. Hamedi et al. investigated the effect of pulsed laser parameters on in-situ Tic synthesis in laser surface treatment [39]. Here, effects of irradiated energy per unit length and pulse duration on mircostructure and hardness were investigated. Results showed improved hardness and fine dendritic morphology to cellular grain structure. The Marangoni convection flows in the melt pool allowed uniform distribution of carbon in liquid Ti because of decreasing peak power and increasing heat input. In order to achieve high volume fraction in laser-alloyed zone, the energy input to the melt pool is increased. Hardness value produced reached 1700 Hv, which is 10 times harder than the substrate.

9.1. Laser power

Laser power is a very important factor in laser cladding as it is responsible for energy density transferred on the powder-substrate system. Optimum parameters during laser cladding are needed to yield microstructure and clad that are free from defects. Wu et al. carried out the effect of process parameters on the microstructure of laser-deposited Ti-6Al-4V [40]. It was observed that long columnar grains dominated the microstructure of the laser-deposited Ti-6Al-4V alloy for a large range of laser powers and is formed for all. The scale of columnar grains also increases with decrease of the laser scan speed and when other parameters are maintained constant. The microstructure of the laser- deposited Ti-6Al-4V is influenced by laser power, scan speed, or powder feed rate.

9.2. Laser scanning speed

Laser scanning speed generally refers to the speed at which laser beam travels along the working piece. Higher laser speeds lead to subsequent reduction in the amount of clad material particulates onto working piece leading to a thin clad layer. Hemmati et al. investigated into the effect of scanning speed on phase constituents and properties of laser-deposited coatings [41]; the investigation showed a significant refinement of dendritic structure that stabilized the phases and reduced the volume fraction of other phases at higher speeds. Laser cladding technology was used to deposit Co–Ti alloy on mild steel using different scanning velocities [42]. This resulted in a different rate of cooling of the clad variations in the microstructural characteristics and hardness of the clad layer. Good-quality clad layer was evident with no visible porosity or cracks. A fine microstructure was achievable with higher laser velocities while lower velocities yielded higher hardness due to a large fraction formation of hard intermetallic phase of $TiCo_3$. The higher hardness of the clad layer compared to the substance is due to the formation of intermetallic compound $TiCO_3$.

9.3. Powder flow rate

Flow rate determines the thickness of the alloyed layer and dilution rate required. This tends to affect microstructural changes, homogeneity, cracking, porosity, and surface finish. Rajaram et al. studied the effect of feed speed and power on laser cut quality of 4130 steel [43]. Power and feed rate had a major effect on kerf width and size of HAZ. In addition, the surface roughness and striation frequency were affected most by feed rate. It was observed that at low power levels, the smallest kerf width and HAZ are obtained and the effect of feed rate is moderate. Low feed rates gave good surface roughness and low striation frequency.

10. Conclusion

Laser generates intense beam energy, which offers outstanding advantages over other conventional surface modification techniques. Laser techniques have become the choice for many industrial applications involving corrosion, wear, oxidation, and general repairs. Laser surface modification techniques overcome limitations such as poor adherence, lower bonding strength, serious crack propagation at the interface, and limitation of the thickness of coatings, which are associated with other forms of surface modification processes [44]. These techniques have known to modify surface composition and microstructure without altering the bulk substrate [25]. Laser cladding technique, a type of laser technique, has advantages over the traditional coating techniques, which include high precision, automation control with choices of clad thickness from about 0.1 millimeters to several centimeters, metallurgical bonding of the cladding material with the base material, lower deposition rate, minimal heat effect on the substrate due to controlled laser energy and rapid cooling, a wide selection of homologous and non-homologous powder materials, and the ability to process virtually any type of metal alloy. Also, the energy input of laser cladding is low, resulting in finer microstructures with superior properties and minimal distortion [32].

Author details

Olanrewaju Adesina, Patricia Popoola and Olawale Fatoba

*Address all correspondence to: popoolaapi@tut.ac.za

Department of Chemical, Metallurgical and Materials Engineering, Tshwane University of Technology, Pretoria, South Africa

References

[1] Li, S., Sun, B., Imai, H. & Kondoh, K.., Powder metallurgy Ti–TiC metal matrix composites prepared by in situ reactive processing of Ti-VGCFs system. Carbon, 2013. 61(0): 216–228.

[2] Bansal, D.G., O.L. Eryilmaz, and P.J. Blau, Surface engineering to improve the durability and lubricity of Ti–6Al–4V alloy. Wear, 2011. 271(9–10): 2006–2015.

[3] Ezugwu, E.O. and Z.M. Wang, Titanium alloys and their machinability—a review. Journal of Materials Processing Technology, 1997. 68(3): 262–274.

[4] Razavi, R. S., Salehi, M., Monirvaghefi, M. & Gordani, G. R.Corrosion behaviour of laser gas-nitrided Ti–6Al–4V alloy in nitric acid solution. Journal of Materials Processing Technology, 2008. 203(1–3): 315–320.

[5] Dong, H., 3 - Tribological properties of titanium-based alloys, in Surface Engineering of Light Alloys, H. Dong, Editor. 2010, Woodhead Publishing. UK. pp. 58–80.

[6] Baker, T.N., 12 - Laser surface modification of titanium alloys, in Surface Engineering of Light Alloys, H. Dong, Editor. 2010, Woodhead Publishing. UK. pp. 398–443.

[7] Polmear, I.J., 6 - Titanium alloys, in Light Alloys (Fourth Edition), I.J. Polmear, Editor. 2005, Butterworth-Heinemann: Oxford. pp. 299–365.

[8] Lütjering, G. and J.C. Williams,. 2003. Titanium: engineering materials and processes. Springer Verlag: Berlin, Heidelberg, New York.

[9] Peters, M. and C. Leyens. 2003: Non-Aerospace Applications of Titanium and Titanium Alloys. Titanium and Titanium Alloys. WILEY-VCH Verlag GmbH & Co. KGaA, Weinheim ISBN: 3-527-30534-3.

[10] Donachie, M., 1988. Titanium: A Technical Guide. ASM Metals international Park, OH44073. OHIO. USA. ISBN 0-81770-309-2.

[11] Liu, X., P.K. Chu, and C. Ding, Surface modification of titanium, titanium alloys, and related materials for biomedical applications. Materials Science and Engineering: R: Reports, 2004. 47(3–4): 49–121.

[12] Momin, O., S.Z. Shuja, and B.S. Yilbas, Laser heating of titanium and steel: Phase change at the surface. International Journal of Thermal Sciences, 2012. 54(0): 230–241.

[13] Balla, V. K., Soderlind, J., Bose, S. & Bandyopadhyay, A.Microstructure, mechanical and wear properties of laser surface melted Ti6Al4V alloy. Journal of the Mechanical Behavior of Biomedical Materials, 2014. 32(0): 335–344.

[14] Dutta Majumdar, J. and I. Manna, 21 - Laser surface engineering of titanium and its alloys for improved wear, corrosion and high-temperature oxidation resistance, in Laser Surface Engineering, J.L.G. Waugh, Editor. 2015, Woodhead Publishing. UK. pp. 483–521.

[15] Adebiyi, D.I. and A.P.I. Popoola, Mitigation of abrasive wear damage of Ti–6Al–4V by laser surface alloying. Materials & Design, 2015. 74(0): 67–75.

[16] Chikarakara, E., S. Naher, and D. Brabazon, High speed laser surface modification of Ti–6Al–4V. Surface and Coatings Technology, 2012. 206(14): 3223–3229.

[17] Bloyce, A., Qi, P. Y., Dong, H. & Bell, T. Surface modification of titanium alloys for com bined improvements in corrosion and wear resistance. Surface and Coatings Technology, 1998. 107(2–3): 125–132.

[18] Weng, F., C. Chen, and H. Yu, Research status of laser cladding on titanium and its alloys: A review. Materials & Design, 2014. 58(0): 412–425.

[19] Tian, Y. S., Chen, C. Z., Li, S. T. & Huo, Q. H.Research progress on laser surface modification of titanium alloys. Applied Surface Science, 2005. 242(1–2): 177–184.

[20] Liu, X.-B., Meng, X.-J., Liu, H.-Q., Shi, G.-L., Wu, S.-H., Sun, C.-F., Wang, M.-D. & Qi, L.-H Development and characterization of laser clad high temperature self-lubricating wear resistant composite coatings on Ti–6Al–4V alloy. Materials & Design, 2014. 55(0): 404–409.

[21] Majumdar, J.D. and I. Manna. 2003. Laser Processing of Materials. Sadhana. Printed in India. Vol28(4): pp. 495–562.

[22] Avadhanulu, M.N. 2009. An Introduction to Lasers: Theory and Applications. S. Chand and Company Ltd.: New Delhi.

[23] Steen, W.M. and K. Watkins. 2003. Laser Materials Processing, Vol. 1, Springer Verlag: New York.

[24] Dahotre, N.B and S.P. Harimkar. 2008. Laser fabrication and machining of materials. Springer: New York.

[25] Liu, Z., 4.09 - Laser Applied Coatings, in Shreir's Corrosion, B.C.G.L.L.R.S. Stott, Editor. 2010, Elsevier: Oxford. pp. 2622–2635.

[26] Wirth, P. 2004. Introduction to industrial laser materials processing. Humburg: Rofin, Germany.

[27] Tseng, W.C. and J.N. Aoh, Simulation study on laser cladding on preplaced powder layer with a tailored laser heat source. Optics & Laser Technology, 2013. 48: 141–152.

[28] Gupta, N. and N. Nataraj, A dual weighted residual method for an optimal control problem of laser surface hardening of steel. Mathematics and Computers in Simulation, 2014. 103: 12–32.

[29] Bojinović, M., N. Mole, and B. Štok, A computer simulation study of the effects of temperature change rate on austenite kinetics in laser hardening. Surface and Coatings Technology, 2015. 273: 60–76. 30.

[30] Li, H. C., Wang, D. G., Chen, C. Z. & Weng, F Effect of CeO2 and Y2O3 on microstructure, bioactivity and degradability of laser cladding CaO–SiO2 coating on titanium alloy. Colloids and Surfaces B: Biointerfaces, 2015. 127(0): 15–21.

[31] Yakovlev, A., P. Bertrand, and I. Smurov, Laser cladding of wear resistant metal matrix composite coatings. Thin Solid Films, 2004. 453–454(0): 133–138.

[32] Ion, J.C., 12 - Cladding, in Laser Processing of Engineering Materials, J.C. Ion, Editor. 2005, Butterworth-Heinemann: Oxford. pp. 296–326.

[33] Kusinki, J., S. Kac, A. Kopia, A. Radziswewska, M. Rozmus-Gornikowska, B. Major, L. Major, J. Marckar, and A. Lisiecki. 2012. Laser modification of materials surface layer-review paper. Bulleting of the polish academy of sciences, 60(4): pp. 711–728.

[34] Morris, D.G., Glass-forming conditions during laser surface melting. Materials Science and Engineering, 1988. 97(0): 177–180.

[35] Poulon-Quintin, A., Watanabe, I., Watanabe, E. & Bertrand, C., Microstructure and mechanical properties of surface treated cast titanium with Nd:YAG laser. Dental Materials, 2012. 28(9): 945–951.

[36] Ana Sofia C.M., D'Oliveira, P. Se'rgio C.P. da Silva, Rui M.C. Vilar (2002). Microstructural features of consecutive layers of Stellite 6 deposited by laser cladding. Surface and Coatings Technology,153, 203–209.

[37] Arlt, A.G., R. Muller. 1994. Technology for Wear Resistant Inside Diameter Cladding of Tubes, Proe. ECLAT '94, pp. 203–212.

[38] Vaziri, S. A., Shahverdi, H. R., Torkamany, M. J. & Shabestari, S. G.Effect of laser parameters on properties of surface-alloyed Al substrate with Ni. Optics and Lasers in Engineering, 2009. 47(9): p. 971–975.

[39] Hamedi, M.J., M.J. Torkamany, and J. Sabbaghzadeh, Effect of pulsed laser parameters on in-situ TiC synthesis in laser surface treatment. Optics and Lasers in Engineering, 2011. 49(4): 557–563.

[40] Wu, X., Liang, J., Mei, J., Mitchell, C., Goodwin, P. S. & Voice, W. Microstructures of laser-deposited Ti–6Al–4V. Materials & Design, 2004. 25(2): 137–144.

[41] Hemmati, I., V. Ocelík, and J.T.M. De Hosson, The effect of cladding speed on phase constitution and properties of AISI 431 stainless steel laser deposited coatings. Surface and Coatings Technology, 2011. 205(21–22): 5235–5239.

[42] Alemohammad, H., S. Esmaeili, and E. Toyserkani, Deposition of Co–Ti alloy on mild steel substrate using laser cladding. Materials Science and Engineering,, 2007. 456(1–2): 156–161.

[43] Rajaram, N., J. Sheikh-Ahmad, and S. Hossein Cheraghi Parametric study of the effect of feed speed and power on laser cut quality of 4130 steel. Wichita State University: Wichita, Kansas 67260–0035

[44] Yang, Y., N. Guo, and J. Li, Synthesizing, microstructure and microhardness distribution of Ti–Si–C–N/TiCN composite coating on Ti–6Al–4V by laser cladding. Surface and Coatings Technology, 2013. 219: 1–7.

[45] Polmear J.J. 1981: 6 - Titanium alloys, in Light Alloys, Edward Arnold Publications: London.

[46] Zhao, X.M., Chen, J., Lin, X. and Huang, W.D. (2008). Study on microstructure and mechanical properties of laser rapid forming Inconel 718. Materials Science and Engineering A-Structural Materials Properties Microstructure and Processing. 478(1-2): 119-124.

Chapter 17

Laser Engineering Net Shaping Method in the Area of Development of Functionally Graded Materials (FGMs) for Aero Engine Applications - A Review

Patricia Popoola, Gabriel Farotade, Olawale Fatoba and Olawale Popoola

Additional information is available at the end of the chapter

http://dx.doi.org/10.5772/61711

Abstract

Modern aero engine components are subjected to extreme conditions were high wear rate, excessive fatigue cycles, and severe thermal attack are inevitable. These aggressive conditions reduce the service life of components. Its generic effect is magnified in the light of understanding the fact that aero engine parts are highly sensitive to functional and dimensional precision; therefore, repair and replacement are great factors that promote downtime during operation. Hard thermal barrier coatings have been used in recent times due to their optimized properties for maximum load bearing proficiency with high temperature capability to meet performance and durability required. Nevertheless, less emphasis is being given to the coating-substrate interaction. Functionally graded structures have better synergy and flexibility in composition than coatings, giving rise to controlled microstructure and improved properties in withstanding acute state of affairs. Such materials can be fabricated using Laser Engineered Net Shaping (LENS™), a laser-based additive manufacturing technique. LENS™ offers a great deal in rapid prototyping, repair, and fabrication of three-dimensional dense structures with superior properties in comparison with traditionally fabricated structures. The manufacture of aero engine components with functionally graded materials, using LENS™, can absolutely mitigate the nuisance of buy-to-fly ratio, lost time in repair and maintenance, and maximize controlled dimension and multi-geometric properties, enhanced wear resistance, and high temperature strength. This review presents an extensive contribution in terms of insightful understanding of processing parameters and their interactions on fabrication of functionally graded stainless steel, which definitely influence the final product quality.

Keywords: Functionally graded materials, LENS™, processing parameters

1. Introduction

1.1. The aerospace industry

The urge for advanced materials and faster processes for manufacture of products and delivery of services is critical in the aerospace industry due to the increasing rate of air travel and stringent environmental regulations. Researchers and manufacturers, under constant hand-to-hand interactions, are strained to meet demands and maintain supply chains. However, research-driven technologies have, relatively, created a platform for non-stop affordable flights with destinations as far as the North Pole, belting over to the other flank of the world. These advances suppress the underlying challenges in the past, a major percentage of aero engines parts are made of super alloys and hard alloys coated with thermal barrier coatings (TBCs) to improve their elevated temperature strengths. The use of super-alloys is a valuable consideration, withstanding high temperatures, mitigating the concern for premature part failure and limited air travel-engine use duration but to a reduction in application as a result of low service strength at ultra-high temperatures for a very long time. In turn, repair and replacement of intricate regions of the aero engine, such as compressors, turbine blades, pistons, and cylinders, may have adverse effects in terms of loss time during service. Another issue is the need to reduce the usage of expensive rare elements in the manufacture of high strength-high temperature components in the engine, as this will pull a net positive effect on the customer's effective evaluation and interest [1]. Based on its high precision and sensitivity to minute anomalies, engines account for approximately 30% of the life cycle cost of modern airlines [2]. As a result of this, materials engineers are constantly on the frontline of developing highly multi-functional materials with less production time, which will maintain their operating properties over a wide range of extreme temperatures for extended flight hours.

2. Functionally Graded Materials (FGMs)

2.1. History

The concept and processing of FGMs seem to be a new generational novelty but only the processing techniques, invented by humans, can be considered as innovative. FGMs have been existing as a result of nature. Human bones, skin, and bark of trees are examples of naturally occurring FGMs. The first humanly created FGM was industrialized in Japan in the early 1980s with the idea of fabricating a space craft during a space plane project. Eventually, enormous studies have been performed on advancing the viability of these materials [3].

2.2. Definition

FGMs are composites possessing continuous and coherent variation in composition, micro-structure [4], and even mechanical properties [5] from a region to another along the build axis, in an effect to attain improved performance and reliability. Traditional composites are associated with frequent thermal stresses, stress singularities, and residual stresses due to solidification history while the FGMs offer new generational components that can endure these critical thermal and mechanical stresses that are typical with aircraft engines and aerospace assembly during operation [6]. The rationale for the resistance to these stresses is the gradient effect of properties in FGMs and also the ability to integrate materials with contradicting properties such as high wear resistant ceramic and a tough metal in a single structure [7]. In addition, the feasibility and ease of tailoring material properties to match the desired requirements is a tip of the numerous benefits of these advanced composites. Patterns made with FGMs are presented in Figure 1.

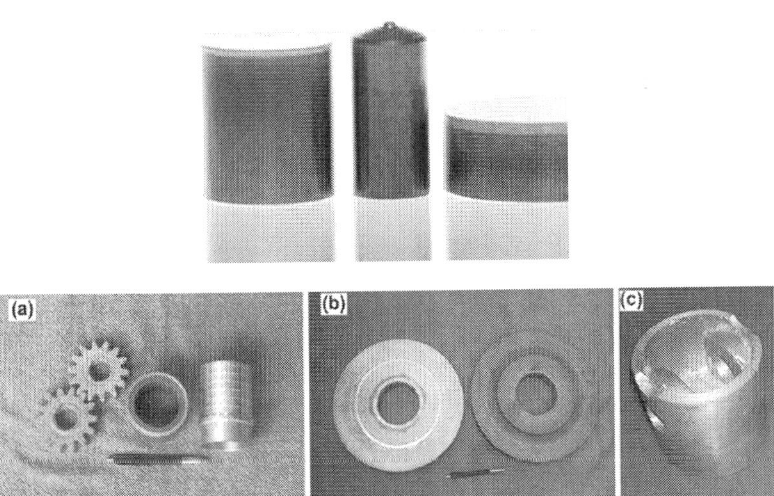

Figure 1. Functionally Graded Components; (a) Cylinder liners and gears; (b) Brake rotor discs; and (c) Piston fabricated by centrifugal casting method [8].

There are two major ways by which FGMs are synthesized, the first is an incessant formation, whereby compositional change seems invisible with no distinct discrete gradation of layers. Secondly, a step-wise synthesis is achieved by stacking layers with near-composition as a fractional material's composition dominates while the other declines with position. In the latter, the inhomogeneity in composition can physically be detected. These methods of synthesizing FGMs consist of similar preparation modes namely: gradation and consolidation. Gradation enables the formation of layers of different percentage compositions with little or no significance in the composition change along the axis of build up, while consolidation takes care of the elimination of sharp discontinuous interfaces present between these successive layers and conversion of these layers into continuous or single gradient structure through material transport. Optimal achievement of a successful FGM structure is highly dependent on the chemistry of bonding between successive layers, therefore materials scientists have combined different techniques to achieve suitable functionally graded structures that are functions of the component materials [9, 10] and prior knowledge of the uneven shrinkage of sintered layers, as a result of dependence of sintering pattern to the particle size, porosity, and composition of the elemental powders [11]. However, numerous theoretical efforts on fabricating and analyzing the adherence of FGMs to required expectations have been reported four decades ago, but industrial applications have been restricted due to little knowledge of processing techniques of these materials until recent times [12].

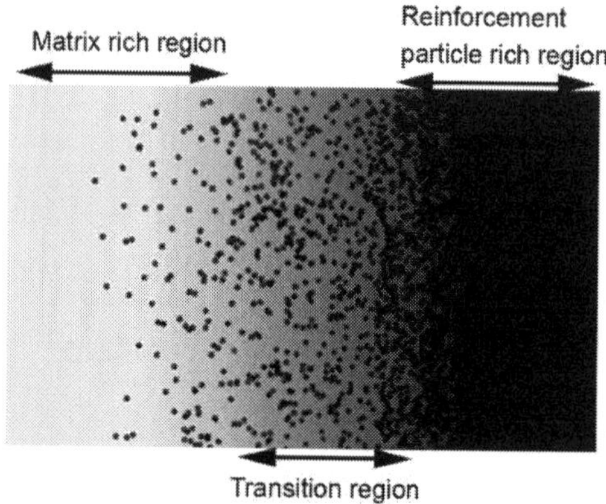

Figure 2. Representation of a typical functionally graded material [8].

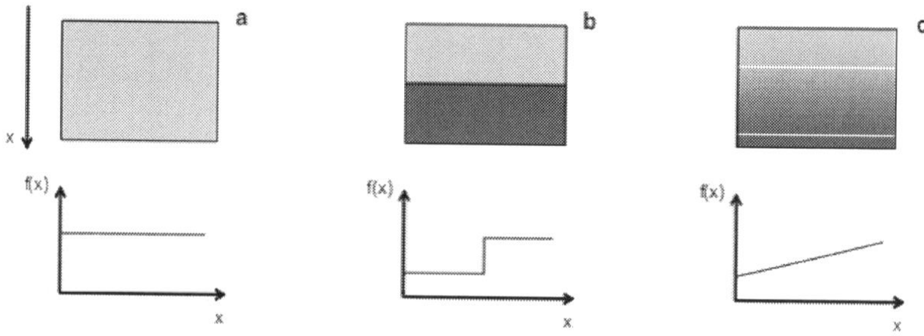

Figure 3. Diagrammatic representation of (a) homogeneous material; (b) layer-wise and; (c) FGM [13].

3. Operational properties of FGMs

In an FGM, the presence of different phases account for the various properties possessed. These phases make the material behave as a heterogeneous material unlike alloys and composites that have unique properties. An effect of this is the difficulty of enumerating material parameters [14]. In recent times, several studies have been performed to produce models that account for the effective material properties. Application of the Mori-tanaka method to establish directives for mixing of ceramic-metal composites [15], synthesis of experimental and analytical methods to determine elastic behavior of graded materials using micro-indentation approach [16], the investigation of properties of graded materials by inverse analysis, and instrumented indentation using kalman filter technique [14]. Also, the thermal residual stresses at the layer interface that also influence the mechanical properties of these materials have been studied [17]. Different approaches are adopted to investigate and determine the effective material parameters in order to understand the elastic-plastic behavior of FGMs when subjected to thermo-mechanical loading.

4. Processing techniques of FGMs

With increasing development of automated materials processing techniques, suitable techniques have been constructed in order to achieve various graded structures irrespective of component geometry and size. The use of these techniques is based on the position of the desired gradation. There are two types of gradation in a component. The thin gradation is mostly situated at the top part of the component while the bulk gradation is effective in the wholeness of the component. Thin gradation can be achieved by plasma spraying, self-developing high temperature synthesis, and vapor depositions (Chemical or Physical). Centrifugal method, powder metallurgy, hot and cold pressing, sintering method, infiltration method, and solid free form techniques are typical bulk gradation techniques [3]. The current and typical processing techniques adopted in fabricating FGMs are discussed below.

5. Thin FGM processing techniques

5.1. Vapor deposition

The vapor deposition technique is of importance due to its concept of iso static or vacuum hot pressing of desired alloys to form a composite component. With this technique it is possible to achieve a controlled deposition and deposition thickness, also, layer spacing precision for laminated grading. Sputtering [18], thermal deposition, and chemical vapor deposition are forms of direct vapor deposition that have been reported to be of good consideration for microlaminates [19]. Jet Vapor Deposition (JVD™) is a later form that deposits concentrated levels of different materials at reduced vacuum situations using an unreactive gas jet. The unreactive gas jet is combined with a resistive evaporation supply that concentrates the deposits [20]. Vapor deposition can be adopted to deposit functionally graded coatings with exceptional fine microstructure on surfaces of aero engine components. Mitsubishi Miracle inserts were functionally graded with chemical vapor deposition of a three-layer structure on a carbide substrate. This tremendously reduced the challenge of plastic deformation and damage [21].

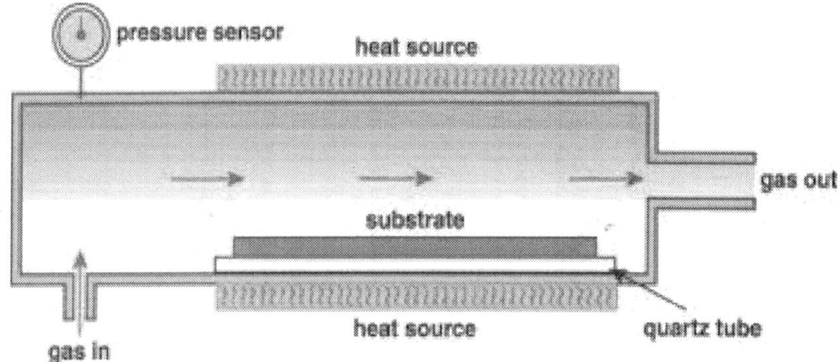

Figure 4. A schematic view of the chemical vapor deposition process [22].

5.2. Plasma spraying

Thermal spraying processes are typically called different names such as plasma spraying, metal spraying, high velocity oxygen fuel (HVOF) spraying, arc spraying, etc. Considering a porous ceramic coating that is applied to a low corrosion resistant component, a typical composite for the hot regions in aero engines, high velocity powder particles possessing elevated melting point is melted and accelerated by the heat generated by the plasma cloud towards the substrate. Required coating thickness is achieved by successive actions of the plasma cloud. Hard, low friction, and improved fatigue resistant coatings can be achieved for applications, such as airplane landing gear piston surfaces, applying this technique [23].

Conventional thermal-sprayed TBCs have been reportedly stated to spall when subjected to mechanical loading at ultra-high temperatures due to their low bonding strengths and residual stresses. Functionally graded TBCs are suitable materials to resist such challenges. In the work performed by Khor and Gu, functionally graded coating of yttrium-stabilized ZrO_2/NiCo-CrAlY was used as a TBC. The coating was applied using thermal spray technique. Highly deposited coating with enhanced coating density and chemical homogeneity was achieved compared to the results from duplex coatings, in addition it was observed that the oxidation of the FGM coating was impeded. Zirconia-based coatings are commonly used in thermal spraying due to the formation of non-transformable tetragonal phase that does not undergo martensitic transformation during cooling by quenching [24]. A schematic illustration of thermal spraying is shown in Figure 5.

5.3. Ion Beam-assisted Deposition (IBAD)

Ion Beam Assisted Deposition (IBAD) can be referred to as deposition of thin film by the blending of evaporation with simultaneous bombardment in high vacuum environments. The simultaneous bombardment involved in this technique makes it different from other deposition techniques. Effective modification can be achieved by the bombardment of developing

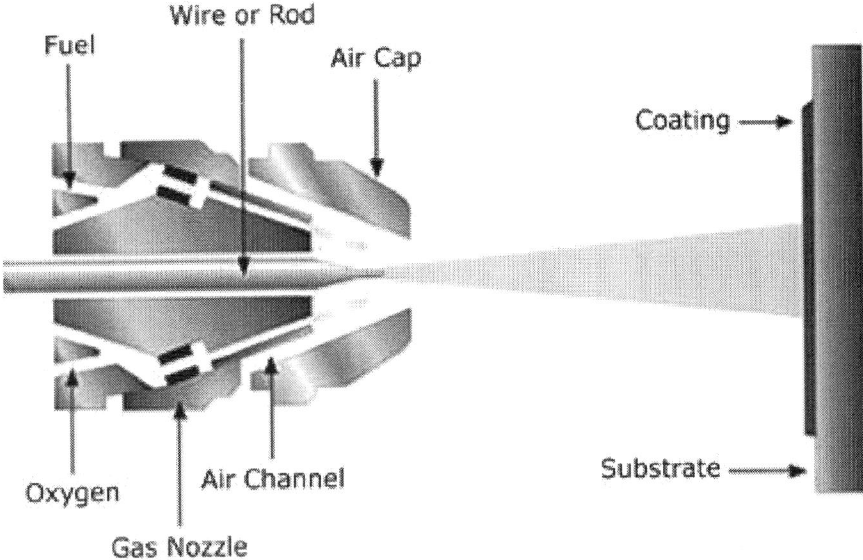

Figure 5. Concept of thermal spraying [25].

films with high velocity particles in various morphologies that is crucial to the activity of thin coatings. These can be achieved by: change in orientation, densification of developing films at low temperatures, and alteration of grain size and mechanical properties. This technique can be used to deposit coatings of high bond strength, varying the concentration of particles for each successive layer through simultaneous bombardment. Functionally graded coatings fabricated by IBAD are suitable for applications at low temperatures, having a high control of developed film/coating, while the chemical mixture of the substrate with the film provides better adhesion compared to vapor deposited counterparts [26]. Ion beam assisted deposition can be limited by substrate geometry and low mechanical strength of deposition.

5.4. Electrodeposition

Functionally graded deposits of metal-ceramic composites and bi-metals have been achieved by electrodeposition in recent times. This is attained by co-deposition of ceramic particles and metallic particles from electrolytes containing the metal ion, varying either the particle ratio or the current density with respect with time. Pulsed ElectroDeposition (PED) is a novel technique of depositing nanocrystalline materials on numerous metallic substrates.

With electrodeposition, single-layered coatings have proven to project lower expectations compared to their graded counterparts in terms of corrosion, mechanical, and wear properties. Deposition with just a single electrolytic bath has also been performed with less additive involved to achieve controlled, excellent graded coatings.

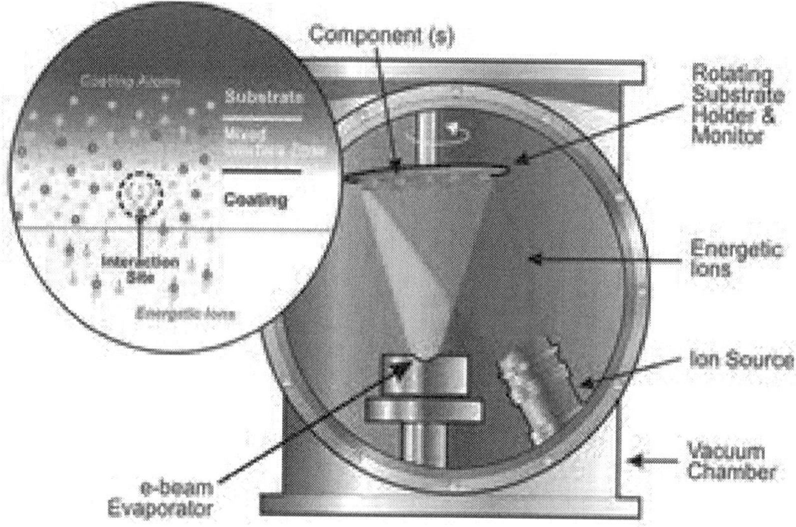

Figure 6. Ion Beam Assisted Deposition (IBAD) Process [27].

Coatings are graded perpendicular to the surface providing properties parallel to surfaces where composition is uniform, hence functionality cannot be tailored in any direction with these techniques [28].

6. Bulk FGM processing techniques

6.1. Powder metallurgy

Powder metallurgy is one fabrication technique that is obviously adopted for FGMs that makes use of solid materials (powders). Powder preparation, material processing, and forming and sintering processes are stages that appear as principal characteristics of this technique. The powder preparation entails methods such as grinding, deposition or chemical reactions, and producing a massive rate of powder. Also, the rate of production can be controlled to the desired size. Considering the processing stage of the powder, attention is placed on the sampling and distribution of the powders while those of forming (piling and pressing stage) and sintering (consolidation of powders) stages are influenced by the working environment, which must be achieved at room temperature and high pressure, consecutively. Powder metallurgy possesses the feature of premixing powders and piling layers with graded compositions that fuses the layers together with either hot pressing or cold pressing [8]. During the sintering stage, preservation must be incorporated in order to achieve good product quality because some metals that are highly reactive can be oxidized during hot pressing. However, the limitation in fabricating FGMs with powder metallurgy is the complexity in densification mode because densification is based on the dominant fraction of powder (positional matrix) which, thus, results in distortion and variation of shrinkage of the layers [29].

6.2. Centrifugal casting

FGMs fabricated by centrifugal force can undergo either centrifugal, centrifugal slurry, centrifugal pressurization, or centrifugal casting method [30]. Gradation occurs when a homogeneous molten metal, dispersed with ceramic particles, is stirred and left to settle based on differences in density. Studies have shown suitable centrifugal methods for different materials; for instance, centrifugal solid-particle method is suitable for Al/SiC and Al/Al$_3$Ti FGMs and centrifugal in situ method used for the processing of Al/Al$_3$Ni FGMs [31]. Unlike stepwise gradation, which occurs in powder metallurgy technique, FGMs processed with this technique possess continuous gradation [32]. Centrifugal casting of FGMs is dependent on the molten matrix content, rotation speed, and the rate of settling conditions of the particle to achieve a controlled and desired production in mass [33]. However, this technique is limited in producing FGM components with regular geometrical shapes.

Other common techniques that are not elaborated are infiltration [4], spray casting [8], and ultrasonic separation [34].

7. Laser-based fabrication of FGMs

FGMs by laser-based techniques in the aerospace industry is rapidly sky-rocketing because they offer unique solutions to industrial problems. Aero engines experience limited off-flight time in order to maximize their service life and this requires cyclic rapid heating and cooling of the engine parts. As a result of this effect, materials may fail due to fatigue and thermo-mechanical stresses. As stated earlier, FGMs are applicable to mitigate these situations but most techniques used in fabricating these materials fail to produce consistent and desired properties at ultrahigh temperatures and extreme stresses that are typical with aero engine components. Therefore, laser-based techniques are outstandingly appropriate to fabricate these materials with better flexibility of controlled fabrication to produce complex components having superior hardness/toughness compatibility and minute differential shrinkage. Rapid manufacturing is another advantage of fabricating bulk FGM components with laser-based techniques, whereby functionally graded engine parts are produced within hours. Composition and properties can be highly monitored within the structure during fabrication by either pre-mixing or combining various elemental powders using multiple feeding systems, depositing the powders in a melt pool created by a laser. The different forms of laser-based techniques used for FGMs are: laser surface melting, laser cladding, laser surface alloying, direct laser deposition, LENS™, and so on. Powder feed systems have better adhesion and metallurgical properties than powder bed systems, and also the manufacture of FGMs with powder bed laser based systems is limited to layer-wise gradation and not localized functionality. This is due to the fact that positional distribution of elemental powders cannot be accounted for in every layer. Examples of powder feed systems are direct laser metal deposition, laser surface melting, laser surface alloying, laser cladding, and LENS™ techniques.

Laser surface melting, alloying, and laser cladding are mostly referred to as laser-based surface modification techniques. These techniques are used to apply heat- and wear-resistant coatings on surfaces of alloys and metal matrix composites. Based on the direct localized laser energy density, limited zones are affected by heat so grains are refined due to rapid cooling as heat is conducted away from the modified surface regions. In the case of laser surface melting and alloying, mixing between the deposited coatings and the substrate occurs at regions close to the surface. This allows proper metallurgical cohesion between the coating and substrate while the bulk material is unaffected. Laser cladding possesses a very low dilution of coating in the substrate surface. Although these techniques provide improved graded properties to components, the presence of sharp discontinuities between the substrate composition and the coating may not withstand extreme thermal stress conditions components are exposed to in aero engine compartments. Furthermore, smart structures that possess functionality in any direction as desired for respective applications cannot be achieved with laser surface modification techniques as gradation of properties is limited to directions perpendicular to layer surface. For instance, complex components and intricate regions of components cannot be altered satisfactorily with laser surface modification techniques. On the other hand, the LENS™ technique can be used to scale through these challenges because of its flexibility to pre-design the desired component with a CAD software and convert it to an STL (Standard Triangulation Language) file that slices the bulk design to thousands of layers. The composition of each layer can be fabricated by using multiple powder feeding systems and controlling their speed of deposition. The realization of the final product can be achieved within weeks irrespective of the size of the engine component.

Here, in this chapter, we will look into the fabrication of FGMs using the LENS™ technique.

8. LENS™ technique

Highly dense three-dimensional FGMs are produced using a CAD model that is then fed into the laser system through an STL file (slicing of the designed model into layers). The sliced image is interpreted by the laser system. The laser system can be connected to a CO_2 laser or an Nd:YAG (Neodymium Yttrium-Aluminium Garnet) laser source. Powder feeding systems are connected to a delivery head of the laser system, whereby an inert gas serves as the carrier gas of the material powders from the feeding system to the work area. The delivery head is made up of a lens that concentrates the laser beam to a focal point to create a molten pool. As the deposition of the powder is done co-axially with the laser beam into the molten pool, the powder melts and fuses, thus, creating a layer along the direction of deposition. This process can be seen in the illustration given in Figure 7.

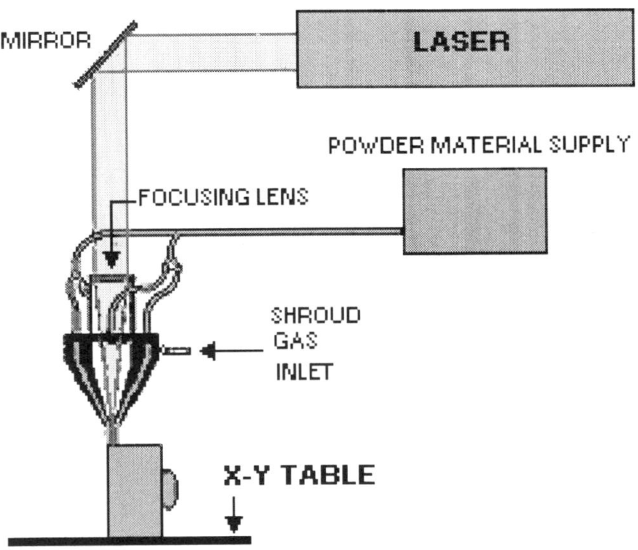

Figure 7. Set up of a LENS™ process [35].

9. Benefits of LENS™ over other techniques

The manufacture of FGMs using LENS™ have been realized efficaciously [36]. The process is carried out in an inert environment having an oxygen concentration that is less than 10 ppm, as this protects the melt from oxidation. The fabricated part may undergo heat treatment, hot iso-static pressing, or light machining to fit the required specification. Fine microstructures develop while consolidating metal powders due to rapid cooling of preceding layers. This results in improved tensile strength and toughness of metallic components than convention‐ally fabricated ones.

LENS™ is not restricted to proprietary material formulations, which encourages the formation of numerous novel alloys and composites, as is typically the case with most other processes. The powder size generally used ranges from 20 to 100 μm. Either pre-alloyed powders or suitably blended elemental powders can be used. Elemental powders can be delivered in precise amounts to the melt zone using separate feeders to generate various alloys and composite materials in situ. With the adoption of this technique, countless number of FGMs can be fabricated into complex shapes as the rate of elemental powder deposition can be controlled for each feeder during the fabrication for each layer and the final product can be achieved within hours. The following is a summary of the advantages of LENS™ over other laser-based additive manufacturing techniques:

- Fully dense fabricated parts with no compositional degradation

- Possibility of repair and overhaul, rapid manufacturing, and limited run manufacturing

- Reduced manufacturing time and cost in realization of functional metal parts [37]

- Additional cost savings is realized through increased material utilization as compared with bulk removal processes

- Components can be fabricated with reduced or eliminated micro-segregation, refined microstructure [38], and graded compositions

- Closed loop control of process for accurate part fabrication

- Potential to significantly reduce manufacturing costs, reduce the time from design to market, and simultaneously improve component performance

- Ability to tailor deposition parameters to feature size for speed, accuracy, and property control (i.e., the possibility to create parts where the composition and properties can be tailored to best meet the needs of the application)

- Composite and functionally-graded material deposition

- Mechanical properties similar or better than traditional processing methods

- Environmental compatibility based on controlled containment of expensive and hazardous materials during processing using inert conditions [39]

Figure 8. Fabrication of impeller pump using LENS™ [40].

10. Processing parameters of LENS™

The desired properties and outstanding performances of aero engine components made of FGMs using LENS™ can only be achieved if the various parameters and their influences on the final outcome are properly understood. For this reason, the primary processing parameters that are dominant are briefly discussed below.

10.1. Feed rate

The powder feed rate is the speed at which the carrier gas (inert gas that is mostly argon) conveys the elemental powders through the tubing to the delivery head, depositing the powders into the melt pool created by the focused beam on the substrate. Associated problems of powder delivery in LENS™ are the constraint in sustaining retention capability, poor deposition rate, and high surface roughness in the finished part [41]. The layer thickness is influenced by the corresponding feed rate. If the feed flow rate is high, its outcome is a highly thick layer formation. The deposited layer thickness is a function of dimension, geometry, and chemistry between layers, i.e., the thicker the deposited layer, the poorer the adhesion between layers [37]. Feed rate is key in the development of FGMs. Light components are needed for aerospace engines, therefore, amount of deposited layers should be minimum as possible to achieve the desired functionality. Higher feed rate will cause more deposition of powders, with this, the material/component thickness and density are increased, having a net increase in the weight of the engine. In addition, increasing the feed rate attracts higher beam energy to fuse the powder deposited, resulting in an increased thermal stress and distortion in the fabricated part.

10.2. Laser power

As the intensity of the laser beam increases, with time the energy injected into the melt pool also increases. This time rate of energy from the laser beam is called the laser power. Mahamood and Akinlabi [42] studied the effect of laser power on the surface finish of the Ti6Al4V substrate with deposited Ti6Al4V powder using laser metal deposition. The scanning speed, powder flow rate, and the gas flow rate were kept constant. According to their analysis, surface roughness decreased linearly with increasing laser power, thereby improving the surface finish. A study by Alimardani et al. [43] on the effect of scanning speed and melt pool dimension increase on surface finish quality showed that the melt pool geometry control and temperature had a positive influence on surface finish.

Considering the laser surface modification techniques mentioned earlier, laser surface melting and laser surface alloying require high energy input in order to achieve a thorough dilution of the coating in the surface region of the substrate. High laser power encourages large heat affected zones that may eventually alter the properties of the bulk substrate. For materials with low thermal conductivities, such as titanium, formation of large grain structures is inevitable, consequently, diminishing high temperature strength during service. This problem can be corrected with secondary heat treatment but the technique becomes less economical.

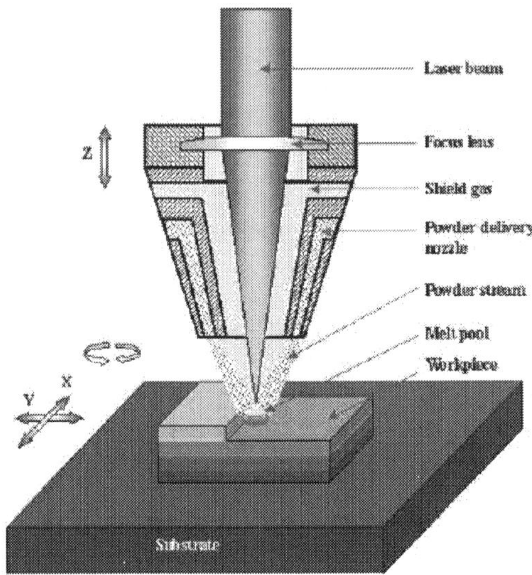

Figure 9. Schematic illustration of the LENS™ process [44]

10.3. Scan speed

The speed at which the laser beam carries out its deposition along a track path, as the component is being built, is known as the scanning speed. An increase in the scanning speed can result to a reduction in the expected time for completion of the proposed component. Nonetheless, this can be deleterious to the final product in such a way that the difference in layer height between the edges and the midmost regions may occur as the delivery head decelerates, at constant powder flow rate, depositing more materials at the edges, as in the case of a reverse deposition for each layer.

11. FGMs and LENS™

Co-axial deposition capability of LENS™ enables the feasibility of fabricating smart and functional components, allowing maximum precision in composition and improved properties compared to conventional methods. This technique takes care of the issue of oxidation during material processing by working in an enclosed compartment. The compartment is stripped of oxygen to a level lower than 10 ppm, while further prevention is realized by the use of an inert carrier gas from the powder feed system, through the delivery head to the work area, and also a shielding gas. This makes the process much effective in recycling the unmelted powders.

New innovation can be introduced into the aerospace technology with a reduced trepidation over ultrahigh temperature effects on engine parts when fabricated with LENS™. The use of metallic bond coat to prevent migration of delicate elements either from substrate to coatings or otherwise can be eliminated because such migration leaves the region bare and exposes the coating/substrate to oxidation/corrosion as the case may be, thus, functionally graded coatings blended with substrate will achieve higher reliability.

Author details

Patricia Popoola[1], Gabriel Farotade[1], Olawale Fatoba[1] and Olawale Popoola[2]

*Address all correspondence to: popoolaapi@tut.ac.za

1 Department of Chemical, Metallurgical and Materials Engineering, Tshwane University of Technology, Pretoria, South Africa

2 Centre for Energy and Electrical Power, Department of Electrical Engineering, Tshwane University of Technology, Pretoria, South Africa

References

[1] Schafik R. E. and Watson S., (2008). Challenges for High Temperature Materials in the New Millennium, the Minerals, Metals and Materials Society, 2008.

[2] Ackert S., (2011). Engine Maintenance Concept for Financiers, Aircraft Monitor, www.aircraftmonitor.com.

[3] Aysha C. P. M. S., Varghese B., and Baby A., (2014). A Review on Functionally Graded Materials, the International Journal of Engineering and Science, 3 (6):90-101, ISSN: 2329-1805.

[4] Jamaludin S. N. S., Mustapha F., Nuuzzaman D. M., and Basri S. N., (2013). A Review on the Fabrication Techniques of Functionally Graded Ceramic-Metallic Materials in Advanced Composites, Academic Journals, 8 (21):828-840, ISSN 1992-2248.

[5] Cooley W. G., (2005). Application of Functionally Graded Materials in Aircraft Structures, Airforce Institute of Technology.

[6] Wood M. and Ward-Close M., (1995). Fibre-Reinforced Intermetallic Compounds by Physical Vapor Deposition, Materials Science and Engineering, A 192/193, 590-596.

[7] Liu W. and Dupont J. N., (2003). Fabrication of Functionally Graded TiC/Ti Composites by Laser Engineered Net Shaping, Journal of Scripta Materialia, 48:1337-1342.

[8] Rajan T. P. D. and Pai B. C., (2014). Developments in Processing of Functionally Gradient Metals and Metal-Ceramic Components: A Review, The Chinese Society for Metals and Springer-Verlag Berlin Heidelberg, Acta Metallugica Sinica (English Letters), 27(5):825-838.

[9] Miyamoto Y., Kaysser W. A., Rabin B. H., Kawasaki A., and Fod R. G., (1999), Functionally Graded Materials: Design, Processing and Applications (Materials Technology Series) 1st ed., (Springer, 1999), pp 352.

[10] Kieback B., Neubrand A., and Riedel H., (2003). Processing Techniques for Functionally Graded Materials, Material Science and Engineering, A362:81-105.

[11] Schatt W., (1992). Sintervorgange, VDI-Verlang, Dusselderf, pp. 275.

[12] Bever M. B. and Duwez P. F., (1972). Gradient in Composite Materials, Materials Science Engineering, 10:1-8.

[13] El-Wazery M. S. and El-Desouky A. R., (2015). A Review of Functionally Graded Ceramic-Metal Materials, Journal of Materials and Environmental Science, 6 (5): 1369-1376, ISSN: 2028-2508.

[14] Nakamura T., Wang T., and Sampath S., (2000). Determination of Properties of Graded Materials by Inverse Analysis and Instrumented Indentation, Journal of Acta Materialia, 48:4293-4306.

[15] Weissenbek E., Pettermann H. E., and Suresh S., (1997). Elastic-Plastic Deformation of Compositionally Graded Metal-Ceramic Composites, Journal of Acta Materialia, 45:3401-3434.

[16] Giannakopoulos A. E. and Suresh S., (1997). Indentation of Solids with Gradients in Elastic Properties. Part I: Point free and Part II: Axisymmetric Indentors, International Journal of Solid Structures, 34 (19):2357-2428.

[17] Williamson R. L., Rabin B. H., and Drake J. T., (1993). Finite Element Analysis of Thermal Residual Stresses at Graded Ceramic-Metal Interfaces, Part I, Model Description and Geometrical Effects, Journal of Applied Physics, 74(2):1310-1320.

[18] Bunshah R. F., (ed.). Handbook of Deposition Technologies for Films and Coatings, Second Edition, Noyes Publications, Park Bridge, NJ, 1994.

[19] Groves J. F. and Wadley H. N. F., (1997). Functionally Graded Materials Synthesis via Low Vacuum Directed Vapor Deposition.

[20] Hsiung L. M., Zang J. Z., McIntyre D. C., Golz J. W., Halpern B. I., Schmitt J. J., and Wadley H. N. G., (1993). Structure and Properties of Jet Vapour Deposited Aluminium-Aluminium Oxide Nanoscale Laminates, Journal of Scripta Met., 29:293-298.

[21] Mitsubishi Carbides (www.mitsubishicarbides.com).

[22] Azonano Website: www.azonano.com/article.aspx?ArticleID=3423.

[23] www.progressivesurface.com/thermalsprayingprocess.php

[24] Khor K. A., Dong Z. L., and Gu Y. W., (1999). Plasma Sprayed Functionally Graded Thermal Barrier Coatings, Materials Letters, 38:437-444.

[25] Advanced Coating Activities, www.advanced-coating.com/english/spraying-flame.htm.

[26] Nakatani M., Shimizu S., and Harada Y., (2014). Fretting Fatigue Behaviour of Titanium Alloy Coated with Functionally Graded Ti/TiN Film, Fatigue 2014 presentations, www.fatigue2014.com/presentations/monday-3march-2014/36594.pdf.

[27] NASA SBIR Success, Oxidation Resistant Ti-6Al-4V-SiC Composite Materials by Ion-Beam Processing, Spire Corporation, Bedford, MA. https://sbir.gsfc.nasa.gov/SBIR/successes/ss/3-011text.html.

[28] Knoppers G., Gunnink J. W., Van der Hout J., and Van Vliet W. The Reality of Functionally Graded Material Products, TNO Science and Industry, The Netherlands, pp. 38-43.

[29] Zhang B. S. and Gasik M. M., Computational Material Science, 25, 264 (202).

[30] Watanabe Y., Inaguma Y., Sato H., and Miura-Fujiwara E., (2009). A Novel Fabrication Method for Functionally Graded Materials under Centrifugal Force: The Centrifugal Mixed-Powder Method, Materials, 2(4):2510-2525, EISSN 1996-1944.

[31] Watanabe Y., Kim I. S., and Fukui Y., (2005). Microstructures of Functionally Graded Materials Fabricated by Centrifugal Solid-particle and in-situ Methods, Metals and Materials International, 11(5):391-399.

[32] Bohidar S. K., Sharma R., and Mistra P. R., (2014). Functionally Graded Materials: A Critical Review, International Journal of Scientific Footprints, ISSN 2310-4090.

[33] Jaworska L., Rozmus M., Krolicka B., and Twardowska A., (2006). Functionally Graded Cermets, Journal of Aci. Materials, 17(1-2):73-76.

[34] Zhongtao Z., Tingju L., Hongyun Y., Jian Z., and Jie L., (2008). Preparation of Al/Si Functionally Graded Materials Using Ultrasonic Separation Method, China Foundry, 5(3)(A):1672-6421, 03-0194-05.

[35] Castle Island's Worldwide Guide to Rapid Prototyping, Laser Powder Forming, www.additive3d.com/len_int.htm.

[36] Rangaswamy P., Holden T. M., Rogge R. B., and Griffith M. L., (2003). Residual Stresses in Components formed by the Laser Engineered Net Shaping (LENS) Process, Journal of Strain Analysis for Engineering Design, 38(6):519-527.

[37] Ludovico A. D., Angelastro A., and Camparelli S. L., (2013). Experimental Analysis of the Direct Laser Metal Deposition Process, www.intechopen.com.

[38] Lewis G. K. and Schlienger E., (2000), Practical Considerations and Capabilities for Laser Assisted Direct Metal Deposition, Materials and Design, 21(4):417-423, ISSN: 0261-3069.

[39] Milewski J. O., Lewis G. K., Thoma D. J., Nemec R. B., and Renert R. A., (1998). Directed Light Fabrication of a Solid Metal Hemisphere using 5-axis Powder Deposition, Journal of Materials Processing Technology, 75(1-3):165-172, ISSN: 0924-0136.

[40] Sciammerella F., (2014). Fabricating the Future, Layer by Layer, The Fabricator, www.thefabricator.com/article/metalsmaterials/fabricating-the-future-layer-by-layer.

[41] Syed W. U. H., Pinkerton A. J., and Li L., (2006). Simultaneous Wire- and Powder-Feed Direct Metal Deposition: An Investigation of the Process Characteristics and Comparison with Single-Feed Methods, Journal of Laser Applications, 18(1).

[42] Mahamood R. M. and Akinlabi E. T., (2014). Effect of Laser Power on Surface Finish during Laser Metal Deposition Process, Proceedings of the World Congress on Engineering and Computer Science 2014, vol. II, San Francisco, USA.

[43] Alimardani M., Fallah V., Iravani-Tabrizipour M., and Khajepour A., (2012). Surface Finish in Laser Solid FreeForm Fabrication of an AISI 303L Stainless Steel Thin Wall, Journal of Materials Processing Technology, 212, 113-119.

[44] Julice M. Schoenung Research Group, Department of Chemical Engineering and Materials Science, University of California-Davis, https://chms.ucdavis.edu/research/web/schoenung/research.html.

Chapter 18

Fiber Lasers in Material Processing

Catherine Wandera

Additional information is available at the end of the chapter

http://dx.doi.org/10.5772/62014

Abstract

The economic aspects of laser usage in manufacturing that form important criteria in the choice of a suitable laser system for thick-section metal cutting include: high processing speeds, high processing depths, high cut edge quality, and high wall-pug efficiency of the laser system. Consequently, the performance of the high brightness ytterbium fiber laser system in thick-section metal cutting is evaluated based on the maximum achievable cutting speeds, maximum cutting depths possible, and cut edge quality attainable. The maximum processing speeds, maximum processing depths, and resulting cut edge quality are governed by a number of parameters related to the laser system, workpiece specification, and the cutting process. The effects of the processing parameters in the cutting of thick-section stainless steel and mild steel and medium-section aluminium have been reported; optimization of the processing parameters for enhancement of the cut edge quality has been discussed.

Keywords: Ytterbium Fiber Laser, Thick-section Metal, Cutting

1. Introduction

1.1. The high-power fiber laser

The development in the output power of solid-state fiber laser source has resulted in the increasing interest in the use of the high brightness fiber laser in macro laser material processing applications, especially cutting and welding of metal [1]. The essence of the design of the high-power fiber laser (a solid-state laser) is the improved cooling of the laser-active medium which enables attainment of higher output power with high beam quality. Among the rare earth ions used in fiber lasers, ytterbium is highly absorbing of pump radiation and is preferred as doping material for the high-power fiber laser operating at 1,060–1,080 nm spectral range and delivering kilowatt output power suitable for material processing [2].

1.2. The structure of the ytterbium fiber laser

The structure of the high-power ytterbium fiber laser includes a double-clad glass fiber having a core region in which the laser-active ytterbium ions are deposited. The core region of highest refractive index is surrounded by two cladding regions of progressively decreasing refractive index which serve to confine the pump light within the core region. The ytterbium ion-doped core region is surrounded by an inner cladding of lower refractive index than the core and the inner cladding is in turn surrounded by an outer cladding of still lower refractive index forming a step index fiber (see Figure 1a). The pump light from high-power laser diode arrays is guided into the inner cladding (referred to as pump cladding), and this pump light is confined in the inner cladding by the lower refractive index outer cladding (see Figure 1b). The confinement of pump light rays within the fiber core region – subject to some losses through absorption or scattering – maintains the pump light intensity propagating in the fiber over a fiber length of several meters. Subsequently, the pump light propagating in the fiber is absorbed by the laser-active ytterbium ions in the core region of the fiber, resulting in the lasing action of the ytterbium ions. The stimulated emission resulting from the lasing action is guided inside the core region building up to high intensities before it finally emerges as a high-power laser beam at near-infrared spectral range of 1,060–1,080nm [3, 4, 5]. The small core diameter of single-mode fibers (3–10 µm) ensures that the power density of the output beam is very high. The use of cladding-pumping in the high-power ytterbium fiber laser limits the thermal issues (i.e., the variation of the refractive index with temperature) that affect the stability of high-power Nd: YAG lasers [6]. The temperature in the fiber core is determined primarily by heat transport through the outer surface of the fiber [7]; the geometry of the fiber laser exposes a large surface area per unit volume, which aids the cooling of fiber lasers [8]. Therefore, the improved cooling mechanism in the ytterbium fiber laser has enabled the solid-state laser to achieve near diffraction limited beam quality at high output power [3].

1.3. Fiber laser features beneficial to materials processing

The high brightness ytterbium fiber laser operating at 1,060–1,080 nm spectral range has a unique combination of high-power output, high beam quality, and near-infrared wavelength, and these characteristics offer increased performance flexibility in materials processing applications. Consequently, the solid-state ytterbium fiber laser has introduced solid-state laser sources to the material processing applications – such as thick-section metal cutting and welding – that had previously been considered impractical for the traditional solid-state Nd: YAG laser due to its limited beam quality at high output power [3, 4, 9, 10, 11, 12].

1.3.1. Output power

The availability of high-power diode lasers as pumping sources for fiber lasers and improvements in fiber laser design have enabled power scaling of ytterbium (Yb^{3+})-doped fiber lasers to output powers beyond 1 kW in *cw* operation with near diffraction-limited beam quality [8, 13]. Additionally, laser beam combination techniques are also used for power scaling of the ytterbium fiber laser by bundling multiple single-mode fiber laser elements to provide up to

(a) Structure of the Fiber Laser

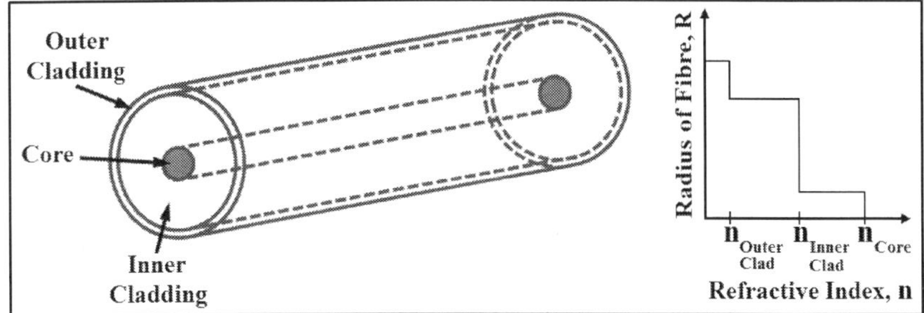

(b) Pump light propagating in the fibre

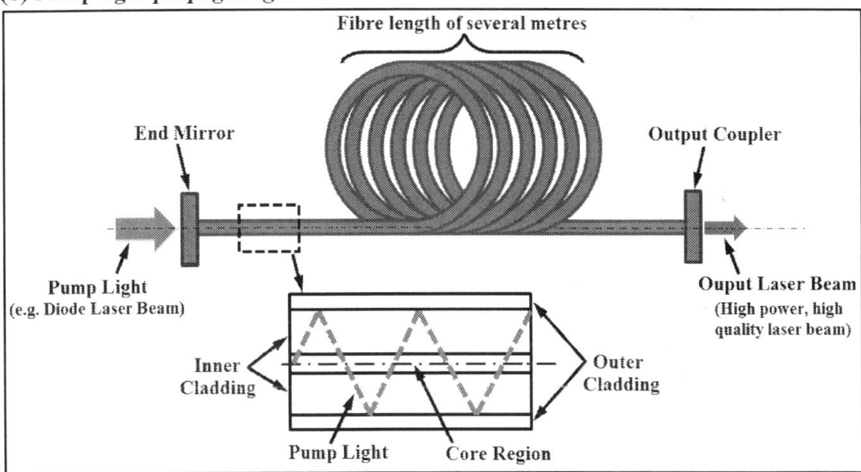

Figure 1. A schematic illustration of the double-clad fiber

10 kW multi-mode power output. Consequently, ytterbium fiber lasers delivering over 10 kilowatt power output are now available for materials processing applications [14].

1.3.2. Laser beam quality and focusing

The performance of the high-power ytterbium fiber laser in materials processing applications is enhanced by its high beam quality at high output power, which results in high brightness. The quality of a laser beam is characterized by the beam parameter product (BPP), which is the standard measure of beam quality that incorporates the wavelength effects. The beam parameter product (BPP) – defined by the relationship in Equation (1) – is an important parameter for the comparison of the beam quality of laser beams of different wavelengths from different laser sources. In the relation given in Equation (1), λ is the wavelength of the laser beam and M^2 is the beam quality factor (i.e., times diffraction limit factor) which tells how much larger the BPP of the laser beam under consideration is compared to the lowest value of λ / π for the basic Gaussian TEM_{00} mode (diffraction limit) [3, 15].

$$BPP = \frac{\lambda}{\pi} \cdot M^2 \tag{1}$$

Focusability of the laser beam is an important requirement for its utilization in cutting and welding applications. The laser beam quality is the critical parameter that influences the focusability of the laser beam; Equation (2) gives the relationship between the BPP of the laser beam and the minimum focused spot size, d_f for focusing optics of focal length, f, and a raw beam diameter on the focusing optics, D. [3, 15]. Near diffraction limited beam quality (i.e., high beam quality denoted by low BPP) is essential for focusing of the laser beam to a small focal spot size to give very high power intensity necessary for processing of thick-section metals at the required high processing speed. The high power intensity enables melting of the workpiece at the laser-material interaction zone at a high processing speed [16].

$$d_f = \frac{4f}{D} \cdot BPP \tag{2}$$

The depth of focus – i.e., the effective distance over which the minimum focused beam diameter is maintained – is the distance over which the power intensity is maximum and satisfactory cutting can be achieved. A longer depth of focus is essential for good cut edge quality like in thick-section metal cutting. For a given laser beam quality, Equation (3) gives the relationship between the focal length of the focusing optics, f, and the depth of focus, d_z, which shows that use of a longer focal length focusing optics increases the depth of focus [3]. However, the minimum focused spot size is also directly proportional to the focal length (see Equation (2)) such that use of longer focal length optics for focusing of the laser beam results in a larger focused spot size with reduced power intensity. Therefore, the low BPP of the high brightness fiber laser beam allows the use of long focal length focusing optics for achievement of a long depth of focus and long working distance without much compromise on the minimum focused spot size.

$$d_z = 4\left(\frac{f}{D}\right)^2 \times BPP \tag{3}$$

where: d_z is the depth of focus, f is the focal length of the focusing optics, D is the raw beam diameter on the focusing optics, and BPP is the beam parameter product of the incident laser beam.

1.3.3. Wavelength

Absorption of the laser beam by metals increases for wavelengths toward the visible and ultraviolet regions and decreases towards the longer infrared wavelengths. The near infrared wavelength of the high brightness ytterbium fiber laser offers a higher absorptivity by metals than the CO_2 laser. The more energetic photons of the shorter wavelength radiation of the ytterbium fiber laser can be absorbed by a greater number of electrons in the metal structure such that the reflectivity of the metal surface falls and absorptivity is greatly increased [15]. Additionally, the fiber laser wavelength offers lower sensitivity to laser-induced plasmas during materials processing when compared to the CO_2 laser. The inverse bremsstrahlung absorption by the plasma produced during laser welding is lower with the 1.06 μm wavelength radiation of the solid-state Nd: YAG laser than with the 10.6 μm radiation of the CO_2 laser [15]. Furthermore, fiber laser beam wavelength offers more flexible beam handling through use of narrow optical fibers.

2. Absorption of the laser beam by metals

The focused high-intensity laser beam radiation that is incident on the surface of a metal workpiece is partly absorbed and partly reflected by the metal surface. In a laser cutting or welding process, a sufficient amount of the focused laser beam incident on a metal workpiece must be absorbed by the workpiece so as to cause melting of the material in the laser beam–material interaction zone at the desired processing speed.

2.1. Mechanisms of laser beam absorption

Laser cutting and welding of metals require absorption of high power intensities to enhance cutting and welding at high processing speeds. The two absorption mechanisms that prevail during cutting and welding of metal include *Fresnel absorption* and *plasma absorption* (inverse bremsstrahlung effect); these laser beam absorption mechanisms are explained in the following sections.

2.1.1. Fresnel absorption mechanism

The direct absorption of the beam by the workpiece takes place through the Fresnel absorption mechanism (i.e., absorption during reflection from the surface). Fresnel absorption occurs during direct interaction of the beam and the material in which the photons of the incident laser beam radiation are absorbed by the free electrons in the metal structure. The absorbed energy sets the electrons in forced vibration motion which can be detected as heat. Absorption increases with increase in temperature of the material due to an increase in the phonon population causing more phonon–electron energy exchanges and more tendencies for the electrons to interact with the material structure with the resultant fall in reflectivity [15].

2.1.2. Plasma absorption mechanism

Plasma absorption mechanism (i.e., inverse bremsstrahlung effect) occurs when there is presence of laser-induced plasma during the process. Plasma absorption occurs through absorption of the laser beam by the free electrons in the plasma (i.e., hot metal vapor) leading to plasma re-radiation [15]. The two methods of welding of metal using a laser beam include: *conduction limited welding* and *keyhole welding*. Conduction limited welding occurs when the power density is not sufficient to cause evaporation of part of the melt at the given welding speed. Keyhole welding occurs when the energy is sufficient to cause boiling and evaporation of part of the melt creating a hole (referred to as keyhole) in the melt pool and plasma (metal vapor) [17, 18]; the keyhole is stabilized by the pressure from the vapor generated. The plasma in the keyhole consists of both vapor from the evaporated melt of the metal being welded and the shroud gas sucked into the hot vapor due to the pulsation of the keyhole. The absorption of the laser beam within the keyhole is through both Fresnel absorption (i.e., absorption directly by the material) and plasma absorption (i.e., inverse bremsstrahlung effect) by the free electrons in the metallic plasma. Consequently, the presence of plasma interferes with the laser beam delivery to the interaction zone by blocking the beam through beam-scattering effects caused by changes in refractive index and particles caught up in the plasma [15].

2.2. Laser beam absorption during metal cutting

The melt film in the cutting front is generated by the melting action of the absorbed laser beam power and the oxidation reaction power (in the case of oxygen or compressed air assist gas). Initiation of laser cutting of metals by piercing of the workpiece with a focused incident laser beam to generate a melt surface throughout the workpiece thickness is affected by metal surface reflectivity. The surface reflectivity limits the amount of laser energy coupled to the workpiece; therefore, metals with high surface reflectivity – e.g., aluminum – require higher power intensity for cut initiation. After the initiation of cutting, the cutting process progresses by the laser beam absorption on the steeply sloped cut front by the two absorption mechanisms, namely Fresnel absorption and plasma absorption and re-radiation (15). However, the plasma buildup is not very significant in cutting due to the assist gas which blows it away; therefore, plasma absorption mechanism is very limited in laser cutting. With absorption of sufficient laser energy, the thermal vibrations in the metal become so intense that the molecular bonding is stretched and is no longer capable of exhibiting mechanical strength, resulting in melting of the metal at the interaction zone. Olsen [19, 20], in his description of the mechanisms of the cutting front formation, identified the *melt surface, melt film*, and *melt front* as the three zones that comprise the cutting front. After cut initiation, the laser cutting process proceeds through absorption of the incident laser beam at the melt surface; the absorbed laser beam is transmitted to the melt front through the melt film. There is a minimum melt film thickness necessary for transmission of the absorbed energy from the melt surface to the melt front. The melt surface propagates through the material with a velocity that depends on the energy input, thermal properties of the workpiece material, and the molten material removal mechanisms. The melt front velocity increases with increasing laser power intensity which enhances the penetration speed [21]. Multiple reflections of the incident laser beam inside thick-section cut kerfs result in increased absorption of the beam inside the cut kerf; consequently, the maximum temperature in the cutting front occurs below the material surface. Multiple reflections of the laser beam inside the cut kerf increase with increasing workpiece thickness and cutting speed because the multiple reflections are a function of the cutting depth and cutting front inclination [22].

3. Laser cutting of thick-section metals

Laser cutting of metals requires very high power intensities (intensities of the magnitude 10^{10} W/m^2) to melt the metallic material to the required penetration depth at a high cutting speed. Therefore, the material's surface reflectivity, thermal conductivity, and workpiece thickness are the material parameters that critically affect the efficiency of the laser cutting process. The speed of penetration of a metal workpiece during laser cutting depends on the absorbed incident laser power intensity. Therefore, the ease with which a metallic material can be cut depends on the absorptivity of the material to the incident laser beam, and the melting temperature of the material or oxide formed when a reactive assist gas is used [15].

3.1. Applicable laser cutting methods

Laser fusion cutting using an inert gas or an active gas is commonly used in thick-section metal cutting. Due to the high power requirement for vaporization cutting, this cutting method is not applicable to thick-section metal cutting. Figure 2 schematically illustrates the cut kerf generated and the volume of material removed during laser fusion cutting of thick-section metal. The cut kerf is the opening that is created during through-thickness penetration of a workpiece; therefore, the cut kerf width, w, shown in Figure 2 is the separation distance between the two cut surfaces of the cut kerf which represents the amount of material removed during the laser cutting process.The laser power absorbed at the cutting front is utilized in melting the kerf volume at the rate of cutting and part of the absorbed laser power is lost from the cutting zone through heat conduction to the substrate metal. The maximum cutting speed that can be applied on a given metal thickness is influenced by the laser power intensity that is available for melting of the kerf volume at the applied cutting speed. Consequently, penetration of a metal workpiece of a given thickness may not be achievable at cutting speeds beyond the maximum cutting speed for the applied laser power intensity.

3.1.1. Laser fusion cutting using an inert assist gas jet

The principle role of the inert assist gas jet – e.g., nitrogen – during laser cutting of a metal workpiece is to eject the molten metal to create the cut kerf. Metallic materials like stainless steel and aluminium are often laser cut using an inert assist gas jet (usually nitrogen) to give clean unoxidized cut edges which do not require any cleaning operation after cutting. The inert gas-assisted laser fusion cutting process utilizes a focused high-intensity laser beam to melt the kerf volume and a coincident high-pressure inert gas jet to blow out the molten metal to form a cut kerf. In this cutting process, the melt temperature is not raised to boiling point; and the melt removal process is solely dependent on the drag force supplied by the high pressure assist gas jet. During inert gas-assisted laser fusion cutting of metal, the absorbed laser power is the only incoming power contribution to the cutting zone. The absorbed laser power is utilized in both melting the workpiece material equivalent to the kerf volume and accounting for the inevitable power losses from the cutting zone. Therefore, in a pure laser fusion cutting process – typical for thick-section metal cutting – where the kerf volume is melted but

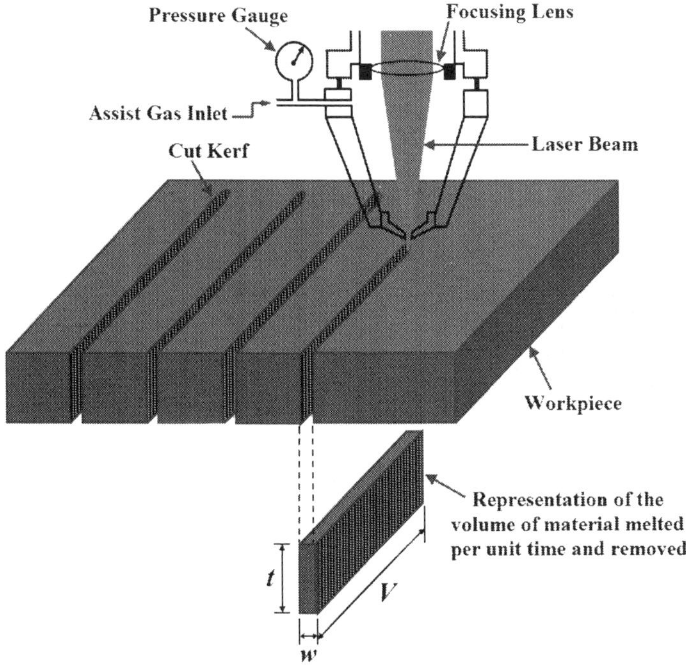

Figure 2. Cut kerf generated and the volume of material removed

negligibly vaporized and the conduction power losses from the cutting zone are significant, the power balance at the cutting front is given in Equation 4 [15].

$$A P_L = P_{Melt} + P_{Loss} \tag{4}$$

where A is the absorptivity of the workpiece to the incident laser radiation, P_L is the incident laser power, P_{Melt} is the power utilized in melting the kerf volume, and P_{Loss} is the inevitable power loss from cutting zone.

In examining the power requirement during laser cutting of a metal workpiece using inert and oxidizing assist gas jets, Wandera et al. [23] developed a theoretical model to estimate the power requirement for melting the kerf volume and the inevitable conduction power losses. Schulz et al. [24] developed an analytical approximation of the heat conduction losses during laser cutting of metals and provided an expression that can be used to estimate the temperature change in the substrate metal during laser cutting; the temperature change in the substrate metal is inversely proportional to the Peclet number which is directly proportional to the cutting speed.

3.1.2. Reactive fusion cutting

Reactive fusion cutting utilizes an active assist gas jet (usually oxygen or compressed air) that is capable of reacting exothermically with the molten metal and the reaction generates an additional heat source to the cutting process [15, 25]. The two major roles of an active assist gas jet (oxygen or compressed air) during laser cutting of a metal are to influence the energy balance at the cutting zone through the exothermic oxidation reaction and also eject the oxidized molten metal. The exothermic oxidation reaction contributes up to 40% of the energy used in the laser cutting of mild steel and stainless steel using oxygen assist gas jet [15].

During reactive fusion cutting, the incident laser beam melts the workpiece and also ignites and carries on the exothermic reaction between the molten metal and the active gas jet. Therefore, the active assist gas jet passing through the cut kerf plays two important roles which include exerting the necessary drag force to blow the molten material out the cut kerf and providing additional heat to the cutting process. Consequently, for the same incident laser power, cutting speeds of the reactive fusion cutting using an active (oxygen or compressed air) assist gas are usually higher compared to the inert gas-assisted cutting process. The incoming power contributions to the cutting zone during active gas-assisted laser cutting of metal include the absorbed laser power and the power from the exothermic oxidation reaction. In reactive fusion cutting of thick-section metal, the proportion of the kerf volume that is vaporized is considered to be negligible due to the high conduction losses which scale up with increase in workpiece thickness. The conduction heat transfer from the cutting front through the kerf walls to the substrate metal is the significant means of power loss from the laser–material interaction zone (i.e., cutting zone).

Iron-oxide (FeO) generated in oxygen-assisted laser cutting of mild steel does not boil, but would dissociate when heated to high temperatures; the iron-oxide dissociation process which consumes much energy could lead to a collapse of the cutting process [26]. Therefore, it is sufficient to assume that molten metal oxide is removed through the bottom of the cut kerf without vaporization. The oxidation of the metal melt during laser cutting is sustained in as much as the reactants - i.e., O_2 and molten metal – are available in the laser material interaction zone. Therefore, the power contribution to the cutting process by the oxidation reaction (i.e., reaction power) is estimated from either the oxygen flow into the interaction zone or the molten iron flow into the interaction zone [27]. The extent of the exothermic reaction in the cutting front is limited by the flow rate of the rarer type of reactant (either oxygen or iron). Only a small proportion of the oxygen jet is consumed in the oxidation reaction as part of the oxygen jet is utilized as drag to accelerate the melt out of the cut kerf while part of it is lost across the top workpiece surface or down the kerf; consequently, only about 50% of the molten iron reacts with the oxygen in the cut kerf. The cutting speeds in reactive fusion cutting are much higher than in fusion cutting with an inert assist gas because of the additional heat added by the exothermic reaction. Due to the high productivity of the reactive fusion cutting, this method is often used in industry for cutting of mild steel (i.e., low alloy steel). The presence of the oxide layer on the cut edge is the downside of this process as the oxide layer on the cut edge influences the final quality of the part; this oxide layer may require to be removed in a cleaning operation prior to further processing of the part in welding and painting operations.

3.2. Absorptivity (coupling coefficient) and melting efficiency

Absorptivity, A (also known as absorption coefficient), of the metal surface to the laser radiation is defined as the ratio of the laser power absorbed at the surface to the incident laser power. Absorptivity depends on the wavelength of laser radiation, plane of polarization of the light beam, angle of incidence, material type, and temperature and state of the material (solid, liquid, or gas). For an opaque material such as a metal, the absorptivity is given as $A = 1 - R$, where R is the reflectivity of the workpiece surface. Absorptivity of the light beam by the metal workpiece generally increases with increase in the temperature of the metallic material. When the angle of incidence is zero (i.e., vertical incidence), the parallel polarized laser beam (R_P) and the perpendicularly polarized laser beam (R_s) are absorbed equally. However, the absorption coefficient of the parallel polarized light (R_P) increases with increase in angle of incidence and is highest at the Brewster angle while the absorption coefficient of the perpendicularly polarized light (R_s) decreases with increase in angle of incidence [25, 27, 28].

The absorptivity (coupling coefficient), A, of the workpiece to the incident laser beam can be estimated by considering that the absorbed laser power is utilized to account for both the melting of the kerf volume and the inevitable conduction power losses. For a pure fusion cutting process utilizing only the incident laser power as the incoming power source, the absorptivity (coupling coefficient) of the workpiece to the incident laser radiation can be estimated using Equation 6.

$$A = \frac{Absorbed\ Laser\ Power\left(AP_L\right)}{Incident\ Laser\ Power\left(P_L\right)} = \frac{\left(P_m + P_{Losss}\right)}{P_L} \tag{5}$$

$$A = \frac{\left[\rho w t V\left(C_p \Delta T + L_m\right)\right] + \left[2\left(\rho C_p \Delta T_{Loss} V L t\right)\right]}{P_L} \tag{6}$$

where A is the absorptivity of the workpiece to the incident laser radiation; P_L is the incident laser power, (W); P_m is the laser power for melting the kerf volume, (W); P_{Loss} is the inevitable power loss from cutting zone (W); w is the kerf width, (m); t is the workpiece thickness, (m); V is the cutting speed, (m/s); ρ is the metal workpiece density, (kg/m³); C_p is the specific heat capacity of the material, (J/kg/K); ΔT is the temperature rise to cause melting of kerf volume, (K) ; L_m is the latent heat of melting of the material (J/kg); ΔT_{Loss} is the temperature change in the substrate metal (K).

3.3. Melt removal during laser cutting

In laser cutting of metal, the cut kerf is created by continuously shearing and acceleration of the molten metal out of the cutting front by action of the drag force of the high pressure assist gas jet acting coaxially with the laser beam. The efficiency of melt removal from the cut kerf

during laser cutting of a metal workpiece plays a very important role on the cutting performance and the resulting cut edge quality. A minimum melt film thickness should be maintained
at the cutting front to enhance efficient energy coupling to the unmelted metal in the line of
traverse of the laser beam for continued cutting. Molten metals that have high surface tension
and high viscosity are more difficult to remove from the cutting front and have a high tendency
to attach on the underside of the cut as dross. Thus, the quality of the cut depends on the
quantity of the melt which builds up in the cut kerf and causes dross on the cut surface.

3.3.1. Melt flow velocity and melt film thickness

The molten metal ejection is mainly driven by the shear force at the assist gas–melt interface
and the pressure gradient created in the cut kerf. With the coaxial arrangement of the assist
gas nozzle with the incident laser beam, the entire melt surface in the cut kerf is in contact with
the assist gas jet. The cut kerf width is usually a fraction of a millimeter and the molten metal
has a high viscosity such that the melt flow can be assumed to cover the entire cut kerf.
Consequently, the expression for the melt flow can be analyzed by applying the principles of
conservation of mass and momentum to the boundary layer flow of the melt film.

Wandera and Kujanpää [29] modeled the maximum melt flow velocity, U, at the gas–melt
interface and the melt film thickness, t_{melt} under high pressure inert assist gas processing
conditions as presented in Equations 7 and 8, respectively. The characteristic gas velocity,
U_g, inside the cut kerf is estimated using the Bernoulli equation as presented in Equation 9.
During laser fusion cutting using an inert assist gas jet, the molten metal has a high viscosity
such that high assist gas pressure is required to facilitate high melt flow velocity, ensuring a
minimal melt film thickness and achievement of high cut edge quality.

$$U = \left(\frac{1}{\eta + \mu}\right)\left(\frac{\rho_g U_g^{\,2}}{16\,t}\right) w^2 \tag{7}$$

$$t_{melt} = \frac{24\,V\,d^2\,(\eta + \mu)}{\rho_g U_g^{\,2} w^2} \tag{8}$$

$$P = \frac{\rho_g U_g^{\,2}}{2} \tag{9}$$

where U is the maximum melt flow velocity; η is the gas viscosity; μ is the melt kinematic
viscosity ; ρ_g is the gas density; U_g is the characteristic gas velocity inside the cut kerf; t is the
workpiece thickness ; w is the kerf width; t_{melt} is the melt film thickness; V is the cutting speed;
d is the workpiece thickness; P is assist gas pressure.

The variation of melt flow velocity and melt film thickness along the cut depth is shown in
Figure 3. There is retardation of the melt as the melt flow progresses down the cut kerf,

resulting in the melt buildup at the lower section of the cut kerf and subsequently dross attachment on the lower cut edge. The melt flow velocity increases with increase in assist gas pressure and increase in the cut kerf width, resulting in a reduction in the melt film thickness.

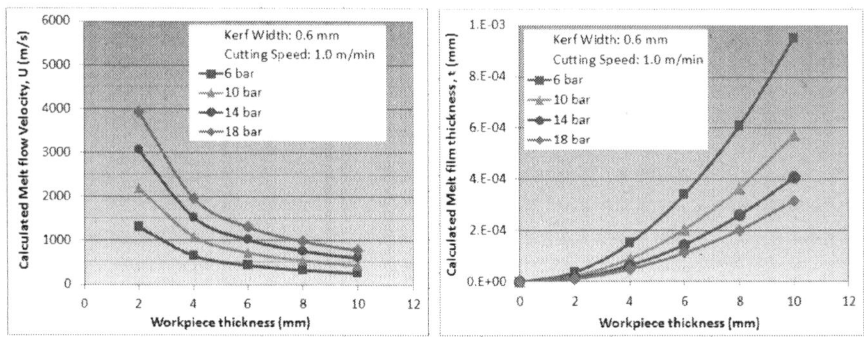

Figure 3. Variation of melt flow velocity and melt film thickness with cut depth for different assist gas pressure.

3.3.2. Separation and transition of melt flow

The retardation of the viscous melt streamlines during laser cutting of a thick-section metal using an inert assist gas jet can result in flow separation as the melt layer thickens rapidly in order to satisfy continuity within the boundary layer. The point along the cut edge where the flow separation occurs is referred to as the boundary layer separation point (BLS), which is shown in Figure 4. Downstream from the boundary layer separation point, there is a back-flow of the melt adjacent to the kerf wall and the boundary layer flow transitions from a laminar flow into a turbulent flow in which the melt particles move in random paths. The transition to turbulent boundary layer flow can also be caused by the disturbances in the laser cutting process – e.g., fluctuations in processing parameters – which may become amplified until turbulence is developed [30].

4. Performance of fiber lasers in metal cutting

The evaluation of the performance of the high brightness ytterbium fiber laser system in thick-section metal cutting is based on the maximum achievable cutting speeds, maximum cutting depths possible, and cut edge quality attainable. The maximum processing speeds, maximum processing depths, and resulting cut edge quality are governed by a number of parameters related to the laser system, workpiece specification, and the cutting process [15]. For cutting of a specific metal of a given thickness, the cutting process parameters can be altered by the operator so as to optimize the cutting process and obtain high cut quality at an optimal cutting speed for high productivity.

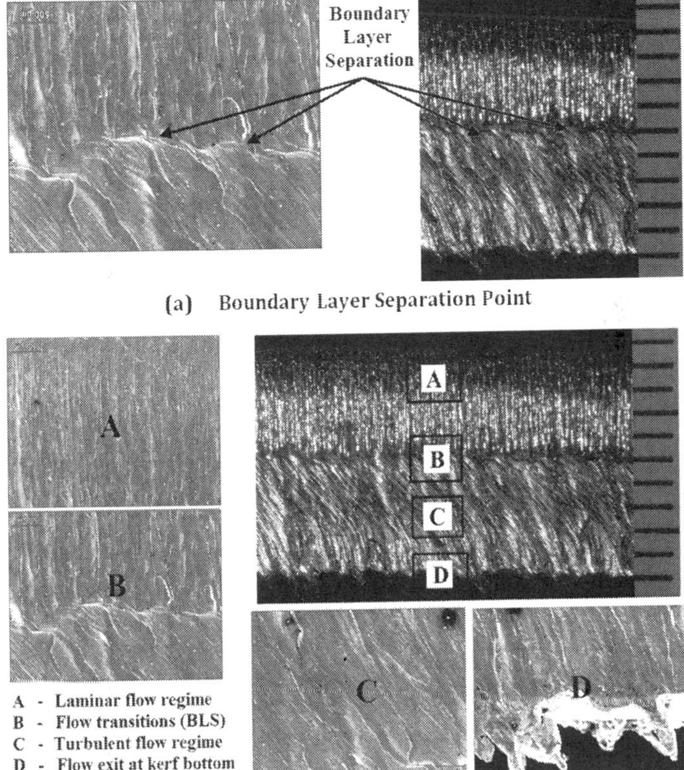

Figure 4. SEM image of the boundary layer separation on a 10-mm stainless steel cut edge

4.1. Maximum achievable cutting speed

The maximum achievable cutting speed for a given laser power level is the maximum cutting speed at which the cut edges are separated. Wandera et al. [23, 31] tested the maximum cutting speeds with the corresponding required laser power levels for the cutting of 10-mm stainless steel, 15-mm mild steel, and 4-mm aluminium using the fiber laser as presented in Figure 5. Using 5 kW fiber laser power, stainless steel of 10-mm thickness can be cut at a maximum cutting speed of 1.5 m/min with nitrogen assist gas jet and mild steel of 15-mm thickness can be cut at a cutting speed of 1.8 m/min with oxygen assist gas jet. Aluminium of 4-mm workpiece thickness can be cut at 10.2 m/min using 5 kW fiber laser power and nitrogen assist gas jet. A comparison of the fiber laser and CO_2 laser cutting speeds in cutting of 15-mm mild steel and 10-mm stainless steel showed that the fiber laser cutting speeds were over 1.3 times higher than the CO_2 laser cutting speeds for the same power level (see Figure 6). And in the cutting of 6–10-mm stainless steel, Sparkes et al. [32] reported an increase in fiber laser cutting speeds of up to 1.5 times higher than the CO_2 laser cutting speeds. Wandera et al. [33] also investigated the cutting of 1–6-mm stainless steel using a fiber laser of beam quality 2.5mm.mrad and reported more than double increase in cutting speeds in 1–4-mm workpiece thickness compared to the CO_2 laser cutting speeds. The cutting speed difference between the fiber laser and the CO_2 laser reduces with increase in metal workpiece thickness to the thick-section domain. The increased cutting speeds for fiber laser is an indication of a higher absorption of the fiber laser beam by the metal workpiece compared to the absorption of the CO_2 laser beam. The melting efficiency of a given laser increases with increase in the absorptivity since a larger proportion of the incident laser radiation is absorbed by the material and utilized in melting of the kerf volume during cutting. Due to the reduction in the proportion of the absorbed laser beam that is lost through conduction to the substrate metal with higher cutting speeds, the potential increase in cutting speed when using the high-power ytterbium fiber laser increases the melting efficiency because the conduction energy losses from the cutting front decrease with increase in cutting speed.

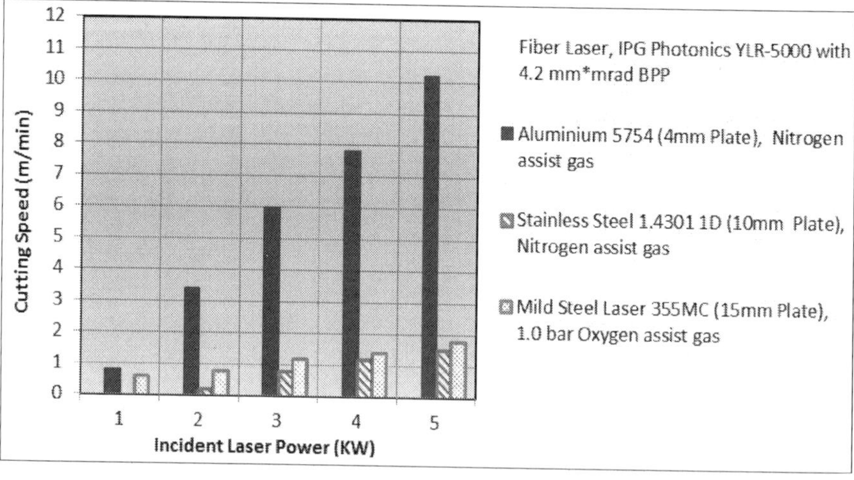

Figure 5. Maximum achievable cutting speeds using the ytterbium fiber laser

4.2. Cut quality characterization

According to the ISO standard for classification of thermal cuts (SFS-EN ISO 9013:2002) [34], the cut edge characteristics that are used to classify thermal cuts include: surface roughness and perpendicularity (squareness) deviation. Dross adherence and presence of the boundary layer separation on the laser cut edges which affect the surface roughness characteristic of a laser cut edge are critical quality aspects that need to be considered in thick-section metal laser cutting.

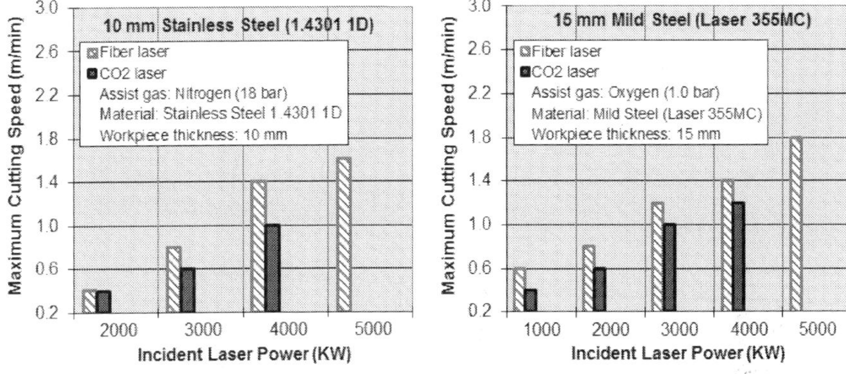

Figure 6. Maximum cutting speeds for cutting 10-mm stainless steel and 15-mm mild steel

4.2.1. Boundary layer separation and dross adherence

The presence of the boundary layer separation and dross adherence on the laser cut edge are characteristics of poor cut edge quality. The tendency of boundary layer separation and dross adherence on cut edges during thick-section metal cutting using an inert assist gas is influenced by the efficiency of melt removal from the cut kerf. The striations observed on the stainless steel cut edges during fiber laser cutting with an inert assist gas are associated with the melt flow mechanisms. Therefore, the cutting process parameters that affect the power intensity at the cutting front and the gas dynamics in the narrow thick-section cut kerf have a great bearing on boundary layer separation, dross adherence, and the resultant cut edge surface roughness.

Surface roughness is the unevenness of the cut surface profile which is observed as striations and adherent dross on the cut edge caused by the dynamical behavior of the laser cutting process. The dynamics of the laser cutting process affects the shape of the cutting front and the melt flow mechanism. Inefficient melt removal and the geometrical shape of the lower edge of the melting front have a strong bearing on the occurrence of adherent dross and is closely related to the properties of the melt flow [30]. High surface tension and viscosity of the molten metals affect the melt ejection and can cause dross adherence on the lower cut edge.

The kerf size and assist gas pressure influence the melt velocity and melt film thickness. Dross adherence on the cut edge as indicated in Figure 7 is caused by the inefficient melt ejection at the lower cut edge because the molten metal is thicker at the bottom of the kerf due to the deceleration of the melt film. Dross adherence in cutting of thick-section stainless steel using fiber laser with an inert assist gas jet is prevented by use of high assist gas pressures and large kerf width shown in Figure 7. The cutting process parameters that influence the kerf size and gas flow dynamics in the cut kerf – such as cutting speed, focal point position, and assist gas pressure – need to be optimized for improved melt removal. A large kerf size, high assist gas pressure, and large nozzle diameter enhance gas flow rates in the thick-section cut kerfs.

Common cutting process parameters:
Stainless steel: 10 mm, Fiber Laser power: 4 kW, Cutting speed: 1.0 m/min,
Nozzle: 2.5 mm, Focal length 190 mm, Focal position: -8, Nitrogen assist gas

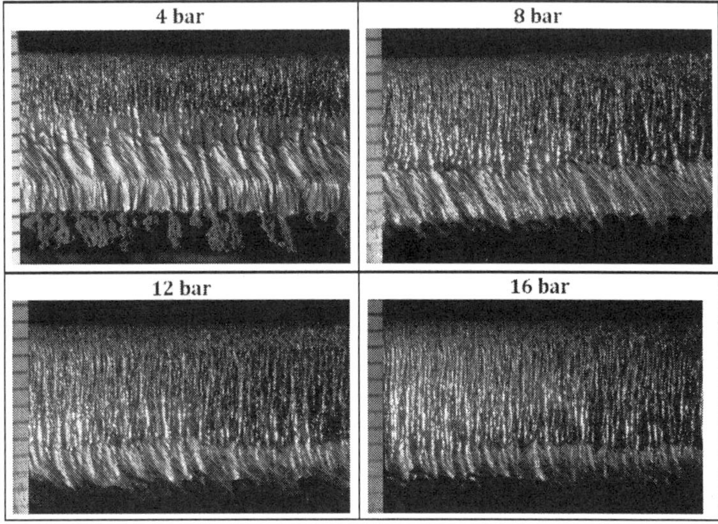

Figure 7. Effect of assist gas pressure on the location of the boundary layer separation point and dross attachment on cut edges in inert gas-assisted laser cutting

4.2.2. Categorization of the fiber laser cut edges

Wandera et al. [31] presented the different categories of the fiber laser cut edges in thick-section steel and medium-section aluminum for different cutting speeds at different power levels. There was existence of adherent dross on cut edges obtained at the slow cutting speeds and at the cutting speeds close to the maximum achievable cutting speed for a given laser power level. Wandera and Kujanpää [35] established the cutting process parameter combinations for optimization of the cut edge quality in 10-mm stainless steel workpiece. The poor cut edge quality – dross attachment and boundary layer separation – observed in thick-section stainless steel cut edges obtained using the high brightness fiber laser and an inert assist gas jet is caused by the difficulty in melt ejection through the narrow thick-section cut kerfs [32, 36]. Therefore, the rate of melt removal from the narrow thick-section cut kerf may be a potential factor limiting the maximum workpiece thickness that can be cut using the high brightness fiber laser rather than the required laser power.

4.3. Effects of laser cutting process parameters

Cutting process parameters that can be altered for the improvement of the cutting process and the resulting cut edge quality include: used laser power, cutting speed, type and pressure of assist gas, nozzle diameter and nozzle stand-off distance, focal point position relative to the workpiece, and focal length of focusing optics. Maximum cutting speed at a given laser power level and efficient melt removal for prevention of the undesired dross adherence on the cut edge can be achieved through optimization of the cutting process parameters. Sparkes et al. [32] experimentally investigated the effects of different cutting process parameters in the cutting of 6–10 mm 304 stainless steel using a high brightness ytterbium fiber laser. Wandera et al. [23, 31] and Wandera and Kujanpää [29], also experimentally investigated the effects of process parameters in the cutting of 10-mm stainless steel, 15-mm mild steel, and 4-mm aluminum using the high brightness fiber laser. They observed that the cut edge quality in 6–10-mm stainless steel improved with increase in cutting speeds, higher nitrogen assist gas pressures, and wider cut kerfs. The cut kerf width depends on the focused spot size, laser power absorbed, and the applied cutting speed.

4.3.1. Effect of laser power and cutting speed

The maximum achievable cutting speeds increase with increase in the incident laser power used for cutting. Wandera et al. [23, 31] and Sparkes et al. [32] reported that for the cut edge quality in 6–10- mm stainless steel, the location of the boundary layer separation point moves closer to the bottom cut edge with increase in cutting speed. A lower dross attachment is experienced with high cutting speeds than when slow cutting speeds are used. The higher energy density loss from the cutting front when cutting at low cutting process increases the tendency of dross adherence on the cut edge. The cut edge quality improved with increase in cutting speeds.

There is a tendency of dross adherence on the cut edge when cutting at the maximum cutting speeds using the high brightness ytterbium fiber laser; therefore, there is need to define the acceptable cut edge quality. Typically, the cutting speed giving the best cut edge quality is lower than the maximum achievable speed for cutting through a given material at the given laser power level. Therefore, the cutting speed giving the best cut quality is the optimum cutting speed, especially for applications where cut edge quality is of paramount importance [35]. Application of cutting speeds that are beyond or below the optimum cutting speed results in dross adherence on the lower cut edge. To improve the cut quality in metal cutting using the high brightness and short wavelength lasers, Olsen et al. [37] developed a multi-beam approach to control the melt flow out of the cut kerf. Their approach involves splitting up the beams from two single mode fiber lasers and positioning the beams in a pattern in the cut kerf in such a way that there is a melt beam that performs melting and the melt ejection beam.

In reactive fusion cutting using an active assist gas jet, there is a significant variation in kerf width with cutting speed because the exothermic reaction is very erratic at slow cutting speeds, resulting in increased sideways burning and widening of the kerf width (see Figure 8). There is a reduction in the occurrence of the irregular deep grooves on the cut edge with a reduction of assist gas pressure and increase in cutting speed to optimum levels. Increased cutting speeds also result in poor melt ejection at the bottom of the cut kerf, causing dross attachment and in worst cases the eventual resealing of the lower cut edge by the resolidified melt.

Common cutting process parameters:

Fiber Laser 2 kW, 1.0 bar Oxygen, 2.0 mm Nozzle, -10 focal position, 190 mm focal length, 15 mm workpiece thickness

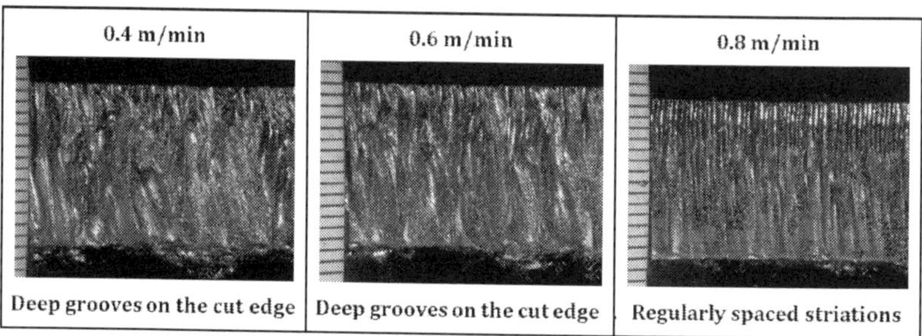

Figure 8. Effect of cutting speed on striation pattern in 15- mm mild steel

4.3.2. Effect of type and pressure of assist gas

Wandera et al. [29] showed that there was a good correlation between the calculated melt flow velocity and melt film thickness with the location of the boundary layer separation point on the 10-mm stainless steel cut edges made using a high-power fiber laser with an inert assist gas jet. The melt flow velocity increases with increasing assist gas pressure so that the melt film thickness decreases and the boundary layer separation point moves closer to the kerf bottom with increase in assist gas pressure. Consequently, in the cutting of the 10-mm stainless steel plate using the high brightness fiber laser, the cut edge quality was optimized with increase in assist gas pressure to over 16 bars so as to eliminate the boundary layer separation on the cut edge (see Figure 9). The dross attachment on the cut edge and surface roughness is also reduced with increase in assist gas pressure as shown in Figure 10. It is shown in Figure 9 that the assist gas pressure does not have an effect on the kerf width in inert gas-assisted laser cutting.

The dynamic nature of the exothermic oxidation reaction that occurs during reactive fusion laser cutting of mild steel is favored by high oxygen gas pressure resulting in wider non-uniform cut kerfs as shown in Figure 11. The nature of the erratic exothermic oxidation reaction at high oxygen pressure produces irregular deep striations on the cut edge. The ridges observed on the cut edge in laser oxygen-assisted cutting of 15-mm mild steel using the fiber laser is typical of the mechanism of the sideways burning, as originally explained by Arata et al. and Miyamoto and Maruo [38, 39].

4.3.3. Effect of focal point position

The focal point position is the location of the minimum focused spot size relative to the workpiece top surface. The focal point position which affects the laser power intensity on the workpiece influences the cut kerf size and affects the melt removal process, thus affecting the

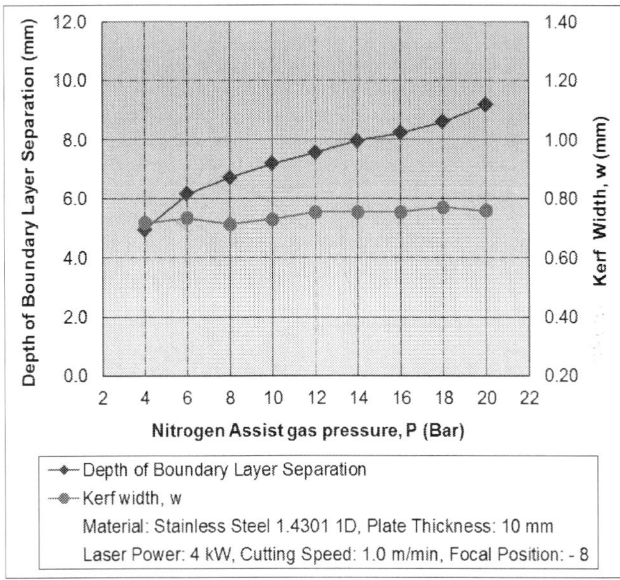

Figure 9. Effect of assist gas pressure on the location of boundary layer separation point and kerf width in inert gas-assisted laser cutting

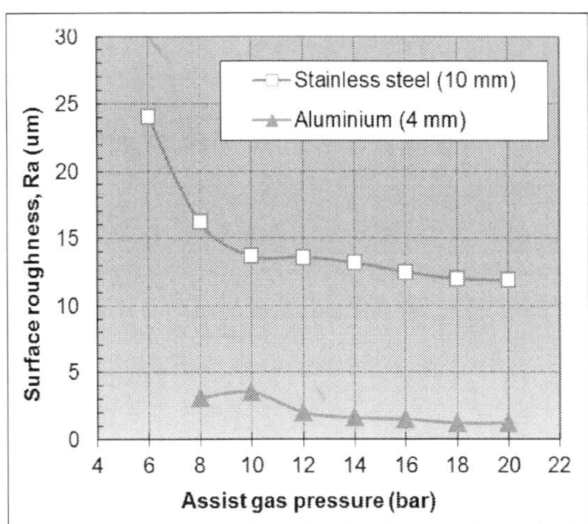

Figure 10. Variation of surface roughness with assist gas pressure for the fiber laser cut edges in 10-mm stainless steel and 4-mm aluminum

Common cutting process parameters:

Fiber Laser 3 kW, 0.8 m/min, 2.0 mm Nozzle, -10 focal position, 190 mm focal length, 15 mm workpiece thickness

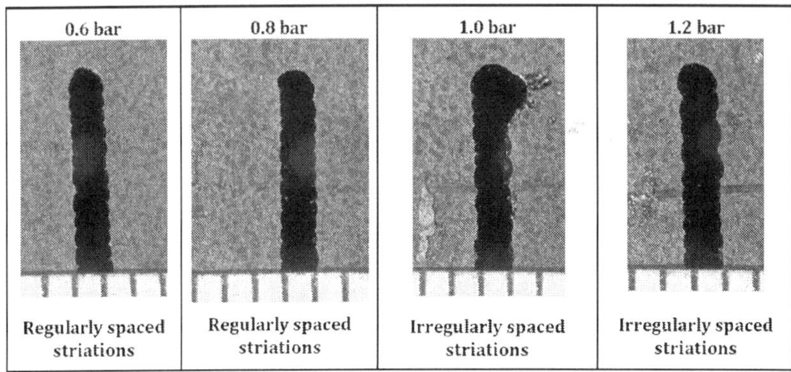

Figure 11. Effect of oxygen assist gas pressure on the kerf width and striations

cut edge quality. The highest incident power intensity necessary for penetration of a thick-section metal workpiece at the highest cutting speed is achieved when the focal position is at the workpiece top surface. However, for fiber laser cutting, the focal position at the workpiece top surface results in narrow cut kerfs and the poor melt removal mechanism in the narrow thick-section laser cut kerfs causes dross attachment on the lower cut edge. Usually, the lower section of the cut edge has a higher surface roughness than the upper section due to the melt build-up at the lower cut section, resulting in inefficient melt removal.

Efficient melt removal is obtained when wider cut kerfs are created with the focal position located below the workpiece top surface. For improved cut edge quality, the focal positions below the workpiece top surface are essential in thick-section metal cutting using the fiber laser so as to obtain wider cut kerfs for efficient melt removal as long as the power intensity at the workpiece top surface is sufficient to obtain complete penetration of the workpiece. When cutting thick-section stainless steel using the fiber laser, optimum cut edge quality is obtained when the focal position is located on the lower surface of the workpiece as long as the power intensity is sufficient to penetrate the workpiece at the applied cutting speed; in this case, the applied cutting speed should be lower than the maximum achievable speed for the applied laser power. Complete penetration of the thick-section stainless steel workpiece cannot be achieved at the maximum cutting speed for a given power level when the focal point position is far into the lower half section of the workpiece thickness because of the reduced power intensity on the workpiece surface. On the cutting of a 10-mm stainless steel workpiece using the fiber laser, the tendency for dross attachment on the lower cut edge is more significant with focal point positions located on the upper half section of the workpiece thickness (see Figure 12). Focal positions close to the workpiece bottom surface produce clean dross-free cuts because of the wider cut kerfs formed with these focal positions.

Common cutting process parameters:

Stainless steel: 10 mm, Fiber Laser power: 4 kW, Cutting speed: 1.0 m/min, Nozzle: 2.5 mm, Focal length 254 mm, 16 bar Nitrogen assist gas pressure

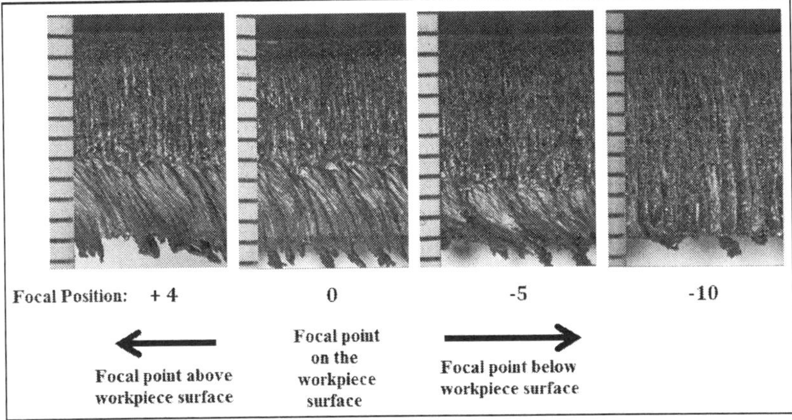

Figure 12. Effect of focal point position on the dross adherence on 10-mm stainless steel

4.3.4. Effect of nozzle diameter and stand-off distance

The efficiency of melt ejection is influenced by the nozzle diameter which determines the amount of the cutting gas jet available at the cutting front. In the case of thick-section metal cutting with an inert assist gas jet, the nozzle diameter significantly affects the location of the boundary layer separation point on the cut edge. The location of the boundary layer separation point moves toward the bottom cut edge with increase in the nozzle diameter because of the enhanced melt removal process facilitated by the increased amount of assist gas provided by the large-size nozzle. There is no boundary layer separation on the cut edge with the 2.5-mm nozzle diameter [29].

In the oxygen-assisted laser cutting, a smaller nozzle size enhances achievement of a good cut quality with a finer uniform striation pattern on the cut edge by limiting the extent of the sideways burning oxidation reaction. In fiber laser cutting of 15-mm mild steel using oxygen assist gas, the best cut quality was obtained with a 1.5-mm nozzle diameter and the 2.5-mm nozzle diameter produced the worst cut edge quality [23].

The nozzle stand-off distance – distance between the nozzle and the workpiece top surface – influences the gas flow dynamics at the entrance of the cut kerf and consequently affects the gas flow patterns at the cutting front. The gas flow patterns at the cutting front have a strong effect on the resulting cut edge quality especially during high pressure inert gas-assisted laser cutting [40, 41, 42]. The nozzle stand-off distances (0.5–1.2 mm) tested by Wandera and Kujanpää [29] in the cutting of 10-mm stainless steel workpiece showed that the nozzle stand-off distance did not have an effect on the location of boundary layer separation [29].

4.3.5. Effect of focal length of focusing optics

The effects of cutting speed and focal point position on the surface roughness of the fiber laser cut edges are more significant when a short focal length of the focusing optics is used than when longer focal length optics is used. In cutting of 10-mm stainless steel using the 4 kW fiber laser power at cutting speed of 1.0 m/min and focal position located on the bottom workpiece surface, dross-free cut edges with lower surface roughness were obtained with the 254-mm focal length optics than when the 190.5-mm focal length optics was used. This is because wider cut kerfs were obtained with the 254-mm focal length optics compared to the 190.5-mm focal length optics [35].

5. Conclusion

Performance of the high brightness ytterbium fiber laser system in metal cutting has been extensively evaluated in stainless steel, mild steel and aluminium cutting. The maximum cutting speeds and maximum workpiece thickness that can be cut using the ytterbium fiber laser have been shown to be largely governed by a number of cutting process parameters that affect the melt removal process and influence the resultant cut edge quality. Exceedingly higher cutting speeds in thin-section to medium-section (less than 6 mm sheet thickness) metal cutting have been realized with the ytterbium fiber laser compared to the CO_2 laser. However, a drastic reduction in maximum cutting speed has been reported in thick-section (above 6 mm sheet thickness) metal cutting using the ytterbium fiber laser due to deterioration in cut edge quality at the maximum cutting speed. Optimization of the cutting process parameters for enhancement of the cut edge quality in thick-section metal cutting at high cutting speeds using the ytterbium fiber laser has shown that the maximum applicable cutting speed is influenced by the melt removal process.

Acknowledgements

The company HT Laser Oy is acknowledged for funding the fiber laser experimental work which formed a basis for analysis of the performance of the high brightness ytterbium fiber laser in metal cutting that is presented in the Chapter on Fiber Lasers in Materials Processing. Lappeenranta Laser Processing Center at Lappeenranta University of Technology in Finland is acknowledged for facilitation of the fiber laser experimentation. Professor Veli Kujanpää of VTT Technical Research Centre of Finland and Professor Antti Salminen of Lappeenranta University of Technology are acknowledged for the invaluable discussions that inspired the formulation of the ideas presented in this chapter. I also acknowledge Busitema University, Faculty of Engineering, for providing a conducive environment during the preparation of this book chapter.

Author details

Catherine Wandera*

Address all correspondence to: cathywandera@yahoo.com

Busitema University, Tororo, Uganda

References

[1] Petring D, Schneider F, Wolf N, Nazery V. The relevance of brightness for high pow‐ er laser cutting and welding. Proc. 27th Int. Congr Applic Lasers Electro Optics, ICA‐ LEO 2008 (October 20-23), Temecula, California, USA, paper 206, 2008;95–103.

[2] Müller H-R, Kirchhof J, Reichel V, Unger S. Fibers for high-power lasers and amplifi‐ ers. Comptes Rendus Physique 2006;7(2):154 –62.

[3] Hügel H. New solid-state lasers and their application Potentials. Optics Lasers Eng 2006;34(4–6):213 –29.

[4] Canning J. Fiber lasers and related technologies. Optics Lasers Eng 2006;44(7):647 – 76.

[5] Nilsson J, Clarkson WA, Selvas R, Sahu JK, Turner PW, Alam S-U, Grudinin AB. High-power wavelength-tunable cladding-pumped rare-earth-doped silica fiber la‐ sers. Optical Fiber Technol 2004;10(1):5 –30.

[6] Jarman RH. Novel optical fiber lasers. Curr Opin Solid State Mater Sci 1996;1(2):199 – 203.

[7] Limpert J, Schreiber T, Liem A, Nolte S, Zellmer H, Peschel T, Guyenot V., Tünner‐ mann A. Thermo-optical properties of air-clad photonic crystal fiber lasers in high power operation. Optics Express 2003;11(22):2982 –90.

[8] Hecht J. Photonic Frontiers: Fiber Lasers: Fiber lasers ramp up the power. [In OptoIQ www-pages], [retrieved May 5, 2009]. From: http://www.optoiq.com/index/photon‐ ics-technologies-applications/lfw-display/lfw-article-display/371319/articles/laser-fo‐ cus-world/volume-45/issue-12/features/photonic-frontiers-fiber-lasers-fiber-lasers-ramp-up-the-power.html).

[9] O'Neill W, Sparkes M, Varnham M, Horley R, Birch M, Woods S, Harker A. High power high brightness industrial fiber laser technology. Proc 23rd Int Congr Applic Lasers Electro Optics, ICALEO 2004 (October 4–7), San Francisco, California, USA, paper 301, 2004;1–7.

[10] Thomy C, Seefeld T, Vollertsen F. High-power fiber lasers – application potentials for welding of steel and aluminium sheet material. Adv Mater Res 2005;6–8:171–8, available online at http://www.scientific.net/

[11] Kancharla V. Applications review: materials processing with fiber lasers under 1kW. Proc 25th Int Congr Applic Lasers and Electro Optics, ICALEO 2006 (October 30–November 2), Scottsdale, Arizona, USA, Paper 1301, 2006;579–85.

[12] Quintino L, Costa A, Miranda R, Yapp D, Kumar V, Kong CJ. Welding with high power fiber lasers – a preliminary study. Mater Design 2007;28(4):1231 –7.

[13] Jeong Y, Sahu JK, Payne DN, Nilsson J. Ytterbium-doped large-core fiber laser with 1.36 kW continuous-wave output power. Optics Express 2004;12(25):6088 –92.

[14] IPG Photonics. Materials Processing: Single mode fiber lasers, Multi mode fiber lasers. [In IPG Photonics www-pages], [retrieved May 10, 2010]. From http://www.ipgphotonics.com/apps_materials.htm

[15] Steen WM. Laser Material Processing, 3rd edn. Springer-Verlag, London, 2003, pp. 83–90, 111–122, 126, 163–165.

[16] Powell J. CO_2 Laser Cutting, 2nd edn, Chapter 1: Section 1.2. Springer, London Ltd, 1998.

[17] Rémy F, Sonia S, Frédéric C, Francis B, Bruno D, Guillaume L. Analysis of basic processes inside the keyhole during deep penetration Nd-Yag Cw laser welding. Proc 25th Int Congr Applic Lasers Electro-Optics ICALEO 2006, 30th October –2nd November 2006, Scottsdale, Arizona, USA.

[18] Salminen A, Fellman A. The effect of laser and welding parameters on keyhole and melt pool behavior during fiber laser welding. Proc 26th Int Conf Lasers Electro Optics ICALEO2007, Orlando, FL, USA, October 29–November 1, 2007.

[19] Olsen FO. Fundamental mechanisms of cutting front formation in laser cutting. Proc SPIE, 2207, 1994;235–47.

[20] Olsen FO. Cutting front formation in laser cutting. Annal CIRP 1989;38(1):215–8.

[21] Yilbas BS. Laser heating process and experimental validation. Int J Heat Mass Transfer 1997;40(5):1131 –43.

[22] Duan J, Man HC, Yue TM. Modeling the laser fusion cutting process: I. Mathematical modelling of the cut kerf geometry for laser fusion cutting of thick metal. J Physics D Appl Phys 2001;34(14):2127 –34.

[23] Wandera C, Kujanpää V, Salminen A. Laser power requirement for cutting of thick-section steel and effects of processing parameters on mild steel cut quality. Proc IMechE B, J Eng Manufacture 2011;225:651–61.

[24] Schulz W, Becker D, Franke J, Kemmerling R, Herziger G. Heat conduction losses in laser cutting of metals. J Phys D Appl Phys 1993;26(9):1357 –63.

[25] Ion JC. Laser Processing of Engineering Materials: Principles, Procedure and Industrial Application. Elsevier Butterwort-Heinemann, 2005.

[26] Powell J, Petring D, Kumar RV, Al-Mashikhi SO, Kaplan AFH, Voisey KT. Laser-oxygen cutting of mild steel: the thermodynamics of the oxidation reaction. J Phys D Appl Phys 2009;42(1):1 –11.

[27] Ready JF, Farson DF. (Eds). LIA Handbook of Laser Materials Processing. Laser Institute of America, 2001.

[28] Xie J, Kar A, Rothenflue JA, Latham WP. Temperature-dependent absorptivity and cutting capability of CO_2, Nd: YAG and chemical oxygen-iodine lasers. J Laser Applic 1997;9(2):77–85.

[29] Wandera C, Kujanpää V. Characterization of the melt removal rate in laser cutting of thick-section stainless steel. J Laser Applic 2010;22(2):62–70.

[30] Schulz W, Kostrykin V, Nießen M, Michel J, Petring D, Kreutz EW, Poprawe R. Dynamics of ripple formation and melt flow in laser beam cutting. J Phys D Appl Phys 1999;32(11):1219 –28.

[31] Wandera C, Salminen A, Kujanpää V. Inert gas cutting of thick-section stainless steel and medium-section aluminium using a high power fiber laser. J Laser Applic 2009;21(3):154–61.

[32] Sparkes M, Gross M, Celotto S, Zhang T, O'Neill W. Practical and theoretical investigations into inert gas cutting of 304 stainless steel using a high brightness fiber laser. J Laser Applic 2008;20(1):59 –67.

[33] Wandera C, Salminen A, Kujanpää V, Olsen F. Cutting of stainless steel with fiber and disk laser, 2006, 10 pages, Proc 25th Int Congr Applic Lasers Electro-Optics ICALEO 2006, 30th October – 2nd November 2006, Scottsdale, Arizona, USA.

[34] European Committee for Standardization. Thermal cutting – Classification of thermal cuts, Geometrical product specification and quality tolerances. EN ISO 9013: 2002 (SFS-ISO EN 9013:2002.

[35] Wandera C, Kujanpää V. Optimization of parameters for fiber laser cutting of a 10 mm stainless steel plate, Proc IMechE Part B, J Eng Manufacture 2011;225:641–9.

[36] Sparkes M, Gross M, Celotto S, Zhang T, O'Neill W. Inert cutting of medium section stainless steel using a 2.2 KW high brightness fiber laser. Proc 25th Int Congr Applic Lasers Electro Optics, ICALEO 2006 (October 30 – November 2), Scottsdale, Arizona, USA, paper 402, 2006;197–205.

[37] Olsen FO, Hansen KS, Nielsen JS. Multibeam fiber laser cutting. J Laser Applic 2009;21(3):133 –8.

[38] Arata Y, Maruo H, Miyamoto I, Takeuchi S. Dynamic behavior in laser gas cutting of mild steel. Transac JWRI 1979;8:15–26.

[39] Miyamoto I, Maruo, H. The mechanism of laser cutting. Weld World (UK) 1991;29:283 –94.

[40] Fieret J, Terry MJ, Ward BA. Overview of flow dynamics in gas-assisted laser cutting. Paper presented to the Fourth International Symposium on Optical and Optoelectronic Applied Science and Engineering, Topical Meeting on High Power Lasers: Sources, Laser-Material Interactions, High Excitations, and Fast Dynamics in Laser Processing and Industrial Applications, 1987 (March 30 – April 3), The Hague, The Netherlands.

[41] Zefferer H, Petring D, Beyer E. Investigation of the gas flow in laser beam cutting. Lectures and Posters of the 3rd International Beam Technology Conference, Karlsruhe, 1991 (March 13–14).

[42] Man HC, Duan J, Yue TM. Behaviour of supersonic and subsonic gas jets inside laser cut kerfs. Proc of ICALEO 1997, San Diego, USA, B27-36.